Lecture Notes in Computer Science 1091

Edited by G. Goos, J. Hartmanis and J. van Leeuwen

Advisory Board: W. Brauer D. Gries J. Stoer

Springer
Berlin
Heidelberg
New York
Barcelona
Budapest
Hong Kong
London
Milan
Paris
Santa Clara
Singapore
Tokyo

Jonathan Billington   Wolfgang Reisig   (Eds.)

Application and Theory of Petri Nets 1996

17th International Conference
Osaka, Japan, June 24-28, 1996
Proceedings

Springer

Series Editors

Gerhard Goos, Karlsruhe University, Germany

Juris Hartmanis, Cornell University, NY, USA

Jan van Leeuwen, Utrecht University, The Netherlands

Volume Editors

Jonathan Billington
University of South Australia, Institute for Telecommunications Research
Warrendi Road, The Levels, Pooraka, South Australia, 5095 Australia

Wolfgang Reisig
Humboldt-University of Berlin, Computer Science Department
Unter den Linden 6, D-10099 Berlin, Germany

Cataloging-in-Publication data applied for

Die Deutsche Bibliothek - CIP-Einheitsaufnahme

Application and theory of Petri nets ... : ... international
conference ; proceedings. - Berlin ; Heidelberg ; New York ;
Barcelona ; Budapest ; Hong Kong ; London ; Milan ; Paris ;
Santa Clara ; Singapore ; Tokyo : Springer.
Früher begrenztes Werk in verschiedenen Ausg.

17. 1996. Osaka, Japan, June 24 - 28, 1996. - 1996
  (Lecture notes in computer science ; 1091)
  ISBN 3-540-61363-3 (Berlin ...)
NE: GT

CR Subject Classification (1991): F.1-3, C.1-2, G.2.2, D.4, J.4

ISSN 0302-9743
ISBN 3-540-61363-3 Springer-Verlag Berlin Heidelberg New York

© Springer-Verlag Berlin Heidelberg 1996
Printed in Germany

Typesetting: Camera-ready by author
SPIN 10513152   06/3142 – 5 4 3 2 1 0   Printed on acid-free paper

Preface

This volume contains the proceedings of the 17th International Conference on Application and Theory of Petri Nets. The aim of the Petri net conference is to create a forum for the dissemination of the latest results in the application and theory of Petri nets. Typically the conferences have 150-200 participants and usually one third of these come from industry while the rest are from universities and research institutions. The conferences always take place in the last week of June.

The previous conferences (1980-1995) were held in Strasbourg (France), Bad Honnef (Germany), Varenna (Italy), Toulouse (France), Aarhus (Denmark), Espoo (Finland), Oxford (UK), Zaragoza (Spain), Venice (Italy), Bonn (Germany), Paris (France), Aarhus (Denmark), Sheffield (UK), Chicago (USA), Zaragoza (Spain), and Turin (Italy).

The conferences and a number of other activities are coordinated by a steering committee with the following members: G. Balbo (Italy), J. Billington (Australia), G. De Michelis (Italy), C. Girault (France), K. Jensen (Denmark), T. Murata (USA), C.A. Petri (Germany; honorary member), W. Reisig (Germany), G. Roucairol (France), G. Rozenberg (The Netherlands; chairman), M. Silva (Spain).

We are delighted that the 17th conference has been organized for the first time in Asia, and on this occasion in Osaka, Japan. In addition to the traditional conference, a smorgasbord of other activities provided includes: an exhibition and presentation of Petri net tools; extensive introductory tutorials for those commencing in the field; two advanced tutorials, one on scheduling and the other on distributed algorithms; and two workshops held during the first two days: Object-Oriented Programming and Models of Concurrency; and Manufacturing and Petri Nets. The tutorial notes, tool presentations, and proceedings of the workshops are not published in this proceedings but limited copies are available from the organizers.

We have received 78 submissions from 19 countries covering five continents (still none from Antarctica where you would expect there to be fascinating research into Petri net models of the mating habits of emperor penguins!) and 26 have been accepted for presentation. Invited lectures are given by G. Agha (USA), A. Valmari (Finland), and T. Murata (USA).

The submitted papers were evaluated by a program committee (PC) with the following members: P. Azéma (France), J. Billington (Australia; co-chair), S. Christensen (Denmark), J.-M. Colom (Spain), F. De Cindio (Italy), J. Esparza (Germany), G. Franczeschinis (Italy), J. Kleijn (Netherlands), S. Kodama (Japan), M. Koutny (UK), C. Lakos (Australia), T. Murata (USA), L. Ojala (Finland), K. Onaga (Japan), L. Petrucci-Dauchy (France), W. Reisig (Germany; co-chair), D. Simpson (UK), D. Stotts (USA), M. Woodside (Canada). The program committee meeting took place at the Humboldt-University of Berlin in Germany and by email link to Adelaide Australia, the first step towards distributed PC meetings over the internet.

We should like to express our gratitude to all authors of submitted papers, to the members of the program committee, and to the referees who assisted them. The names of the referees are listed on the following page. For the local organization of the conference, we greatly appreciate the enormous efforts of the organizing committee: S. Kumagai (Chair), K. Hasagawa, M. Yamamoto, I. Shirakawa, S. Shinoda, T. Matsumoto, T. Ushio, K. Takahashi, H. Hasegawa, K. Hiraishi, K. Tsuji, T. Miyamoto. The support of The Commemorative Association for the Japan World Exposition is gratefully acknowledged.

Finally, we would like to acknowledge excellent cooperation with Alfred Hofmann of Springer-Verlag in the preparation of this volume.

Adelaide, Australia   Jonathan Billington
Berlin, Germany   Wolfgang Reisig

April 1996

List of Referees

M. Ajmone Marsan
Y. Al-Salqan
T. Aura
P. Azéma
J. A. Bañares
T. Basten
E. Battiston
V. Baudin
C. Bellettini
L. Bernardinello
E. Best
A. Bianco
C. Birkinshaw
A. Bondavalli
R. Bosworth
U. Buy
J. Campos
L. Capra
A. Cheng
B. Chezalviel-Pradin
G. Chiola
A. Chizzoni
S. Christensen
P. Chrzastowski-Wachtel
G. Ciard
J. M. Colom
M. Combacau
J. Dayao
F. de Cindio
R. de Lemos
G. de Michelis
H. Demmou
J. Desel
M. Dickson
S. Donatelli
K. Drira
J. Engelfriet
J. Esparza
J. Ezpeleta
M. Farrington
D. Floreani
G. Franceschinis
R. Gaeta
L. Gallon
F. García-Vallés
T. Gelsema
C. Girault
H. J. M. Goeman
D. Gomm
L. P. J. Groenewegen

J. Hage
J. Hall
X. He
A. Heise
K. Hiraishi
M. Holcombe
L. Holloway
S. Honiden
H. J. Hoogeboom
R. P. Hopkins
S. Huang
N. Husberg
J. Jeffrey
K. Jensen
J. B. Jørgensen
E. Y. T. Juan
G. Juanole
P. Kemper
A. Kiehn
E. Kindler
H. C. M. Kleijn
S. Kodama
M. Koutny
A. Kovalyov
L. Kristensen
D. Kuske
M. Z. Kwiatkowska
C. Lakos
D.J. Leu
J. Lilius
C. Lin
J. Martinez
R. Mayr
S. Melzer
J. Menden
F. Michel
T. Miyamoto
N. Miyoshi
M. Molloy
K. H. Mortensen
M. Mukund
T. Murata
M. Nielsen
L. Ojala
K. Onaga
M. Paludetto
C. Pérez
L. Petrucci
J. F. Peyre
D. Poitrenaud

L. Pomello
L. Portinale
T. Pyssysalo
M. D. Radola
J. Ranicki
L. Recalde
J. Rekers
M. Ribaudo
S. Römer
N. Sabadini
W. H. Sanders
P. Senac
M. Sereno
S. Shatz
H. Shiizuka
M. Silva
C. Simone
D. Simpson
V. Sliva
R. Sloan
R. S. Sreenivas
D. Stotts
I. Suzuki
K. Takahashi
S. Takai
T. Takine
S. Tavares
E. Teruel
P. S. Thiagarajan
V. Thomas
M. Tiusanen
S. Tu
N. Uchihira
T. Ushio
R. Valette
W. M. P. van der Aalst
K. Varpaaniemi
F. Vernadat
R. F. Vidale
W. Vogler
M. Voorhoeve
F. Wallner
R. Walter
W. Weitz
M. Woodside
A. Yakovlev
G. Yee
D. Zhang
W. H. Zuberek
R. Zurawski

Table of Contents

Invited Papers

Full Papers

Modeling Concurrent Systems: Actors, Nets, and the Problem of Abstraction and Composition

Gul A. Agha

Open Systems Laboratory
Department of Computer Science
1304 W. Springfield Avenue
University of Illinois at Urbana-Champaign
Urbana, IL 61801, USA

Email: agha@cs.uiuc.edu   http://www-osl.cs.uiuc.edu

Abstract. This paper reviews the state of the art in building and reasoning about concurrent system using actors. We first provide a brief definition of actors and discuss the status of actor theory. We then describe a number of programming abstractions that are useful in developing and maintaining complex concurrent systems. Defining such abstractions requires a sort of system decomposition that is not supported by standard models of concurrency, including actors and nets. Rather a suitable meta-architecture is needed and its satisfactory formal definition remains elusive. We currently have only rudimentary semantics for the different programming abstractions that we have developed.

1 Introduction

Systems in the real-world consist of many distributed, asynchronous components which are *open* to interaction with their environment. We will call such systems *open (distributed) systems.*

Open systems are reconfigurable and extensible: they may allow components to be dynamically replaced – an essential requirement in developing, for example, fault-tolerant applications. The functionality of open systems is not defined by the result of evaluating an expression; instead the relative state of components, the relative timing of actions, locality and distribution of the computation, etc., are all critical to the correctness of the system.

Concurrency leads to complex interactions between different modules. To simplify the task of implementing real-world systems, we must abstract over patterns of interaction between components. Models of concurrency are generally based on message passing as the mechanism to support interaction between components. Unfortunately, programming using only message passing is somewhat worse than programming in assembler: sending a message is not only a jump, it is a concurrent one! The goal of our research is to find the right kinds of abstractions which capture interactions between concurrent modules.

We use actors as our model of concurrent computation. But the questions I will pose are equally applicable to any of the current models of concurrency. Consider requirements such as security, availability, and atomicity. How can we, for example, define a module which imposes a certain security policy on an arbitrary application? A security policy may be implemented by encrypting messages; after we fix the desired implementation, current techniques require that we modify the code for each actor to implement the encryption. We cannot simply define a module for encryption and compose it with an arbitrary collection of actors. But such hand coding is not satisfactory: to address the complexity of real-world systems we need to decompose systems into simpler parts which address distinct design concerns. I will begin with an introduction to actors and then return to the problem of decomposition.

2 Actors

The Actor model provides a flexible method for representing computation in real-world systems. Actors extend the concept of objects to concurrent computation [1]. Recall that objects encapsulate a state and a set of procedures that manipulate the state; actors extend this by also encapsulating a thread of control. Each actor potentially executes in parallel with other actors and may send messages to actors it knows the addresses of. Actor addresses may be communicated in messages, allowing dynamic interconnection. Finally, new actors may be created; such actors have their own unique address.

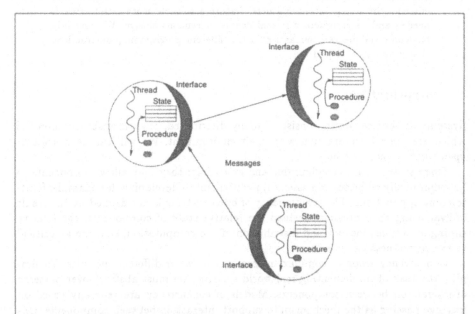

Fig. 1. Actors encapsulate a thread and state. The interface is comprised of public methods which operate on the state.

It is possible to extend any sequential language with the actor constructs. For example, the call-by-value λ-calculus is extended in [2]. Specifically, the following operators are added to expressions:

$\text{send}(a, v)$ creates a new message:
  - with receiver a, and
  - contents v

$\text{newactor}(e)$ creates a new actor:
  - which is evaluating the expression e, and
  - returns its address

$\text{ready}(b)$ captures local state change:
  - alters the behavior of the actor executing the ready expression to b
  - frees that actor to accept another message.

A let construct may be used to allow mutual reference of newly created actors. The behavior of an actor is represented by a lambda abstraction, the acceptance of a message by function application, where app(b,m) represents b applied to message m.

Instantaneous snapshots of actor systems are called *configurations*; actor computation is defined by a transition relation on configurations. The notion of open systems is captured by defining a dynamic interface to a configuration, i.e. by explicitly representing a set of *receptionists* which may receive messages from actors outside a configuration and a set of actors *external* to a configuration which may receive messages from the actors within.

Definition (Actor Configurations): An *actor configuration* with actor map, α, multi-set of messages, μ, receptionists, ρ, and external actors, χ, is written

$$\left\langle\!\!\left\langle \alpha \mid \mu \right\rangle\!\!\right\rangle_{\chi}^{\rho}$$

where ρ, χ is a finite set of actor addresses, α maps a finite set of addresses to their behavior, μ is a finite multiset of (pending) messages, and let $A = \text{Dom}(\alpha)$, i.e., the domain of α, then:

(0) $\rho \subseteq A$ and $A \cap \chi = \emptyset$,
(1) if $a \in A$, then $\text{FV}(\alpha(a)) \subseteq A \cup \chi$, where $\text{FV}(\alpha(a))$ represents the free variables of $\alpha(a)$; and if $<v_0 \Leftarrow v_1>$ is a message with content v_1 to actor address v_0, then $\text{FV}(v_i) \subseteq A \cup \chi$ for $i < 2$.

An actor may be in one of two kinds of states: busy or ready to accept a message. For an actor with address a, we indicate these states as follows:

- $(b)_a$ ready to accept a message, where b is its behavior, a lambda abstraction;
- $[e]_a$ busy executing e, e represents the actor's current (local) processing state.

Now we can extend the local transitions defined for a sequential language ($\overset{\lambda}{\mapsto}$), by providing transitions for the actor program as follows (assume that R is the reduction context in which the expression currently being evaluated occurs):

4

Definition ($\mapsto$):

$$e \xrightarrow{\lambda}_{\text{Dom}(\alpha) \cup \{a\}} e' \Rightarrow \left\langle\!\!\left\langle \alpha, [e]_a \mid \mu \right\rangle\!\!\right\rangle^{\rho}_{\chi} \mapsto \left\langle\!\!\left\langle \alpha, [e']_a \mid \mu \right\rangle\!\!\right\rangle^{\rho}_{\chi}$$

$$\left\langle\!\!\left\langle \alpha, [R[\texttt{newactor}(e)]]_a \mid \mu \right\rangle\!\!\right\rangle^{\rho}_{\chi} \mapsto \left\langle\!\!\left\langle \alpha, [R[a']]_a, [e]_{a'} \mid \mu \right\rangle\!\!\right\rangle^{\rho}_{\chi} \qquad a' \text{ fresh}$$

$$\left\langle\!\!\left\langle \alpha, [R[\texttt{ready}(v)]]_a \mid \mu \right\rangle\!\!\right\rangle^{\rho}_{\chi} \mapsto \left\langle\!\!\left\langle \alpha, (v)_a \mid \mu \right\rangle\!\!\right\rangle^{\rho}_{\chi}$$

$$\left\langle\!\!\left\langle \alpha, [R[\texttt{send}(v_0, v_1)]]_a \mid \mu \right\rangle\!\!\right\rangle^{\rho}_{\chi} \mapsto \left\langle\!\!\left\langle \alpha, [R[\texttt{nil}]]_a \mid \mu, m \right\rangle\!\!\right\rangle^{\rho}_{\chi} \qquad m = \texttt{<}v_0 \Leftarrow v_1\texttt{>}$$

$$\left\langle\!\!\left\langle \alpha, (v)_a \mid \texttt{<}a \Leftarrow cv\texttt{>}, \mu \right\rangle\!\!\right\rangle^{\rho}_{\chi} \mapsto \left\langle\!\!\left\langle \alpha, [\texttt{app}(v, cv)]_a \mid \mu \right\rangle\!\!\right\rangle^{\rho}_{\chi}$$

$$\left\langle\!\!\left\langle \alpha \mid \mu, m \right\rangle\!\!\right\rangle^{\rho}_{\chi} \mapsto \left\langle\!\!\left\langle \alpha \mid \mu \right\rangle\!\!\right\rangle^{\rho'}_{\chi}$$
$$\text{if } m = \texttt{<}a \Leftarrow cv\texttt{>}, \ a \in \chi, \text{ and } \rho' = \rho \cup (\text{FV}(cv) \cap \text{Dom}(\alpha))$$

$$\left\langle\!\!\left\langle \alpha \mid \mu \right\rangle\!\!\right\rangle^{\rho}_{\chi} \mapsto \left\langle\!\!\left\langle \alpha \mid \mu, m \right\rangle\!\!\right\rangle^{\rho}_{\chi \cup (\text{FV}(cv) - \text{Dom}(\alpha))}$$
$$\text{if } m = \texttt{<}a \Leftarrow cv\texttt{>}, \ a \in \rho \text{ and } \text{FV}(cv) \cap \text{Dom}(\alpha) \subseteq \rho$$

Based on a slight variant of the transition system described above, a rigorous theory of actor systems is developed in [2]. Specifically, we define and study various notions of testing equivalence on actor expressions and configurations. The model we have developed provides fairness, namely that any enabled transition eventually fires. Fairness is an important requirement for reasoning about eventuality properties. It is particularly relevant in supporting modular reasoning: if we compose one configuration with another which has a nonterminating computation, computation in the first configuration may nevertheless proceed as before, for example, if actors in the two configurations do not interact.

An important result is that the three forms of equivalence, namely, convex, must, and may equivalences, collapse to two in the presence of fairness. We have further developed methods for proving laws of equivalence and have developed proof techniques that simplify reasoning about actor systems. Finally, note that the composition of configurations defines an algebra.

A more concrete way to think of actors is that they represent an abstraction over concurrent architectures. An actor runtime system provides the interface to services such as global addressing, memory management, fair scheduling, and communication. It turns out that these services can be efficiently implemented, thus raising the level of abstraction while reducing the size and complexity of code on concurrent archictectures [8].

Note that the Actor model is, like the theory of higher order nets, general and inherently parallel. Specifically, asynchronous communication preserves the available potential for parallel activity: an actor sending a message need not block until the recipient is ready to receive (or process) a message. More complex communication patterns, such as remote procedure calls, can easily be expressed as a series of asynchronous messages [3]. In fact, it is possible to model a system of actors by means of a higher order net and vice versa [12, 9].

3   Communication Abstractions

It is often possible to encode the computation and communication behavior defined in one model of concurrency in terms of another while preserving the intuitive properties of the behavior. This suggests that there is some sort of equivalence between many different, relatively powerful models of concurrency, although we do not have a consensus about what defines a universal model for concurrency. However, from a programming language point of view, such generality is not enough.

Consider the fact that although asynchronous communication is a natural way of representating primitive interaction in distributed systems, other forms of communication, such as remote procedure calls, are often useful. In actor terms, an rpc-like communication is represented by an actor entering a "wait" state after sending a request to another actor; in this state all messages other than a reply from the "called" actor are deferred. In other words, expressing rpc-like communication in terms of primitive actors (as well as in higher order nets), requires unfolding the continuation behavior into a separate actor (place). The need for such transformation places an unnecessary burden of book-keeping on the programmer.

No model of concurrency can, and perhaps should, allow all communication abstractions to be directly expressed; in fact, different models make some forms easier to express than others. For example, a communication abstraction that appears to be quite easy to express in Petri Nets is invocation by a set of messages. The communication represents a common schema where an actor carries out some useful actions only after some arbitrary set of messages has been received (*input synchronization*).

In a high level actor language, a new programming abstraction must be introduced to allow an abstract specification of a behavior that is activated by a set of messages. In [6], we defined an abstraction, called *activators*, to allow specification of both input synchronization and *reply synchronization*; reply synchronization generalizes rpc-like communication to support the concurrent invocation of a group of actors. A compiler can translate activators into a collection of actors which maintain the local state necessary to capture the partial set of messages received.

4   Interaction Policies

An interaction policy may be expressed in terms of the interfaces of actors and implemented by using appropriate protocols to coordinate between actors. A protocol imposes a certain role on each actor governed by the protocol: in essence it mediates the interactions between actors to ensure that each relevant actor implements its end of the interaction policy.

Consider a common interaction policy, namely, atomicity. Atomicity may be realized by using a protocol such as *two phase commit*. Notice that the implementation of a two phase commit is quite involved: it involves exchanging a number of messages between different actors. Current techniques for developing distributed software require developers to implement interaction policies and application behavior together, significantly complicating code. The lack of modularity not only makes it hard to reason about code; it limits its reusability and portability. Moreover, the resulting code is brittle: modifying an interaction policy to satisfy changing requirements requires modifying the code of each relevant component and then reasoning about the entire system, essentially from scratch. One cannot, for example, simply pull out a two phase commit and replace it with a three phase commit.

Composition of configurations is not sufficient to capture the kind of composition we need here. In the first place, in standard actor semantics, we cannot even express a two-phase protocol by a configuration of actors; what we need to express a two phase commit as an abstraction is the ability to write meta-programs with distributed scope. A two phase commit meta-program imposes a role for each actor, specifically, trapping and tagging incoming and outgoing messages to implement the protocol . Such customization of an individual actor's mail system may be further limited only for the duration of an interaction.

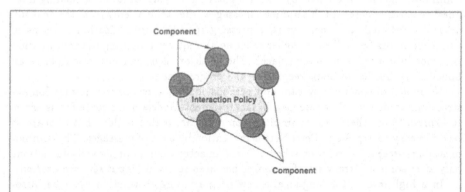

Fig. 2. A distributed system consists of a set of components carrying out local computations and interacting in accordance with a set of *policies*.

We have developed a language for describing and implementing interaction policies [13]; using our language, a protocol abstraction, such as one for the two phase commit protocol, may be instantiated by specifying a particular group of actors and other initialization parameters. The runtime system must then support specific forms of reflection which are sufficient to enable dynamic modification of the mail system, to store and retrieve actor states, or other parts of the meta-architecture.

Now notice that the semantics of actor systems in the presence of protocols is quite different from the semantics of ordinary (the so-called base-level) actor systems. Our pragmatic experience suggests that reasoning about distributed applications is simplified by our meta-programming system; after all, code size is reduced by at least an order of magnitude, and the application is decomposed into more intuitive units corresponding to the requirements specification. However, the semantics of meta-level operations remains poorly understood. Recent research based on actors has made progress on the problem of reasoning in the presence of meta-actors, specifically, by defining a reasoning system and using it to prove the correctness of a meta-level algorithm for taking a global snapshot of a running distributed system of actors [14].

5 Declarative Coordination Constraints

If we want to further abstract over possible mechanisms to implement interaction policies, we need to focus on coordination constraints. As a gross simplification, interaction policies determine when actions take place rather than what individual actors do. It turns out that two types of coordination are often useful.

One coordination scheme imposes precedence constraints on otherwise asynchronous events at different actors, and the other requires such events to be atomic (loosely speaking, to co-occur). By providing a language abstraction, called a *synchronizer*, to express these two types of coordination constraints, we are able to show that the task of distributed programming may be further simplified [5]. Figure 3 shows an example of a synchronizer. Frølund has defined a transition system for actors in the presence of a collection of synchronizers and showed that synchronizers can be implemented efficiently. Because synchronizers may be superimposed, and may be dynamically added or removed, implementing such a system proves to be a fairly challenging task [7].

```
AllocationPolicy =
{
 init prev := 0

 enter(adm1,adm2,max)
    prev = max  disables (adm1.request or adm2.request),
    (adm1.request or adm2.request) updates prev := prev + 1,
    (adm1.release or adm2.release) updates prev := prev - 1,
}
```

Fig. 3. Synchronizer that enforces a global bound, max, on the number of resources allocated by two actors, adm1 and adm2. We assume that each request uses a single resource from a pool of resources.

As one might expect, the semantics of synchronizers fundamentally alters the nature of transitions in actor systems. Essentially, synchronizers provide a declarative mechanism for customizing the behavior of meta-level actors that collectively represent the scheduler. Observe that synchronizers do not add to the set of possible events in an actor system; they merely rule out certain interleavings of events.

6 Real-Time Requirements

Many real world systems need to respond in real-time. Linguistic abstractions can simplify distributed real-time system design, implementation and reasoning. Specifically, we developed a method for modularizing timing properties, separate from each actors functionality. Instead, timing properties are specified as constraints between actors [11]. Because real-time constraints are separately specified, using generic software in real-time systems becomes feasible.

Semantics for concurrent programming languages usually focus on qualitative aspects, which is unsuitable for real-time programming languages. It is critical for real-time applications that quantitative aspects are analyzed and explained. We have shown how an operational semantics, together with the timing constraints, may be translated into an underlying real-time formalism, namely, timed graphs, to provide the real-time semantics of an actor language [10]. A goal of this research is to make reasoning about real time systems compositional.

7 Naming and Groups

Groups of actors are an important unit of representation; for example, in defining protocols we can assign roles to a group of actors rather than an individual actor. The Actorspace model allows an abstract specification of a group of actors [4]. An actorspace associates an actor with specific attributes; the sender of a message specifies a destination pattern which is pattern matched against the attributes of actors in the actorspace. A simple analogy with set theory illustrates the difference between naming in actors and actorspaces. A set may be defined by enumerating its elements, or by specifying a characteristic function which defines a subset in a domain. The first method is analogous to actor communication (where an explicit collection of mail addresses of actors must be specified), whereas the second method corresponds to actorspace communication. Of course, in conventional mathematics the two ways of characterizing sets are equivalent since the properties of mathematical objects are static; by contrast, actors may dynamically change their attributes. Actorspace provides a transparent way of managing groups of actors. It generalizes the notion of ports in process calculi, where object identity is not uniquely defined.

8 Conclusions

We have discussed a number of ways in which the development of concurrent systems may be simplified by abstractions. Such abstractions are useful only if new methods are developed for composing modules, as well as for reasoning about such composition. A key idea is that it is not sufficient to think of composition as a way of plugging together different modules; rather the implementation of services in a concurrent system must itself be thought of as a system of actors that is composed with a (base-level) application and such composition may itself be subject to reconfiguration. One may abstract over patterns of interaction between meta-level actors as well as the mechanisms they use to manage base-level actors by defining higher-level coordination structures.

Acknowledgements

Research described in this paper has been made possible by intense interaction with the past and present members of the Open Systems Laboratory including Mark Astley, Christian Callsen, Chris Houck, Svend Frølund, Shingo Fukui, Nadeem Jamali, WooYoung Kim, Brian Nielsen, Rajendra Panwar, Anna Patterson, Shangping Ren, Masahiko Saito, R. K. Shyamasundar, Daniel Sturman, Takuo Watanabe, and Nalini Venkatasubramanian. I have also particularly benefitted from discussions with Ian Mason, Scott Smith and Carolyn Talcott. The research described has been supported in part by the Office of Naval Research (ONR contract numbers N00014-90-J-1899 and N00014-93-1-0273), the Digital Equipment Corporation, Hitachi, and the National Science Foundation (NSF CCR 93-12495).

References

1. G. Agha. *Actors: A Model of Concurrent Computation in Distributed Systems*. MIT Press, Cambridge, Mass., 1986.
2. G. Agha, I. A. Mason, S. F. Smith, and C. L. Talcott. A foundation for actor computation. *Journal of Functional Programming*, 1996. to appear.
3. Gul Agha. Concurrent Object-Oriented Programming. *Communications of the ACM*, 33(9):125–141, September 1990.
4. C. J. Callsen and G. A. Agha. Open Heterogeneous Computing in ActorSpace. *Journal of Parallel and Distributed Computing*, pages 289–300, 1994.
5. S. Frølund and G. Agha. A language framework for multi-object coordination. In *Proceedings of ECOOP 1993*, volume 707 of *Lecture Notes in Computer Science*. Springer Verlag, 1993.
6. S. Frølund and G. Agha. Abstracting interactions based on message sets. In *Object-Based Models and Languages for Concurrent Systems*, volume 924, pages 107–124. Springer-Verlag, 1996. Lecture Notes in Computer Science.
7. Svend Frølund. *Coordinating Distributed Objects: An Actor-Based Approach to Synchronization*. MIT Press, 1996.
8. W. Kim and G. Agha. Efficient Support of Location Transparency in Concurrent Object-Oriented Programming Languages. In *Supercomputing '95*. IEEE, 1995.
9. S. Miriyala, G. Agha, and Y. Sami. Visiualizing actor programs using predicatae transition nets. *Journal of Visual Languages and Computation*, 3(2):195–220, June 1992.
10. B. Nielsen and G. Agha. Semantics for an actor-based real-time language. In *Fourth International Workshop on Parallel and Distributed Real-Time Systems*, Honolulu, April 1996. (to be published).
11. S. Ren and G. Agha. Rtsynchronizers: Language support for real-time specifications in distributed systems. In *Proceedings of ACM SIGPLAN 1995 Workshop on Languages, Compilers, and Tools for Real-time Systems*, pages 55–64, 1995.

12. Y. Sami and Vidal-Naquet. Formalization of the behavior of actors by colored petri nets and some applications. In *Conference on Parallel Architectures and Languages Euorpe, PARLE'91*, 1991.
13. D. Sturman and G Agha. A protocol description language for customizing failure semantics. In *The 13th Symposium on Reliable Distributed Systems, Dana Point, California.* IEEE, October 1994.
14. N. Venkatasubramanian and C. L. Talcott. Reasoning about Meta Level Activities in Open Distributed Systems. In *Principles of Distributed Computation*, 1995.

Temporal Uncertainty and Fuzzy-Timing High-Level Petri Nets[1]

Tadao Murata

Department of Electrical Engineering and Computer Science
University of Illinois at Chicago
Chicago, Illinois 60607-7053 USA

Abstract This paper discusses an approach to deal with temporal uncertainty and introduces fuzzy timing in a high-level Petri net model. The main features of the present model are the four fuzzy set theoretic functions of time called fuzzy timestamp, fuzzy enabling time, fuzzy occurrence time and fuzzy delay, all of which capture temporal uncertainty in a form not violating the axiom of measurement recently proposed by Dr. Petri. Fuzzy-timing nets are suitable for time-critical applications since fuzzy time functions can be computed very fast.

1. Introduction

During initial deliberations on what topic to present as an invited talk at this conference, the author learned that Dr. Carl Adam Petri was planning to attend and present a special lecture entitled "Nets, Time and Space: an Introduction to Operative Topology." This motivated the author to prepare this paper on a topic related to time and nets. In addition, time is one of the most important considerations in designing *practical* systems, particularly time-critical systems. The notion of time plays a vital role in performance evaluation and specification of dynamic concurrent systems. Many readers have probably wondered at one time or another, as the author did, why Dr. Petri did not include time explicitly in his original definition of nets [Pet66]. We understand that net theory "features proper concurrency, stressing the partial independence of system parts and the priority of causality over temporal order" [Pet96]. Other reasons may be found in Petri's 1987 paper [Pet87]. For example, he provides the following statements: "When I look at timed and stochastic nets, I see some basic problems which have to do with the conceptualization of time and chance." "The first main problem is that you can specify what you wish. But the implementer cannot have equality, he can have only similarity." "Treat time as the state of clocks." In his recent paper [Pet96], he presents many axioms (rules), among which the axioms of measurement and of control are related to time and nets. The axiom of measurement "defines measurement as a set of atomic judgments of the form, $t_0 < t < t_1$. It characterizes time t by the two transitions t_0 and t_1 (not necessarily neighbored ones) which

---
[1] This work was supported by the NSF under Grant CCR-93-21743.

delimit a segment on the scale, ... It establishes a law of uncertainty in continuous/discrete modeling." The axiom of control "defines control cycles as sets of atomic couplings $t_0 < t_a < t_1 < t_b < t_0$, and fixes the details of coupling between Boolean functions of time."

Dr. Petri 's reference to similarity and uncertainty in the above statements reminded the author of fuzzy set theory or fuzzy logic [Zad65, Zad73, Zim91, DP80], which has been applied successfully in modeling and designing many *real-world* systems in environments of uncertainty and imprecision. Both Petri nets and fuzzy logic were conceived about the same time, during the 1960s, but the latter has grown into a more widely-known field of study. (It is said that in Japan the word "fuzzy" is used and applied by the general public and has become a household word.) In light of this background, it seems useful to take a look at a way to model time explicitly in terms of fuzzy set theory (as opposed to the implicit form $t_0 < t < t_1$ characterized by the two transitions t_0 and t_1).

This paper presents a proposed high-level Petri net model (or semantics) which features the fuzzy set theoretic notion of time using the following four fuzzy time functions: fuzzy timestamp $\pi(\tau)$, fuzzy enabling time $e(\tau)$, fuzzy occurrence time $o(\tau)$, and fuzzy delay $d(\tau)$. A fuzzy time function, or possibility distribution, is a function from the time scale *TS*, the set of all non-negative reals, to the real interval [0, 1]. It indicates the more or less possible values (degrees of possibilities) of a date or a point of time τ. It is a subjective way of representing the available, *uncertain* knowledge about time τ. The fuzzy timestamp $\pi(\tau)$ gives the numerical estimate of the possibility that a token *arrives* at time τ in a place. When $\pi(\tau) = 1$, it is completely possible that the token will arrive at time τ; when $\pi(\tau) = 0.5$, there is a '50-50 chance' that it will arrive at time τ; and when $\pi(\tau) = 0$, there is little or no possibility that it will arrive at time τ. For simplicity, we often describe a fuzzy time function $\pi(\tau)$ in the trapezoidal or triangular possibility distribution specified by the 4-tuple ($\pi 1$, $\pi 2$, $\pi 3$, $\pi 4$) as shown in Fig. 1, where the triangular form in Fig. 1(b) is a special case ($\pi 2 = \pi 3$) of the trapezoidal form in Fig. 1(a). As explained in the sequel, fuzzy timestamps, fuzzy

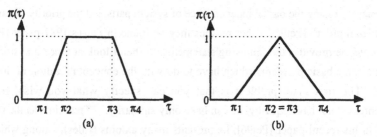

Fig. 1 Trapezoidal and triangular possibility distributions

enabling times and fuzzy occurrence times are computed and updated from the current fuzzy timestamps and initially specified fuzzy delays, *each time when a firing (atomic action) occurs*. The timestamp $\pi(\tau)$ specified by a trapezoidal possibility distribution ($\pi1$, $\pi2$, $\pi3$, $\pi4$) captures temporal uncertainty by the "fuzzy time interval," $\pi1 < \tau < \pi4$, which is in agreement of the form, $t_0 < t < t_1$, stated in the axiom of measurement. Further, in our model a transition occurrence (firing) is atomic. So, delay is not associated with a transition but with an arc directed into a place. The detailed semantics of our model are presented first informally in Section 2 and then more formally in Section 3. Related work is given in Section 4, and concluding remarks in Section 5. It is assumed that the reader is familiar with basic terminology of Petri nets [Mur89].

2. Informal Introduction to Fuzzy-Timing High-Level Petri Nets

In this section we present an informal introduction to our model called fuzzy-timing high-level Petri nets (FTHN). A token in place p of a FTHN has three attributes denoted by $<<p$, $v>$, $\pi(\tau)>$, indicating that this token is in place p with value (or color) v, and fuzzy timestamp $\pi(\tau)$. The fuzzy timestamp $\pi(\tau)$ gives a possibility distribution that the token with value v arrives at time τ in place p. The value of a token is often referred to as the token color. Each place has a color set which specifies the set of allowed values. The FTHN shown in Fig. 2 represents a job shop, where jobs arrive in place p_{in}, and leave the system via place p_{out}. The job shop has a number of machines. Each machine is represented by a

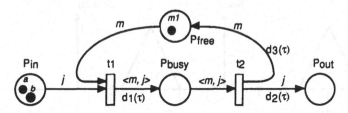

Fig. 2 A FTHN model of a job shop

token which is either in place p_{free} or p_{busy}. There are three color sets: $M = \{m1, m2, \ldots \}$, $J = \{a, b, \ldots \}$, and $M \times J = \{<mi, j > \mid mi \in M \text{ and } j \in J\}$. Color set J (job types) is attached to places p_{in} and p_{out}, color set M (machine types) is attached to place p_{free}. Color set $M \times J$ is attached to place p_{busy}. Transition t_1 has two input places (p_{in}, p_{free}) and one output place (p_{busy}). Transition t_2 has one input place (p_{busy}) and two output places (p_{free} and p_{out}). As in any Petri net, a firing (or occurrence) of an enabled transition consumes tokens from its input places and produces tokens in its output places. The relation between the multiset of the consumed tokens and the multiset of produced tokens is described by the transition function F. For example, the domain of transition function $F(t_1)$ in the job shop model shown in Fig. 2 is:

$$dom(F(t_1)) = \{<<p_{in}, j>, \pi_j(\tau) > + <<p_{free}, m>, \pi_m(\tau) > \mid j \in J \text{ and } m \in M\}$$

and the range is:

$$range(F(t_1)) = \{ << p_{busy}, <m, j>>, \pi_{mj}(\tau) > \mid j \in J \text{ and } m \in M \}.$$

That is, for $j \in J$ and $m \in M$, we have:

$$F(t_1)(<<p_{in}, j> , \pi_j(\tau) > + <<p_{free}, m>, \pi_m(\tau) >) = << p_{busy}, <m, j>>, \pi_{mj}(\tau) >.$$

This means that if t_1 fires or occurs, it consumes one token with value j (a waiting job) from p_{in} and another token with value m (a free machine) from p_{free}, and it produces one token with the attribute, $<< p_{busy}, <m, j>>, \pi_{mj}(\tau) >$, in p_{busy}.

Similarly, for $j \in J$ and $m \in M$, we have:

$$F(t_2)(<<p_{busy}, <m, j>>, \pi_{mj}(\tau) >) = <<p_{out}, j>, \pi_{2j}(\tau)> + <<p_{free}, m>, \pi_{3m}(\tau)>.$$

Transition t_2 represents the completion of a job, and when t_2 fires, it consumes one token $<m, j>$ from p_{busy} and it produces one token $<<p_{out}, j>, \pi_{2j}(\tau)>$ in p_{out} and another $<<p_{free}, m>, \pi_{3m}(\tau)>$ in p_{free}. Computation of these fuzzy timestamps will be explained shortly.

The initial marking M_0 shown in Fig. 2 can be described by the following formal sum of attributes of the three tokens:

$$M_0 = << p_{in}, a>, \pi_{0a}(\tau) > + <<p_{in}, b>, \pi_{0b}(\tau)> + <<p_{free}, m1>, \pi_0(\tau)>$$

where $\pi_{0a}(\tau), \pi_{0b}(\tau)$ and $\pi_0(\tau)$ are the fuzzy timestamps. For the present illustration, let $\pi_{0a}(\tau) = (1,3,3,5), \pi_{0b}(\tau) = (3,5,5,7)$ and $\pi_0(\tau) = (0,0,0,0)$ which are depicted in Fig. 3.

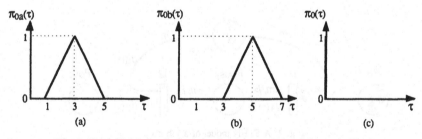

Fig. 3 The initial fuzzy timestamps of the three tokens of the net in Fig. 2

These fuzzy timestamps indicate that machine $m1$ is available starting at time $\tau = 0$, job a arrives in place p_{in} most likely at $\tau = 3$ but possibly between $\tau = 1$ and $\tau = 5$, and job b arrives most likely at $\tau = 5$ but possibly between $\tau = 3$ and $\tau = 7$, with the possibility distributions shown in Fig. 3 (a) and (b), respectively. There are two possible occurrences of transition t_1: one corresponding to event e_{1a} in which machine $m1$ takes job a first, and the other to event e_{1b} in which machine $m1$ takes job b first. We denote an event by a 3-tuple $<t, b_{in}, b_{out}>$, where b_{in} and b_{out} are the multisets of tokens to be consumed and produced, respectively, by an occurrence of t. Thus the above two events e_{1a} and e_{1b} are denoted by $e_{1a} = <t_1, b_{in-a}, b_{out-a}>$ and $e_{1b} = <t_1, b_{in-b}, b_{out-b}>$, where:

$$b_{in-a} = <<p_{free}, m1>, \pi_0(\tau)> + <<p_{in}, a>, \pi_{0a}(\tau)>,$$

$b_{out\text{-}a} = <<p_{busy}, <m1, a>>, \pi_{1a}(\tau)>$

$b_{in\text{-}b} = <<p_{free}, m1>, \pi_0(\tau)> + << p_{in}, b>, \pi_{0b}(\tau)>$, and

$b_{out\text{-}b} = <<p_{busy}, <m1, b>>, \pi_{1b}(\tau)>.$

The possibility distribution of the time at which both tokens, $<<p_{free}, m1>, \pi_0(\tau)>$ and $<<p_{in}, a>, \pi_{0a}(\tau)>$, become available for an occurrence of t_1 is given by the *latest* operation (explained later) on fuzzy timestamps of the tokens to be consumed. This is given for our job shop example by

$$e_{1a}(\tau) = latest\{\pi_0(\tau), \pi_{0a}(\tau)\} = latest\{(0,0,0,0), (1,3,3,5)\} = (1,3,3,5).$$

Similarly , the possibility distribution of the time at which both tokens, $<<p_{free}, m1>, \pi_0(\tau)>$ and $<<p_{in}, m1>, \pi_{0b}(\tau)>$ become available for an occurrence of t_1 is given by

$$e_{1b}(\tau) = latest\{\pi_0(\tau), \pi_{0b}(\tau)\} = latest\{(0,0,0,0), (3,5,5,7)\} = (3,5,5,7).$$

$e_{1a}(\tau)$ and $e_{1b}(\tau)$ are called fuzzy enabling times. A fuzzy enabling time is a possibility distribution of the latest arrival time among arrival times of all the tokens necessary for an occurrence of a transition. As seen in Fig. 3, token a arrives at p_{in} sometime between $\tau = 1$ and $\tau = 5$ and token b sometime between $\tau = 3$ and $\tau = 7$. Thus after $\tau = 3$, there is a *conflict* between the two events e_{1a} and e_{1b} of transition t_1. This conflict is depicted as the conflict between the two transitions e_{1a} and e_{1b} in the uncolored Petri net of Fig. 4, which is obtained by 'unfolding' the FTHN shown in Fig. 2.

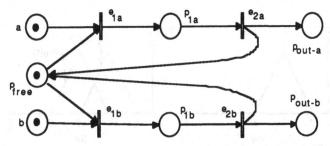

Fig. 4 The uncolored job shop net obtained by unfolding the net shown in Fig. 2

Now, we need to find possibility distributions of the times at which two events e_{1a} and e_{1b} start. To this end, we adopt a 'first-come *possibly*-first-serve' policy that gives a higher priority to a possibly earlier enabled event. (As discussed in Remark 3 in Section 3, this *'fuzzy firing policy'* does not make partial-ordered events totally ordered but adds information on the degrees of possibilities for which event may occur first.) To find the possibly earliest enabled event among all events which are possibly involved with some conflicts, we perform the *earliest* operation (explained later) on the fuzzy enabling times of all quasi-enabled (defined later) events. In this example, we have only two quasi-enabled events and we get:

$earliest\{e_{1a}(\tau), e_{1b}(\tau)\} = e_{1a}(\tau)$.

The *earliest* operator finds a possibility distribution of the earliest time at which an event gets enabled.

Next, to find a possibility distribution of the time at which a specific event e_t occurs and in order not to violate the monotonic increasing property of time, we perform the intersection (minimum operation in fuzzy arithmetic) of the following two fuzzy time functions: the fuzzy enabling time of event e_t and the result of an earliest operation. For our job shop example, we find the possibility distributions of occurrences of the two events as follows:

$$o_{1a}(\tau) = min\{e_{1a}(\tau), earliest\{e_{1a}(\tau), e_{1b}(\tau)\}\}$$
$$= min\{e_{1a}(\tau), e_{1a}(\tau)\} = e_{1a}(\tau) = (1,3,3,5)$$
$$o_{1b}(\tau) = min\{e_{1b}(\tau), earliest\{e_{1a}(\tau), e_{1b}(\tau)\}\}$$
$$= min\{e_{1b}(\tau), e_{1a}(\tau)\} = 0.5(3,4,4,5).$$

$o_{1a}(\tau)$ and $o_{1b}(\tau)$ are called fuzzy occurrence times. A fuzzy occurrence time of an event is a possibility distribution of the time at which the event occurs. From the fuzzy occurrence times $o_{1a}(\tau)$ and $o_{1b}(\tau)$ depicted in Fig. 5, it can be seen that event e_{1a} may occur between $\tau = 1$ and $\tau = 5$ and event e_{1b} may occur between $\tau = 3$ and $\tau = 5$. Thus it is possible that either event occurs first between $\tau = 3$ and $\tau = 5$. However, the possibility that event e_{1a} occurs before event e_{1b} is greater than the possibility that event e_{1b} occurs first.

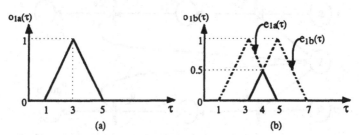

Fig. 5 Fuzzy occurrence times of events e_{1a} and e_{1b}

In any case, there are two possible firing (occurrence) sequences in this example given by

$$\sigma_1 : M_0[e_{1a} > M_1[e_{2a} > M_2[e'_{1b} > M_3[e'_{2b} > M_4$$
$$\sigma_2 : M_0[e_{1b} > M_5[e_{2b} > M_6[e'_{1a} > M_7[e'_{2a} > M_8$$

where $(e_{1a}, e_{2a}, e'_{1b}, e'_{2b})$ and $(e_{1b}, e_{2b}, e'_{1a}, e'_{2a})$ denote the events of σ_1 and σ_2 for transition t_1 or t_2 (shown by the first subscript) with token a or b (shown by the second subscript), respectively. These events correspond to the occurrences (firings) of the unfolded transitions shown in Fig. 4.

Next we explain the fuzzy delay, $d(\tau)$. A fuzzy delay is a fuzzy time function associated with each arc from a transition t to each of its output places p and indicates the possibility distribution of the *relative time* for a token to arrive at place p, i.e., delay from the initiation of an event of transition t to the arrival of a token at place p. Thus, the fuzzy timestamp $\pi(\tau)$ of a token produced at an output place p is computed as the extended addition of the fuzzy occurrence time $o(\tau)$ and the fuzzy delay $d(\tau)$. The extended addition $\oplus$ is based on the extension principle [DP80], a most basic concept of fuzzy set theory that can be used to generalize a crisp mathematical concept to fuzzy sets. The addition $\oplus$ is defined in general by Equation (4) in Section 3. For the trapezoidal forms of possibility distributions, the extended addition simply reduces to the additions of the four parameters [DP80, DP89]:

$$\pi(\tau) = o(\tau) \oplus d(\tau) = (o1,o2,o3,o4) \oplus (d1,d2,d3,d4)$$
$$= (o1+d1,\ o2+d2,\ o3+d3,\ o4+d4). \qquad (1)$$

For the FTHN model shown in Fig. 2, let us assume that $d_1(\tau) = (0,0,0,0)$ and $d_2(\tau) = d_3(\tau) = (4,5,7,9)$, as is depicted in Fig. 6. Note that unlike the other three fuzzy time

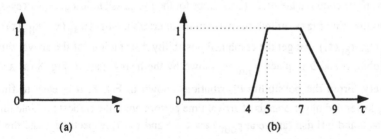

(a)   (b)

Fig. 6 The fuzzy delays d_1, d_2 and d_3 of the net shown in Fig. 2

functions ($\pi(\tau)$, $e(\tau)$, and $o(\tau)$), a fuzzy delay $d(\tau)$ is concerned with the *relative time measured* from the occurrence of an event. Thus, $d_2(\tau) = (4,5,7,9)$ means that the possible delay from the starting time of an event (firing t_2) to the arrival time of a token at p_{out} is between 4 and 9 sec.

Now we illustrate how to compute the fuzzy timestamps of tokens in the two firing sequences: σ_1 and σ_2, which result in the following markings, respectively:

$$M_4 = <<p_{free},\ m1>,\ \pi_4(\tau)> + << p_{out},\ a>,\ \pi_{4a}(\tau)> + << p_{out},\ b>,\ \pi_{4b}(\tau)>$$
and $$M_8 = <<p_{free},\ m1>,\ \pi_8(\tau)> + << p_{out},\ a>,\ \pi_{8a}(\tau)> + << p_{out},\ b>,\ \pi_{8b}(\tau)>.$$

The fuzzy timestamp $\pi_{4a}(\tau)$ (for the arrival time of token a at p_{out} in σ_1) can be computed from $o_{1a}(\tau) = (1,3,3,5)$, $d_1(\tau) = (0,0,0,0)$ and $d_2(\tau) = (4,5,7,9)$, as follows:

$$\pi_{4a}(\tau) = \pi_{2a}(\tau) = \pi_{1a}(\tau) \oplus d_2(\tau) = [o_{1a}(\tau) \oplus d_1(\tau)] \oplus d_2(\tau) = (5,8,10,14),$$

where $\pi_{2a}(\tau)$ and $\pi_{1a}(\tau)$ are the fuzzy timestamps of token a at M_2 and M_1, respectively. The timestamp $\pi_{4b}(\tau)$ (the arrival time of token b at p_{out} in σ_1) can be computed as follows:

$$\pi_2(\tau) \quad = [o_{1a}(\tau) \oplus d_1(\tau)] \oplus d_3(\tau) = \pi_{2a}(\tau) = (5,8,10,14)$$

$$o'_{1b}(\tau) = \textit{latest}\{\pi_{0b}(\tau), \pi_2(\tau)\} = \pi_2(\tau) = (5,8,10,14), \text{ and}$$

$$\pi_{4b}(\tau) = [o'_{1b}(\tau) \oplus d_1(\tau)] \oplus d_2(\tau) = (5,8,10,14) \oplus (4,5,7,9) = (9,13,17,23),$$

where $\pi_2(\tau)$ is the fuzzy timestamp of token $m1$ at M_2 (for the arrival time of machine $m1$ at p_{free} after completing job a in σ_1).

Similarly, for the firing sequence σ_2, $\pi_{8b}(\tau)$ and $\pi_{8a}(\tau)$ (the arrival times of tokens b and a at p_{out} in σ_2) can be computed as follows:

$$\pi_{8b}(\tau) = o_{1b}(\tau) \oplus d_1(\tau) \oplus d_2(\tau)$$

$$= 0.5(3,4,4,5) \oplus (0,0,0,0) \oplus (4,5,7,9) = 0.5(7,9,11,14)$$

$$\pi_{8a}(\tau) = 0.5(7,9,11,14) \oplus d_2(\tau) = 0.5(11,14,18,23).$$

The four timestamps, $\pi_{4a}(\tau)$, $\pi_{4b}(\tau)$, $\pi_{8b}(\tau)$ and $\pi_{8a}(\tau)$ are depicted in Fig. 7(a), (b), (c) and (d), respectively.

Finally, if we combine the fuzzy timestamps for the two possible firing sequences σ_1 and σ_2 by taking the fuzzy union or maximum operation, $max\{\pi_{4a}(\tau), \pi_{8a}(\tau)\}$ and $max\{\pi_{4b}(\tau), \pi_{8b}(\tau)\}$, we get the combined possibility distributions of the arrival times of tokens (jobs), a and b in place p_{out}, as shown by the heavy lines in Fig. 8 (a) and (b), respectively. From the possibility distributions shown in Fig. 8, it is easy to find the following lower and upper bounds of arrival time performance: the earliest possible times at which jobs a and b finish (arrive at p_{out}) are $\tau = 5$ and $\tau = 7$, respectively; and the latest (worst) possible times at which jobs a and b finish (arrive at p_{out}) are $\tau = 23$, in both cases.

Further, we see from the possibility distributions shown in Fig. 8 that job a finishes most likely (the possibility = 1) between $\tau = 8$ and $\tau = 10$, while there is a 50-50 chance (the possibility = 0.5) that it finishes between $\tau = 14$ and $\tau = 18$. Similar observations and statements can be made about job b.

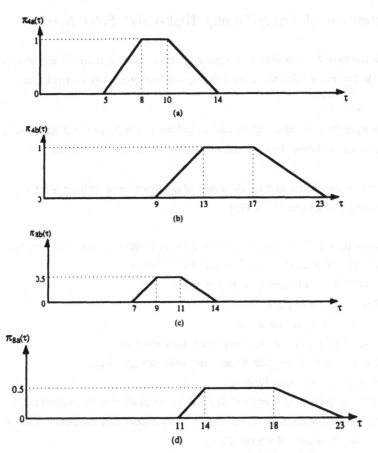

Fig. 7 Fuzzy timestamps of tokens a and b in p_{out} for σ_1 and σ_2

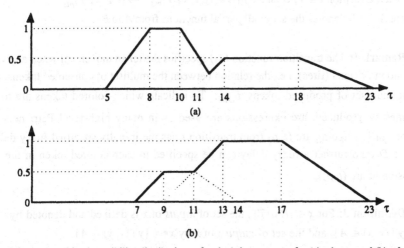

Fig. 8 σ_1-σ_2 combined possibility distributions of arrival times at p_{out} for (a) token a and (b) token b

3. Definition of Fuzzy-Timing High-Level Petri Nets

Definition 1: A *multiset* is a generalized set where multiple appearances of an element in the set are allowed. A multiset B_{ms} can be expressed as a formal sum:
$$\sum_{a \in B_{ms}} \#(a)a$$
where $\#(a)$ is the number of appearances of element a in B_{ms} and subscript ms is used to indicate a multiset but is often omitted when it is obvious from the context.

The reader unfamiliar with the definition of operations on a multiset is referred to one of two excellent articles [Jen92, vdA93].

Definition 2: The static structure of a *Fuzzy-Timing High-Level Petri Net (FTHN)* is a nine tuple, $N = (\Sigma, P, T, A, C, D, CT, FT, F)$ where:

(i) Σ is a finite set of types, called color sets.

(ii) P is a finite set of places.

(iii) T is a finite set of transitions.

(iv) $A \subseteq (P \times T) \cup (T \times P)$ is a set of arcs (flow relation).

(v) D is a set of all fuzzy delays associated with arcs $\subseteq (T \times P)$.

(vi) $C \in P \rightarrow \Sigma$ is a color function.

(vii) $CT = \{<p,v> \mid p \in P \text{ and } v \in C(p)\}$ is the set of all possible colored tokens.

(viii) FT is a set of all fuzzy timestamps. A fuzzy timestamp is a function from the time scale TS to the real interval $[0, 1]$.

(ix) F is the transition function defined from T into functions.
For each transition $t \in T$, we have $F(t) \in (CT \times FT)_{ms} \xrightarrow{\circ} (CT \times FT)_{ms}$
where $A \xrightarrow{\circ} B$ denotes the set of all partial functions from A to B.

Remark 1: The transition function $F(t)$ specifies the token movement when an event of transition t occurs (fires), i.e., the relation between the multiset of consumed tokens, b_{in} and the multiset of produced tokens, b_{out}. To indicate which colored tokens are to be consumed or produced, arc expressions are used as in many high-level Petri nets. In addition, each outgoing arc (t, p) from transition t has the initially specified fuzzy delay, $d_{tp}(\tau) \in D$ (and further a fuzzy delay can be specified to each colored token in the arc expression of arc (t, p)).

Definition 3: For $x \in (P \cup T)$, the set of *inputs* of x is defined and denoted by $\bullet x = \{y \mid (y, x) \in A\}$, and the set of *outputs* of x by $x\bullet = \{y \mid (x, y) \in A\}$.

Definition 4:. A *marking* M of an FTHN is defined as the multiset and expressed as the formal sum of token attributes, $<<p,v>, \pi(\tau)>$, for all tokens present in all places $p \in P$, where $<p,v> \in CT$ and $\pi(\tau) \in FT$. In particular, the initial marking is a marking specified initially, including the initial fuzzy timestamps. All fuzzy timestamps except for the initial fuzzy timestamps are computed and updated each time when an event occurs. We define the *untimed marking* M^u of a marking M of an FTHN as the multiset (formal sum) of token colors, $<p,v>$, for all tokens present in all places $p \in P$ under the marking M. That is, M^u is M with the timestamp attribute $\pi(\tau)$ disregarded.

Definition 5: An *event* is a triple $<t, b_{in}, b_{out}>$, where $F(t)(b_{in}) = b_{out}$, which represents a possible occurrence (firing) of transition t. An occurrence of an event removes the tokens specified by the multiset b_{in} from $\bullet t$ and adds the tokens specified by the multiset b_{out} to $t\bullet$. We denote the set of all possible events given by

$$E = T \times (CT \times FT)_{ms} \times (CT \times FT)_{ms}$$

Definition 6: An event $<t, b_{in}, b_{out}>$ is said to be *quasi-enabled* at M if $(b_{in})^u \leq M^u$, where M^u denotes the untimed marking of M and $(b_{in})^u$ is the b_{in} under the untimed marking M^u. That is, an event is quasi-enabled if there are 'enough, right-colored' tokens in each of its input places.

Definition 7: The *fuzzy enabling time* $e_t(\tau)$ of an event $<t, b_{in}, b_{out}> \in E$ is the possibility distribution of the latest arrival time among arrival times of all tokens in b_{in}, and is computed as follows: If b_{in} has n tokens with n fuzzy timestamps $\pi_i(\tau)$, $i = 1, 2, ..., n$, respectively, we have

$$e(\tau) = latest\{\pi_i(\tau), i = 1, 2, ..., n\}, \qquad (2)$$

where *latest* is the operator that picks the latest arrival distribution among n distributions (a fuzzy-set notation of this operator is under investigation).

Remark 2: The fuzzy time Petri net (FTPN) defined in [FPC94] has a "fuzzy enable interval" associated with each transition, during which the transition remains enabled. We think that this fuzzy enable interval is not necessary in our FTHN model, since it can be included in the fuzzy delay defined in Definitions 2 and 9.

Example 1: As an illustration of (2), consider the two fuzzy timestamps, $\pi_1(\tau) = 0.5(0,1,5,6)$ and $\pi_2(\tau) = (1,3,3,4)$ shown in Fig. 9 (a). Then $\pi_4(\tau) = latest\{\pi_1(\tau), \pi_2(\tau)\} = 0.5(1,2,5,6)$ and is shown by the heavy line in Fig. 9 (a). As another illustration, let's add the 3rd fuzzy timestamp, $\pi_3(\tau) = (6,7,7,8)$ to $\{\pi_1(\tau), \pi_2(\tau)\}$. Then we get $\pi_5(\tau) =$

$latest\{\pi_1(\tau), \pi_2(\tau), \pi_3(\tau)\} = 0.5(6, 6.5, 7.5, 8)$, which is shown by the heavy line in Fig. 9 (b).

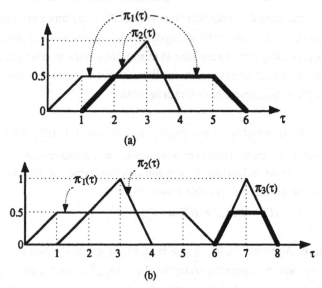

Fig. 9 The heavy lines show (a) *latest* $\{\pi_1(\tau), \pi_2(\tau)\}$ and (b) *latest* $\{\pi_1(\tau), \pi_2(\tau), \pi_3(\tau)\}$

Definition 8: The *fuzzy occurrence time* $o_t(\tau)$ of an event $e_t = <t, b_{in}, b_{out}> \in E$ is the possibility distribution of the time at which the event e_t occurs (or fires) and is computed as follows: Suppose that there are m events e_i, $i = 1, 2, ..., t, ..., m$, which are quasi-enabled. Let their fuzzy enabling times computed by (2) be $e_i(\tau)$, $i = 1, 2, ...t, ..., m$. Then the fuzzy occurrence time of an event $<t, b_{in}, b_{out}>$ with its fuzzy enabling time, $e_t(\tau)$, is given by

$$o_t(\tau) = min\{e_t(\tau), earliest\{e_i(\tau), i = 1, 2, ...t, ..., m\}\} \qquad (3)$$

where *earliest* is the operator that picks up the 'earliest enabling' time among all quasi-enabled events in a net (a fuzzy-set notation of this operator is under investigation).

Remark 3: How the fuzzy occurrence times should be computed depends on what kind of 'firing policy' to adopt or not to adopt for selecting which transition(s) to fire when more than one is enabled. This is one of the most difficult or controversial semantics to consider when time is introduced in net models of concurrent systems with partial-ordered events, since introducing time means adding some (temporal) ordering to partial ordering. Note that in our FTHN model, fuzzy timing does not make partial-ordered events totally ordered but adds some information on degrees of possibilities for event occurrence times. How the fuzzy occurrence times should be computed in our FTHN model depends on specific applications. For time-critical applications, it is justifiable to adopt a policy that gives a possibly higher priority on 'earliest enabling' events, such as the policy given by (3). But, there may be other firing policies that are perfectly justifiable for particular application domains.

Example 2: As an illustration of the computation by (3), consider the three fuzzy enabling times, $e_1(\tau) = 0.5(0,1,6,7)$, $e_2(\tau) = (1,3,3,5)$, and $e_3(\tau) = (6,7,7,8)$ shown in Fig. 10 (a). The result of the earliest operation, $earliest\{e_1(\tau), e_2(\tau), e_3(\tau)\}$ is shown by the

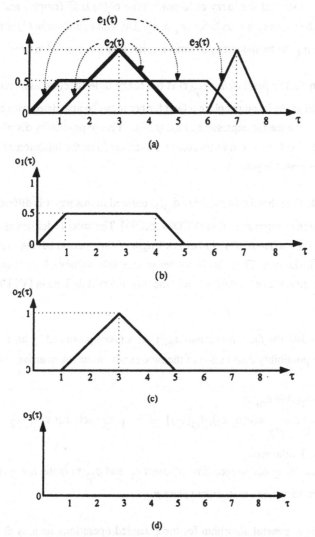

Fig. 10 (a) $e'(\tau) = earliest\{e_1(\tau), e_2(\tau), e_3(\tau)\}$ shown by the heavy line; (b) $o_1(\tau) = min\{e_1(\tau), e'(\tau)\}$; (c) $o_2(\tau) = min\{e_2(\tau), e'(\tau)\}$; and (d) $o_3(\tau) = \varnothing$

heavy line in Fig. 10 (a). Then for i = 1, 2, and 3, the fuzzy occurrence time $o_i(\tau)$ can be found as the intersection of two fuzzy time functions: the fuzzy enabling time $e_i(\tau)$ and the result of the earliest operation. That is,

$$o_1(\tau) = min\{e_1(\tau), earliest\{e_1(\tau), e_2(\tau), e_3(\tau)\} = 0.5(0,1,4,5)$$
$$o_2(\tau) = min\{e_2(\tau), earliest\{e_1(\tau), e_2(\tau), e_3(\tau)\} = (1,3,3,5)$$
$$o_3(\tau) = min\{e_3(\tau), earliest\{e_1(\tau), e_2(\tau), e_3(\tau)\} = \varnothing$$

$o_1(\tau)$, $o_2(\tau)$, and $o_3(\tau)$ are shown in Fig. 10 (b), (c), and (d), respectively. As seen from Fig. 10 (a), *earliest* is the operator that picks up the 'earliest enabling' time of all quasi-enabled events. Note that the fuzzy occurrence time $o_3(\tau)$ is $\varnothing$ (empty) and there is no possibility that the event e_3 occurs before e_1 or e_2. The min operation in (3) is necessary to ensure not violating the monotonicity of time, i.e., not going backward in time.

Definition 9: The *fuzzy delay* $d_{tp}(\tau)$ is the fuzzy time function associated with arc (t, p) from transition t to its output place p (and further it can be associated with each colored token with color v in the arc expression of arc (t, p)). It is the possibility distribution of the length of time period for a token (with color v) to travel from the initiation of an event of transition t to the arrival in place p.

Remark 4: Note that the fuzzy delay $d_{tp}(\tau)$ defined in this paper is different from the fuzzy time duration or interval used in [VCD89. deF94]. The latter is defined as an 'interval' between *two ill-known dates*. The former is a length of time period having *only one value* which may be ill-known. Thus the fuzzy set of possible values of our fuzzy delay is considered to be disjunctive [DP89], whereas the time interval defined in [VCD8, deF94] is conjunctive.

Definition 10: The *fuzzy timestamp* $\pi_{tp}(\tau)$ of a token produced by an event $e_t = \langle t, b_{in}, b_{out}\rangle$ is the possibility distribution of the time at which the token arrives in $p \in t\bullet$ and is given by

$$\pi_{tp}(\tau) \quad = o_t(\tau) \oplus d_{tp}(\tau)$$
$$= \sup_{\tau=\tau_1+\tau_2} min\{o_t(\tau_1), d_{tp}(\tau_2)\}, \text{ if } \exists\ \tau_1, \tau_2 \text{ such that } \tau=\tau_1+\tau_2 \quad (4)$$
$$= 0 \text{ otherwise,}$$

where $o_t(\tau)$ is the fuzzy occurrence time of event e_t and $d_{tp}(\tau)$ is the fuzzy delay of the outgoing arc from transition t to its output place p.

Remark 5: A general algorithm for the extended operations such as $\oplus$ is found in [DP80]. For the trapezoidal forms of possibility distributions, the above extended addition

$\oplus$ is reduced to Equation (1) in Section 2. The four fuzzy time functions: $\pi(\tau)$, $e(\tau)$, $o(\tau)$ and $d(\tau)$ discussed in this paper are concerned with a *date* or a *point of time* τ (or a length of time in the case of $d(\tau)$), each of which may be ill-known but represents *one distinct number* in the time scale *TS*. Thus the fuzzy set of possible values of an ill-known time τ in the four fuzzy time functions is a *disjunctive* set [DP89], and many concepts for modeling and processing fuzzy temporal knowledge discussed in [DP89] can be applied.

Definition 11: If an event $e_1 = \langle t_1, b_{in}, b_{out} \rangle$ occurs at a marking M_1, the resulting marking M_2 is given by:
$$M_2 = (M_1 - b_{in}) + b_{out} \qquad (5)$$
where the fuzzy timestamps in b_{out} is computed by using (2), (3) and (4). M_2 is said to be *directly reachable* from M_1 via event e_1 and we write $M_1[e_1 \rangle M_2$. A *firing sequence* σ is an alternating sequence of markings and events, and we write $\sigma = M_1[e_1 \rangle M_2[e_2 \rangle M_3[e_3 \rangle M_4 \cdots [e_n \rangle M_n$. Marking M_n is said to be *possibly reachable* from marking M_1 if there is a possibility of a finite firing sequence that transforms M_1 into M_n.

4. Related Work

Although it appears that there exist many papers that combine the concepts of fuzzy logic or fuzzy set theory with Petri nets, only two works that introduce fuzzy timing in Petri nets have come to our attention. One is due to Valette et al [VCD89] and the other is due to Figueiredo et al [deF94, FPC94]. In the context of imprecise markings for modeling real-time control and monitoring manufacturing systems, Valette et al [VCD89] presents a brief description of 'fuzzy-time Petri nets' and outlines a procedure to compute fuzzy markings and fuzzy firing dates. In essence, they have extended the crisp time interval of Merlin's time Petri net [Mer76] to a fuzzy time interval using a class of high-level nets, called Petri nets with objects. Merlin's time Petri net is said to be the first model to consider (non-fuzzy) temporal uncertainty, where the time interval $[t_0, t_1]$ associated with each transition models an uncertain occurrence (firing time) of the transition. In his Ph.D. thesis [deF94] and also in a conference paper [FPC94], Figueiredo presents a comprehensive fuzzy time Petri net model and introduces new concepts such as a fuzzy reachability graph. Then the model is extended to a modular high-level Petri net called fuzzy time G-net system and it is applied to the timing analysis of real-time software systems [FPC94].

In addition to the difference pointed out in Remark 4, there is another major difference between our FTHN model and the models introduced by Valette et al and Figueiredo. In these models, the fuzzy time function or interval associated with place p is concerned with the time that a token *exists* in the place p (as opposed to the arrival time of a token in p for our model). This interpretation causes some difficulties, for example, in computing upper bound performance such as latest (worst case) arrival times, since some tokens may exist in a place all the time, while some of other tokens are moving in and out of the place.

While fuzzy time functions are not considered, deterministic timestamps are used in many net models, e.g., van der Aalst [vdA92, vdA93], Ghezzi et al [GMMP91], and [Jen92]. In particular, our model is closely related to and can be viewed as a generalization of the interval timed colored Petri net (ITCPN) model proposed by van der Aalst [vdA93]. For this reason and to ease comparison, in Sections 2 and 3 we have chosen to present and illustrate our model using a running example and formalism that are similar to those used in [vdA93]. In the ITCPN model, the timestamp is defined first as a point of time at which a token becomes available and then extended to a time interval in order to define transition enabling and to reduce the size of a reachability graph for analysis of ITCPNs. This 'extended' ITCPN model can be considered as a special case of our FTHN model, where the four fuzzy time functions, $\pi(\tau)$, $e(\tau)$, $o(\tau)$ and $d(\tau)$ are of the crisp trapezoidal form ($\pi 1$, $\pi 1$, $\pi 4$, $\pi 4$), i.e., an interval [$\pi 1$, $\pi 4$]. In this sense and as discussed in [vdA92], our FTHN can also be used to model other existing semantics of time or timed nets and for analysis of performance measures such as throughput time, response time, occupation rate, etc.

5. Concluding Remarks

The computations involved in FTHNs are basically repeated additions and comparisons of real numbers and are necessary only for certain finite firing sequences, such as those in which each transition in an uncolored (unfolded) net fires at most once. Thus these computations can be done very fast, making FTHNs suitable for time-critical applications [MSS96]. As mentioned in [VCD89], it is well known in the field of fuzzy logic that allowing some degree of uncertainty in modeling tends to simplify the modeling and analysis of very complex systems. Although results obtained from these simplified models are less precise, their relevance to the original systems is fully maintained. The FTHN model is consistent with this, since it provides fast derivation of relevant information without generating the entire state space. FTHNs are considered to be complementary to existing timed or stochastic net models.

As discussed in Remark 3 in Section 3, FTHNs provide additional information on partial ordered events in terms of their degrees of possibilities (instead of transforming them into a total ordering). Thus we should take advantage of temporal information as long as it is reliable, since it can provide additional useful information about the behavior of systems.

As is evident from Dr. Petri's recent work [Pet96], the subject of time and nets contains profound issues that are fundamental for development of the Petri net field. In the present paper, we have presented some initial thoughts on a method for modeling time explicitly and yet in a form not violating the axioms concerning time [Pet96]. Hopefully, this paper, together with the recent work of Dr. Petri [Pet96], will stimulate further research and development in the area of time and nets. This can help make new needed progress into *real-world* applications of Petri nets.

Acknowledgment

Being an invited paper, this paper did not get the benefits of the refereeing process. However, the following people kindly read and provided helpful suggestions to improve the paper: Sol Shatz, Takeshi Suzuki, Shawn Huang and Eric Juan. Also, I'd like to thank Carl Adam Petri, Robert Valette, Jorge C.A. de Figueiredo and Wil van der Aalst for providing their papers and theses.

References

[Pet66] C.A. Petri, "Kommunikation mit Automaten." Bonn: Institut fur Instrumentelle Mathematik, Schriften des IIM Nr. 3, 1962, also, English translation, "Communication with Automata", New York: Griffiss Air Force Base, Tech. Rep. RADC-TR-65-377, Vol. 1, Suppl. 1, 1966.

[Pet87] C.A. Petri,. "'Forgotten Topics' of Net Theory," in Procs. of the 1987 Advanced Course in Petri Nets - Part II, Lecture Notes in Computer Science, Vol. 255, Springer-Verlag, New York, pp. 500-514, 1987.

[Pet96] C.A. Petri, "Nets, Time and Space," *Theoretical Computer Science*, 153 (1996), pp. 3-48.

[Mur89] T. Murata, "Petri Nets: Properties, Analysis and Applications," *Proceedings of the IEEE*, Vol. 77, No 4, April, 1989, pp. 541-580.

[Zad65] L.A. Zadeh, "Fuzzy Sets," *Information and Control*, Vol. 8, pp. 338-353, 1965.

[Zad73] L.A. Zadeh, "Outline of a New Approach to the Analysis of Complex Systems and Decision Processes," *IEEE Trans. on Systems, Man, and Cybernetics*, Vol.-SMC-3, No. 1, Jan. 1973.

[Zim9] H. J. Zimmermann, *Fuzzy Set Theory and its Applications*, Kluwer Academic Publishers, Norwell, Massachusetts, 1991.

[DP80] D. Dubois and H. Prade, *"Fuzzy Sets and Systems: Theory and Applications,"* Academic Press, Inc., New York, 1980.

[MSS9] T. Murata, T. Suzuki, and S. Shatz, "A Fuzzy-Timing Petri Net Model and its Application to a Real-Time Network Protocol," in preparation, 1996.

[VCD89] Valette, R., Cardoso, J., Dobois, D., "Monitoring Manufacturing Systems by Means of Petri Nets with Imprecise Markings", *IEEE International Symposium on Intelligent Control*, September 25-26, 1989, Albany, N.Y., USA.

[DP88] D. Dubois and H. Prade, "Possibility Theory: an Approach to Computerized Processing of Uncertainty," Plenum Press, New York, 1988.

[DP89] D. Dubois and H. Prade, "Processing Fuzzy Temporal Knowledge," *IEEE Trans.. on Systems, Man, and Cybernetics*, Vol. 19, No. 4, July/August 1989, pp. 729-744.

[Mer76] P.M. Merlin, "A Methodology for the Design and Implementation of Communication Protocols," *IEEE Trans. on Communications*, Vol. COM-24, No. 6, pp. 614-621, June 1976.

[deF94] J. C. A. de Figueiredo, "Fuzzy Time Petri Net", *Ph.D Dissertation*, Universidade Electrical Engineering Department, Federal da Paraíba, August, 1994.

[FPC94] J. C. A. de Figueiredo, A. Perkursich, and S.K. Chang, "Timing Analysis of Real-Time Software Systems Using Fuzzy Time Petri Nets," In Procs. of the 6th International Conference on Software Engineering and Knowledge Engineering, Riga, Latvia, June 1994.

[vdA92] W. van der Aalst, "Timed Coloured Petri Nets and their Applications to Logistics," *Ph.D. thesis*, Eindhoven Univ. of Technology, Eindhoven, 1962.

[vdA93] W. van der Aalst, "Interval Timed Colored Petri Nets and their Analysis," Application and Theory of Petri Nets 1993 (Procs. of the 14th Int. Conf. on Application and Theory of Petri Nets, Chicago, IL USA, June 21-25, 1993), pp. 453-472, LNCS 691, Springer-Verlag.

[Jen92] K. Jensen, *"Coloured Petri Nets,"* Vol. 1 and Vol. 2, EATCS Monographs on Theoretical Computer Science, Springer - Verlag, 1992 and 1994.

[GMMP91] C. Ghezzi, D. Mandrioli, S. Morasca, and M. Pezze, "A Unified High-Level Petri Net Formalism for Time-Critical Systems," *IEEE Trans. on Software Engineering*, Vol. 17, No. 2, Feb. 1991, pp. 160-172.

Compositionality in State Space Verification Methods

Antti Valmari

Tampere University of Technology
Software Systems Laboratory
PO Box 553, FIN-33101 Tampere
FINLAND
ava@cs.tut.fi

Abstract. The purpose of this article is to introduce the concepts and ideas that are necessary for understanding computerised process-algebraic compositional verification. Furthermore, by describing two recent case studies an attempt is made to demonstrate the power of the compositional approach and to describe some advanced ways of using the basic techniques. The case studies are given first so that the basic concepts may be introduced in an informal manner. Then the article attempts to elucidate the process-algebraic way of modelling systems and their behaviours. The idea of compositionality is explained. The most important process-algebraic semantic models are described in detail and related to each other, paying special emphasis on algorithmic issues. The representation is at the semantic and state space level; process-algebraic languages are not discussed. Therefore, the techniques presented should be immediately applicable to a wide range of formalisms that can be given semantics in terms of state spaces and transition occurrences.

1 Introduction

The goal of this article is to present a "road map" of the land of process-algebraic compositional state space verification. Although process-algebraic theories were developed mostly in the context of process-algebraic specification languages, the verification algorithms operate essentially at the level of state spaces. They are thus almost completely independent of the languages and can be applied to Petri net reachability graphs. It is necessary to use some language operators to obtain compositionality, though. The operators may be chosen in many ways, but they usually include some versions of *parallel composition* and *hiding*. Typical parallel composition operators correspond closely to transition fusion in Petri nets, and hiding affects only the names of transitions. Furthermore, the syntactical appearance of operators and expressions composed of them is not important. Therefore, the operators do not cause much problems to Petri net applications of the methods described in the sequel.

Process-algebraic compositionality is only one, although very important, approach to compositionality. The author regrets that others could not be covered in this presentation. Furthermore, although the goal was to cover the most important ideas, this presentation is no doubt biased towards the preferences of its author. Even so, because this article introduces many general concepts and terminology, it should be beneficial also for studying the forgotten topics from original papers.

The next section describes two verification examples that give some idea of the kinds of things facilitated by the techniques presented in the later sections. They also informally introduce many key concepts that will be needed later. The way systems and their behaviours are modelled in the process-algebraic world and the notions of black-box semantics and compositionality are discussed in Section 3. Section 4 introduces the most important process-algebraic equivalences, plus the author's favourite, and the basic algorithms used in manipulating them. Some more advanced ideas that could not be explained in detail in this presentation are mentioned briefly in the concluding section.

2 Two Motivating Examples

2.1 A Modified Alternating Bit Protocol

The *alternating bit protocol* [BSW69] is a simple link layer communication protocol used in many verification case studies. In it, if the sender process does not receive an acknowledgement for a message from the receiver process within a specified period of time, it re-transmits the message. If necessary, the message is transmitted for a third time and for a fourth time and so on without limit, until an acknowledgement arrives. Each message is augmented by one bit (called *alternating bit*), with which the protocol receiver and sender can distinguish previously unseen messages and acknowledgements from retransmissions of previous messages and acknowledgements.

Unlimited retransmission is not what practical protocols are expected to do. Instead, after a specified number of futile transmission attempts, the sender should give up and inform the client that the connection is probably broken. When the client later issues a new transmission request, the connection may be operational again, and if it is, the protocol should deliver the message normally.

Motivated by this reasoning, in [VKS96] the present author and two of his students used their *LTS abstraction, reduction and visualisation* technique [VaS96] to find out what would happen if, after giving up, the sender just gave an error indication and returned to its initial state. They wrote a corresponding model of the protocol and then commanded the *ARA* computer tool [VK+93, Sav95] to show its global behaviour as seen by its clients. The outcome is shown in Figure 1. The figure has been faithfully redrawn according to a computer-generated original. The only important difference in Figure 1 and the original is that in the original, names of edges were denoted by showing the edges in different colours, which would of course have made the names indistinguishable in this black-and-white printout.

In the figure, the initial state of the protocol is the one in the middle with a short incoming arrow starting from nowhere. The edges labelled $send\langle m_1 \rangle$ and $rec\langle m_1 \rangle$ denote transmission requests by the sending client and deliveries to the receiving client of the message m_1. The use of only one kind of messages is, of course, a restriction, but that problem will be solved later in this section. Error indications are denoted by *err*.

The τ-labelled edges specify who makes the choices between alternative transitions, the protocol or its clients. They can be thought of as invisible actions made by the protocol (although we will see in Section 4.7 that the issue is not that simple). For instance, at the bottommost state either a *rec*-transition may be taken, or a τ-transition

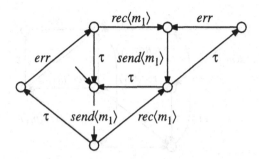

Fig. 1. External behaviour of Alternating bit protocol
with bounded number of retransmissions

followed by *err*. This means that the protocol may deliver the most recently sent message, or it may choose not to deliver it and give an error indication instead. That there is a τ-edge preceding the *err*-edge means that the clients have no way of preventing the *err*-branch from being taken. That is, they cannot prevent the protocol from losing the message. This was expectable, because losses of messages are ultimately determined by the lower-level channels used by the protocol, whose operation is hidden in the figure. That there is no similar τ-edge before the *rec*-edge tells us that if the receiving client refuses to receive the message, then the protocol is guaranteed to issue an error indication sooner or later. The notion of "refusal from receiving a message" might seem strange at first, but it is a possibility allowed by most process-algebraic theories (it is a consequence of the assumption that all communication is synchronous). The readers who feel uneasy with it may at this stage just ignore it and read the picture as saying that it is the protocol, not the clients, who decides whether the *rec*- or *err*-branch is taken.

We can see from the figure that a transmission request issued when the protocol is in its initial state can generate four different sequences of transitions. The simplest to explain is the *send-rec-τ*-triangle taking the protocol back to the initial state: it corresponds to a successful transmission. Also the *send-τ-err-τ*-cycle at the left side of the figure was expectable, since in it the protocol fails to deliver the message and returns to the initial state. But the *send-τ-err-rec*- and *send-rec-τ-err*-sequences may seem surprising. First, in them the protocol delivers the message, but still gives the error indication. The explanation is that it is possible that the original transmission or some of its retransmissions gets through, but all acknowledgements are lost. Then the protocol receiver delivers the message, but the protocol sender does not know that the message got through and can thus only issue the error indication. Second, these sequences of actions do not return the protocol to its initial state, but to the middle state in the top row. We can read from the figure that the next sending request is bound to not lead to the delivery of a message. Furthermore, it is either not followed by an error indication, or it leads the protocol back to the problematic state. This is not acceptable behaviour.

The reason for the above-described bad behaviour is that if the message gets through but all acknowledgements are lost, then the protocol receiver reverses its alternating bit, but the protocol sender does not. Therefore, the sender and receiver have different ideas as to what the alternating bit of the next original transmission should be.

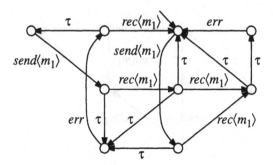

Fig. 2. External behaviour after modifications

Consequently, the receiver rejects the next new message as a retransmission of the previous message.

However, even without knowing the cause of the bad behaviour, an interesting observation can be made. After an error indication, the protocol may lose the next message but no further messages, unless it gives a new error indication. This suggests that the bad behaviour might be fixed by introducing a "void" or "synchronisation" message which is otherwise like an ordinary message (in particular, it contains the alternating bit), but the receiver should never deliver it to the client. Whenever the previous transmission resulted in an error indication, the protocol sender should send the synchronisation message before the actual message. The synchronisation message is either accidentally or intentionally not delivered to the client, but the net effect is the same: it is not delivered. The attempt to transmit the synchronisation message either returns the protocol to the normal operation, so that the actual message will be handled correctly, or leads to an error indication, in which case transmission of the actual message is not even attempted.

The behaviour of the protocol with the above-mentioned modification is shown from the clients' viewpoint in Figure 2, again faithfully redrawn according to a computer-generated original. Most of the behaviour is perfectly acceptable, such as the sequences *send-rec-τ*, *send-rec-τ-err*, *send-τ-err-rec* and *send-τ-err-τ-send-rec-τ* which lead from the initial state back to itself. The only suspicious detail is the doubled reception in the middle row of the figure, because it seems as if the same message could be delivered twice. However, it is easy to check that a double delivery has at least one message loss in its history, so the extra delivery might actually be a delayed delivery of the earlier message believed be lost, and the protocol might actually be correct. If the possibility of message duplication could be ruled out, then the protocol would be known correct.

Figure 3 shows the global behaviour of the protocol when two different messages, m_1 and m_2, may be transmitted. The view is restricted so that only the transmission requests and deliveries of m_2-messages are shown. Figure 3 reveals that the very first m_2-message is delivered at most once, even if an arbitrary number of m_1-messages has been handled before and after it. The protocol processes never look at the contents of the message. Therefore, for an arbitrary positive integer n, the fact that the nth message of the sequence consisting of $n-1$ m_1-messages, one m_2-message, and zero or more

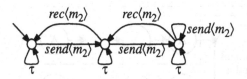

Fig. 3. Checking message duplications

m_1-messages is delivered at most once implies that the nth message of any sequence is delivered at most once. In brief, the protocol does not duplicate messages. (This kind of reasoning appealing to the insensitivity of the model to the contents of messages is called *data-independency* [Wol86].)

Figure 3 contains one more detail worth an explanation, namely the τ-loops attached to each of its states. A τ-loop denotes the possibility of an infinite sequence of actions that are not visible from the chosen point of view. In Figure 3 they represent the possibility that from some point on, only m_1-messages, and an infinite number of them, are given to the protocol for transmission. Because the chosen view shows only the sending and reception of m_2-messages, an infinite sequence of transmissions of m_1-messages appears as a τ-loop. We can even reason that the τ-loops are not caused by livelocks within the protocol. Otherwise Figure 2 would contain a τ-loop.

This example shows that visualisation of behaviour that has been restricted to a suitable point of view can be a very efficient and pleasant way of obtaining information of a concurrent system, in particular when playing with new design ideas. The theoretical notions and algorithms facilitating the generation of restricted views will be discussed in later sections. (More details of this case study, including accurate specifications of the final protocol sender and receiver processes, are presented in [VKS96].)

2.2 Sliding Window Protocol with Arbitrarily Long Channels

The alternating bit protocol is, in the end, a small system, and well within the capacity of today's verification tools. In order to investigate the power of certain verification approaches in more complicated problems, Roope Kaivola conducted a detailed case study [Kai96] on the *sliding window protocol* [Ste76].

The sliding window protocol is a generalisation of the alternating bit protocol where the alternating bit is replaced by a *sequence number*. The sequence number is an integer between 0 and $w-1$ for some positive integer w, and successive original messages are given successive sequence numbers modulo w. An acknowledgement with sequence number x confirms the reception of all messages up to the one numbered by x. Therefore, it is not necessary to acknowledge all messages separately. The maximum number of unacknowledged messages is given as a parameter tw to the protocol. Furthermore, the receiver is allowed to accumulate at most rw messages that have not yet been delivered to the client, where rw is another parameter. This is useful if, say, message number 3 was lost but 4 and 5 did arrive, because then the receiver could retain the messages 4 and 5 until a retransmission of message 3 arrives, and then deliver all three to the customer and acknowledge all three in one acknowledgement. It is required that $tw + rw \leq w$.

The sliding window protocol is a challenge to verification tools, because its behav-

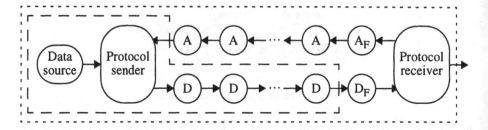

Fig. 4. Verification model of the sliding window protocol

iour is complicated, and the size of its state space grows rapidly when the parameters w, tw and rw or the capacities of the low-level transmission channels are increased. Computer-aided verification of it has been attempted at least once before Kaivola's work, namely in [RR+88], but that analysis was restricted to so-called *safety properties* (that is, liveness and progress were left out), and the parameters and channel capacities were given only small values. As a matter of fact, the version analysed in [RR+88] was slightly different from the one in [Ste76]. Kaivola analysed also the [RR+88] version and found a liveness error in it. (Also in [BrJ82] safety properties of the sliding window protocol were verified with the aid of a computer, but the approach used there was semi-automated theorem proving heavily guided by the human verifier.)

Kaivola modelled the protocol as is shown in Figure 4. The low-level channel for data messages was modelled by connecting together $n-1$ reliable channel cells D and one unreliable cell D_F. The capacity of each cell is one, so the channel as a whole modelled a fifo queue with capacity n which may lose messages. The low-level channel for acknowledgements was constructed in a similar way from $m-1$ reliable cells A and one unreliable cell A_F. Protocol sender and Protocol receiver modelled the actual protocol. Finally, Data source was added for verification purposes. Its task was to generate a certain sequence of messages for transmission by the protocol. Two different data sources were used, one (DSsafe) for verifying the safety properties of the protocol and another (DSlive) for liveness. The data sources are shown in Figure 5.

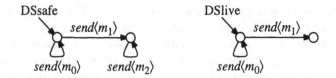

Fig. 5. The data sources used in verification

To verify safety properties of the protocol, Kaivola restricted the view such that only the deliveries of messages were visible, and visualised the resulting LTSs with the ARA toolset [VK+93]. Then he used data-independency for checking the correctness of the protocol from the visualised LTSs. For instance, he reasoned that if the protocol could duplicate a message, then the visualised LTS obtained with DSsafe would contain a sequence with two deliveries of m_1-messages. But the visualised LTS did not show any such sequence, so he reasoned that the protocol was correct in this respect. To verify liveness properties Kaivola added a *tester process* to the system such that violations of the properties emerged as infinite sequences of a special action of the tester. Altogether he verified a fairly complete set of safety and liveness properties.

Kaivola's way of constructing an LTS for the protocol was an elaboration of the following. First an LTS for the parallel composition of Data source and Protocol sender is computed. Then it is *reduced* with ARA in such a way that its externally observable behaviour does not change. Let us call the reduced LTS DP. Next a parallel composition of DP and D is constructed, and the result is reduced again. Continuing in this way, the data channel cells D are added one by one, then the acknowledgement channel cells A, and finally D_F, A_F and Protocol receiver, each time reducing the result before computing the next parallel composition. In this way it is possible to work with reasonably small LTSs all the time; it is never necessary to construct the huge full LTS of the protocol. This kind of *compositional LTS construction* is a very useful technique for attacking the state explosion problem.

Actually, Kaivola's approach was even more clever. Instead of using Protocol sender, he used a process (call it PS′) that is "worse" than Protocol sender in a certain precise sense (namely *NDFD-preorder*, Section 4.7). Roughly speaking, PS′ was equivalent to Protocol sender together with arbitrarily long data and acknowledgement channels. It so happened that the externally observable behaviour of the parallel composition of Data source, PS′ and D proved equivalent to the parallel composition of Data source and PS′ only. But this implies that the parallel composition of Data source, PS′ and any number of D-processes is equivalent to the parallel composition of Data source and PS′. Furthermore, the same happened with the acknowledgement cells A. Therefore, the parallel composition of Data source and PS′ actually represented the parallel composition of Data source, PS′ and any arbitrary numbers of A- and D-processes. Consequently, the behaviour he obtained when he added also D_F, A_F and Protocol receiver to the system was valid for arbitrary channel capacities.

Of course, the behaviour obtained with PS′ is not the behaviour of the real protocol, because PS′ is not the real Protocol sender, but its approximation. On the other hand, the techniques used by Kaivola guaranteed that because PS′ is "worse" than the real Protocol sender, the behaviour of the whole protocol obtained using PS′ is "worse" than the behaviour of the real protocol. Because Kaivola was able to verify that the "worse" behaviour is correct, also the behaviour of the real protocol is correct. In brief, using a behaviourally "worse" approximation of Protocol sender, and observing that it yields global behaviour that satisfies the requirements for the protocol and is independent of channel capacities, he proved the behaviour of the real protocol correct independently of the channel capacities.

Although ingenuity was required to devise this verification approach and it took some trial and error to find a suitable "worse" approximation of Protocol sender, Kaivola's verification was mostly automatic. Now that the approach has been tried once it can be re-used more easily. We may conclude that techniques such as visualisation, data-independency, approximation with a process-algebraic preorder relation (i.e. "behaviourally worse"), and compositional LTS construction have potential for solving very demanding verification tasks.

3 Black-Box Behaviour and Compositionality

3.1 Labelled Transition Systems, Parallel Composition and Hiding

In order to talk about the behaviour of a system or process, it is necessary to describe how it communicates with its environment. With few exceptions, communication is synchronous in process algebras. There are slight syntactical variations in different process algebras, but we may assume in this article that each process has a set of named *gates*. A process communicates with its environment by performing *actions* in the gates. An action consists of the name of the corresponding gate together with zero or more data parameters. Actions are instantaneous. The environment may observe them and even prevent them from occurring. Therefore, an action may occur only if both the process and its environment are ready for it. Actions are symmetric in the sense that an action does not (necessarily) have a sender and receiver. Instead, it has one or more *participants*.

For instance, the gates of the modified alternating bit protocol of Section 2.1 are *send*, *rec* and *err*. As was discussed in Section 2.1, it is perhaps unreasonable for the environment to prevent the actions *err* and *rec⟨msg⟩* (where *msg* ranges over the set of different messages the protocol may deliver) from occurring. The power of preventing *send*-actions from occurring is, on the other hand, important. With it the environment specifies what messages and when it wants to be transmitted. When the protocol is ready for a new sending request, it allows the actions *send⟨msg⟩* for all possible messages *msg*. The environment issues the sending request for message m_2 by simultaneously allowing the action *send⟨m_2⟩* and preventing *send⟨msg⟩* for all messages *msg* other than m_2. We see that although actions as such are symmetric, the ability of preventing actions allows the imitation of the roles of sender and receiver.

Most of the methods and algorithms discussed in later sections assume that the behaviour of a process or system is represented as a *labelled transition systems (LTS)*. An LTS is an edge-labelled graph with an alphabet for labels and a designated initial vertex. The alphabet is the set of actions that may occur in the gates of the process. For instance, if the set of possible messages is $\{m_1, m_2\}$, then the alphabet of the modified alternating bit protocol is $\{$ *send⟨m_1⟩*, *send⟨m_2⟩*, *rec⟨m_1⟩*, *rec⟨m_2⟩*, *err* $\}$. There is a special symbol τ which does not belong to the alphabet of any LTS but which can be used as a transition label. It is used to mark those actions of the process that the environment cannot observe. We have already seen graphical representations of LTSs in Figures 1, 2, 3 and 5.

Definition 3.1 A *labelled transition system (LTS)* is a 4-tuple (S, Σ, Δ, is) where S is the set of *states*, Σ is a set of symbols called the *alphabet* such that $\tau \notin \Sigma$, $\Delta \subseteq S \times (\Sigma \cup \tau) \times S$ is the *transition relation*, and $is \in S$ is the *initial state*. $\square$

An execution of an LTS consists of a sequence of successive transitions starting at the initial state. Each time the LTS performs a transition labelled with a symbol $a \in \Sigma$, the environment observes the action a. The environment cannot directly observe an execution of a τ-transition, but it is often possible to reason indirectly that a τ-transition has been executed. For instance, if the environment of the LTS in Figure 1 has observed the actions *send⟨m_1⟩* and *err*, it can reason that one τ-transition was executed before the *err*-transition.

The parts of an LTS not reachable from its initial state via transitions do not affect its behaviour, and may thus usually be removed.

It is customary to model systems as collections of processes connected in parallel. Several different parallel composition operators have been defined in the process algebra literature. For the purpose of this presentation, we define our operator as follows. It is equivalent to the parallel composition operator of the CSP language [Hoa85].

Definition 3.2 Let $L_1 = (S_1, \Sigma_1, \Delta_1, is_1), ..., L_n = (S_n, \Sigma_n, \Delta_n, is_n)$ be LTSs. Their *parallel composition* is the LTS $(S, \Sigma, \Delta, is) = L_1 \parallel ... \parallel L_n$ defined as follows:

- $\Sigma = \Sigma_1 \cup ... \cup \Sigma_n$, $S = S_1 \times ... \times S_n$ and $is = (is_1, ..., is_n)$.
- $((s_1, ..., s_n), \tau, (s'_1, ..., s'_n)) \in \Delta$ if and only if there is $1 \le i \le n$ such that $(s_i, \tau, s'_i) \in \Delta_i$ and $s'_j = s_j$ whenever $1 \le j \le n$ and $i \ne j$.
- If $a \in \Sigma$, then $((s_1, ..., s_n), a, (s'_1, ..., s'_n)) \in \Delta$ if and only if for every $1 \le i \le n$ either $a \in \Sigma_i$ and $(s_i, a, s'_i) \in \Delta_i$, or $a \notin \Sigma_i$ and $s'_i = s_i$. $\square$

That is, the states of a parallel composition are vectors consisting of states of the component processes. Each component can perform a τ-action independently of other components. A non-τ action is executed synchronously by all processes which have the action in their alphabets. The joint alphabet of the parallel composition is the union of the alphabets of the components.

This definition of parallel composition tends to introduce lots of states that are not reachable from the initial state. As we already mentioned, they do not affect the behaviour of the composition. Therefore, algorithms for parallel composition usually construct only the reachable states and transitions.

It is often the case that an action which is needed for communication between component processes of a system is not intended to be observable by the environment of the system. Such actions may be made unobservable or *invisible* by the *hiding operator*, which converts them into τ-actions. The hiding operator below is the same as in Lotos [BoB87].

Definition 3.3 Let $L_1 = (S_1, \Sigma_1, \Delta_1, is_1)$ be an LTSs, and A some set not containing τ. The LTS $(S, \Sigma, \Delta, is) = \textbf{hide } A \textbf{ in } L_1$ is defined as follows:

- $\Sigma = \Sigma_1 - A$, $S = S_1$ and $is = is_1$.
- $(s, \tau, s') \in \Delta$ if and only if there is $(s, a, s') \in \Delta_1$ such that $a = \tau$ or $a \in A \cap \Sigma_1$.
- If $a \in \Sigma_1 - A$, then $(s, a, s') \in \Delta$ if and only if $(s, a, s') \in \Delta_1$. $\square$

Parallel composition is often shown in drawings by connecting together all gates with the same name. Hiding is denoted by drawing a box around the system such that the gates corresponding to non-hidden actions extend to the exterior of the box.

To give an example, the internal process structure of the modified alternating bit protocol is shown in Figure 6. We see that it consists of four component processes, two modelling the logic of the protocol, and two the underlying lower-level links. There are many invisible actions such as $sdata\langle msg, nr\rangle$, with which Protocol sender sends the message *msg* with the alternating bit value *nr* to Data channel; and *losedata*, which corresponds to loss of a message in Data channel. The total behaviour of the protocol is obtained as the result of the parallel composition of the behaviours of its component processes, with the internal actions converted to τ-actions.

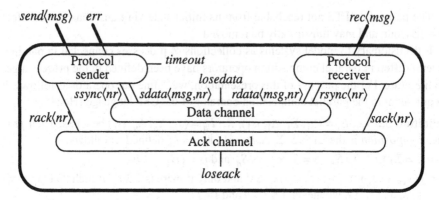

Fig. 6. Process structure of the modified alternating bit protocol

3.2 Black-Box Behaviour

One of the main sources of the power of process-algebraic verification methods is the emphasis of process algebras on *black-box* views to systems. A black-box view is what users of a system see of the behaviour of the system; it is the *externally observable* behaviour of the system. To take an example, the users of the modified alternating bit protocol of Section 2.1 are the clients who issue sending requests and receive the messages delivered by the protocol. They see the sending requests and deliveries, but they do not see the actual transmissions of messages in the low-level channels, alternating bits, losses of messages in the channels, or retransmissions. However, if losses of messages lead to failure of delivery, then they see the absence of delivery.

The hiding of internal actions is one step towards a black-box view. However, it is not sufficient alone, because it removes the names of certain actions, but not the actions themselves. Therefore, it does not make the number of states of a system any smaller. Furthermore, the view produced by hiding is not abstract enough. Although it is not possible to see what invisible actions occur, it is possible to count how many of them occur. For instance, hiding converts the sequence *send*⟨m_1⟩-*sdata*⟨m_1,0⟩-*rdata*⟨m_1,0⟩-*rec*⟨m_1⟩ to *send*⟨m_1⟩-τ-τ-*rec*⟨m_1⟩, from which it is possible to count that there are two internal actions between the sending request and delivery. This kind of information is usually not relevant.

The following notation is useful for abstracting from invisible actions.

Definition 3.4 Let $L = (S, \Sigma, \Delta, is)$ be an LTS. Let $s, s' \in S$, $a \in \Sigma \cup \{\tau\}$, and $a_1, a_2,$ $..., a_n \in \Sigma$.

- $s -a\rightarrow_L s'$ iff $(s, a, s') \in \Delta$.
- $s =\varepsilon\Rightarrow_L s'$ iff there are $n \geq 0$ and $s_0, s_1, ..., s_n$ such that $s = s_0$, $s_n = s'$ and $s_0 -\tau\rightarrow_L$ $s_1 -\tau\rightarrow_L ... -\tau\rightarrow_L s_n$, i.e. s' can be reached from s by zero or more τ-transitions.
- $s =a\Rightarrow_L s'$ iff $a \neq \tau$ and there are s_1 and s_2 such that $s =\varepsilon\Rightarrow_L s_1 -a\rightarrow_L s_2 =\varepsilon\Rightarrow_L s'$.
- $s =a_1a_2...a_n\Rightarrow_L s'$ where $n \geq 0$ iff there are $s_0, s_1, ..., s_n$ such that $s = s_0$, $s_n = s'$ and $s_0 =a_1\Rightarrow_L s_1 =a_2\Rightarrow_L ... =a_n\Rightarrow_L s_n$.
- $s -a\rightarrow_L$ iff there is s' such that $s -a\rightarrow_L s'$, and similarly with $s =a_1a_2...a_n\Rightarrow_L$.

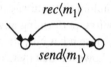

$rec\langle m_1\rangle$

$send\langle m_1\rangle$

Fig. 7. Externally observable behaviour in the absence of transmission errors

In all of the above, the subscript "$_L$" may be omitted if the LTS is obvious from the context. $\square$

One possible way of throwing all information about invisible actions away is the restriction of attention to the *traces* of the system. A trace is any finite prefix of any sequence of visible actions an execution of the system may generate. In particular, the empty sequence ε is a trace of any system. In automata-theoretic terms, the set of traces of a system is the *language* generated by the system.

Definition 3.5 The set of *traces* of an LTS L is $tr(L) = \{\ \sigma \in \Sigma^* \mid s = \sigma \Rightarrow_L \}$. $\square$

The set of traces of a system can be thought of as a very abstract black-box view to the system. It is called the *trace semantics* of the system. The trace semantics induces natural ways for comparing the behaviours of two systems.

Definition 3.6 Let L, L_1 and L_2 be LTSs with a common alphabet (i.e. $\Sigma = \Sigma_1 = \Sigma_2$).

- The *trace semantics* of L is $tr(L)$.
- L_1 and L_2 are *trace equivalent*, denoted by $L_1 =_{tr} L_2$, iff $tr(L_1) = tr(L_2)$.
- The *trace preorder* "$\leq_{tr}$" is defined by $L_1 \leq_{tr} L_2$ iff $tr(L_1) \subseteq tr(L_2)$. $\square$

If *Sys* is the LTS of the modified alternating bit protocol in Figure 6, and *Serv* is another LTS representing the specification of the service the protocol should provide, then it would be natural to require that $Sys =_{tr} Serv$. Furthermore, if the trace semantics did capture all the essential behavioural aspects of a system (which it does not), then checking that $Sys =_{tr} Serv$ would be sufficient for checking that the protocol is correct with respect to the service specification.

To understand the need for a preorder (such as "$\leq_{tr}$") assume for a moment that the low-level channels used by the modified alternating bit protocol are reliable, and timeouts do not occur prematurely. Let us call the resulting LTS *RelSys*. It is very simple, as can be seen from Figure 7. Because the service specification allows failure of transmission as long as an error indication is given, *Serv* has the trace "$send\langle m_1\rangle\ err$". (For the sake of this example we may assume that *Serv* is the LTS given in Figure 2.) However, because messages are never lost or delayed in the channels, the protocol will never fail in message transmission, and *RelSys* does not contain any transitions labelled with the error indication action *err*. In particular, it does not have the trace "$send\langle m_1\rangle\ err$". As a consequence, $RelSys \neq_{tr} Serv$. If trace equivalence were our notion of "satisfies the specification", then we would have to reject the system with reliable channels! Of course, that would be stupid. The reliable system is not equivalent with the service specification because it is better.

Preorders are intended to capture the notion of "equivalent or better". Also the words "implements" and "refines" are in use. Indeed, if Figure 2 is taken as the service specification *Serv* and *RelSys* is as in Figure 7, then $RelSys \leq_{tr} Serv$ holds. The trace preorder is not a fully satisfactory notion of "equivalent or better", because according to it the system which deadlocks before performing any visible actions is "equivalent or better" than any other system, because it has only the trace "ε" that all systems have. We will see in Sections 4.6 and 4.7, however, that there are preorders which do not suffer from this problem.

Although the trace semantics is abstract enough, it is inappropriate for many process-algebraic verification tasks, because it throws away some essential information. Most importantly, the trace semantics does not reveal whether the choice between two actions is made by the system, its environment, or jointly by both. For instance, the modified alternating bit protocol may initially perform the actions $send\langle m_1\rangle$ and $send\langle m_2\rangle$ corresponding to transmission requests of two different messages m_1 and m_2. It is of course important to know whether it is the client or the protocol which chooses the message to be transmitted, but the trace semantics does not separate the two cases.

It is thus necessary to preserve some, but not full, information on invisible actions. Opinions as to how much information should be preserved differ among researchers. This has led to the development of numerous semantic models with related behavioural equivalences and preorders. We will discuss some of the best known in Section 4.

Black-box behaviours are not useful in computer-aided verification unless they can be represented within a computer and manipulated by algorithms. It is seldom reasonable or even possible to represent a black-box semantics as such. For instance, the set of traces of a system is usually infinite and cannot thus be enumerated within a computer. Instead, behaviour is usually represented as an LTS, and the semantic model is taken into account in the manipulation of LTSs. For example, Section 4.5 describes an algorithm for checking trace equivalence of two finite LTSs. (The problem is, of course, just a slightly simplified case of the well-known problem of checking the equivalence of the languages accepted by two finite automata.)

One more thing regarding black-box behaviour is worth mentioning here. Namely, black-box behaviour can be analysed without putting a system into any particular environment. Therefore, in process-algebraic verification it is often the case that the customers who generate impulses to the system and receive its responses are not modelled at all. To give an example, the clients of the modified alternating bit protocol were not given in Figure 6.

3.3 Compositional LTS Construction

For many black-box semantics one or more *reduction* algorithms have been invented. The task of a reduction algorithm is to produce a small equivalent LTS from a given LTS. A reduction algorithm is not required to produce the (or a) smallest possible equivalent LTS, i.e. *minimise* the LTS, because for some black-box equivalences minimisation is computationally quite expensive. Consequently, there may be different reduction algorithms with different reduction capabilities for the same equivalence.

Some black-box semantic equivalences are *congruences*. That is, they guarantee that if a component of a system is replaced by an equivalent one, the behaviour of the resulting system remains equivalent to the behaviour of the original system. The notion of "congruence" will be defined formally in Section 4.1.

Reduction algorithms for congruences render possible *compositional* construction of an LTS for a big system. Let the application of the reduction algorithm to LTS L be denoted by $red(L)$. Assume that we have to construct an LTS for the system

$$L = \textbf{hide } a_1, a_2, b_1, c_1, c_2 \textbf{ in } (L_1 \parallel L_2 \parallel L_3 \parallel M_1 \parallel M_2 \parallel N_1)$$

where a_1 and a_2 belong only to the alphabets of L_1, L_2 and L_3; b_1 to the alphabets of M_1 and M_2; and c_1 and c_2 to the alphabet of N_1. It can be proven that an isomorphic (and thus equivalent under any reasonable notion of equivalence) LTS would be obtained if hiding and parallel composition were rearranged as follows:

$$L' = \textbf{hide } a_1, a_2 \textbf{ in } (L_1 \parallel L_2 \parallel L_3) \parallel \textbf{hide } b_1 \textbf{ in } (M_1 \parallel M_2) \parallel \textbf{hide } c_1, c_2 \textbf{ in } N_1$$

Because the operator $red(L)$ preserves the semantics and the equivalence is a congruence, an LTS equivalent to L' (and thus L) would be obtained by computing in the following way:

$$\begin{aligned} R = red(\ &red(\ \textbf{hide } a_1, a_2 \textbf{ in } (L_1 \parallel L_2 \parallel L_3)\ ) \\ &\parallel\ red(\ \textbf{hide } b_1 \textbf{ in } (M_1 \parallel M_2)\ ) \\ &\parallel\ red(\ \textbf{hide } c_1, c_2 \textbf{ in } N_1\ ) \\ &) \end{aligned}$$

It is often the case that all LTSs encountered when computing R are much smaller than L or L'. However, because R is equivalent to L, it can be used instead of L as long as the semantic model that is used preserves everything that is considered essential. The computation of R instead of L is known as *compositional LTS construction*.

We saw an example of compositional LTS construction already in Section 2.2. The earliest explicit mentions of compositional LTS construction the author is aware of are [SLU89, MaV90]. The prerequisites of compositional LTS construction were analysed in [Val93].

Although compositional LTS construction is usually effective, the results depend on the way the system is decomposed to subsystems for reduction. Furthermore, there are systems for which the basic compositional approach does not work at all [GrS91]. This happens when all components of a system restrict each others' behaviour efficiently enough. For instance, in a complete token-ring there is only one token in the ring at any instant of time, which efficiently reduces the number of states. However, in an open-ended segment of the ring, new tokens can be received from the open end without limit, introducing a big LTS.

It is important to notice that not all black-box equivalences found in the literature are congruences. Therefore, the choice of the equivalence requires some care. In the next section we analyse several equivalences from the point of view of compositional LTS construction.

4 Equivalence Notions and Algorithms

4.1 Basic Concepts

Before discussing the various black-box behavioural equivalences and preorders defined in the literature, it is necessary to introduce some basic concepts. An *equivalence* is, of course, a reflexive, transitive and symmetric relation. A *preorder* is reflexive and transitive, but it need not be symmetric. Essentially same equivalences and preorders may almost always be defined both for expressions in process-algebraic languages and for LTSs. When this is possible, it is even possible to compare an expression to an LTS using the equivalence or preorder, or to obtain an equivalent (possibly infinite) LTS from an expression. For simplicity, only the LTS versions of definitions are given in the sequel.

An equivalence "$\approx$" is a *congruence* with respect to some function f, iff $L_1 \approx L_1'$ and $L_2 \approx L_2'$ and ... and $L_n \approx L_n'$ imply $f(L_1, L_2, ..., L_n) \approx f(L_1', L_2', ..., L_n')$. Correspondingly, a preorder "$\leq$" is a *precongruence* with respect to f, iff $L_1 \leq L_1'$ and $L_2 \leq L_2'$ and ... and $L_n \leq L_n'$ imply $f(L_1, L_2, ..., L_n) \leq f(L_1', L_2', ..., L_n')$. If a relation is a precongruence or congruence with respect to some given functions, it is a precongruence or congruence with respect to all functions composable of the given functions. We say that an equivalence or preorder is a congruence or precongruence, if the set of functions with respect to which it is a congruence or precongruence is obvious from the context.

In particular, the process composition operators "$\|$" and "**hide**" defined in Section 3.1 can be thought of as functions on LTSs. If an equivalence is a congruence with respect to them, and a system is composed of LTSs without using any additional operators, then the replacement of a component LTS of the system by an equivalent LTS yields a system that is equivalent to the original one. This observation remains valid for other operators and even if the components are not given as LTSs, but as process-algebraic expressions, as long as it is possible to obtain equivalent LTSs from the expressions and the operators can be interpreted as functions on LTSs. (Most process-algebraic languages contain *fixed-point operators* or recursive process definitions which cannot always easily be thought of as functions on LTSs.)

Every preorder "$\leq$" induces an equivalence, namely the one where $L_1 \approx L_2$ iff $L_1 \leq L_2$ and $L_2 \leq L_1$. The equivalence induced by a precongruence is a congruence.

Equivalences can be compared according to their capability of making distinctions between systems. Equivalence "$\approx_2$" is *weaker* or *coarser* than "$\approx_1$", if $L_1 \approx_1 L_2$ implies $L_1 \approx_2 L_2$ for all LTSs L_1 and L_2. Furthermore, if "$\approx_1$" and "$\approx_2$" are not the same equivalence, then "$\approx_2$" is *strictly* weaker (or strictly coarser) than "$\approx_1$".

In the remainder of this section several equivalences and preorders will be introduced, and algorithms for manipulating LTSs according to them will be discussed. The algorithms are intended for the following tasks:

- **Equivalence comparison.** The checking whether two LTSs are equivalent according to the semantics.
- **Preorder comparison.** The checking whether a given preorder relation holds between two LTSs.
- **Reduction.** The construction of an LTS that is semantically equivalent to the input LTS, but (if possible) smaller. The result needs not be the smallest possible equivalent LTS.
- **Minimisation.** The construction of the (or a) smallest possible LTS that is semantically equivalent to the input LTS.
- **Normalisation.** The construction of an LTS that is semantically equivalent to the input LTS and in a form that makes semantical comparisons easy.

4.2 Strong Bisimilarity

Strong bisimilarity (sometimes called also *bisimulation equivalence*) is not a black-box semantics in the sense of Section 3.2, because it preserves almost full information on invisible actions. It is, however, useful as an "intermediate" semantics, because it is strictly stronger than any of the black-box equivalences discussed later in this section, and it is easy to manipulate both theoretically and algorithmically. The need for an intermediate semantics may arise, for instance, when constructing LTSs from process-algebraic expressions; it defines the sense in which the LTS is equivalent to the expression. Furthermore, some important truly black-box semantic models and algorithms for manipulating them resemble strong bisimilarity and its algorithms.

The basic idea of strong bisimilarity is that two LTSs are strongly bisimilar if and only if they can simulate each other in a certain way. Actually, the notion of simulation is brought to the level of individual states, and it is required that the initial states of the LTSs can simulate each other. A state s_1 can simulate the state s_2, if for every output transition of s_2, there is an output transition of s_1 such that the labels of the transitions are the same, and the end states of the transitions simulate each other. As such this description would be a circular definition. The circularity may be broken, however, by not using it as a definition, but as a check for a potential simulation relation that has been given from the outside. Any relation that passes the check is called a *strong bisimulation*. Strong bisimilarity of two LTSs may now be defined simply by requiring the existence of some strong bisimulation relation that makes their initial states to simulate each other.

Strong bisimilarity is a useful concept also for comparing two states of the same LTS. Therefore, the definition is formulated as follows.

Definition 4.1 [Par81, Mil89] Let $L = (S, \Sigma, \Delta, is)$ be an LTS. A binary relation "$\sim$" $\subseteq$ $S \times S$ over the states of L is a *strong bisimulation*, iff for every s_1, s_2 and $s \in S$ such that $s_1 \sim s_2$ and every $a \in \Sigma \cup \{\tau\}$ the following hold:

- If $s_1 -a \rightarrow s$, then there is $s' \in S$ such that $s \sim s'$ and $s_2 -a \rightarrow s'$.
- If $s_2 -a \rightarrow s$, then there is $s' \in S$ such that $s' \sim s$ and $s_1 -a \rightarrow s'$.

Furthermore,

- The states s_1, $s_2 \in S$ of L are *strongly bisimilar*, iff there is a strong bisimulation "$\sim$" such that $s_1 \sim s_2$.
- Let $L_1 = (S_1, \Sigma, \Delta_1, is_1)$ and $L_2 = (S_2, \Sigma, \Delta_2, is_2)$ be two LTSs such that their alphabets are the same and, furthermore, $S_1 \cap S_2 = \emptyset$. They are *strongly bisimilar*, iff their initial states is_1 and is_2 are strongly bisimilar in their joint LTS $(S_1 \cup S_2, \Sigma, \Delta_1 \cup \Delta_2, is_1)$. $\square$

Strong bisimilarity is denoted in this article by "$=_{sb}$". A simple induction argument shows that if two systems are strongly bisimilar, then they can simulate not only each other's individual transitions, but also transition sequences. More formally, if s_0 and s_0' are strongly bisimilar and $s_0 -a_1 \rightarrow s_1 -a_2 \rightarrow \ldots -a_n \rightarrow s_n$, then there are states s_1', s_2', $\ldots$, s_n' such that each s_i is strongly bisimilar with the corresponding s_i', and $s_0' -a_1 \rightarrow s_1' -a_2 \rightarrow \ldots -a_n \rightarrow s_n'$.

It can be proven that strong bisimilarity indeed is an equivalence (see e.g. [Mil89]). Furthermore, it is a congruence with respect to most (or perhaps all) commonly used process operators, including "∥" and "**hide**".

Let *reachable-parts(L)* denote the result of the removal from L of all states and transitions that cannot be reached from the initial state of L. The relation mapping each state of *reachable-parts(L)* to the corresponding state in L is a strong bisimulation. Therefore, *reachable-parts(L)* $=_{sb} L$. (This formalises the remark after Definition 3.1 that the unreachable parts of an LTS can be removed without modifying its behaviour.)

Regarding the development of algorithms, strong bisimilarity has a pleasant property. Namely, for any given LTS, there is a unique minimal LTS strongly bisimilar with it. It may be characterised as follows.

Theorem 4.2 Let $L = (S, \Sigma, \Delta, is)$ be a finite LTS. Whenever $s \in S$, let $ec(s)$ denote the equivalence class determined by s and "$=_{sb}$", i.e. $ec(s) = \{ s' \in S \mid s' =_{sb} s \}$. Let the LTS $L_m = (S_m, \Sigma, \Delta_m, is_m)$ be obtained from L as follows:

- $S_m = \{ ec(s) \mid s \in S \}$.
- $(ec(s_1), a, ec(s_2)) \in \Delta_m$ where s_1 and $s_2 \in S$ and $a \in \Sigma \cup \{\tau\}$, iff for every $s'_1 \in ec(s_1)$ there is $s'_2 \in ec(s_2)$ such that $(s'_1, a, s'_2) \in \Delta$.
- $is_m = ec(is)$.

Then $L_m =_{sb} L$. Furthermore, every LTS L' such that $L' =_{sb} L$ has at least as many states and transitions as *reachable-parts(L_m)*. □

The LTS L_m of Theorem 4.2 can be constructed quite rapidly by an application of known algorithms for the *relational coarsest partitioning problem* [BoS88, Fer90]. Because *reachable-parts(L_m)* is the minimal LTS equivalent to L and it can be easily constructed from L_m, we have a fast minimisation algorithm for strong bisimilarity. Furthermore, the construction of L_m can also be used for checking strong bisimilarity of two LTSs. This is done by joining them into one LTS like in Definition 4.1, minimising the joint LTS, and checking whether the initial states of the original LTSs end up in the same equivalence class, i.e. the same state of the minimised LTS.

4.3 Weak Bisimilarity or Observation Equivalence

Weak bisimilarity [Mil89], also called *observation equivalence*, is a black-box semantics whose definition resembles a lot the definition of strong bisimilarity. The only difference is that instead of simulating transitions of the form $s -a\rightarrow s'$ by transitions of the same form, transition sequences of the form $s =a\Rightarrow s'$ where either $a = \varepsilon$ or $a \in \Sigma$ are simulated by sequences of the same form. Because of this difference, a τ-transition may now be simulated by a sequence of τ-transitions of an arbitrary length, including zero. So weak bisimilarity does not preserve information of the number of successive τ-transitions. Weak bisimilarity is denoted in this article by "$=_{wb}$".

Just like in the case of strong bisimilarity, a simple induction argument shows that weak bisimilarity guarantees that any sequence of the form $s =\sigma\Rightarrow s'$ (where $\sigma \in \Sigma^*$) can be simulated by a sequence of the same form. As one would expect, $L_1 =_{sb} L_2$ implies $L_1 =_{wb} L_2$.

Weak bisimilarity is not a congruence with respect to all customary process-algebraic operators (although it is with respect to "∥" and "**hide**"). Therefore, a slightly stronger variant called *observation congruence* has been defined [Mil89]. Its details are not important for the current presentation, so we do not go into it.

The result of the existence of a unique minimal LTS extends to weak bisimilarity. However, unlike in the case of strong bisimilarity, the minimisation of the number of states by fusing equivalent states together does not any more automatically minimise the number of transitions. This is because a transition may be *redundant* in the sense that its addition or removal does not affect weak bisimilarity. For instance, the transition $s -a \rightarrow s'$ (where $a \neq \tau$) is redundant if there is a state s_1 such that $s -\tau \rightarrow s_1 -a \rightarrow s'$ and $s_1 \neq s$; even if it were removed, $s =a \Rightarrow s'$ would remain valid. The existence of a unique state- and transition-minimal LTS was proven in [BeK89] and [Elo91], and the latter article gave also an algorithm for the removal of all redundant transitions.

In any case, the fusion of weakly bisimilar states produces a state-minimal LTS even without transition minimisation. The proper handling of the "$=a\Rightarrow$"-relation leads, in essence, to the problem of computing the transitive closure of the relation determined by τ-transitions. Actually, a state-minimal LTS can be constructed by computing and storing the transitive closure and then applying minimisation with respect to strong bisimilarity [BoS88]. Because the number of transitions in the transitive closure may be quadratic in the number of states, this approach sometimes consumes quite a lot of memory. Alternatively, the transitive closure needs not be stored if it is simulated on-the-fly, but then time consumption increases.

The technique of verifying equivalence by checking whether the initial states of the LTSs in question end up in the same state of the minimised LTS is valid also with weak bisimilarity.

Weak bisimilarity allows the simulation of a τ-loop by zero transitions. Therefore, it does not preserve information about *divergence*, that is, infinite sequences of invisible actions. This is sometimes considered as an advantage, but, since divergence is closely related to livelocks, also the opposite view is justified. Divergence information can be added to weak bisimilarity by requiring that a divergent (i.e. immediately capable for an infinite sequence of invisible actions) and non-divergent state may not simulate each other. This modification can be relatively easily taken into account in algorithms for weak bisimilarity. Divergence-preserving weak bisimilarity has been investigated in detail in [Elo94].

Algorithms for strong and weak bisimilarity have been implemented into many tools. Perhaps the most well known is the *Concurrency Workbench* [CPS90]. One of the earliest explicit mentions of compositional LTS construction was in the context of the *AUTO* tool [MaV90]. These and many other tools are listed in [InP91].

4.4 Branching Bisimilarity

Branching bisimilarity [vGW89, vGl90] is another semantic model based on bisimulation and abstraction of τ-transitions. In it, when $s_1 \sim s_2$, each transition of the form $s_1 -a \rightarrow s_1'$ (where $a \neq \tau$) is simulated by a sequence of the form $s_2 =\varepsilon\Rightarrow s_2'' -a \rightarrow s_2'$, where $s_1 \sim s_2''$ and $s_1' \sim s_2'$. Each transition $s_1 -\tau \rightarrow s_1'$ is simulated either in the same way, or by doing nothing, i.e. by staying in s_2. In the latter case it is required that $s_1' \sim s_2$. Transitions $s_2 -a \rightarrow s_2'$ are similarly simulated by sequences $s_1 =\varepsilon\Rightarrow s_1'' -a \rightarrow s_1'$; or perhaps by staying in s_1, if $a = \tau$ and $s_1 \sim s_2'$.

Branching bisimilarity is strictly stronger than weak bisimilarity and strictly weaker than strong bisimilarity. Algorithms for branching bisimilarity resemble those for weak bisimilarity, but there is no need to compute the transitive closure of τ-transitions. As a consequence, minimisation with respect to branching bisimilarity is sometimes significantly faster in practice than minimisation with respect to weak bisimilarity [GrV90]. Another important property of branching bisimilarity is that it and its variants have interesting connections with the well-known logic $CTL*-X$ (i.e. $CTL*$ [CES86, Eme90] without the "next state" operator) [DNV90].

4.5 Trace Semantics

The trace semantics, trace equivalence and trace preorder were described already in Definition 3.6. The trace semantics can be used for the verification of safety properties, and some other important semantics can be thought of as enrichments of the trace semantics, so the basic algorithms for trace semantics are worth an introduction.

Because a finite LTS can be thought of as a finite automaton whose states are all accepting, and the set of traces is the same as the language accepted by the corresponding finite automaton, the trace semantics can be manipulated with algorithms for finite automata. The problem of checking whether an automaton accepts the language $\Sigma*$ (i.e. all words in its alphabet) is polynomial space complete (e.g. [GaJ79], "regular expression non-universality", p. 174). This problem can be converted into LTS form where all states are accepting by adding an extra symbol δ to the alphabet, an extra state s_δ, a δ-transition from each original acceptance state to s_δ, and an a-transition from s_δ to itself for every $a \in \Sigma \cup \{\delta\}$. Because $\Sigma*$ is accepted by the one-state LTS $L_{\Sigma*}$ which has an a-transition from its only state to itself for every $a \in \Sigma$ (Figure 8), the problem "does L accept $\Sigma*$" can be solved by checking if $L =_{tr} L_{\Sigma*}$ and by checking if $L_{\Sigma*} \leq_{tr} L$. Therefore, the latter two problems are polynomial space complete. Furthermore, because $L_{\Sigma*}$ is clearly minimal, the LTS minimisation problem for trace equivalence is polynomial space hard. We cannot thus expect to find algorithms for these problems that run efficiently with all inputs. As a matter of fact, it can be shown with an example that a (possibly nondeterministic) minimal automaton for a given language is not always unique. Therefore, minimisation with respect to trace semantics is not feasible.

It is well known that a deterministic automaton (or LTS) $det(L)$ can be constructed that accepts the same language as a given finite automaton (or LTS) L, but the result may be exponentially bigger than the input automaton. (Some versions of determinisation may create one non-accepting state $s_\emptyset$ that is reached by all rejected words, but it and its incoming edges can be discarded.) Furthermore, a unique minimal deterministic automaton exists and can be constructed quite fast with the aid of known algorithms for the ordinary (functional) coarsest partitioning problem. (Also minimisation w.r.t. strong bisimilarity produces it, when applied to a deterministic automaton.) Let us denote the result of the minimisation of $det(L)$ by $mindet(L)$. It so happens that $mindet(L_1)$ is isomorphic to $mindet(L_2)$ if and only if $L_1 =_{tr} L_2$, and isomorphism of deterministic LTSs is easy to check. So we have a normal form, a normalisation algorithm and an equivalence checking algorithm for the trace semantics. The normalisation algorithm consumes exponential time in the worst case, but runs reasonably fast when the determinisation stage does not make the LTS grow. It can, therefore, be used in practice.

The question "does $L_1 \leq_{tr} L_2$ hold?" may be answered by computing $L_1 \parallel$ *mindet*(L_2) and seeking for states where L_1 is ready to perform an a-transition for some $a \neq \tau$ but *mindet*(L_2) refuses a. Alternatively, one may keep in *mindet*(L_2) the rejection state $s_\emptyset$ mentioned above, and seek for states where *mindet*(L_2) is in it. Actually, any deterministic LTS that is trace equivalent to L_2 can be used in the place of *mindet*(L_2). It is worth noticing that this algorithm does not require the determinisation of L_1 and can thus be used even if L_1 is big, as long as L_2 can be determinised.

What is more, the above algorithm may even be combined to the construction of L_1 from its component processes. That is, if L_1 is of the form **hide** $a_1, ..., a_m$ **in** ($L_{11} \parallel ... \parallel L_{1n}$), then the question "does $L_1 \leq_{tr} L_2$ hold?" may be answered by computing $L_{11} \parallel ... \parallel L_{1n} \parallel$ *mindet*(L_2) and seeking for states where an action would otherwise be enabled, but *mindet*(L_2) refuses it (or states where *mindet*(L_2) is in $s_\emptyset$). Of course, the computation may be terminated when such a state $(s_{11}, ..., s_{1n}, s_2)$ and action a are found for the first time, because then it is already known that $L_1 \leq_{tr} L_2$ does not hold. An action sequence σ can easily be found in the already computed part such that if ρ is obtained from σ by removing all occurrences of τ and $a_1, ..., a_m$, then $\rho a \in tr(L_1)$ and $\rho a \notin tr(L_2)$. This premature termination may save lots of memory and time, if $L_1 \leq_{tr} L_2$ does not hold. Furthermore, L_2 may have been designed to represent some particular behavioural property instead of the full specification. Then savings may be obtained also when checking a correct L_1, because L_2 may prevent L_1 from entering those branches of behaviour that are irrelevant for the property. The idea of detecting violations against a specification (such as a trace of L_1 that L_2 forbids) during the computation of a state space or LTS for the system is known as *on-the-fly* verification. It has been applied to much more complicated verification tasks than trace preorder checking (see later sections and [VaW86]).

Because *mindet*(L) is in practice often smaller than L, the construction of *mindet*(L) can be used as a reduction algorithm, as long as the result is discarded if it is bigger than the input LTS. Furthermore, several heuristic reduction algorithms for trace semantics may be developed. For instance, one simple but not very powerful algorithm removes all states other than the initial state and the end states of transitions labelled by visible actions, and adds a transition (s, a, s') whenever s and s' are not removed, $a \neq \tau$, and there is a (possibly removed) s_1 such that $s =\varepsilon\Rightarrow s_1 -a\rightarrow s'$ holds in the input LTS. Also the minimisation algorithm w.r.t. weak bisimulation can be used as a reduction algorithm for the trace semantics, because weak bisimilarity is strictly stronger than the trace semantics. Finally, these two algorithms can be combined to obtain better reduction than either of them provides alone.

4.6 The Failures-Divergences Model of CSP

One of the most important goals in the development of the standard semantic model of CSP [BrR85, Hoa85] (earlier version in [BHR84]) was to find a model that preserves information on deadlocks or, more generally, the ability of refusing or preventing actions, but is otherwise as weak as possible. The importance of refusal information was discussed in Section 3.1. Refusal information may be formalised by the notion of *stable failures*.

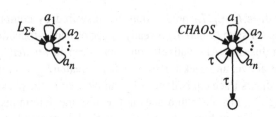

Fig. 8. L_{Σ^*} and *CHAOS* in LTS form, when $\Sigma = \{a_1, a_2, ..., a_n\}$

Definition 4.3 Let $L = (S, \Sigma, \Delta, is)$ be an LTS, $\sigma \in \Sigma^*$, and $A \subseteq \Sigma$. The pair (σ, A) is a *stable failure* of L, iff there is $s' \in S$ such that $is =\!\sigma\!\Rightarrow s'$, $s' \not\!\!-\tau\!\!\rightarrow$, and $s' \not\!\!-a\!\!\rightarrow$ for every $a \in A$. The set of stable failures of L is denoted by *sfail(L)*. $\square$

That is, $(\sigma, A) \in$ *sfail(L)*, if and only if L has a state s' such that L cannot perform τ or any actions from A when it is in s', and L has an execution that ends up in s' and produces the trace σ. Any state from which a process cannot execute τ is called *stable*, because the process is guaranteed to stay in it until the environment executes or observes some visible action.

It can be proven that the *stable failures equivalence* "$=_{sf}$" defined by $L_1 =_{sf} L_2$ if and only if *sfail(L_1)* = *sfail(L_2)* is the weakest possible congruence with respect to "‖" and "**hide**" that keeps deadlocking and non-deadlocking systems apart [Val95]. It has, however, a surprising and unpleasant feature: it does not imply trace equivalence, i.e. it does not preserve complete information about the traces a system may execute. For instance, the LTS in Figure 3 has no stable states. Therefore, it has no stable failures, and it would remain equivalent according to "$=_{sf}$" even if all of its edges were relabelled by τ.

If $\sigma \in tr(L)$ and L may reach a stable state after executing σ, then at least $(\sigma, \emptyset)$ is a stable failure of L. Therefore, "$=_{sf}$" loses information on a trace σ only if it is not possible to reach a stable state after executing σ. But then all states reachable by σ have an outgoing τ-transition, and it is possible to execute an infinite sequence of invisible actions after σ. The trace σ is thus a *divergence trace* in the sense of the following definition.

Definition 4.4 Let $L = (S, \Sigma, \Delta, is)$ be an LTS. The sequence $\sigma \in \Sigma^*$ is a *divergence trace* of L iff there are states $s_0, s_1, s_2, ...$ such that $is =\!\sigma\!\Rightarrow s_0 -\tau\!\rightarrow s_1 -\tau\!\rightarrow s_2 -\tau\!\rightarrow ....$. The set of divergence traces of L is denoted by *divtr(L)*. $\square$

For instance, all traces of the LTS in Figure 3 are divergence traces, while the LTS in Figure 2 has no divergence traces. It is possible that $\sigma \in$ *divtr(L)* even if $(\sigma, \emptyset) \in$ *sfail(L)*, if it is possible to reach both a stable state and an infinite sequence of τ-actions by σ.

The developers of CSP-semantics considered divergence as undesirable. They built their model in such a way that it does not preserve any information whatsoever about the behaviour of any process after it has executed a divergence trace. (This was a natural decision from the point of view of the mathematical methods they used in defining the model. They did not use LTSs.) They gave the name *CHAOS* to the process that behaves like the LTS in Figure 8, i.e. has all possible traces, stable failures and diver-

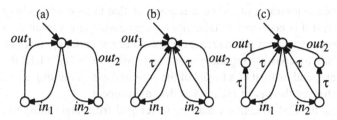

Fig. 9. A reliable and two unreliable channels

gence traces. In their model any process that can execute a divergence trace starts to behave like *CHAOS* after executing a divergence trace. When ρ and σ are strings, let ρ ≤ σ denote that ρ is a (not necessarily proper) prefix of σ. The CSP model can be defined as follows in the present framework.

Definition 4.5 Let L, L_1 and L_2 be LTSs with a common alphabet.

- The *CSP-divergences* of L are $CSPdiv(L) = \{ \sigma \in \Sigma^* \mid \exists \rho \in divtr(L): \rho \leq \sigma \}$.
- The *CSP-failures* of L are $CSPfail(L) = sfail(L) \cup (CSPdiv(L) \times 2^\Sigma)$.
- The *CSP model* of L is the pair $(CSPfail(L), CSPdiv(L))$.
- $L_1 =_{CSP} L_2$ iff $CSPfail(L_1) = CSPfail(L_2) \wedge CSPdiv(L_1) = CSPdiv(L_2)$.
- $L_1 \leq_{CSP} L_2$ iff $CSPfail(L_1) \subseteq CSPfail(L_2) \wedge CSPdiv(L_1) \subseteq CSPdiv(L_2)$. □

In the absence of divergences, "$=_{CSP}$" collapses to "$=_{sf}$" and is thus the weakest deadlock-preserving congruence.

The preorder "$\leq_{CSP}$" has proven very natural and useful in practice. In the absence of divergences, $L_1 \leq_{CSP} L_2$ implies that $L_1 \leq_{tr} L_2$. Furthermore, stable failures make it possible to specify that a system has to have some traces and, better still, execute certain actions if the environment so desires. For instance, if $L_1 \leq_{CSP} L_2$ is required, and assuming that $L_2 \neq_{CSP} CHAOS$, then L_1 has the liberty to initially refuse the action $send\langle m_1 \rangle$ if and only if $(\varepsilon, \{send\langle m_1 \rangle\}) \in sfail(L_2)$. These features of "$\leq_{CSP}$" make "implementation $\leq_{CSP}$ specification" a reasonable notion of "implementation satisfies specification", as long as divergence is considered as forbidden. Interestingly, *CHAOS* is a maximal element with respect to "$\leq_{CSP}$", that is, it corresponds to a specification that all systems satisfy. If $\Sigma \neq \emptyset$, then no LTS is minimal, that is, there is no system that satisfies all specifications.

To give another example, Figure 9 (a) shows a reliable channel of capacity one, and (b) and (c) show two erroneous channels which may lose messages. We have (a) $\leq_{CSP}$ (c) and (b) $\leq_{CSP}$ (c), but not (a) $\leq_{CSP}$ (b) nor the other way round, because $(in_1, \{in_1\}) \in CSPfail((a))$ and $(in_1 in_1, \emptyset) \notin CSPfail((a))$, but $(in_1, \{in_1\}) \notin CSPfail((b))$ and $(in_1 in_1, \emptyset) \in CSPfail((b))$. This implies that if a protocol is verified using (c) as the model of its low-level channels, then the protocol is correct also with reliable channels (a) or faulty channels that behave like (b). However, correctness with (b)-channels does not imply correctness with (a)-channels. Indeed, if the protocol may accidentally reach a state where the recipient insists on receiving out_2, the sender insists on sending in_2, and the channel is ready for out_1, then (a)-channels may lead to a deadlock, but (b)-channels may not.

Usually the environment of a system is not modelled in process-algebraic verification. However, if it is necessary to take into account assumptions about the behaviour of the environment, such as it can refuse certain actions at certain points of execution, this can be done by constructing the greatest (according to "$\leq_{CSP}$") LTS that satisfies the requirements and using it as a model of the environment. A missing environment is equivalent to having the L_{Σ^*} of Figure 8 as the environment.

Algorithms for CSP-semantics may be developed from algorithms for the trace semantics by adding information about failures and divergences to $det(L)$ and $mindet(L)$. Failures may be represented by adding so-called *acceptance sets* to the states of the deterministic LTS [Hen85]. To represent divergences it suffices to represent *CHAOS*, and that can be done by having a special state that denotes *CHAOS* and has no outgoing transitions. During minimisation of $det(L)$ states that have different acceptance sets or different divergence information have to be kept apart, but this can be taken care of in the initial partitioning of coarsest partitioning algorithms. The on-the-fly algorithm for "$\leq_{tr}$" presented in Section 4.5 can be adapted to "$\leq_{CSP}$" by adding a suitable comparison of acceptance sets, or by taking "mirror images" of acceptance sets according to [Bri88].

Perhaps the most important tool supporting CSP-semantics is *FDR* (*Failures-Divergence Refinement*) by Formal Systems (Europe) Ltd. [Ros94]. In the *Concurrency Workbench* [CPS90] algorithms for strong bisimilarity are adapted to some failure-based semantic models [ClH90].

Although the CSP model is one of the most important process-algebraic semantic models, there are good reasons to dislike the way divergence is handled in it. For instance, Figure 3 was useful, but it could not have been produced with CSP-semantics. Indeed, several other semantic models based on failures have been presented in the literature. In the next section one of them will be discussed in detail.

4.7 CFFD-Equivalence

The *CFFD model* [VaT91, VaT95] was developed as an attempt to remove *CHAOS* from the CSP model. It develops further some ideas in [BKO87], and may be defined as follows:

Definition 4.6 Let L, L_1 and L_2 be LTSs, where $L = (S, \Sigma, \Delta, is)$, and L_1 and L_2 have the same alphabet.

- *stable(L)* is a predicate that holds iff *is* is stable, i.e. $is \not\xrightarrow{\tau}$.
- The set of *infinite traces* of L is $inftr(L) =$
  $\{ a_1a_2a_3\ldots \mid \exists s_0, s_1, s_2, \ldots \in S: is = s_0 \land s_0 = a_1 \Rightarrow s_1 = a_2 \Rightarrow s_2 = a_3 \Rightarrow \ldots \}$.
- The *CFFD model* of L is the 4-tuple ($stable(L)$, $sfail(L)$, $divtr(L)$, $inftr(L)$).
- $L_1 =_{CFFD} L_2$ iff $stable(L_1) = stable(L_2) \land sfail(L_1) = sfail(L_2) \land divtr(L_1) = divtr(L_2) \land inftr(L_1) = inftr(L_2)$.
- $L_1 \leq_{CFFD} L_2$ iff ($stable(L_1) \lor \neg stable(L_2)$) $\land sfail(L_1) \subseteq sfail(L_2) \land divtr(L_1) \subseteq divtr(L_2) \land inftr(L_1) \subseteq inftr(L_2)$. $\square$

The component *stable(L)* was added to the model to ensure that "$=_{CFFD}$" is a congruence with respect to some process composition operators that lack from the CSP language. It is not needed when only "$\|$" and "**hide**" are used in building up systems. It is, however, only one bit of information and it can be easily handled in algorithms, so keeping it unnecessarily does not cost a lot.

When LTSs are finite, the infinite traces are not needed in the model, because they can be derived from traces, and traces can be derived from stable failures and divergence traces.

Theorem 4.7 Let L be an LTS.

- $tr(L) = divtr(L) \cup \{ \sigma \mid (\sigma, \varnothing) \in sfail(L) \}$.
- If L is finite, then $inftr(L) = \{ a_1a_2a_3... \mid \forall i: a_1a_2a_3...a_i \in tr(L) \}$. $\square$

Without the (fairly inessential) stability predicate CFFD-equivalence is strictly weaker than divergence-preserving weak bisimilarity, and strictly stronger than CSP-equivalence. The LTS *CHAOS* in Figure 8 is maximal also with respect to CFFD-preorder. In the absence of divergences CFFD-equivalence is essentially the same as CSP-equivalence. Therefore, "$=_{CFFD}$" and "$\leq_{CFFD}$" can be used like their CSP counterparts, except that also behaviour after the first divergence is taken into account in the comparisons.

The CFFD model has an interesting connection to the classic linear-time temporal logic [MaP92] from which the "next state" operator has been removed. Namely, a slight variant of CFFD-equivalence known as *NDFD-equivalence* [KaV92, VaT95] is the weakest possible congruence that preserves the validity of formulae in the logic. Furthermore, CFFD-equivalence is the weakest possible congruence that preserves both deadlocks and the validity of formulae in the logic [KaV92]. If $L_1 \leq_{CFFD} L_2$, then L_1 satisfies all the formulae that L_2 satisfies and perhaps some more. The exact formulation of these results required the building of a formal "bridge" between the state-oriented logic and action-oriented equivalence. In [Val95b] an alternative "bridge" was given that is very relevant for compositional verification. The equivalence in [DaG90] is essentially the same as NDFD-equivalence, although it is presented in a different formal framework.

Algorithms for the CFFD model can be developed along the same lines as in the case of CSP. Divergence information has to be handled differently, of course. It is sufficient to add one bit for each state of $det(L)$ and $mindet(L)$ that specifies whether an infinite sequence of invisible actions can be started at the state. Not surprisingly, the polynomial space hardness results mentioned in the context of trace semantics apply here, too (and in the case of CSP [BrR83, KaS90]). A theory of acceptance graphs and a normalisation algorithm for the CFFD model were given in [VaT91]. The algorithm was implemented in the *ARA* toolset [VK+93]. Tester processes for on-the-fly checking of "$\leq_{CFFD}$" were defined in [Val93b].

The case studies in Section 2 were conducted using CFFD- and NDFD-semantics and ARA. The LTSs shown in Figures 1, 2 and 3 were obtained by the determinisation and minimisation approach discussed above, and by representing acceptance set and divergence information by putting additional states and τ-transitions into suitable places. Therefore, there is no direct mapping between the τ-transitions in the figures and invisible actions made by the systems, and the interpretation of τ-transitions as arising from particular system transitions is sometimes doomed to failure. This issue is discussed in more detail in [VaS96].

4.8 What Equivalence to Choose?

Many, many equivalences have been defined in the literature in addition to the ones discussed above. For instance, [vGl93] lists 155 different equivalences. The problem of choosing a suitable equivalence for a verification task may thus seem formidable.

When choosing an equivalence for compositional verification, two things have to be kept in mind:

- The equivalence must preserve all the behavioural properties that are essential for the verification task.
- The equivalence must be a congruence with respect to the operators used in building up the system under analysis from its components.

The present author suggests a third criterion:

- The equivalence should preferably be as weak as possible without violating the first two criteria.

The basic reason for this recommendation is that the weaker the equivalence is, the more reduction is obtainable. Indeed, the collection of available reduction algorithms and methods is larger for weaker equivalences, because reduction methods for all strictly stronger equivalences are valid also for the weaker equivalence. The smaller the reduced LTS is, the less states it contributes when it is used as a component in a bigger system. A smaller LTS is also easier to analyse using the visualisation technique discussed in Section 2.1.

Although two arbitrarily given equivalences may be incomparable because of details such as the preservation of divergences, bisimulation-based equivalences are generally stronger than failure-based. On the other hand, we saw that the equivalence checking problem for failure-based equivalences is polynomial space hard, while fast algorithms are known for bisimulation-based equivalences. Implementors of failure-based equivalences tend to announce at this stage that the exponential blow-up of states involved in their algorithms occurs with some artificially constructed systems, but not very often with systems arising from real-world verification tasks, so the algorithms run fast in practice at least most of the time [ClH90, VK+93, Ros94]. It is also important that equivalence checking is only one step in the verification process, and not necessarily the one that determines the overall complexity. After all, even with bisimulation-based equivalences there is always at least one potentially exponential step, namely the computation of parallel compositions. Reduction with respect to failure-based equivalences is not inherently exponential, as long as normalisation or minimisation is not attempted. Altogether, it is not at all obvious whether the expensive equivalence checking algorithm outweighs the benefits of better reduction results. This issue was discussed in detail in [Val95b].

Classic linear-time temporal logic is widely accepted as a sufficiently strong specification formalism for many tasks. The link between it and the CFFD model mentioned in Section 4.7 implies that also CFFD-semantics is strong enough for many tasks. On the other hand, if deadlocks have to be preserved, then at least the stable failures equivalence "$=_{sf}$" is needed. With finite-state systems, the only essential difference between "$=_{sf}$" and CFFD-equivalence is the divergence component. It thus seems that CFFD-equivalence is not far too strong even when its full power is not needed, as long as deadlocks have to be preserved. This makes CFFD-equivalence a good candidate for a general-purpose equivalence.

If only so-called *safety properties* are considered interesting, then the trace semantics is a good choice. If so-called *branching time properties* are analysed (for instance, using a logic such as *CTL–X* [CES86, Eme90]), then CFFD-equivalence does not suffice and a branching-bisimulation-based equivalence has to be used.

5 Concluding Remarks

The purpose of this article was to give a quick introduction to the basic concepts and algorithms in process-algebraic compositional verification, and to highlight some of the advantages it has to offer. Disadvantages have not been emphasised so far. Compared to other automatic verification approaches based on state enumeration, there is perhaps only one major problem. Namely, it is not known how to take so-called *fairness assumptions* into account in a reasonable way and still maintain compositionality.

Unfortunately, there is no space to discuss advanced techniques in any detail. So we only briefly mention some below.

Many algorithms and methods for alleviating the state explosion problem have been suggested in the literature, and some of them may be used in process-algebraic compositional verification. The report [Val92] presents a stubborn set (or partial-order) algorithm that can be used during the computation of LTSs of the form **hide** $a_1, ..., a_m$ **in** ($L_1 \parallel ... \parallel L_n$). The algorithm reduces the size of the result while preserving CSP- and CFFD-semantics. It was applied to on-the-fly verification of "$\leq_{CFFD}$" in [Val93b]. A less powerful version of the algorithm can be used for weak bisimilarity [GK+94].

It was mentioned in Section 3.3 that sometimes the compositional LTS construction approach fails, because isolated subsystems may exhibit behaviour that they do not have when they are within the full system. For instance, an isolated open-ended segment of a token ring may contain several tokens, although the full system contains only one. In order to remedy this problem, in [GrS91] a technique was developed that allows the user to present some invariant information to the compositional verification process. For instance, the user may specify that the open-ended segment contains at most one token at any instant of time. If both the system and the invariant given by the user are correct, the method uses the invariant for reducing the number of states and finally declares success in verification. If the user gave an invariant that does not actually hold, then the method declares failure. Of course, failure is declared also with an incorrect system. The method was developed a bit further in [ChK93].

The methods and models discussed in this article concentrate on actions and synchronous communication. In some cases it would be more natural to analyse states and possibly also use shared variable communication. This was made possible in [Kai96] by adding state information to the semantic model. In [Val94] the same problem was solved with ordinary CFFD-equivalence by defining new system composition operators.

54

Acknowledgements

The time I had available for writing this text was not abundant. Fortunately Jaana Eloranta and Roope Kaivola helped me to reduce the number of errors and omissions in it.

References

[BSW69] Bartlett, K. A., Scantlebury, R. A. & Wilkinson, P. T.: *A Note on Reliable Full-Duplex Transmission over Half-Duplex Links*. Communications of the ACM 12 (5) 1969, pp. 260–261.

[BeK89] Bergstra, J. A. & Klop., J. W.: *Process Theory Based on Bisimulation Semantics*. Linear Time, Branching Time and Partial Order in Logics and Models for Concurrency, Lecture Notes in Computer Science 354, Springer-Verlag 1989, pp. 50–122.

[BKO87] Bergstra, J. A., Klop, J. W. & Olderog, E.-R.: *Failures Without Chaos: A New Process Semantics for Fair Abstraction*. Formal Description of Programming Concepts III, North-Holland 1987, pp. 77–103.

[BoB87] Bolognesi, T. & Brinksma, E.: *Introduction to the ISO Specification Language LOTOS*. Computer Networks and ISDN Systems 14 1987 pp. 25–59.

[BoS88] Bolognesi, T. & Smolka, S. A.: *Fundamental Results for the Verification of Observational Equivalence: a Survey*. Proc. Protocol Specification, Testing and Verification VII, North-Holland 1988, pp. 165–179.

[BrJ82] Brand, D. & Joyner, W. H.: *Verification of HDLC*. IEEE Transactions on Communications, Vol. 30, no. 5., 1982, pp. 1136–1142.

[Bri88] Brinksma, E.: *A Theory for the Derivation of Tests*. Proc. Protocol Specification, Testing and Verification VIII, North-Holland 1988, pp. 63–74.

[BHR84] Brookes, S. D., Hoare, C. A. R. & Roscoe, A. W.: *A Theory of Communicating Sequential Processes*. Journal of the ACM, 31 (3) 1984, pp. 560–599.

[BrR85] Brookes, S. D. & Roscoe, A. W.: *An Improved Failures Model for Communicating Sequential Processes*. Proc. NSF-SERC Seminar on Concurrency, Lecture Notes in Computer Science 197, Springer-Verlag 1985, pp. 281–305.

[BrR83] Brookes, S. D. & Rounds, W. C.: *Behavioural Equivalence Relationships Induced by Programming Logics*. Proc. 10th International Colloquium on Automata, Languages, and Programming, Lecture Notes in Computer Science 154, Springer-Verlag 1983, pp. 97–108.

[ChK93] Cheung, S. C. & Kramer, J.: *Enhancing Compositional Reachability Analysis with Context Constraints*. Proc. ACM SIGSOFT '93: Symposium on the Foundations of Software Engineering, ACM Software Engineering Notes Vol. 18 Nr 5, 1993, pp. 115–125.

[CES86] Clarke, E. M., Emerson, E. A. & Sistla, A. P.: *Automatic Verification of Finite State Concurrent Systems Using Temporal Logic*. ACM Transactions on Programming Languages and Systems, 8 (2) 1986, pp. 244–263.

[ClH90] Cleaveland, R. & Hennessy, M.: *Testing Equivalence as a Bisimulation Equivalence*. Proc. Workshop on Automatic Verification Methods for Finite State Systems, Lecture Notes in Computer Science 407, Springer-Verlag 1990, pp. 11–23.

[CPS90] Cleaveland, R., Parrow, J. & Steffen, B.: *The Concurrency Workbench*. Proc. Workshop on Automatic Verification Methods for Finite State Systems, Lecture Notes in Computer Science 407, Springer-Verlag 1990, pp. 24–37.

[DaG90] Darondeau, P. & Gamatie, B.: *Infinitary Behaviours and Infinitary Observations*. Fundamenta Informaticae XIII (1990) pp. 353–386.

[DNV90] De Nicola, R. & Vaandrager, F.: *Three Logics for Branching Bisimulation*. Report CS-R9012, Centrum voor Wiskunde en Informatica, Amsterdam 1990.

[Elo91] Eloranta, J.: *Minimizing the Number of Transitions with Respect to Observation Equivalence*. BIT 31 (1991) pp. 576–590.

[Elo94] Eloranta, J.: *Minimal Transition Systems with Respect to Divergence Preserving Behavioural Equivalences*. PhD Thesis, University of Helsinki, Department of Computer Science, Report A-1994-1, Helsinki, Finland 1994, 162 p.

[Eme90] Emerson, E. A.: *Temporal and Modal Logic*. Handbook of Theoretical Computer Science, Volume B: Formal Models and Semantics, Elsevier Science Publishers 1990, pp. 995–1072.

[Fer90] Fernandez, J.-C.: *An Implementation of an Efficient Algorithm for Bisimulation Equivalence*. Science of Computer Programming 13 (1989/90) pp. 219–236.

[GaJ79] Garey, M. R. & Johnson, D. S.: *Computers and Intractability: A Guide to the Theory of NP-Completeness*. W. H. Freeman and Company, 1979, 340 p.

[GK+94] Gerth, R., Kuiper, R., Peled, D. & Penczek, W.: *A Partial Order Approach to Branching Time Model Checking*. Proc. Third Israeli Symposium on the Theory of Computing Systems, 1994.

[GrS91] Graf, S. & Steffen, B.: *Compositional Minimization of Finite State Processes*. Proc. Computer-Aided Verification '90, AMS-ACM DIMACS Series in Discrete Mathematics and Theoretical Computer Science, Vol. 3, American Mathematical Society 1991, pp. 57–73.

[GrV90] Groote, J. F. & Vaandrager, F.: *An Efficient Algorithm for Branching Bisimulation and Stuttering Equivalence*. Proc. 17th International Colloquium on Automata, Languages, and Programming, Lecture Notes in Computer Science 443, Springer-Verlag 1990, pp. 626–638.

[Hen85] Hennessy, M.: *Acceptance Trees*. Journal of the ACM 32 (4) 1985, pp. 896–928.

[Hoa85] Hoare, C. A. R.: *Communicating Sequential Processes*. Prentice-Hall 1985, 256 p.

[InP91] Inverardi, P. & Priami, C.: *Evaluation of Tools for the Analysis of Communicating Systems*. EATCS Bulletin 45, October 1991, pp. 158–185.

[Kai96] Kaivola, R.: *Equivalences, Preorders and Compositional Verification for Linear Time Temporal Logic and Concurrent Systems*. PhD Thesis, University of Helsinki, Department of Computer Science, Report A-1996-1, Helsinki, Finland 1996, 185 p.

[KaV92] Kaivola, R. & Valmari, A.: *The Weakest Compositional Semantic Equivalence Preserving Nexttime-less Linear Temporal Logic*. Proc. CONCUR '92, Lecture Notes in Computer Science 630, Springer-Verlag 1992, pp. 207–221.

[KaS90] Kanellakis, P. C. & Smolka, S. A.: *CCS Expressions, Finite State Processes, and Three Problems of Equivalence*. Information and Computation 86 (1990) pp. 43–68.

[MaV90] Madelaine, E. & Vergamini, D.: *AUTO: A Verification Tool for Distributed Systems Using Reduction of Finite Automata Networks*. Proc. Formal Description Techniques II (FORTE '89), North-Holland 1990, pp. 61–66.

[MaP92] Manna, Z. & Pnueli, A.: *The Temporal Logic of Reactive and Concurrent Systems: Specification*. Springer-Verlag 1992, 427 p.

[Mil89] Milner, R.: *Communication and Concurrency*. Prentice-Hall 1989, 260 p.

[Par81] Park, D.: *Concurrency and Automata on Infinite Sequences*. 5th GI Conference on Theoretical Computer Science, Lecture Notes in Computer Science 104, Springer-Verlag 1981, pp. 167–183.

[RR+88] Richier, J. L., Rodriguez, C., Sifakis, J. & Voiron, J.: *Verification in Xesar of the Sliding Window Protocol*. Proc. Protocol Specification, Testing and Verification VII, North-Holland 1988, pp. 235–248.

[Ros94] Roscoe, A. W.: *Model-Checking CSP.* A Classical Mind: Essays in Honour of C. A. R. Hoare, Prentice-Hall 1994, pp. 353–378.

[SLU89] Sabnani, K. K., Lapone, A. M. & Uyar, M. Ü.: *An Algorithmic Procedure for Checking Safety Properties of Protocols.* IEEE Transactions on Communications, Vol. 37, no. 9, 1989, pp. 940–948.

[Sav95] Savola, R.: *A State Space Generation Tool for LOTOS Specifications.* VTT Publications 241, Technical Research Centre of Finland (VTT), Espoo, Finland 1995, 107 p.

[Ste76] Stenning, N. V.: *A Data Transfer Protocol.* Computer Networks, vol. 11, 1976, pp. 99–110.

[Val92] Valmari, A.: *Alleviating State Explosion during Verification of Behavioural Equivalence.* Department of Computer Science, University of Helsinki, Report A-1992-4, Helsinki, Finland 1992, 57 p.

[Val93] Valmari, A.: *Compositional State Space Generation.* Advances in Petri Nets 1993, Lecture Notes in Computer Science 674, Springer-Verlag 1993, pp. 427–457.

[Val93b] Valmari, A.: *On-the-fly Verification with Stubborn Sets.* Proc. Computer-Aided Verification '93, Lecture Notes in Computer Science 697, Springer-Verlag 1993, pp. 397-408.

[Val94] Valmari, A.: *Compositional Analysis with Place-Bordered Subnets.* Proc. Application and Theory of Petri Nets 1994, Lecture Notes in Computer Science 815, Springer-Verlag 1994, pp. 531-547.

[Val95] Valmari, A.: *The Weakest Deadlock-Preserving Congruence.* Information Processing Letters 53 (1995) 341-346.

[Val95b] Valmari, A.: *Failure-based Equivalences Are Faster Than Many Believe.* Proc. Structures in Concurrency Theory, Springer-Verlag "Workshops in Computing" series 1995, pp. 326–340.

[VKS96] Valmari, A., Karsisto, K. & Setälä, M.: *Visualisation of Reduced Abstracted Behaviour as a Design Tool.* Proc. PDP'96, the Fourth Euromicro Workshop on Parallel and Distributed Processing, IEEE Computer Society Press 1996, pp. 187-194.

[VK+93] Valmari, A., Kemppainen, J., Clegg, M. & Levanto, M.: *Putting Advanced Reachability Analysis Techniques Together: the "ARA" Tool.* Proc. Formal Methods Europe '93, Lecture Notes in Computer Science 670, Springer-Verlag 1993, pp. 597–616.

[VaS96] Valmari, A. & Setälä, M.: *Visual Verification of Safety and Liveness.* Proc. Formal Methods Europe '96, Lecture Notes in Computer Science 1051, Springer-Verlag 1996, pp. 228–247.

[VaT91] Valmari, A. & Tienari, M.: *An Improved Failures Equivalence for Finite-State Systems with a Reduction Algorithm.* Protocol Specification, Testing and Verification XI, North-Holland 1991, pp. 3–18.

[VaT95] Valmari, A. & Tienari, M.: *Compositional Failure-Based Semantic Models for Basic LOTOS.* Formal Aspects of Computing (1995) 7: 440–468.

[vGl90] van Glabbeek, R.: *Comparative Concurrency Semantics and Refinement of Actions.* PhD Thesis, Centrum voor Wiskunde en Informatica, Amsterdam 1990.

[vGl93] van Glabbeek, R.: *The Linear Time — Branching Time Spectrum II: The Semantics of Sequential Systems with Silent Moves.* Proc. CONCUR '93, Lecture Notes in Computer Science 715, Springer-Verlag 1993, pp. 66–81.

[vGW89] van Glabbeek, R. & Weijland, W.: *Branching Time and Abstraction in Bisimulation Semantics (Extended Abstract).* Proc. IFIP International Conference on Information Processing '89, North-Holland 1989, pp. 613–618.

[VaW86] Vardi, M. Y. & Wolper, P.: *An Automata-theoretic Approach to Automatic Program Verification.* Proc. IEEE Symposium on Logic in Computer Science, 1986, pp. 322–331.

[Wol86] Wolper, P.: *Expressing Interesting Properties of Programs in Propositional Temporal Logic.* Proc. 13th ACM Symposium on Principles of Programming Languages, 1986, pp. 184–193.

On Liveness and Controlled Siphons in Petri Nets

Kamel Barkaoui and Jean-François Pradat-Peyre

Conservatoire National des Arts et Métiers
Laboratoire CEDRIC
292 rue Saint-Martin, PARIS 75141 Cedex 03, FRANCE
barkaoui@cnam.fr, peyre@cnam.f

Abstract. Structure theory of Petri nets investigates the relationship between the behavior and the structure of the net. Contrary to linear algebraic techniques, graph based techniques fully exploit the properties of the flow relation of the net (pre and post sets). Liveness of a Petri net is closely related to the validation of certain predicates on siphons. In this paper, we study thoroughly the connections between siphons structures and liveness. We define the controlled-siphon property that generalizes the well-known Commoner's property, since it involves both traps and invariants notions. We precise some structural conditions under which siphons cannot be controlled implying the structural non-liveness. These conditions based on local synchronization patterns cannot be captured by linear algebraic techniques. We establish a graph-theoretical characterization of the non-liveness under the controlled-siphon property. Finally, we prove that the controlled-siphon property is a necessary and sufficient liveness condition for simple nets and asymmetric choice nets. All these results are illustrated by significant examples taken from literature.

1 Introduction

Place/Transition nets [12] are a mathematical tool well suited for the modeling and analyzing systems exhibiting behaviors such as concurrency, conflict, and causal dependency between events. However, the high degree of complexity of the analysis is considered as the key obstacle that limits the applicability of Petri nets to real-world problems. The reachability graph of such systems is actually unmanageable, thus it is crucial to enforce the analysis power of techniques based on the net structure. Two paradigms have been proposed to tackle with this explosion problem: the relevant properties of the entire system are gained from the corresponding properties of its smaller and simpler components; the structure theory [4] consists in investigating the relationship between the behavior of the net and its flow relation while the initial marking is considered as a parameter. This paper presents new results in this second direction.

Background

Liveness is an important behavioral property of nets. It corresponds to the absence of global or local deadlock situations. The liveness of a Petri net is closely related to the satisfiability of some predicates on siphons. A siphon is a subset of places once "unsufficiently marked", will never again get new tokens. A siphon is said to be controlled if for each reachable marking the siphon remains "sufficiently" marked. When all siphons are controlled, the net is said to be satisfying the controlled-siphon property (cs-property for short). One major goal of structure theory is to propose necessary and sufficient structural conditions ensuring the controlled-siphon property.

It is showed that the cs-property is a necessary and sufficient liveness condition for some classes of Petri nets. Moreover, under the boundedness hypothesis, liveness, i.e. cs-property, of these sub-classes can be checked in polynomial time. For all algorithms proposed, the controlled-siphon property has been expressed either using graph theoretical structures such as traps, conflict-free paths in [3, 1], or using linear algebraic properties such as conservativeness, consistency and rank condition in [6, 8, 11].

Contribution of the paper

The paper is organized as follows: we present in the next section, the basic concepts and notations used, we introduce in section 3 the notions of min and max controlled siphon. After defining the controlled siphon property (cs-property), we show how this property can be checked structurally using in a refined manner both traps [9] and invariants [13]. Hence we extend the decision power of the Commoner's property. In section 4, we highlight new structural non-liveness conditions related to the close structure of siphons containing no trap. We show through a significant example how this characterization escapes to linear algebraic techniques. In section 5, we state a graph-theoretical characterization of the structural non-liveness under the cs-property hypothesis. From these results we prove in section 6 that the cs-property is a necessary and sufficient liveness condition for simple and asymmetric choice nets. We can then determine for such nets initial markings under which the net is live. Finally, we discuss worthwhile future work and conclude.

2 Basic Definitions and Notations

We briefly define Petri nets. A complete definition can be found in [12].

Definition 1. A Petri net N is a 4-tuple $N = \langle P, T, F, V \rangle$ where:

- $G = \langle P, T, F, V \rangle$ is a weighted bipartite digraph:
  P is the set of node places and T is the set node transitions with
  $P \cup T \neq \emptyset$ and $P \cap T = \emptyset$
  F is the flow relation: $F \subseteq (P \times T) \cup (T \times P)$
  V is the weight application (valuation): $V \in [F \to \mathbb{N}^+]$

Definition 2. A marking M of a Petri net $N = \langle P, T, F, V \rangle$ is a mapping from P to $\mathbb{N}$ where $M(p)$ denotes the number of tokens contained in place p.

A marked Petri net is a couple (N, M_0) where N is a Petri net and M_0 a marking of N called the initial marking.

We denote for a node $s \in P \cup T$, ${}^\bullet s$ (resp. $s^\bullet$) the set of nodes s' such that $(s', s) \in F$ (resp. $(s, s') \in F$).

Given a place p, we denote $Max_{t \in p^\bullet} \{V(p, t)\}$ by $max_{p^\bullet}$, $Min_{t \in p^\bullet} \{V(p, t)\}$ by $min_{p^\bullet}$, $Max_{t \in {}^\bullet p} \{V(t, p)\}$ by $max_{{}^\bullet p}$ and $Min_{t \in {}^\bullet p} \{V(t, p)\}$ by $min_{{}^\bullet p}$.

In all the paper, we denote by N a Petri net $N = \langle P, T, F, V \rangle$ as defined above and by (N, M_0) a marked Petri net.

Basic Structural Objects of Petri Nets

Definition 3. Let N be a Petri net. The subnet induced by (P', T') with $P' \subseteq P$, $T' \subseteq T$, is the net $N' = \langle P', T', F', V' \rangle$ where : $F' = F \cap ((P' \times T') \cup (T' \times P'))$ and V' is the restriction of V on F'.

Definition 4. Let N' be a subnet of a Petri net N. The valuation V' of N' is said to be :

- *homogeneous* if and only if : $\forall p \in P', \forall t_1, t_2 \in p^\bullet, V'(p, t_1) = V'(p, t_2)$
- *non-blocking* if and only if : $\forall p \in P', p^\bullet \neq \emptyset \Rightarrow min_{{}^\bullet p} \geq min_{p^\bullet}$
- *strongly non-blocking* if and only if : $\forall p \in P', p^\bullet \neq \emptyset \Rightarrow min_{{}^\bullet p} \geq max_{p^\bullet}$

The valuation V of a Petri net N can be extended to the application W from $(P \times T) \cup (T \times P) \rightarrow \mathbb{N}$ defined by : $\forall u \in (P \times T) \cup (T \times P), W(u) = V(u)$ if $u \in F$ and $W(u) = 0$ otherwise. The matrix C indexed $P \times T$ and defined by $C(p, t) = W(t, p) - W(p, t)$ is called the incidence matrix of the net.

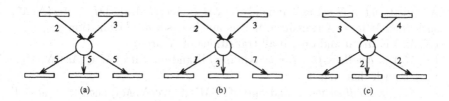

(a)　　　　　　(b)　　　　　　(c)

Fig. 1. A subnet with : (a) homogeneous valuation, (b) non-blocking (but not strongly non-blocking) valuation, (c) strongly non-blocking valuation

Definition 5. A integer vector f ($f \neq 0$) indexed by P ($f \in \mathbb{Z}^P$) is a place invariant if it satisfies ${}^t f.C = 0_T$. The positive support of f is the set of places defined by : $||f||^+ = \{p \in P \mid f(p) > 0\}$. and the negative support of f is the set of places defined by : $||f||^- = \{p \in P \mid f(p) < 0\}$.

Definition 6. Let N be a Petri net. Let $A \subseteq P, A \neq \emptyset$.

- A is called a *siphon* if and only if $^\bullet A \subseteq A^\bullet$. The siphon A is *minimal* if and only if it contains no other siphon as a proper subset and A is *maximal* if and only if none other siphon includes it.
- A is called a *trap* if and only if $A^\bullet \subseteq {}^\bullet A$. The trap A is *minimal* if and only if it contains no other trap as a proper subset and A *maximal* if and only if none other trap includes it.
- A is said to be *max-marked* (resp. *min-marked*) at a marking M if and only if $\exists p \in A$ such that $M(p) \geq max_{p^\bullet}$ (resp. $M(p) \geq min_{p^\bullet}$).
- we denote by $M(A)$ the sum of tokens contained in A at the marking M : $M(A) = \sum_{p \in A} M(p)$

Some Behavioral Properties of Petri Nets

Definition 7. Let (N, M_0) be a marked Petri net.

- A transition $t \in T$ is enabled at a marking M if and only if $\forall p \in {}^\bullet t, M(p) \geq W(p,t)$. The marking M' reached by firing t at M is defined by :

$$\forall p \in P, M'(p) = M(p) - W(p,t) + W(t,p)$$

(denoted by $M[t > M']$)
- By extension, a marking M' is said to be reachable from a marking M if there exists a sequence of transitions $s = t_0.t_1.....t_n$ and a series of marking $m_1, \ldots, m_n$ such that $M[t_0 > m_1, m_1[t_1 > m_2, \ldots, m_n[t_n > M'$ (denoted $M[s > M']$).
- The set of the markings of N reachable from a marking M is denoted by $Acc(N, M)$.

Definition 8. Let (N, M_0) be a marked Petri net.

- A transition t of N is *live* if and only if : $\forall M \in Acc(N, M_0), \exists M' \in Acc(N, M)$ such that $M'[t >$. A transition that is not live is a *dead* transition.
- (N, M_0) is *live* if and only if all transitions of N are live.
- (N, M_0) is *deadlock-free (or weakly-live)* if and only if : $\forall M \in Acc(N, M_0)$, $\exists t \in T$ such that $M[t >$
- (N, M_0) is *deadlockable* if and only if : $\exists M^* \in Acc(N, M_0)$ such that $\not\exists t \in T$ with $M^*[t >$ (M^* is called a "dead marking").
- N is *structurally non-live* if and only if : $\not\exists M_0 \in \mathbb{N}^P$ such that (N, M_0) is a live Petri net.

3 The Controlled-Siphon Property

In this section we first establish two propositions linking liveness or weak-liveness of a net to the initial marking of its siphons. We deduce of these propositions the notion of controlled siphons. We then give two structural ways ensuring the cs-property.

Definition 9. A siphon S of a marked Petri net (N, M_0) is said to be *min-controlled* if and only if S is min-marked at any reachable marking;

$$\forall M \in Acc(N, M_0), \exists p \in S \text{ such that } M(p) \geq min_{p^\bullet}$$

N satisfies the *min cs-property* when all siphons of N are min-controlled.

Definition 10. A siphon S of a marked Petri net (N, M_0) is said to be *max-controlled* if and only if S is max-marked at any reachable marking:

$$\forall M \in Acc(N, M_0), \exists p \in S \text{ such that } M(p) \geq max_{p^\bullet}$$

N satisfies the *max cs-property* when all siphons of N are max-controlled.

Definition 11. A Petri net (N, M_0) is said to be satisfying the *controlled-siphon property* (cs-property) if and only if each minimal siphon of (N, M_0) is min or max controlled.

Remark. Obviously, a max-controlled siphon is also a min-controlled siphon and, when the valuation of the net is homogeneous, a min-controlled siphon is also a max-controlled siphon.

Proposition 12. *If a marked Petri net (N, M_0) is live then it satisfies the min cs-property.*

Proof. We prove that each minimal siphon is min-controlled.

Let (N, M_0) be a live Petri net and S a siphon of N. Suppose that there exists a reachable marking M with $\forall p \in S, M(p) < min_{p^\bullet}$.

At this marking, all transitions of $S^\bullet$ are not enabled. Since S is a siphon ($^\bullet S \subseteq S^\bullet$), all transitions of $S^\bullet$ will never be enabled at any marking reachable from M, which contradicts the liveness of N. □

Proposition 13. *If a marked Petri net (N, M_0) satisfies the max cs-property, then it is weakly-live.*

Proof. If (N, M_0) is not weakly-live then there exists a reachable dead marking M. At M, $\forall t \in T, \exists p_t \in {}^\bullet t \mid M(p_t) < V(p_t, t) \leq max_{p_t^\bullet}$. Let $A = \{p_t\}_{t \in T}$ be the set of such places. A is by construction not a max-controlled siphon. qed. □

Two structural conditions can ensure that a siphon is min or max controlled. For the first one, the control is internal to the siphon and involves trap structure and, for the second one, the control is external to the siphon and is related to place invariants.

Proposition 14. *Let (N, M_0) be a marked Petri net and S a siphon of N. If one of the two following conditions holds then S is min-controlled.*

1. *there exists a trap R included in S such that: R is min-marked at M_0 and the valuation of the subnet induced by $(R, R^\bullet)$ is non blocking.*
2. *there exists a place invariant f such that: $\forall p \in (\|f\|^- \cap S), min_{p^\bullet} = 1$, $\|f\|^+ \subseteq S$, and $\sum_{p \in P}[f(p).M_0(p)] > \sum_{p \in S}[f(p).(min_{p^\bullet} - 1)]$*

Proof. Suppose that point 1 of the proposition holds but that S is not min-controlled. In this case, there exists a marking M reachable from M_0 for which $\forall p \in S$, $M(p) < min_{p^\bullet}$. In particular, the trap R included in S would be not min-marked ($\forall p \in R$, $M(p) < min_{p^\bullet}$). Since the valuation of the subnet induced by $(R, R^\bullet)$ is non blocking, and R is min-marked at M_0, the last conclusion is not possible. Then S is necessarily min-controlled.

Suppose now that point 2 holds. If S is not min-controlled, then there exists a marking M reachable from M_0 for which $\forall p \in S$, $M(p) \le min_{p^\bullet} - 1$. Since $\forall p \in \|f\|^- \cap S$, $min_{p^\bullet} = 1$, we obtain $\sum_{p \in S} f(p).M(p) \le \sum_{p \in S} f(p).(min_{p^\bullet} - 1)$. Moreover as $f(p) < 0$ for $p \in P \backslash S$, we have $\sum_{p \in P} f(p).M(p) \le \sum_{p \in S} f(p).M(p)$ and thus $\sum_{p \in P} f(p).M(p) \le \sum_{p \in S} f(p).(min_{p^\bullet} - 1)$. This inequality can not be satisfied because f is a place invariant ($\sum_{p \in P} f(p).M(p) = \sum_{p \in P} f(p).M_0(p)$) and f min-controls S ($\sum_{p \in P}[f(p).M_0(p)] > \sum_{p \in S}[f(p).(min_{p^\bullet} - 1)]$). $\qquad\square$

Proposition 15. *Let (N, M_0) be a marked Petri net and S a siphon of N. If one of the following condition holds then S is max-controlled.*

1. *there exists a trap R included in S such that: R is max-marked at M_0 and the valuation of the subnet induced by $(R, R^\bullet)$ is strongly non blocking.*
2. *there exists a place invariant f such that: $\forall p \in (\|f\|^- \cap S)$, $max_{p^\bullet} = 1$, $\|f\|^+ \subseteq S$, and $\sum_{p \in P}[f(p).M_0(p)] > \sum_{p \in S}[f(p).(max_{p^\bullet} - 1)]$*

Proof. Suppose that point 1 of proposition holds but that S is not max-controlled. In this case, there exists a marking M reachable from M_0 for which $\forall p \in S$, $M(p) < max_{p^\bullet}$. In particular, the trap R included in S would be not max-marked ($\forall p \in R$, $M(p) < max_{p^\bullet}$). Since valuation of the subnet defined by R is strongly non blocking ($\forall p \in R$, $min_{\bullet p} \ge max_{p^\bullet}$), and initially R is max-marked, the last conclusion is impossible. So S is necessary min-controlled.

Suppose now that point 2 holds. If S is not max-controlled, then there exists a marking M reachable from M_0 for which $\forall p \in S$, $M(p) \le max_{p^\bullet} - 1$. Since $\forall p \in \|f\|^- \cap S$, $max_{p^\bullet} = 1$, we obtain $\sum_{p \in S} f(p).M(p) \le \sum_{p \in S} f(p).(max_{p^\bullet} - 1)$. Moreover as $f(p) < 0$ for $p \in P \backslash S$, we have $\sum_{p \in P} f(p).M(p) \le \sum_{p \in S} f(p).M(p)$ and thus $\sum_{p \in P} f(p).M(p) \le \sum_{p \in S} f(p).(max_{p^\bullet} - 1)$. This inequality can not be satisfied because f is an invariant ($\sum_{p \in P} f(p).M(p) = \sum_{p \in P} f(p).M_0(p)$) and f max-controls S ($\sum_{p \in P}[f(p).M_0(p)] > \sum_{p \in S}[f(p).(max_{p^\bullet} - 1)]$). $\qquad\square$

Remark. When all siphons of a marked Petri net (N, M_0) are min-controlled with respect to condition 1 of proposition 14, (N, M_0) is said to be satisfying the siphon-trap property (Commoner's property) [9], [12].

It is well known that liveness is not a monotonic property (N live under M_0 does not imply N live under $M_0' > M_0$). One must observe that if a siphon S is min or max controlled by a trap under M_0, S remains controlled under $M_0' > M_0$. On the contrary, this control is not necessarily preserved if S is min or max controlled by an invariant. This feature is important since it explains partially the non-monotonicity of the liveness property. Thus, the cs-property is more powerful than the siphon-trap property in the sense that it gives more strict necessary liveness conditions.

Example : Consider the following net N. This net contains two minimal siphons $D_1 = \{p, q\}$ and $D_2 = \{r\}$.

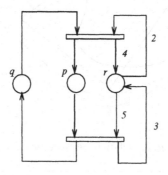

Fig. 2. A Petri net

D_1 is both a trap and the support of a positive invariant. Since valuation of the subnet induced by D_1 is homogeneous D_1 is min or max controlled as soon as p or q is marked.

D_2 is a trap with non blocking valuation (but not strongly non blocking). So D_2 is min-controlled if $M_0(r) \geq 2$. Moreover, $f = r - 2p$ is an invariant such that $||f||^+ = D_2$. So D_2 is min-controlled for initial markings M_0 satisfying $M_0(r) > 1 + 2M_0(p)$ and max-controlled for initial markings M_0 satisfying $M_0(r) > 4 + 2M_0(p)$.

Using the min cs-property, one can conclude that if N is live then necessarily, either $(M_0(q) > 0$ and $M_0(r) > 1)$ (or $M_0(p) > 0$ and $M_0(r) > 3)$. For instance, under $M_0 = p + 2r$, the net is not live although N satisfies the siphon-trap property.

In the same way, if $(M_0(p) > 0$ and $M_0(r) > 4 + 2M_0(p))$ (or $M_0(q) > 0$ and $M_0(r) > 4)$ then N is deadlock free (weakly-live) (for instance, under $M_0 = q + 5r$ or $M_0 = p + 7r$).

Remark. One can note that it may exist a "shadow" interval of initial markings under which cs-property is satisfied but no conclusion about liveness can be deduced.

For instance, under these two following initial markings $M_0 = p + 4r$, or $(M_0 = q + 2r)$ and $M_0' = p + 5r$ or $(M_0' = q + 3r)$ the net of fig.2 satisfies the min cs-property but not max cs. However, one can check that N is deadlockable under M_0 but live under M_0'.

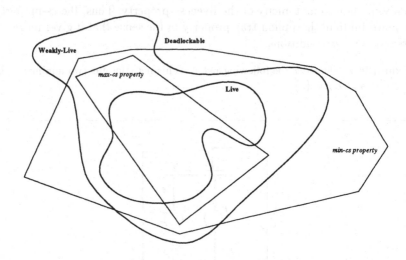

Fig. 3. Relation between properties

4 A Non Controllable Siphons Characterization

The linear algebraic techniques give some structural non-liveness conditions using structural repetitiveness for general nets [5] or using conservativeness [5], consistency [5] and rank theorem [7, 6] for bounded nets.

We establish here a new structural non-liveness condition by exploiting the properties of subnets induced by siphons without trap.

Proposition 16. *Let N be a Petri net and A be a non-empty subset of places which does not contain any trap. There exists a partition $(A_i)_{i=0..I} \subset \mathcal{P}(A)$ of A such that*

$$\forall p \in A_i, \exists t \in p^\bullet \text{ with } t^\bullet \cap \cup_{k \geq i} A_k = \emptyset$$

We denote by $Dec(p) = \{t \in p^\bullet \mid t^\bullet \cap \cup_{k \geq i} A_k = \emptyset\}$

Proof. Let $(S_i) \subset \mathcal{P}(A), (A_i) \subset \mathcal{P}(A), (T_i) \subset \mathcal{P}(T)$ be the series defined by :

- $S_0 = A$,
- $\forall i \geq 0, T_i = S_i^\bullet \setminus {}^\bullet(S_i), A_i = {}^\bullet T_i$ and $S_{i+1} = S_i \setminus A_i$,

Since each S_i is a subset of A, A is finite, and does not contain a trap, then $\exists I > 0$ such that $S_{I+1} = \emptyset$ and $\forall i \leq I, A_i \neq \emptyset, T_i \neq \emptyset$.

Let $i < I, j < I$, and we suppose $i > j$. By construction, $S_i = S_j \setminus \cup_{k=j+1..i} A_i$, and $A_i \subseteq S_i$, so $A_i \cap A_j = \emptyset$. Since $A_i = {}^\bullet T_i$, we also deduce that $T_i \cap T_j = \emptyset$. By construction $S_{I+1} = S_I \setminus \cup_{i=0..I} A_i$, and $S_{I+1} = \emptyset$, so $\cup_{i=0..I} A_i = A$.

Then, $(A_i)_{i=0..I}$ forms a partition of A and $(T_i)_{i=0..I}$ a sub partition of T. Let $p \in A_i$, there exists necessarily $t \in T_i$ such that $t \in p^\bullet$ and $t^\bullet \cap \cup_{k>i} A_k = \emptyset$. Indeed, if such a transition does not exist, either p belongs to A_j with $j > i$, or A contains a trap. This contradicts the hypothesis. $\qquad\square$

Theorem 17. *Let N be a Petri net with homogeneous valuation. If there exists a siphon S of N containing no trap and for which the following condition holds:*

$$\forall p \in S, \forall t \in p^\bullet \setminus Dec(p), \exists t_{dec} \in Dec(p) \text{ such that } {}^\bullet t_{dec} \subseteq {}^\bullet t \qquad [1]$$

then N is structurally not live. Such a siphon is said to be non-controllable.

In order to prove this theorem, we first establish two lemmas.

Lemma 18. *Let A be a finite subset of place and $(A_i)_{i=0..I}$ a partition of A. Let "$\prec_{dec}$" be the order relation on the set of markings of A defined by:*

$$M'_A \prec_{dec} M_A \Leftrightarrow \exists k \in [0, I] \,|\, \forall i > k, M'(A_i) = M(A_i) \text{ and } M'(A_k) < M(A_k)$$

Given an initial (finite) marking of A, M_A, there cannot exist an infinite series of markings of A, starting from M_A and strictly decreasing with respect to $\prec_{dec}$.

Proof. If such an infinite series exists then, either A is infinite, or A is not finitely marked. In both cases, we obtain a contradiction. $\qquad\square$

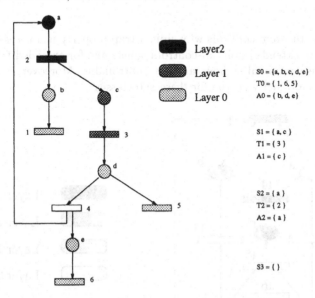

Layer 2
Layer 1
Layer 0

$S0 = \{a, b, c, d, e\}$
$T0 = \{1, 6, 5\}$
$A0 = \{b, d, e\}$

$S1 = \{a, c\}$
$T1 = \{3\}$
$A1 = \{c\}$

$S2 = \{a\}$
$T2 = \{2\}$
$A2 = \{a\}$

$S3 = \{\}$

Fig. 4. Example of construction of the partitions $A = \{a, b, c, d, e\}$

---

[1] this condition means that the siphon has only very locally (according to $Dec(p)$) the free choice pattern synchronization

Lemma 19. *Let M be a reachable marking and $t \in S^{\bullet}$ a transition enabled at M. For all $p \in {}^{\bullet}t \cap S$ there exists a transition $t_{dec} \in S^{\bullet} \cap Dec(p)$ such that t_{dec} is enabled at M.*

Proof. If $t \in Dec(p)$ then the proof is done. Otherwise, by hypothesis, $\exists t_{dec} \in Dec(p)$ such that ${}^{\bullet}t_{dec} \subseteq {}^{\bullet}t$. Since t is enabled at M (${}^{\bullet}t_{dec} \subseteq {}^{\bullet}t$) and the valuation is homogeneous, t_{dec} is also enabled at M. $\qquad\Box$

Proof of theorem 17. Let M_0 be an initial marking of N and suppose that N is live under M_0.

Let $(A_i)_{i<I}$ the partition defined by S (cf. prop 16). Let m_0 be the restriction of M_0 on places of S and let t_0 be the first transition belonging to $S^{\bullet}$ which is enabled at a marking M. Let $p_0 \in {}^{\bullet}t_0 \cap S$.

Consequently to the previous lemma, there exists $t_{dec} \in Dec(p_0)$ such that t_{dec} is also enabled at M. Let then M' be the marking obtained by firing t_{dec} at M. By definition of the order relation " $\prec_{dec}$ ", the restriction of M' on places of S, m_1, is such that $m_1 \prec_{dec} m_0$.

If we suppose that N is live under M_0, we generate an infinite series of markings of S, starting from m_0, strictly decreasing with respect to the order relation $\prec_{dec}$, and that is not possible.

So, N is not live under M_0 and there is no initial marking under which N is live. $\qquad\Box$

Remark. In the context of net analysis, one can only consider minimal siphons of the net.

The previous theorem confirms why siphon-trap property is a necessary liveness condition for extended non self-controlling nets and for extended free-choice nets [1] otherwise minimal siphons are not controllable. Moreover, it explains why liveness monotonicity is true for these sub-classes.

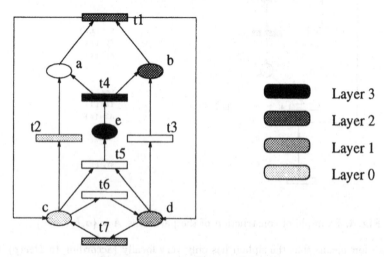

Fig. 5. A structurally non live net

Example : In this net from [7], consider the minimal siphon $S = \{b, c, d, e\}$. One can check that S contains no trap. The partition of S is indicated on figure, and we have $Dec(b) = \{t_1\}, Dec(c) = \{t_2\}, Dec(d) = \{t_7\}, Dec(e) = \{t_4\}$.

We can also check that the "local free-choice" aspect is satisfied : ${}^\bullet t_2 \subseteq {}^\bullet t_5, {}^\bullet t_2 \subseteq {}^\bullet t_6$ for the place c, ${}^\bullet t_7 \subseteq {}^\bullet t_3$ ${}^\bullet t_5 \subseteq {}^\bullet t_3$ for the place d $Dec(b) = b^\bullet$ and $Dec(e) = e^\bullet$. Hence, $S = \{b, c, d, e\}$ is a non-controllable siphon.

So N is structurally non-live, thus N is not a well formed net although N is conservative, consistent, and satisfies the rank condition [6, 7].

The next section is devoted to investigate structural conditions ensuring the liveness once weak-liveness is established by means of the cs-property.

5   A Non-Liveness Structural Characterization

We consider in this section a non live marked Petri net (N, M_0) but for which all siphons are max-controlled $((N, M_0)$ is weakly-live). We suppose also that any shared place p $(Card(p^\bullet) \geq 2)$ satisfies $min\bullet_p \geq max_p\bullet$. This constraint means that the valuation of the net is locally (around shared places) strongly non-blocking.

We are looking for graph-theoretical properties underlying to this pathological behavior (max-cs property but not live).

Proposition 20. *Let (N, M_0) be a marked Petri net satisfying the max-cs property $((N, M_0)$ is weakly-live) but not live. There exists a reachable marking M^* from M_0 and two subsets of transitions, T_L and T_D such that*

- *T_L and T_D is a partition of T $(T_L \neq \emptyset, T_D \neq \emptyset, T_L \cap T_D = \emptyset, T_L \cup T_D = T)$*
- *all transitions of T_L are live and all transitions of T_D are dead.*

Proof. The proof is trivial since the net is not deadlockable (weakly-live) but not live.   □

Remark. This partition is not necessarily unique but there exists at least one. It is important to note that T_D is maximal in the sense that all transitions that do not belong to T_D, will never become dead.

For this particular marking M^*, we characterize now the existence of a particular dead transition, $t^* \in T_D{}^\bullet$.

Theorem 21. *Let (N, M_0) be a non live marked Petri net satisfying the max-cs property. There exists a transition $t^* \in T_D$ such that the set L_P defined by $L_P = \{p \in {}^\bullet(t^*) \mid {}^\bullet p \cap T_L \neq \emptyset$ and $p^\bullet \cap T_L \neq \emptyset\}$ contains at least two items. Furthermore, $\exists M \in Acc(N, M^*)$ such that $\forall p \in {}^\bullet(t^*) \setminus L_P, M(p) \geq V(p, t^*)$.*

Before proving this proposition we establish two lemmas.

Lemma 22. $\forall p \in P, \, {}^\bullet p \cap T_L = \emptyset \Longrightarrow p^\bullet \cap T_L = \emptyset$

Proof. Suppose that this lemma is not true. In this case there exists a place p with all its input transitions in T_D (${}^\bullet p \cap T_L = \emptyset$) and at least one output transition t_v in T_L ($p^\bullet \cap T_L \neq \emptyset$). Since t_v is live, after a finite number of firings place p becomes non min-marked because all its input transitions are dead. So, t_v is dead and denies the maximality of T_D. □

Lemma 23. *There exists $t^* \in T_D$ such that*

1. $\forall p \in {}^\bullet(t^*), \, {}^\bullet p \cap T_L = \emptyset \Longrightarrow \forall M \in Acc(N, M^*), M(p) \geq V(p, t^*)$
2. ${}^\bullet({}^\bullet t^*) \cap T_L \neq \emptyset$

Proof. Suppose that point 1 is false. Since $M^* \in Acc(N, M^*)$, $\forall t \in T_D, \exists p_t \in {}^\bullet t$ with ${}^\bullet p_t \cap T_L = \emptyset$ and $M^*(p_t) < V(p_t, t)$. Let $S = \{p_t\}_{t \in T_D}$. By construction, ${}^\bullet S \subseteq T_D$ and $T_D \subseteq S^\bullet$ (for all p_t in S, ${}^\bullet p_t \cap T_L = \emptyset$). So, S is a siphon. Since $\forall p_t \in S, M^*(p_t) < V(p, t)$, S is non max-marked for M^* and hence, the max-cs property is denied. If S is not minimal, then there exists a minimal siphon included in S which is not max-marked at M^*: that also denies hypothesis.

Using now the previous lemma (lemma 22), (if p has no live input then all outputs of p are dead), we can deduce that the marking of such places does not change for all markings reachable from M^*; that ends the proof of point 1.

If the second point of this lemma is false, then using the point 1 we obtain $\forall p \in {}^\bullet(t^*), M^*(p) \geq V(p, t^*)$. So t^* is enabled at M^* that contradicts hypothesis. □

Proof of theorem 21. Let t^* be a transition fulfilling the conditions of the previous lemma. We first prove that L_p is not empty ($Card(L_p) > 0$).

Suppose that $L_P = \emptyset$: any input place of t^* having a live input transition (there is at least one because of point 2 of the previous lemma) has no live output transition. As the other input places of t^* are such that their pre-conditions on t^* are satisfied at M^* and remain satisfied (point 1 of the previous lemma), we can reach a marking M from M^* such that t^* would be enabled at M. So, $Card(L_P) \geq 1$.

Suppose now that $Card(L_P) = 1$ ($L_P = \{p_1\}$). The valuation is strongly non-blocking around p_1 (p_1 is a shared place), and p_1 has a live input transition. Since the other pre-conditions of t^* are satisfied for M^* (point 1 of lemma 23) then t^* would be enabled at M^*: That is not possible. So, $Card(L_P) \geq 2$.

The last point of the theorem is true for places having no live input transition (point 1 of lemma 23). The other places have a live input transition and are not in T_L. So, these places are not bounded. qed □

We now show how this non-liveness structural characterization allows us to state that the cs-property is a necessary and sufficient liveness condition for asymmetric choice nets with homogeneous valuation.

6 A Necessary and Sufficient Liveness Condition for A.C. Nets

We first establish that the cs-property is a necessary and sufficient liveness condition for simple nets having a modeling power more interesting than free choice nets.

Definition 24. A Petri net N is a simple net if and only if:

$$\forall t \in T, Card(\{p \in {}^\bullet t \,|\, Card(p^\bullet) > 1\}) \leq 1$$

Corollary 25 of theorem 21. *Let (N, M_0) be a simple marked net.*

1. *If (N, M_0) satisfies the max-cs property, then is live.*
2. *Furthermore, when N has an homogeneous valuation, (N, M_0) is live if and only if it satisfies the max-cs property.*

Proof. Suppose that the net is not live. As it is weakly-live since it satisfies the max-cs property, we know (th. 21) that there must exist a transition t^* having at least two input places p_1, p_2 with $Card(p_1{}^\bullet \setminus \{t^*\}) \geq 1$ and $Card(p_2{}^\bullet \setminus \{t^*\}) \geq 1$. So it comes $Card(p_1{}^\bullet) \geq 2$, $Card(p_2{}^\bullet) \geq 2$, $\{p_1, p_2\} \subseteq {}^\bullet(t^*)$ and then the net cannot be a simple net.

The proof of the second point of the corollary is obvious since the max-cs and min-cs properties coincide when the valuation is homogeneous. $\qquad\square$

Example : Consider the following simple net of fig. 6 given in [7]. This net contains six minimal siphons: $D_1 = \{p_1, p_6, p8, p_9\}$, $D_2 = \{p_2, p_9\}$, $D_3 = \{p_6, p_7\}$, $D_4 = \{p_1, p_3, p_4, p_5, p_8\}$ and $D_5 = \{p_2, p_3, p_4, p_5\}$.

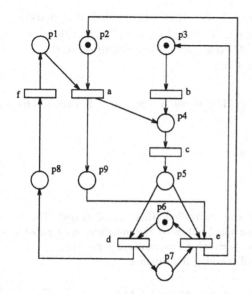

Fig. 6. A simple net

The siphons D_1, D_2, D_3, D_4 are traps and D_5 is controlled by the place invariant $f = p_2 + p_3 + p_4 + p_5 - p_6$.

So, the net is live if the initial marking M_0 satisfies:

1. $\forall i, M_0(D_i) > 0$
2. $M(p_2) + M(p_3) + M(p_4) + M(p_5) > M(p_6)$.

For the marking presented in the figure the net is live but if we e.g. increase by a token the marking of place p_6, the non-liveness holds.

Now we prove that an equivalent result of corollary 1 can be stated for asymmetric choice nets.

Definition 26. A Petri net N is an asymmetric choice net (AC net) if and only if:

$$\forall (p, q) \in P \times P, p^\bullet \cap q^\bullet \neq \emptyset \Longrightarrow p^\bullet \subseteq q^\bullet \text{ or } q^\bullet \subseteq p^\bullet$$

Corollary 27 of theorem 21. *An asymmetric choice net with homogeneous valuation is live if and only if it satisfies the max-cs property.*

Proof. Because min-cs and max-cs properties coincide when the valuation is homogeneous, a live AC net with homogeneous valuation fulfills the max-cs property.

Suppose now that (N, M_0) fulfills the max-cs property (it is weakly-live) but is not live. In this case, there exists (th. 21) a marking M^* and a transition t^* such that the set defined by $L_P = \{p \in {}^\bullet(t^*) \mid {}^\bullet p \cap T_L \neq \emptyset \text{ and } p^\bullet \cap T_L \neq \emptyset\}$ contains at least two items and such that there exists a reachable marking M satisfying $\forall p \in {}^\bullet(t^*) \setminus L_P, M(p) \geq V(p, t^*)$.

The AC net definition implies that the set of places $\{{}^\bullet(t^*)\}$ can be linearly ordered (with the relation $\subseteq$) and then one can derive an ordering on L_P.

Suppose that $L_P = \{p_1, p_2, \ldots, p_k\}$. Without loss of generality we may assume $p_1^\bullet \subseteq \ldots \subseteq p_k^\bullet$.

As $p_1 \in L_P$ ($p_1^\bullet \cap T_L \neq \emptyset$), $\exists t \in p_1^\bullet$ such that t is enabled at a marking M' reachable from M. Since $\forall p \in L_P, p \in {}^\bullet t$ ($p_1^\bullet \subseteq \ldots \subseteq p_k^\bullet$) and valuation is homogeneous, t^* would be enabled at M' that contradicts definition of t^*.

So the max cs-property implies the liveness. $\qquad\qquad\qquad\qquad\qquad\square$

Example : This AC net (given in [14]) is not a simple net. It contains five minimal siphons:

- $D_1 = \{p_{11}, p_1, p_{12}, p_{22}, p_{13}, p_{23}, p_{14}, p_{24}\}$
- $D_2 = \{p_{21}, p_1 + p_{12}, p_{22}, p_{13}, p_{23}, p_{14}, p_{24}\}$
- $D_3 = \{robot, p_{12}, p_{22}, p_{14}, p_{24}\}$
- $D_4 = \{ressources, p_{13}, p_{23}\}$
- $D_5 = \{robot, ressources, p_{14}, p_{24}\}$

The four siphons D_1, D_2, D_3, D_4 are trap-controlled (they are all traps). The siphon D_5 is invariant-controlled. It is controlled either by the invariant $f_1 = robot + ressources - p_1 - p_{11}$ or by the invariant $f_2 = robot + ressources - p_1 - p_{21}$.

So, the net is live as soon as the initial marking M_0 satisfies:

1. $\forall i, M(D_i) > 0$,
2. $M_0(robot) + M_0(ressources) > M_0(p_1) + min(M_0(p_{11}) + M_0(p_{21}))$.

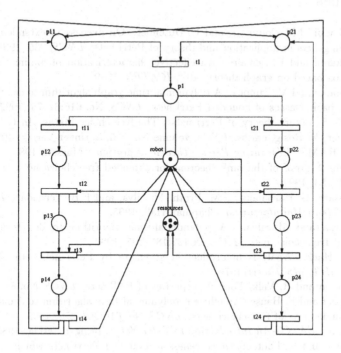

Fig. 7. An asymmetric choice Petri net

7 Conclusion

In this paper we have presented some new results related to structure theory of Petri nets [10, 4] which are more conclusive about liveness than Commoner's property and than algebraic conditions such as consistency, conservativeness and rank theorem. The technical results are based on the controlled-siphon property which uses in a refined manner both traps and invariants in order to "control" siphons. We have proved that this cs-property is a necessary and sufficient liveness condition for asymmetric choice nets. Also, we highlighted a new structural non-liveness condition by investigating the local structure of siphons containing no trap. The cs-property highlights the conditions markings since it provides an explanation of the non-monotonicity of liveness property. In particular, this property allows us to understand why a small modification of the initial marking can make a net non live even if it fulfills the good algebraic properties.

Among the open problems which seem to us relatively approachable in future, we mention the following: How to refine the proposed non-liveness characterization using both a graph theoretical characterization of minimal siphons [2] and the transition invariant notion [13] in order to define a new class of Petri nets having a strong description power and for which the controlled siphon property can be checked structurally in polynomial time.

References

1. K. Barkaoui, J.M Couvreur, and C. Duteilhet. On liveness in extended non self-controlling nets in application and theory of Petri nets. *LNCS*, 935, 1995.
2. K. Barkaoui and B. Lemaire. An effective characterization of minimal deadlocks and traps based on graph theory. 10th *ICATPN*, 1989.
3. K. Barkaoui and M. Minoux. A polynomial time graph algorithm to decide liveness of some basic classes of bounded Petri nets. *LNCS*, No. 616:62–75, 1992.
4. E. Best. Structure theory of Petri nets : The free choice hiatus. In G.Rozenberg W.Brauer, W.Resig, editor, *LNCS*, volume No. 255. Springer-Verlag, 1986.
5. G.W. BRAMS. *Réseaux de Petri : Theorie et pratique*. Masson, 1983.
6. J. Desel. A proof of the rank theorem for extended free choice nets. *LNCS*, No. 616:134–153, 1992.
7. F. Dicesare, G. Harhalakis, J.M. Proth, M. Silva, and F.B. Vernadat. *Practice of Petri Nets in Manufacturing*. Chapman-Hall, 1995.
8. J. Esparza and M. Silva. A polynomial-time algorithm to decide liveness of bounded free-choice nets. *T.C.S*, N 102:185–205, 1992.
9. M.H.T. Hack. Analysis of production schemata by Petri nets. In *Cambridge, Mass.: MIT, MS Thesis*, 1974.
10. M. Jantzen and R. Valk. Formal properties of P/T nets. *LNCS*, No. 84, 1981.
11. P. Kemper and F. Bause. An efficient polynomial-time algorithm to decide liveness and boundedness of free-choice nets. *LNCS*, No. 616:263–278, 1992.
12. W. Reisig. *EATCS-An Introduction to Petri Nets*. Springer-Verlag, 1983.
13. H. Ridder and K. Lautenbach. Liveness in bounded Petri nets which are covered by t-invariants. *LNCS*, No. 815:358–375, 1994.
14. M. Zhou and F. DiCesare. *Petri nets Synthesis for Discrete Event Control of Manufacturing Systems*. Kluwer Academic, 1993.

Behavioural and Structural Composition Rules Preserving Liveness by Synchronization for Colored FIFO Nets

Mohamed-Lyes Benalycherif[1,2] and Claude Girault[1]

[1] Université Pierre et Marie Curie Paris 6, Laboratoire MASI-IBP
4, Place Jussieu. 75252 Paris Cedex 05. France
[2] Institut National des Télécommunications, Département LoR
9, Rue Charles Fourier. 91011 Evry Cedex. France

Abstract. This paper deals with the compositionality of liveness when synchronizing two colored FIFO nets. The composition operator allows to merge transitions as well as some adjacent places or queues.

A behavioural sufficient condition for liveness compositionality relies on a mutual non constraining relation between component nets. A structural sufficient condition for synchronization preserving liveness is then considered in the case of a state machine at the interface of the merged elements with the non merged ones of each component net. It requires that the conflictual colored transitions of the interface state machines satisfy the structural freeing or blocking relations.

Finally an example shows how these conditions simplify the analysis of a protocol within a layered architecture.

Keywords: blocking relation, colored FIFO nets, compositionality, equal conflict, freeing relation, liveness, state machine, synchronization, synchronization medium.

1 Introduction

Compositional specification and validation are fundamental for handling large Petri net models [BC92]. Modular analysis reduces the cost of handling the entire systems. Local properties of each component net can be checked separately and composition rules allow to preserve some global properties.

Communication protocols constitute a privileged field for composition of Petri nets [Dia87] [BWWH88] [BK89]. Protocols which allow to interconnect heterogeneous systems are organized within layered architectures with accurate interface rules between modules. In [BMR83], is proposed a way to construct a live system of sequential machines from smaller ones. [ES91] have studied the synchronization operator which is based on the merging of transitions and of some adjacent places. They have given for this operator a structural characterizations of preserving liveness within the framework of free choice nets. In [Ber86], the addition of a non constraining net is a P/T net transformation defined in behavioural terms which has been proposed to preserve liveness and/or boundedness of the initial net in the resulting net.

High level Petri nets constitute a natural framework for the liveness compositionality. FIFO nets [MF85] are models defined for systems where processes have asynchronous communications by serialized messages. Their queues contain words of colored tokens. Colored FIFO nets defined in Section 2, also have colored transitions and some of their places may contain bags of colored tokens. These nets are very efficient for specification but there is a lack of validation tools because handling the order of tokens is difficult. [Fan91] has defined a process algebra based on FIFO nets. [Sou91] has proposed a structural sufficient condition for liveness compositionality in FIFO nets, which is expressed in terms of connectivity of the non merged queues with the merged transitions. Our paper extends this approach for colored FIFO nets and brings new structural conditions for liveness compositionality.

The notion of a non constraining net is adapted in Section 3 to the synchronization of colored FIFO nets and we show that it constitutes a systematic mean for preserving liveness from component nets to the compound net. We specialize and deepen this behavioural definition to obtain a simpler structural sufficient condition.

In Section 4, we focus on synchronizations where the subnet generated by the merged elements (called the synchronization medium) is a mono-marked state machine. Such a medium is usual in protocol interfaces, because it simplifies the management of messages between communicating entities. We introduce two new asymmetrical behavioural relations involving conflictual colored transitions of the state machine medium: the freeing and the blocking relations. They allow to express a sufficient condition for a component net to be non constraining w.r.t. a second one given the liveness of the merged transitions in the compound net. It is a useful technical result for the composition theorem of Section 5. Then we propose structural sufficient conditions for these behavioural relations which are well suited to colored FIFO nets.

The main part of the paper (Section 5) is devoted to a structural sufficient condition for the non constraining relation. We show that it is sufficient to study separately the two interfaces between the synchronization medium and each component net. They must be mono-marked state machines and any couple of their conflictual colored transitions has to satisfy either the structural blocking relation or the structural freeing relation. The composition theorem states that component nets which are structurally non constraining w.r.t. each other is a sufficient condition for the liveness compositionality. This theorem relaxes the constraints which have limited the use of the result of [Sou91]. At last, this modular specification and validation technique is applied through an example for communication protocols within a layered architecture.

2   Colored FIFO nets and Synchronization

We first give the basic notions of colored FIFO nets and recall the definition of a composition operator between colored FIFO nets, the synchronization.

2.1 Colored FIFO nets

A colored FIFO net is based on a finite valuated tripartite graph where each element x (place, queue or transition) has a finite color domain $C(x)$. The main originality of colored FIFO nets consists in their two types of marking:
- words of $C(q)^*$ for the marking of a queue q,
- multisets on $C(p)$ for the marking of a colored place p.

Straightforward examples show that suppressing the FIFO behaviour of queues may change liveness or boundedness. The decidable problems in P/T nets (liveness, boundedness and reachability) are undecidable in colored FIFO nets. Some classes of FIFO nets with decidable properties have been obtained by restrictions on the valuations of arcs [Mem83], [Fin86], [Cho87].

Definition 1. A multiset over X is a function $f : X \to \mathbb{N}$; the support of f is $Y \subseteq X$ such that $Y = \{x \in X, f(x) \neq 0\}$. The set of multisets over X is denoted by $\mathbb{N}^X$.

Notations. Given a finite set A, $A*$ denotes the free monoid with base A. The empty word is denoted by ϵ. $|x|$ is the length of the word x of A^*. $x \prec y$ means x is a prefix of y i.e. there exists z in A^* such that $y = x.z$ where "." is the concatenation.

Definition 2. (A colored FIFO net) $N = < P, Q, T, C, W >$ is a colored FIFO net where:

- P, Q and T are disjoint finite sets respectively the set of places (represented by circles) the set of FIFO queues (represented by ellipses) and the set of transitions (represented by rectangles)
- C is the color function: $\forall x \in P \cup Q \cup T, C(x)$ is a finite set called the color domain of x. $\forall t \in T, \forall ct \in C(t)$, the couple (t, ct) is called a colored transition. We denote $TC = \{(t, ct) \mid t \in T \text{ and } ct \in C(t)\}$ the set of all colored transitions of N
- W is the colored weight function defined for each arc:
  $W(p, t), W(t, p) \in [C(t) \to \mathbb{N}^{C(p)}], \forall p \in P, \forall t \in T$ and $W(q, t), W(t, q) \in [C(t) \to C(Q)^*], \forall q \in Q, \forall t \in T$.
  For convenience, we denote $W(p, t)((ct, cp))$ by $W(p, t, ct, cp)$, $W(p, t, ct)$ is the function $g : C(p) \to \mathbb{N}$ such that $g(cp) = W(p,t,ct,cp)$ and $W(q, t, ct)$ is a word of $C(q)^*$.

Notations. For any set X of nodes of:
- P the preset (resp. postset) of X is given by $^\bullet X = \{y \in T \mid \exists x \in X, W(y, x) \neq 0\}$ (resp. $X^\bullet = \{y \in T \mid \exists x \in X, W(x, y) \neq 0\}$),
- T the preset (resp. postset) of X is given by $^\bullet X = \{y \in P \mid \exists x \in X, W(y, x) \neq 0\} \cup \{y \in Q \mid \exists x \in X, W(y, x) \neq \epsilon\}$ (resp. $X^\bullet = \{y \in P \mid \exists x \in X, W(x, y) \neq 0\} \cup \{y \in Q \mid \exists x \in X, W(x, y) \neq \epsilon\}$),
- Q the preset (resp. postset) of X is given by $^\bullet X = \{y \in T \mid \exists x \in X, W(y, x) \neq \epsilon\}$ (resp. $X^\bullet = \{y \in T \mid \exists x \in X, W(x, y) \neq \epsilon\}$).

$^\bullet X^\bullet$ denotes the union of the preset and the postset of X.

Marking. A marking of N is a function M defined on $P \cup Q$ where $M(p) \in \mathbb{N}^{C(p)}, \forall p \in P$ and $M(q)) \in C(q)^*, \forall q \in Q$. $|M'(q)|$ is the length of the word $M'(q)$ and $|M'(p)| = \sum_{x \in C(p)} M'(p)(x)$.
The couple (N, M) is a marked colored FIFO net.

Enabling and firing of transitions and sequences. A transition t is enabled at M for the color $ct \in C(t)$, written $M((t, ct) > $ iff $\forall q \in {}^\bullet t \cap Q$, $W(q, t, ct) \prec M(q)$ and $\forall p \in {}^\bullet t \cap P, \forall cp \in C(p)$, $W(p, t, ct, cp) \leq M(p)(cp)$.
If the colored transition (t, ct) is enabled at M, it may fire. Its occurrence leads to a reachable marking M' (written $M((t, ct) > M')$. For each colored place (p, cp), $M'(p)(cp) = M(p)(cp) + W(t, p, ct, cp) - W(p, t, ct, cp)$ and for each queue q, $W(q, t, ct).M'(q) = M(q).W(t, q, ct)$. For the queues of Q, M' is obtained from M by removing from each queue q of ${}^\bullet t$ the word $W(q, t, ct)$ and by appending to each queue q' of $t^\bullet$ the word $W(t, q', ct)$. These notions can be extended in a natural way to sequences of colored transitions $\sigma \in TC^*$.

Reachability set, Language. The reachability set of a marked net (N, M) is denoted by $[N, M >= \{M' \mid \exists \sigma \in TC^*, M(\sigma > M'\}$.
The language of (N, M) is the set $L(N, M) = \{\sigma \in TC^* \mid M(\sigma >\}$.

Boundedness. $x \in P \cup Q$ is bounded in (N, M) iff $\exists n \in \mathbb{N}$ such that $\forall M' \in [N, M >, |M'(x)| \leq n$.
(N, M) is bounded iff $\forall x \in P \cup Q$, x is bounded in (N, M).

Liveness. $t \in T$ is live in (N, M) iff $\forall M' \in [N, M >, \forall ct \in C(t)$, $\exists \sigma \in TC^*$ such that $M'(\sigma.(t, ct) >$.
(N, M) is live (or that M is live) iff every transition of T is live in (N, M).

Subnet. Let P' be a subset of P, Q' be a subset of Q and T' a subset of T. We call subnet of N, generated by (P', Q', T') the colored FIFO net $N(P', Q', T') =< P', Q', T', W', C' >$ defined by: $\forall x, y \in ((P' \cup Q') \times T') \cup (T' \times (P' \cup Q')), W'(x, y) = W(x, y), C'(x) = C(x)$.
The following operators allow to express the relationships between the behaviours of a net and its subnets. Let M be a marking of N, TC' the set of colored transitions of $N(P', Q', T')$, SL a subset of TC^* and σ a sequence of TC^*.
$Proj(\sigma, TC')$ is the sequence obtained by removing from σ the couples (t, ct) which do not belong to TC'.
$Proj(SL, TC') = \{Proj(\sigma, TC') \mid \sigma \in SL\}$.
$M/(P' \cup Q')$ stands for the restriction of M to places of P' and queues of Q'.
$M/(P' \cup Q')$ is therefore a marking of $N(P', Q', T')$.
Let N' be a subnet of N, t a transition of N' and $\hat{\Gamma}_{N'}(t) = \{x \in P' \cup Q' \cup T' \mid \exists x_1, ..., x_n : x_n = x \text{ and } x_1 = t\}$. It denotes the set of descendant nodes of t in the graph N'.

State machine. In a colored FIFO net, all places that only behave as a P/T net place have a color domain that is only the neutral color e which must not appear in the domain of other places and queues. They are represented using the notations of P/T nets, dropping the neutral color indication on their arcs. A state machine is a P/T net with exactly one place in the preset and in the postset of each transition. A mono-marked state machine has only one place marked by one neutral token.

2.2 Synchronization of colored FIFO nets

We consider here colored FIFO nets composed by synchronization. This composition operator has been studied in [ES91] within the framework of P/T nets and in [Sou91] within the framework of FIFO nets.

Synchronization consists of merging distinguished transitions and distinguished adjacent places or queues, so that each one is involved in exactly one merging as shown in Figure 1. A merged transition, place or queue can be interpreted as a communication by respectively rendez-vous, common variables or serialized messages.

A modular view of the synchronization points out two explicit modules, the component nets and an implicit module called the synchronization medium. This medium is a subnet of the compound net generated by all the merged transitions, places and queues and shared by the two component nets. It is thus different from the communication medium [SM89] which is the subnet generated only by the merged transitions.

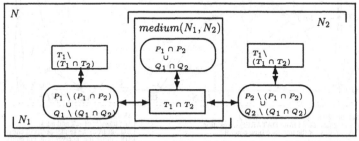

Fig. 1. (N,M) synchronization of (N_1, M_1) and (N_2, M_2)

Definition 3. (Synchronization and synchronization medium)

- Let (N_1, M_1) and (N_2, M_2) be two marked colored FIFO nets. (N_1, M_1) and (N_2, M_2) may be synchronized iff:
  1. ${}^\bullet((P_1 \cap P_2) \cup (Q_1 \cap Q_2))^\bullet \subset T_1 \cap T_2$,
  2. $\forall x, y \in (((P_1 \cap P_2) \cup (Q_1 \cap Q_2)) \times (T_1 \cap T_2)) \cup ((T_1 \cap T_2) \times ((P_1 \cap P_2) \cup (Q_1 \cap Q_2))), W_1(x, y) = W_2(x, y)$,
  3. $M_1/(P_1 \cap P_2) = M_2/(P_1 \cap P_2)$,
  4. $M_1/(Q_1 \cap Q_2) = M_2/(Q_1 \cap Q_2)$.

- The synchronization medium of N_1 and N_2 denoted by $medium(N_1, N_2)$ is the common subnet of N_1 and N_2 generated by $((P_1 \cap P_2), (Q_1 \cap Q_2), (T_1 \cap T_2))$.
- The synchronization (N, M) of (N_1, M_1) and (N_2, M_2) is defined by:
  1. $P = P_1 \cup P_2, Q = Q_1 \cup Q_2, T = T_1 \cup T_2, W = W_1 \cup W_2, C = C_1 \cup C_2,$
  2. $M/(P_i \cup Q_i) = M_i, \forall i \in \{1, 2\}$.

Because synchronization allows communication by rendez-vous through the merged transitions, by projecting sequences of the compound net on the transitions of a component net, we obtain sequences of the component net. The restriction of a global reachable marking to places and queues of a component net also gives a reachable marking of the component net.

Property 4. *Let (N, M) be the synchronization of (N_1, M_1) and (N_2, M_2). We have, $\forall i \in \{1, 2\}$:*

1. $Proj(L(N, M), TC_i) \subset L(N_i, M_i)$
2. $\forall \sigma \in TC^*, \ M(\sigma > M' \Longrightarrow M_i(\sigma_i > M'/(P_i \cup Q_i)$ where $\sigma_i = Proj(\sigma, TC_i)$

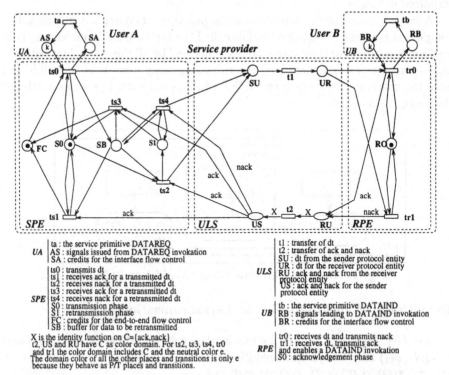

UA	ta : the service primitive DATAREQ AS : signals issued from DATAREQ invocation SA : credits for the interface flow control	**ULS**	t1 : transfer of dt t2 : transfer of ack and nack SU : dt from the sender protocol entity UR : dt for the receiver protocol entity RU : ack and nack from the receiver protocol entity US : ack and nack for the sender protocol entity
SPE	ts0 : transmits dt ts1 : receives ack for a transmitted dt ts2 : receives nack for a transmitted dt ts3 : receives ack for a retransmitted dt ts4 : receives nack for a retransmitted dt S0 : transmission phase S1 : retransmission phase FC : credits for the end-to-end flow control SB : buffer for data to be retransmitted	**UB**	tb : the service primitive DATAIND RB : signals leading to DATAIND invocation BR : credits for the interface flow control
	X is the identity function on C={ack,nack} t2, US and RU have C as color domain. For ts2, ts3, ts4, tr0 and tr1 the color domain includes C and the neutral color e. The domain color of all the other places and transitions is only e because they behave as P/T places and transitions.	**RPE**	tr0 : receives dt and transmits nack tr1 : receives dt, transmits ack and enables a DATAIND invokation S0 : acknowledgement phase

Fig. 2. A colored FIFO net N for a protocol within a layered architecture

Example. The colored FIFO net N represented in Figure 2 models a simple protocol within a layered architecture. The protocol supports a unidirectional flow of data with an error recovery due to negative acknowledgements and a one-sized sliding window for flow control. It will be generalized in Section 5 to an n-sized sliding window.

A functional view of the model brings out three entities: the service provider, user A and user B. Here these user subnets are only represented by single transitions respectively ta and tb with the interface places AS and SA respectively the places BR and RB but in the general case they may be arbitrarily complex nets.

The service provider gathering the subnets SPE (Sender Protocol Entity), ULS (Underlying Layer Service) and RPE (Receiver Protocol Entity) is a natural medium $N_m = SPE \cup ULS \cup RPE$. The underlying layer may corrupt messages and it is assumed that the corruption of messages is reliably detected by the protocol. We consider a very simple service according to which user A invokes a $DATAREQ$ which will be delivered to user B as a $DATAIND$.

Using the synchronization operator, the global colored FIFO net N can be decomposed into two component nets $N_1 = UA \cup N_m$ and $N_2 = UB \cup N_m$ composed of each service user together with the service provider.

3   A behavioural sufficient condition for liveness compositionality by synchronization

Liveness is not preserved by synchronization in a general case whereas boundedness is preserved by synchronization with no restriction (if the component nets are bounded then the compound net is bounded).

A natural composition mechanism preserving liveness is based on non constraining component nets w.r.t. each other. For P/T nets, this notion has first appeared in [Ber86] who has proposed a transformation called the addition of a non constraining net. This transformation which is defined in behavioural terms allows to preserve any sequence of the initial net through a sequence of the resulting net. This notion has been also used in [BRG87] to check the conformity of a protocol w.r.t. its service, both modelled by P/T nets.

We first define the non constraining relation for colored FIFO nets that are synchronized. We then prove that this condition, if verified by each component net w.r.t. the other, is a sufficient condition for the liveness compositionality. This section concludes by a discussion on the conditions for checking the non constraining relation.

Let (N, M) be the synchronization of (N_1, M_1) and (N_2, M_2) with $N_m = medium(N_1, N_2)$.

3.1   The non constraining relation

The non constraining relation net is defined as an asymmetrical relation. A component net is not constraining w.r.t. to a second one if for each reachable marking M' of the compound net N, each transition t which is enabled at the restriction of M' to the second net, is enabled in N after firing at M' a sequence σ of the first net without merged transitions.

Definition 5 (Non constraining relation). (N_1, M_1) is non constraining w.r.t. (N_2, M_2) iff:
$\forall M' \in [N, M >, \forall (t, c) \in TC_2$ such that $M'/(P_2 \cup Q_2)((t, c) >, \exists \sigma \in TC^*$ such that $M'(\sigma.(t, c) >$ and $Proj(\sigma, TC_2) = \epsilon$.

Proposition 6. (N_1, M_1) *is non constraining w.r.t.* (N_2, M_2) *iff:*
$\forall M' \in [N, M >, \forall \sigma \in TC_2^*$ *such that* $M'/(P_2 \cup Q_2)(\sigma >, \exists \sigma' \in TC^*$ *such that* $M'(\sigma' >$ *and* $Proj(\sigma', TC_2) = \sigma$.

Proof.
$\Longleftarrow$ Straightforward.
$\Longrightarrow$ Let $M' \in [N, M >$ and $\sigma \in TC_2^*$ such that $M'/(P_2 \cup Q_2)(\sigma >$. We shall prove that there exists $\sigma' \in TC^*$ such that $M'(\sigma' t >$ and $Proj(\sigma', TC_2) = \sigma$. The proof is constructed by induction over the length n of $Proj(\sigma, TC_1 \cap TC_2)$.

1. $|Proj(\sigma, TC_1 \cap TC_2)| = 0$. $\sigma \in (TC_2 \setminus TC_1)^* \Longrightarrow M'(\sigma >$.
2. Let us suppose that the property is true for any $\sigma \in TC_2^*$ such that $|Proj(\sigma, TC_1 \cap TC_2)| \leq n, \forall n \geq 1$.
   Let $\sigma \in L(N_2, M'/(P_2 \cup Q_2))$ of length n+1.
   Let $\sigma = \alpha.(t', c').\beta$ with $(t', c') \in TC_1 \cap TC_2$ and $\alpha \in (TC_2 \setminus TC_1)^*$.
   Let $M'/(P_2 \cup Q_2)(\alpha.(t', c') > M_2'$.
   As $\alpha \in (TC_2 \setminus TC_1)^*$, we have $M'(\alpha >$.
   Let $M'(\alpha > M''$.
   We can apply the definition of (N_1, M_1) non constraining w.r.t. (N_2, M_2) to M'' and (t', c'). Then there exists a sequence $\delta \in TC^*$ such that $M''(\delta.(t', c') >$ and $Proj(\delta, TC_2) = \epsilon$.
   Let $M''(\delta.(t', c') > M'''$, we have $M'''/(P_2 \cup Q_2) = M_2'$.
   We can apply the induction hypothesis to M''' and β. This leads to the existence of $\zeta \in TC^*$ such that $\zeta = Proj(\beta, TC_2)$.
   Let $\sigma' = \alpha.\delta.(t', c').\zeta$.
   We have $M'(\sigma' >$ and $Proj(\sigma', TC_2) = \sigma$. $\qquad\qquad$ $\square$

3.2 The liveness compositionality

The non constraining relation of a component net w.r.t a second one preserves the liveness of the transitions which belong to the second component net. Moreover if the second net is also non constraining w.r.t. the first one, we obtain a sufficient condition for the liveness of the compound net.

Lemma 7. *If* (N_2, M_2) *is live then:*
(N_1, M_1) *is non constraining w.r.t.* $(N_2, M_2) \Longrightarrow T_2$ *is live in* (N, M).

Proof. Let $M' \in [N, M >$ and $(t, c) \in TC_2$
(N_2, M_2) live and $M'/(P_2 \cup Q_2) \in [N_2, M_2 > \Longrightarrow (N_2, M'/(P_2 \cup Q_2))$ live.
So $\exists \sigma \in TC_2^*$ such that $M'/(P_2 \cup Q_2)(\sigma.(t, c) >$.
Because (N_1, M_1) is non constraining w.r.t. (N_2, M_2), Proposition 6 gives the

existence $\sigma' \in TC^*$ such that $M'(\sigma' >$ and $Proj(\sigma', TC_2) = \sigma.(t, c)$.
Therefore we have the existence of $\alpha, \beta \in TC^*$ such that $\sigma' = \alpha.(t, c).\beta$ and
$M'(\alpha.(t, c) >$. $\qquad\qquad\qquad\qquad\qquad\qquad\qquad\qquad\qquad\qquad\qquad\qquad\square$

Theorem 8. *If* (N_1, M_1) *and* (N_2, M_2) *are non constraining w.r.t. each other
then:*
(N_1, M_1) *and* (N_2, M_2) *are live* $\Longrightarrow$ (N, M) *is live.*

Proof. It follows from lemma 7. $\qquad\qquad\qquad\qquad\qquad\qquad\qquad\qquad\qquad\square$

3.3 How to check the non constraining relation

Practical use of these compositionality results is limited by the difficulty of
checking the non constraining relation. The algorithm deals with the reachability graph of the compound net. The main difficulty is that the finite character
is undecidable for a colored FIFO net as for a FIFO net. Indeed the construction of the reachability graph, if finite, would allow to check the liveness of the
compound net but this would not be in the the spirit of liveness compositionality.

Therefore this relation can be more easily checked for component nets having
a reduced size or some regularities in their structure. The verification of the non
constraining relation of (N_1, M_1) w.r.t. (N_2, M_2) may be simplified if N_1 can be
further decomposed by synchronization into the medium N_m and a new subnet
N_{1m}.

Proposition 9. *Let* (N_1, M_1) *be itself the synchronization of* (N_m, N_m) *and*
(N_{1m}, N_{1m}). *We have:*
(N_{1m}, M_{1m}) *is non constraining w.r.t.* $(N_m, M_m) \Longrightarrow (N_1, M_1)$ *is non constraining w.r.t.* (N_2, M_2)

Proof. Let $M' \in [N, M >$ and $(t, c) \in L(N_2, M'/(P_2 \cup Q_2))$.

1. $(t, c) \in TC_2 \setminus TC_{1m}$. We have $M'((t, c) >$.
2. $(t, c) \in TC_{1m}$
   $(t, c) \in L(N_2, M'/P_2 \cup Q_2) \Longrightarrow (t, c) \in L(N_m, M'/P_m \cup Q_m)$.
   As (N_{1m}, M_{1m}) is non constraining w.r.t. (N_m, M_m), $\exists \sigma \in (TC_{1m} \setminus TC_m)^*$
   such that $M'/(P_1 \cup Q_1)(\sigma.(t, c) >$.
   We have $\sigma \in TC_1 \setminus TC_m$, let us prove that $M'(\sigma.(t, c) >$.
   As $(t, c) \in TC_{1m} \cap TC_m$, σ allows the preconditions of (t, c) in $P_{1m} \cup Q_{1m}$
   to be verified.
   As $M'/(P_2 \cup Q_2)((t, c) >$, the preconditions of (t, c) in $P_m \cup Q_m$ are verified
   at M'.
   As σ does not affect the marking of $P_m \cup Q_m$, the preconditions of (t, c) in
   $P_m \cup Q_m$ remain unchanged after firing σ. Then $M'(\sigma.(t, c) >$. $\qquad\qquad\square$

In Figure 2, $SPE \setminus \{FC\}$ and RPE are mono-marked state machines. Let
$N_{a1} = SPE \setminus \{FC\}$ and $N_{a2} = RPE$. The decomposition of (N_1, M_1) by the
synchronization with (N_{a1}, M_{a1}) as synchronization medium gives (N_m, M_m)

and (N_{1m}, M_{1m}) where $N_{1m} = UA \cup SPE \setminus \{FC\}$ represents user A together with the sender protocol entity. With (N_{a2}, M_{a2}) as synchronization medium, we have the same decomposition for (N_2, M_2) into (N_m, M_m) and (N_{2m}, M_{2m}) where $N_{2m} = UB \cup RPE$ represents user B together with the receiver protocol entity.

The advantage of Proposition 9 is that given a subnet N_{1m} of N_1 non constraining w.r.t. the synchronization medium N_m, we also have N_1 non constraining w.r.t. any component N_2 having the same synchronization medium N_m. Thus it is valuable to create libraries of subnets N_{1m} non constraining w.r.t the synchronization medium N_m.

More particularly state machines appear as interfaces in numerous models, among which communication protocols. They lead to a structural sufficient condition for synchronization preserving liveness in [Sou91]. So in the next section, we consider the case where a component net N_1 may be decomposed into $N_m = medium(N_1, N_2)$ and N_{1m} such that $medium(N_{1m}, N_m)$, allowing us to establish structural conditions for an easy verification of the non constraining relation.

4 A state machine as a synchronization medium

We now assume that the synchronization medium is a mono-marked state machine and that its transitions are live in the compound net. This behaviour is wide spread in interface modules of real systems.

We first present a behavioural sufficient condition for a component net to be non constraining w.r.t. a second one, given the liveness of the state machine transitions in the compound net. This sufficient condition is expressed under the form of two behavioural constraints, the freeing and the blocking relations between the conflictual merged transitions. Finally the well known notions of conflicts [TS93] are exploited and extended to colored FIFO nets to derive structural freeing and blocking relations that imply the behavioural ones. These technical results will be useful for liveness compositionality in Section 5.

4.1 Freeing and blocking relations

As component nets are colored FIFO nets, a conflict in the synchronization medium state machine relates two colored transitions (t, c) and (t', c') where c and c' are respectively colors of transitions t and t' having the same input place with t and t' possibly the same. We introduce two behavioural configurations for the conflicts in the synchronization medium state machine : (t, c) frees (t', c') or (t, c) blocks (t', c') in a given component net N_i. In the last configuration, the non merged part of N_i irreversibly solves the conflict in favour of (t, c) whereas in the former configuration, a resolution of the conflict in favour of (t, c), implies a resolution in favour of (t', c') by the non merged part of N_i. These behavioural conflict relations allow to express a sufficient condition for a component net to be non constraining w.r.t. a second one, provided the state machine transitions are live in the compound net.

Definition 10. (N, M) is a synchronization of (N_1, M_1) and (N_2, M_2) via a mono-marked state machine (N_a, M_a) iff $medium(N_1, N_2) = N_a$ and (N_a, M_a) is a mono-marked state machine.

Definition 11 (Freeness and blocking relations). Let (N, M) be a synchronization of (N_1, M_1) and (N_2, M_2) via a mono-marked state machine (N_a, M_a). Let $p \in P_a$, $t, t' \in p^\bullet$, $c \in C(t)$, $c' \in C(t')$ and $i \in \{1, 2\}$.

1. (t, c) blocks (t', c') in N_i iff:
   $\forall M_i' \in [M_i >, M_i'((t,c) > \implies \nexists \sigma \in (TC_i \backslash \{(t,c)\})^*$ such that $M_i'(\sigma.(t',c') >$,
2. (t, c) frees (t', c') in N_i iff:
   $\forall M_i' \in [M_i >, \exists \sigma \in (TC_i \backslash \tau_b)^*$ such that $M_i'((t,c).\sigma.(t',c') > \implies \exists \sigma' \in (TC_i \backslash (\{(t,c)\} \cup \tau_b)^*$ such that $M_i'(\sigma'.(t',c') >$ and $|Proj(\sigma', \tau)| \leq |Proj(\sigma, \tau)|$
   where $\tau = \{(t'',c'') \in TC_a \mid t'' \in p^\bullet$ and $c'' \in C(t)\}$ and $\tau_b = \{(t'',c'') \in \tau \mid (t',c')$ blocks (t'',c'') in $N_j\}$ with $j \in \{1,2\}$ and $j \neq i$.

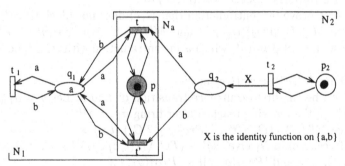

Fig. 3. Example

In the example of Figure 3, the state machine N_a consists only of the transitions t and t' and one place p that contains only a neutral color e. (t, e) and (t', e) are in conflict in N_a.

In N_2, when (t, e) is firable, no sequence without (t, e) allows to fire (t', e). So (t, e) blocks (t', e) in N_2. Similarly, (t', e) also blocks (t, e).

In N_1, for any sequence starting by (t, e) enabling (t', e) without more occurences of (t, e), there exists a sequence which with less conflict transition occurrences which does not include (t, e) and makes (t', e) firable. So (t, e) frees (t', e) in N_1.

The following proposition gives a structural condition for the non constraining relation in the case of a synchronization via a mono-marked state machine provided that the transitions of this state machine are live in the compound net. This rather technical proposition will allow to prove the Lemma 18 of Section 5. The liveness assumption that here appears as limitative, will be a natural requirement for the component nets in order that the composition rules preserve liveness.

Notation. Let $im(\sigma) = \{(t, ct) \in TC \mid n(\sigma, (t, ct)) \geq 1\}$ where $n(\sigma, (t, ct))$ stands for the number of occurrences of (t, ct) in σ.

Proposition 12. *Let (N, M) be a synchronization of (N_1, M_1) and (N_2, M_2) via a mono-marked state machine (N_a, M_a) such that T_a is live in (N, M). If $(\forall p \in P_a, \forall t, t' \in p^\bullet \; \forall c \in C(t), \forall c' \in C(t'), (t, c)$ blocks (t', c') in N_2 or (t', c') frees (t, c) in $N_1)$ then (N_1, M_1) is non constraining w.r.t. (N_2, M_2).*

Proof. Let $M' \in [M >$ such that $M'/(P_2 \cup Q_2)((t, c) >$.

1. $(t, c) \in TC_2 \setminus TC_1$, $M'/(P_2 \cup Q_2)((t, c) > \Longrightarrow M'((t, c) >$
2. $(t, c) \in TC_2 \cap TC_1$

   Because of the liveness of T_a in (N, M), there exists $\sigma \in TC^*$ such that $M'(\sigma.(t, c) >$.

   Let us prove by contradiction that the shortest sequence σ_0 satisfies $Proj(\sigma, TC_2) = \epsilon$.

   Let us assume $(t, c) \notin im(\sigma)$.

   $\forall(t'', c'') \in im(\sigma) \cap \tau$, let $\sigma = \sigma'.(t'', c'').\sigma''$.

   Let us first prove by contradiction that (t, c) does not block (t'', c'') in N_2.

   We have $M'/(P_2 \cup Q_2)((t, c) >$ and $M'/(P_2 \cup Q_2)(Proj(\sigma'.(t'', c''), TC_2))$.

   $(t, c) >$ with $(t, c) \notin im(\sigma')$ which is in contradiction with (t, c) blocks (t'', c'') in N_2.

   So we have (t'', c'') frees (t, c) in N_1.

   Let (t', c') be the first occurrence transition of τ in σ. Let $\sigma = \sigma_1.(t', c').\sigma_2$ and M'' be the marking reached after firing σ_1 at M'.

   We have $M''/(P_1 \cup Q_1)((t', c').Proj(\sigma_2, TC_1).(t, c) >$.

   (t', c') frees (t, c) in N_1 gives $\exists \sigma_1' \in (TC_1 \setminus (\{(t, c\} \cup \tau_b))^*$ such that $M''/(P_1 \cup Q_1)(\sigma_1'.(t, c) >$ and $|Proj(\sigma_1', \tau)| \leq |Proj(\sigma_1, \tau)|$.

   A straightforward reasoning by induction on the length of $Proj(\sigma_1', \tau)$ gives $\exists \sigma_1'' \in (TC_1 \setminus (\{(t, c\} \cup \tau_b))^*$ such that $M''/(P_1 \cup Q_1)(\sigma_1''.(t, c) >$ and $|Proj(\sigma_1'', \tau)| = 0$ or $|Proj(\sigma_1'', TC_a)| = 0$.

   So we have $M'/(P_1 \cup Q_1)(Proj(\sigma_1, T_1).\sigma_1''.(t, c) >$ and $Proj(\sigma_1, T_1).\sigma_1'' \in (TC_1 \setminus TC_a)^*$.

   As $M'/(P_2 \cup Q_2)((t, c) >$ and $Proj(\sigma_1, T_1).\sigma_1''$ does not affect the marking of $P_2 \cup Q_2$, we have $M'(Proj(\sigma_1, T_1).\sigma_1''.(t, c) >$.

   Then $Proj(\sigma_1, T_1).\sigma_1''$ is in contradiction with the minimality of σ. $\qquad\square$

4.2 Structural sufficient condition for the blocking and freeing relations

To obtain structural sufficient condition for the behavioural notions introduced in the former subsection, we use structural conflict relations and connectivity constraints. These relations have been first defined in [TS93] which studied the relationships between the structure and the behaviour of Equal Conflict Systems. These definitions are here naturally extended to colored FIFO nets and a new asymmetrical conflict relation relation is defined.

Definition 13 (Structural conflicts). Let $N = < P, Q, T, C, W >$ be a colored FIFO net, P' a subset of P, Q' a subset of Q and T' a subset of T. Two colored transitions (t, c) and (t', c') of $N(P', Q', T')$ are in:

1. conflict w.r.t. a place or a queue h of $N(P', Q', T')$ iff $h \in {}^\bullet t \cap {}^\bullet t'$,

2. conflict in $N(P', Q', T')$ iff $\exists h \in P' \cup Q'$ such that (t, c) and (t', c') are in conflict w.r.t. h

3. equal conflict in $N(P', Q', T')$ iff (t, c) and (t', c') are in conflict in $N(P', Q', T')$ and $\forall h \in P' \cup Q'$, $W(h, t, c) = W(h, t', c')$

4. asymmetrical conflict in $N(P', Q', T')$ iff (t, c) and (t', c') are in conflict in $N(P', Q', T')$ and $(P \cup Q) \cap {}^\bullet t \subset (P' \cup Q') \cap {}^\bullet t'$ and $\forall h \in (P' \cup Q') \cap {}^\bullet t'$, $W(h, t, c) = W(h, t', c')$.

We first define a structural blocking relation that relies on the order with which words are removed from a FIFO queue. It can be simply expressed by the presence of a common input queue to t and t' such that the words removed by the couples (t, c) and (t', c') are not the prefix of each other. This structural sufficient condition for the blocking relation is easy to check thanks to the specificity of FIFO queues.

Definition 14 (Structural blocking relation). Let $N = < P, Q, T, C, W >$ be a colored FIFO net and P' a subset of P, Q' a subset of Q and T' a subset of T. Let (t, c) and (t', c') be two colored transitions of $N(P', Q', T')$. (t, c) structurally blocks (t', c') in $N(P', Q', T')$ iff $\exists h \in Q'$ such that:

1. (t, c) and (t', c') are in conflict w.r.t. h,

2. $W(h, t, c) \not\prec W(h, t', c)$ and $W(h, t', c) \not\prec W(h, t, c)$,

3. $W(h, t, c) \not\prec W(h, t'', c), \forall t'' \in h^\bullet \setminus \{t, t'\}$.

Proposition 15. Let (N, M) be a synchronization of (N_1, M_1) and (N_2, M_2) via a mono-marked state machine (N_a, M_a). Let $p \in P_a$, $t, t' \in p^\bullet$, $c \in C(t), c' \in C(t')$ and $i \in \{1, 2\}$. We have:
(t, c) structurally blocks (t', c') in $N_i \implies (t, c)$ blocks (t', c') in N_i.

Proof. Let $M_i' \in [M_i >$ such that $M_i'((t, c) >$.
Let us prove by contradiction that $\not\exists \sigma' \in (TC_i \setminus \{(t, c)\})^*$ such that $M_i'(\sigma'.(t', c') >$.
Let $\sigma' \in (TC_i \setminus TC_a)^*$ such that $M_i'(\sigma'.(t', c') >$.
$M_i'((t, c) > \implies W(h, t, c) \prec M_i'(h)$.
As h is a FIFO queue, σ' has to remove $W(h, t, c)$ from h. But the only transition which allows it, is (t, c) which is in contradiction with $\sigma' \in (TC_i \setminus \{(t, c)\})^*$. $\square$

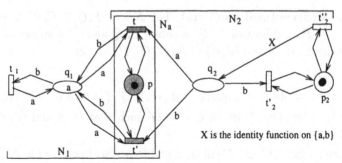

X is the identity function on {a,b}

Fig. 4 Example

In Figure 4, (t,e) and (t',e) are in conflict in N_a and q_2 is the common input queue of (t,e) and (t',e). Firing (t,e) removes a from q_2 whereas firing (t',e) removes b from q_2. But a cannot be removed from q_2 by another transition whereas b can also be removed from q_2 by (t'_2,e). So (t,e) structurally blocks (t',e) in N_2. N_2 can fire (t,e) only when q_2 admits the message a as a prefix of its marking but it cannot fire (t',e) without firing (t,e). So (t,e) blocks (t',e) in N_2.

A threefold structural sufficient condition for the freeing relation is proposed. One feature can be simply expressed by means of the equal conflict relation whereas the other features consist for all the descendant transitions of t' not to be connected to non merged places or queues of N_i.

Definition 16 (Structural freeing relation). Let (N, M) be a synchronization of (N_1, M_1) and (N_2, M_2) via a mono-marked state machine (N_a, M_a). Let $p \in P_a$, $t, t' \in T'$, $c, c' \in C(t) \times C(t')$ and $i \in \{1,2\}$.
(t', c') structurally frees (t, c) in N_i iff:

- (t, c) and (t', c') are in asymmetrical conflict in N_i or
- no transition of $\{t'\} \cup \hat{\Gamma}_{N_a \setminus \{p\}}(t')$ has an adjacent queue or place in $(P_i \setminus P_a) \cup Q_i$ or
- t has no adjacent queue or place in $(P_i \setminus P_a) \cup Q_i$.

Proposition 17. Let (N, M) be a synchronization of (N_1, M_1) and (N_2, M_2) via a mono-marked state machine (N_a, M_a). Let $p \in P_a$, $t, t' \in T_a$, $c \in C(t)$, $c' \in C(t')$ and $i \in \{1,2\}$. We have:
(t', c') structurally frees (t, c) in $N_i \implies (t', c')$ frees (t, c) in N_i.

Proof. Let $\tau = \{(t'', c'') \in TC_a \mid t'' \in p^\bullet$ and $c'' \in C(t'')\}$, $\tau_b = \{(t'', c'') \in \tau \mid (t, c)$ blocks (t'', c'') in $N_j\}$ with $j \in \{1, 2\}$ and $j \neq i$. Let us prove that $\forall M_i' \in [M_i >$ such that $\exists \sigma \in (TC_i \setminus \tau)^*$, $M_i'((t', c').\sigma.(t, c) > \implies \exists \sigma' \in (TC_i \setminus (\{(t', c')\} \cup \tau_b))^*$ such that $M_i'(\sigma'.(t, c) >$. and $|Proj(\sigma', \tau)| \leq |Proj(\sigma, \tau)|$.

1. (t, c) and (t', c') are in asymmetrical conflict in N_i.
   The preconditions of (t, c) are included in those of (t', c'). Then $M_i'((t, c) >$. We have $\sigma' = \epsilon$.

2. No transition of $\{t'\} \cup \hat{\Gamma}_{N_a \setminus \{p\}}(t')$ has an adjacent place or queue in $(P_i \setminus P_a) \cup Q_i$.
   Let $\tau' = \{(t,c) \in TC_a \mid t \in \{t'\} \cup \hat{\Gamma}_{N_a \setminus \{p\}}(t'))$ and $c \in C(t)\}$.
   As $Proj(\sigma, \tau')$ does not affect the marking of places or queues of $(P_i \setminus P_a) \cup Q_i$, let $\sigma' = Proj(\sigma', TC_i \setminus \tau')$.
   We have $M_i'(\sigma'.(t,c) >$ with $|Proj(\sigma', \tau)| < |Proj(\sigma, \tau)|$.
3. t has no adjacent place or queue in $(P_i \setminus P_a) \cup Q_i$.
   The preconditions of (t,c) in N_i are then reduced to p. Moreover at M_i', the only token of N_a is in p. Then $M_i'((t,c) >$. We have $\sigma' = \epsilon$. □

In Figure 4, (t,e) and (t',e) are in equal conflict in N_1. So (t,e) and (t',e) are in asymmetrical conflict and then (t,e) structurally frees (t',e) in N_1. Let M_1 be the marking of N_1. From M_1, N_1 can fire either (t,e) or (t',e) then (t_1,b) to yield M_1'. Therefore at any reachable marking M_i' and for any firable sequence $(t,e).\sigma.(t',e)$ without occurrences of (t,e) in σ, there exists a firable sequence $\sigma'.(t',e)$ such that $(t,e) \notin im(\sigma')$ and $|Proj(\sigma', \tau)| \leq |Proj(\sigma, \tau)|$. So (t,e) frees (t',e) in N_1. Symmetrically (t',e) and (t,e) are in asymmetrical conflict in N_1 so (t',e) frees (t,e) in N_1.

5 A structural sufficient condition for the liveness compositionality

We give in this section a structural sufficient condition for the non constraining relation. This condition relies on the structural freeing and blocking relations which take into account the particularity of FIFO queues. The main result of the paper is a structural sufficient condition for the liveness compositionality which is based on the structurally non constraining relation. This theorem generalizes the one in [Sou91] where the existence of a state machine is required at the interface of the merged transitions with the non merged elements within each component net. Our extension relaxes the constraints concerning the conflicts in this state machine.

5.1 A structural sufficient condition for the non constraining relation

Our structural sufficient condition for the non constraining relation of N_1 w.r.t N_2 requires an adequate decomposition of N_1. It consists in the existence of a mono-marked state machine N_{a1} ensuring the interface with the non merged elements of N_1 (cf. Figure 5). This decomposition may also be expressed by means of a synchronization where N_{a1} is the synchronization medium. The results of the former section which take into account the structural freeing and blocking relations, are applied to express a structural sufficient condition for the non constraining relation.

Let (N, M) be the synchronization of (N_1, M_1) and (N_2, M_2) with $N_m = medium(N_1, N_2)$. Let N_{1m} be a subnet of N_1 such that (N_1, M_1) is the synchronization of (N_{1m}, M_{1m}) and (N_m, M_m) via a mono-marked state machine (N_{a1}, M_{a1}).

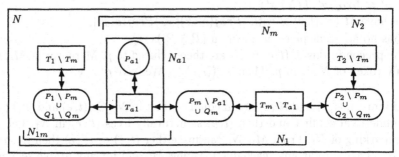

Fig. 5. N_1 structurally non constraining w.r.t. N_2

Definition 18 (Structural non constraining relation). N_1 is structurally non constraining w.r.t. N_2 iff:

1. $^\bullet T_{a1}^\bullet \cap ((P_2 \setminus P_m) \cup (Q_2 \setminus Q_m)) = \emptyset$,
2. $^\bullet((P_1 \setminus P_m) \cup (Q_1 \setminus Q_m)) \cap (T_m \setminus T_{a1}) = \emptyset$,
3. $\forall p \in P_{a1}, \forall t, t' \in p^\bullet, \forall c, c' \in C(t) \times C(t'), (t, c)$ structurally blocks (t', c') in N_m or (t', c') structurally frees (t, c) in N_{1m}.

All the conditions to hold for the structural non constraining relation are easy to check and especially the structural freeing and blocking relations as seen in Section 4.

In the colored FIFO net of Figure 2, N_1 and N_2 are structurally non constraining w.r.t. each other.

Lemma 19. *Let T_{a1} be live in (N_1, M_1). We have:*
N_1 is structurally not constraining w.r.t. N_2 $\Longrightarrow$ (N_1, M_1) is not constraining w.r.t. (N_2, M_2).

Proof. Item 3 of definition 18 allows by applying Propositions 12, 17 and 15 to the decomposition of N_1 into N_{1m} and N_m, to conclude that (N_{1m}, M_{1m}) is non constraining w.r.t. (N_m, M_m). Proposition 9 then gives (N_1, M_1) non constraining w.r.t. (N_2, M_2). $\qquad\square$

5.2 Liveness compositionality theorem

As seen in Section 3, the satisfaction of the mutual non constraining relation by both the component nets is a sufficient condition for liveness compositionality. A structural sufficient condition for the liveness compositionality then follows from the structural non constraining relation. This condition deals with colored

FIFO nets and exploits the specificity of FIFO queues through the basic features of the structural freeing and blocking relations.

Let (N, M) be a synchronization of (N_1, M_1) and (N_2, M_2) with $N_m = medium(N_1, N_2)$. Let N_{1m} be a subnet of N_1 such that (N_1, M_1) is the synchronization of (N_{1m}, M_{1m}) and (N_m, M_m) via a mono-marked state machine (N_{a1}, M_{a1}). Let N_{2m} be a subnet of N_2 such that (N_2, M_2) is the synchronization of (N_{2m}, M_{2m}) and (N_m, M_m) via a mono-marked state machine (N_{a2}, M_{a2}).

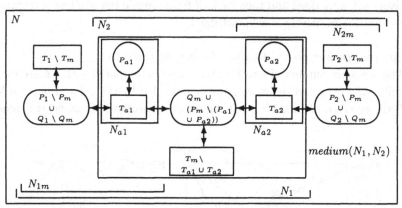

Fig. 6. N_1 and N_2 structurally non constraining w.r.t. each other

Lemma 20. *Let T_{a1} and T_{a2} respectively be live in (N_1, M_1) and (N_2, M_2). If the following conditions hold:*

- *N_1 and N_2 are structurally non constraining w.r.t. each other,*
- *$Q_m \cup (P_m \setminus (P_{a1} \cup P_{a2})) \neq \emptyset$,*
- *${}^\bullet(T_m \setminus (T_{a1} \cup T_{a2}))^\bullet \subset Q_m \cup (P_m \setminus (P_{a1} \cup P_{a2}))$*

then (N_1, M_1) and (N_2, M_2) are non constraining w.r.t. each other.

Proof. It follows from lemma 19. ☐

Theorem 21 (Composition theorem). *If N_1 and N_2 are structurally non constraining w.r.t. each other then:*
(N_1, M_1) and (N_2, M_2) are live $\implies (N, M)$ is live.

Proof. It follows from lemma 20 and theorem 8. ☐

5.3 Example

Protocol specification The colored FIFO net represented in Figure 7 is a generalization of the service provider modelled in Figure 2. The sliding window of the sender protocol entity has now a size of n and all the frames are numbered. The receiver protocol entity always acknowledges dt frames one by one. Functions I, J are used for modelling the management of frames sequence numbers on the arcs.

Within the sender protocol entity, the new features are:

– Two colored places are now necessary for retransmitting dt frames within the sender protocol entity, SB'' shelters the current dt frame to retransmit and SB' shelters all the transmitted but not yet positively acknowledged dt frames.

– A state variable ruling the sequence number of the next dt frame to transmit is modelled by the colored place SN.

– The first dt frame negatively acknowledged is retransmitted until it is positively acknowledged and then each dt frame inside the window is retransmitted until it positively is acknowledged.

Within the receiver protocol entity, the new features are:

– The sequence number of a received dt frame is compared with a state variable ruling the expected sequence number modelled by the colored place RN.

– If the received dt frame has not the expected sequence number, it is ignored.

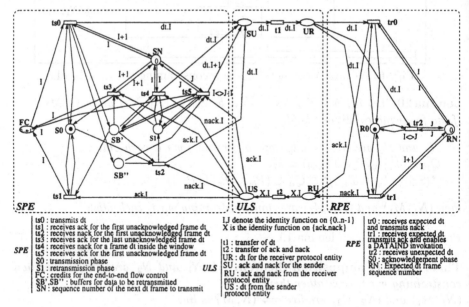

Fig. 7. Service provider model

Protocol verification Let N_1 and N_2 be composed of the service users of Figure 2, each together with the service provider of Figure 7. The service provider is the synchronization medium associated with the synchronization of N_1 and N_2. It has two interface state machines $N_{a1} = SPE \setminus \{FC, SN\}$ and $N_{a2} = RPE \setminus \{RN\}$.

By checking all the couples of conflictual colored transitions of N_{a1} and N_{a2}, we can verify that N_1 and N_2 are structurally non constraining w.r.t. each other. Theorem 21 can then be used to validate the liveness and boundedness of the global colored FIFO net. Indeed, it is sufficient to prove the liveness of each component net in order to conclude that the global colored FIFO net is live. The component nets are bounded by Max(n,k) and live because their marking graphs are strongly connected. We can then conclude that the compound net is bounded and live.

6 Conclusion

In this paper, we have presented a structural sufficient condition for synchronization preserving liveness in colored FIFO nets. This structural condition is a sufficient condition for mutually non constraining component nets which is itself a behavioural sufficient condition for the liveness compositionality. The structural condition requires for each component net the existence of a mono-marked state machine at the interface of the synchronization medium with its non merged places or queues. Within an interface state machine, conflicts are also submitted to two structural relations. These relations which capture the specificity of FIFO queues, are easy to check and may be applied to protocols models. More general behavioural conflict relations have also been defined which, if combined with the component net requirement decomposition, allow a more general sufficient condition so that the synchronization preserves liveness.

This study may be extended in three directions. The first one is to continue relaxing the constraints of our structural sufficient condition for synchronization to preserve liveness. The second direction consists in facing synchronization to the preservation of other properties such as home state. The third direction is to define some classes of colored FIFO nets for which liveness can be checked by polynomial algorithms.

Acknowledgement
We would like to thank Younes Souissi for valuable discussions at the beginnning of the work reported here. We would also like to thank Jean-Michel Couvreur and one anonymous referee for their constructive comments and suggestions which allowed to improve this paper.

References

[BC92]   L. Bernardinello and F. De Cindio. A survey of basic net models and modular net classes. In *Advances in Petri Nets 1992*, volume 609 of *LNCS*. Springer-Verlag, 1992.

[Ber86]   G. Berthelot. Checking Properties of Nets Using Transformations. In *Advances in Petri Nets 1985*, volume 222 of *LNCS*, pages 19–40. Springer Verlag, 1986.

[BK89]   J. Billington and D. Kinny. Computer Aided Protocol Engineering. In *Conference on New Business Applications of Information Technology*, pages 69–73, Melbourne, Australia, 1989.

[BMR83]   G. Berthelot, G. Memmi, and W. Reisig. A Control Structure for Sequential Processes Synchronised by buffers. In *Proc. of the 4th European Workshop on Application and Theory of Petri Nets*, Toulouse, France, 1983.

[BRG87]   A. Bourguet-Rouger and C. Girault. Validation of Parallel System Properties. In *Proc. of International Conference on Parallel Processing and Applications*, L'Aquila, Italy, 1987. North Holland.

[BWWH88]   J. Billington, G.R. Wheeler, and M.C. Wilburn-Ham. PROTEAN: A high-level petri net tool for the Specification and verification of Communication Protocols. *IEEE Transactions on Software Engineering*, 14(3):301–316, 1988.

[Cho87]   A. Choquet. *Analyse et propriétés des processus communiquant par files FIFO : réseaux à files à choix libre topologique et réseaux à files linéaires.* Thèse de doctorat de 3ème cycle, Université Paris XI, 1987.

[Dia87]   M. Diaz. Petri Net Based Models in the Specification and Verification of Protocols. In *Advances in Petri Nets 1986*, volume 255 of *LNCS*, pages 135–170. Springer Verlag, 1987.

[ES91]   J. Esparza and M. Silva. On the Analysis and Synthesis of Free Choice Systems. In *Advances in Petri Nets 1990*, volume 483 of *LNCS*. Springer Verlag, 1991.

[Fan91]   J. Fanchon. Fifo-Net Models for Processes with Asynchronous Communication. In *Proc. of the 12th International Conference on Application and Theory of Petri Nets*, Gjern, Denmak, 1991.

[Fin86]   A. Finkel. *Structuration des systèmes de transitions - Application au contrôle du parallélisme par files FIFO.* Thèse d'état, Université Paris XI, 1986.

[Mem83]   G. Memmi. *Méthodes d'analyse des Réseaux de Petri, Réseaux à files et Application aux Systèmes Temps Réel.* Thèse d'état, Université Paris VI, 1983.

[MF85]   G. Memmi and A. Finkel. An Introduction to FIFO Nets-Monogeneous Net: A subclass of FIFO Nets. *Theoretical Computer Science*, 35:191–214, 1985.

[SM89]   Y. Souissi and G. Memmi. Composition of nets via a communication medium. In *Proc. of the 10th International Conference on Application and Theory of Petri Nets*, Bonn, 1989.

[Sou91]   Y. Souissi. A Modular Approach for the Validation of Comunication Protocols using FIFO Nets. In *Proc. of the XIth International Symposium on PSTV*. North Holland, 1991.

[TS93]   E. Teruel and M. Silva. Liveness and Home States in Equal Conflict Systems. In *Proc. of the 14th International Conference on Application and Theory of Petri Nets*, Chicago, USA, 1993.

High Level Synthesis of Synchronous Parallel Controllers

Krzysztof Biliński, Erik L. Dagless

Advanced Computing Research Center,
Department of Electrical & Electronic Engineering,
University of Bristol, Bristol BS8 1TR, United Kingdom
E-mail:{K.Bilinski, Erik.Dagless}@bristol.ac.uk

Abstract. In this paper the application of Petri nets to high level synthesis of synchronous parallel controllers is presented. A formal specification of a design is given in a form of an interpreted synchronous Petri net. Behavioral properties of the controller are verified using symbolic traversal of its Petri net model. The net state–space explosion problem is managed using binary decision diagrams (BDDs). Once the Petri net specification of a controller is tested, the BDD representation of the net's state–space is used to generate a state assignment with which the controller can be synthesized. The application of the proposed methodology to the design of a MAXbus port controller for a high performance Transputer Framestore and its comparison to the alternative implementation is discussed. The experimental results clearly demonstrate the advantages of the proposed method. Further, the significant increase of the applicability of this approach has also been achieved.

1 Introduction

The customer's desire to increase performance of digital systems may be more easily addressed by applying efficient tools which explore parallelism within the design. In this paper a synthesis method for synchronous parallel controllers is presented. Many of today's digital design methodologies start at the behavioral level. The designer describes how the designed system and its components interact with their environment, specifying *what* the system does, rather than *how* it does it. The design is specified using hardware description languages (HDL) e.g. VHDL or Verilog. Having tested the behavioral description, the design is usually manually translated into a register–transfer level (RTL), and then it is synthesized using automatic synthesis tools.

The target architecture of high level synthesis is a general computing model comprising a data–path and a control unit. To control a complex data path, incorporating many parallel, interacting, quasi–independent subprocesses, the controller needs to support multiple control flow. Such a controller is often called a parallel controller. Applying finite state machine (FSM) techniques to realize parallel controllers, means it must be assembled from a number of linked sequential designs. The FSM technique does not support any explicit representation of a multiple–process system. However, each individual subprocess can be implemented separately as a single FSM. Then all the subcontrollers obtained are linked using semaphore bits and/or by cross–connecting their control lines to realize the entire parallel controller. This approach often yields a non–optimum implementation, places some limits on the amount of parallelism in the design, and can result in parallel synchronization errors in the implementation. A more efficient approach is to use Petri nets to create a formal specification of the design. Behavioral analysis of the controller specification can be performed, using well defined techniques from Petri net theory. Once the Petri net specification of a controller is tested, it can be transformed automatically into a logic–level description.

Recently, several methods which use Petri nets for synchronous parallel controller synthesis have been suggested [3, 9, 11, 13, 16, 18, 22]. However, practical use of Petri-net-based methods in VLSI design is still limited. The main reason is that Petri-net-based models tend to become too large for the analysis of even medium size systems. As a result, methods which operate on a net's reachability graph can only be applied to small examples due to the complexity of the state space explosion [9, 11, 13, 16]. Other methods either suffer from a waste of memory elements [18, 22] or can only be applied efficiently to some specific sub-classes of Petri nets [3]. In this paper, a new synthesis approach based on symbolic traversal of a Petri net is suggested. The net behavior and structure are represented in terms of Boolean functions, which are used to generate a state assignment with which the controller can be synthesized. The complexity of the net state space explosion is successfully managed by using BDDs [4].

The Petri net symbolic traversal techniques have originated from the methods which are used in sequential circuit verification [7, 8]. In [17] an adaptation of these methods to Petri net traversal was reported. The reachability analysis algorithms are modified to represent sets of markings using BDDs. The algorithms are based on simultaneously mapping sets of markings to sets of markings by using net's transition functions. Both marking sets and transition functions are represented with BDDs. Thus, an explicit construction of the net's reachability graph is avoided. In [6, 12, 19] the use of symbolic traversal of Petri nets for synthesis and verification of asynchronous circuits were presented. The reported methods prove to be capable of dealing with large systems, due to the efficient data representation provided by BDDs. Here, we introduce a symbolic traversal method for synchronous interpreted Petri nets.

The remainder of the paper is organized as follows: Sect. 2 gives a brief introduction to parallel controllers, their Petri net specifications, and binary decision diagrams. Modeling of safe Petri nets with BDDs is reviewed in Sect. 3. An overview of the proposed synthesis methodology and its main components are presented in Sect. 4. In Sect. 5 the application of the algorithms presented in the preceding section to the design of a MAXbus port controller is discussed. Experimental results are reported in Sect. 6, and concluding remarks are drawn in Sect. 7.

2 Preliminaries

A Petri net is a bipartite, directed graph, which is defined as 4-tuple [14]: $PN = (P, T, F, M_0)$, where P and T are finite sets of places and transitions, satisfying $P \cap T = \emptyset$ and $P \cup T \neq \emptyset$; $F \subseteq (PxT) \cup (TxP)$ is finite set of arcs; $M_0 : P \rightarrow \{0, 1, 2, ...\}$ is the initial marking. A marking is an assignment of tokens to places. The position and the number of tokens changes during the net execution according to a simple rule for transition enabling and firing. For a formal introduction to Petri nets refer to [14].

2.1 Petri Net Model of Parallel Controllers

A digital system may be viewed as a bipartite network comprising two separate but related modules: a data path and a control unit. Data is manipulated in the data path. However, to ensure that all conversions of the data will be done correctly, the operation of the data path must be controlled by a special unit, able to detect and recognize the state of the system, and produce appropriate control signals, to achieve proper sequencing and synchronization of events in the data path. This unit is called the controller. In complex systems, with many parallel, interacting subprocesses, using an FSM (or a sequential controller) can be awkward. A better solution is to divide each of the global sequential states into a number of concurrently active local states, with each local state controlling a different subprocess. The main difference between a finite state machine and a parallel controller is that whereas a finite state machine has only one state active at any time, a parallel controller can have several states active simultaneously.

When a Petri net is used to model a parallel controller, each of its places represents a local state of the controller. Every marked place represents an active local state, and so the marking of the net represents the entire state of the controller. The movement of the tokens defines the controller behavior, and can be represented using a reachability graph. The use of a Petri net as a hardware description language has resulted in the introduction of some dedicated extensions to ordinary Petri net interpretation.

Firstly, a controller receives signals (inputs or qualifiers) coming from a data path and sometimes from another control unit. Using this information, the controller produces control signals (outputs or instructions) which determine the behavior of the system. Explicit specification of these signals has been incorporated into the interpretation of Petri nets [1]. Input signals affect changes in the state of the system, and are strictly bound to events, and consequently to transitions. Several signals may form a logic function describing a condition for an event to take place. Such a function is called a transition predicate. A transition predicate is a restriction imposed on the transition firing rule. The rule may be defined as follows: a transition is enabled and may fire when all of its input places are marked and a predicate associated with it is asserted.

In addition, a controller should be able to assert output signals both when the system stays in a particular state and when a stated event takes place. This means that outputs should be associated with both places and transitions. Outputs which are associated with places are Moore-type outputs. They only depend on the local states of the system, and are asserted whenever the associated places have tokens. Outputs associated with transitions depend not only on the state of the system, but also on the inputs used in the transition predicate. They are Mealy-type outputs of the controller and they are asserted whenever the associated transitions are enabled. An example of an interpreted Petri net is shown in Fig. 1.

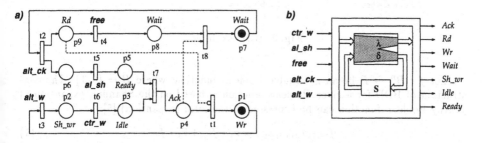

Fig. 1. Example of an Petri net specification (a) of a controller (b).

Secondly, to model systems which are synchronized by a global clock a new transition firing rule has to be introduced. All transitions are synchronized by a global clock, and so all enabled transitions fire simultaneously, and the marking is updated only once per clock cycle. Such a Petri net is said to be a synchronous Petri net.

Finally, the ability to describe priority or synchronization between quasi–independent subprocesses running concurrently in the system is useful and is realized [14] by inhibitor and enabling arcs. An inhibitor arc allows the absence of tokens in a place to be tested, while an enabling arc is used to test for the presence of tokens in a place.

The feature that differentiates inhibitor and enabling arcs from ordinary arcs is that no tokens are moved along the former when transitions fire.

In this paper it is assumed that a parallel controller is synchronized using a global clock, and is described using a synchronous interpreted Petri net.

2.2 Binary Decision Diagrams

A binary decision diagram is a rooted, directed, acyclic graph as defined in [5], which has two sink nodes labeled 0 and 1 representing Boolean functions 0 and 1, and non-sink nodes, each labeled with a Boolean variable x. Each non-sink node has two output edges labeled 0 and 1 and represents the Boolean function $f(x = 0)$ corresponding to its 0 edge or the Boolean function $f(x = 1)$ corresponding to its 1 edge. An ordered binary decision diagram (OBDD) is a BDD with the constraint that all variables are ordered and every path from the root node to a sink node in the OBDD visits the variables in ascending order. A reduced ordered binary decision diagram (ROBDD) is an OBDD where each node represents a distinct logic function.

The ROBDDs are often significantly more compact than traditional forms of representing Boolean functions, such as conjunctive or disjunctive normal forms. The size of the ROBDD strictly depends on variable ordering. Examples of the OBDD and the ROBDD representation of the function $f = a\bar{b} + a\bar{c} + b\bar{c}$ are shown[1] in Fig. 2.

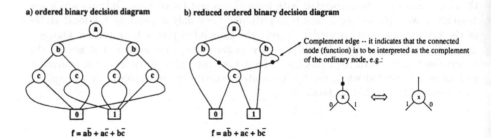

a) ordered binary decision diagram b) reduced ordered binary decision diagram

Complement edge -- it indicates that the connected node (function) is to be interpreted as the complement of the ordinary node, e.g.:

$f = a\bar{b} + a\bar{c} + b\bar{c}$ $f = a\bar{b} + a\bar{c} + b\bar{c}$

Fig. 2. BDD examples.

Let $f : B^n \to B$ be an n-variable Boolean function, where $B = \{0, 1\}$. The construction of BDD for a given Boolean function is based on the Shannon expansion of f. Any Boolean function can be expressed in the form (expansion) [5]:

$$f(x_1, x_2, ..., x_n) = x_i * f_{x_i} + \overline{x_i} * f_{\overline{x_i}} \tag{1}$$

where: $f_{x_i} = f(x_1, ..., x_{i-1}, 1, x_{i+1}, ..., x_n)$ and $f_{\overline{x_i}} = f(x_1, ..., x_{i-1}, 0, x_{i+1}, ..., x_n)$ are called the positive and negative cofactors of f with respect to x_i. Intuitively, equation (1) can be represented using the following fragment of BDD:

$$f(x_1, x_2, ..., x_n) = x_i * f_{x_i} + \overline{x_i} * f_{\overline{x_i}} \implies$$

Applying recursively the Shannon expansion to f_{x_i} and $f_{\overline{x_i}}$ the BDD representation of the function f can be constructed easily.

---

[1] Whenever possible, drawings will be simplified by adopting the following conventions: (i) edges are directed from the top to the bottom; (ii) the 0 edge is the leftmost one.

Boolean operations can be calculated in polynomial time in relation to the size of the BDD. The only other operations which are required for the algorithms that follow are quantifications over Boolean variables and substitution of variables. The existential quantification of f with respect to x_i (i.e. an elimination of the variable x_i from the formula f) is defined as:

$$\exists_{x_i}(f) = f_{x_i} + f_{\overline{x_i}} \qquad (2)$$

The substitution of the variable y_i for the variable x_i in a formula f, denoted $f\langle x_i \leftarrow y_i \rangle$ can be accomplished as:

$$f\langle x_i \leftarrow y_i \rangle = y_i * f_{x_i} + \overline{y_i} * f_{\overline{x_i}} \qquad (3)$$

Henceforth, it will be assumed implicitly that BDDs are reduced and ordered.

3 Modeling Safe Petri Nets with BDDs

In this section, the basic concepts of using BDDs to model structure and behavior of Petri nets are presented. The main theoretical notions involve *characteristic function* of a set of elements, and Petri net's *transition functions*.

3.1 Characteristic Function

The characteristic function χ_A of a set of elements $A \subseteq U$ is a Boolean function: $\chi_A : U \to \{0, 1\}$, that evaluates to 1 on the vertices belonging to A, otherwise it is 0. The function is calculated as a disjunction of all elements of A. Operations on sets are in direct correspondence with operations on their characteristic functions:

$$\chi_{(A \cup B)} = \chi_A + \chi_B; \quad \chi_{(A \cap B)} = \chi_A * \chi_B; \quad \chi_{(\overline{A})} = \overline{\chi_A} \qquad (4)$$

Let PN be a safe Petri net and Ω the set of all markings of PN. A marking of PN can be encoded as $M_k = (\mu_1 \mu_2 \ldots \mu_n)$, where $\mu_i \in \{0, 1\}$ represents the number of tokens in a place p_i. A set of all markings *reachable* from the initial marking (M_0) is denoted $[M_0)$. Any set of markings $M \subseteq \Omega$ can be represented using its characteristic function [17], e.g. given the Petri net depicted in Fig. 1 (a) the characteristic function of the set of markings:

$$M = \{000100001, 001000100, 010000100\}$$

is defined as:

$$\chi_M = \overline{p_1}\,\overline{p_2}\,\overline{p_3}\,p_4\,\overline{p_5}\,\overline{p_6}\,\overline{p_7}\,\overline{p_8}\,p_9 + \overline{p_1}\,\overline{p_2}\,p_3\,\overline{p_4}\,\overline{p_5}\,\overline{p_6}\,p_7\,\overline{p_8}\,\overline{p_9} + \overline{p_1}\,p_2\,\overline{p_3}\,\overline{p_4}\,\overline{p_5}\,\overline{p_6}\,p_7\,\overline{p_8}\,\overline{p_9}$$

We use p_i both to denote a place of PN and a variable of a characteristic function. Note, that there is a one-to-one correspondence between elements μ_i of a set M and variables p_i of its characteristic function χ_M. Figure 3 shows the BDD representation of the set of all reachable markings of the net in Fig. 1 (a).

3.2 Transition Function

The behavior of a Petri net is determined by the net's structure which can be expressed using transition functions. The transition function is used to transform a set of markings M_1 into the set of markings M_2 that can be reached from M_1 in one iteration (i.e. after simultaneously firing of all enabled transitions). The transition function: $\Delta : \Omega \to \Omega$, is defined as a functional vector of Boolean functions:

$$\Delta = [\delta_1(P, X), \delta_2(P, X), \ldots, \delta_n(P, X)] \qquad (5)$$

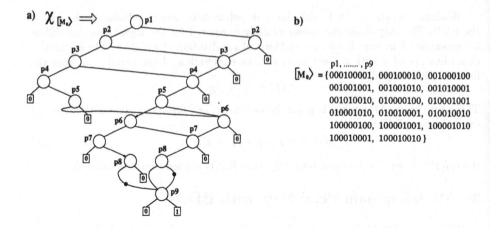

a) $\chi_{[M_\bullet\rangle} \Rightarrow$

b)

$$[M_\bullet\rangle = \begin{matrix} p1, \text{.......}, p9 \\ \{000100001,\ 000100010,\ 001000100 \\ 001001001,\ 001001010,\ 001010001 \\ 001010010,\ 010000100,\ 010001001 \\ 010001010,\ 010010001,\ 010010010 \\ 100000100,\ 100001001,\ 100001010 \\ 100010001,\ 100010010\ \} \end{matrix}$$

Fig. 3. BDD representation of the set of all reachable markings of the net in Fig. 1.

where: $\delta_i(P, X)$ is a transition function of the place p_i; P and X are sets of places and input signals respectively. The function δ_i evaluates to 1 when the place p_i will have a token in the next iteration, otherwise it is 0. It can be efficiently implemented by using the topological information from the Petri net.

Consider again the net shown in Fig. 1(a). Every function δ_i consists of two parts: (a) a part describing a situation when the place p_i will receive a token, and (b) a part describing a situation when the place will keep a token. For example, place p_7 will have a token after one iteration if: (a) places p_4 and p_8 have tokens (transition t_8 will fire) or (b) place p_7 has already got a token and input signal alt_ck is 0 (transition t_2 is disabled), thus the function δ_7 can be defined as follows:

$$\delta_7 = \underbrace{p_4 * p_8}_{(a)} + \underbrace{p_7 * \overline{alt\_ck}}_{(b)} \tag{6}$$

Note, that part (a) of the function δ_7 does not imply that tokens will be removed from places p_4 and p_8 in next iteration if this part is asserted (i.e. places p_4 and p_8 have tokens). The movements of tokens to/from place p_4 are defined by function δ_4:

$$\delta_4 = \underbrace{p_3 * p_5}_{(a)} + \underbrace{p_4 * p_9}_{(b)} \tag{7}$$

An alternative concept of the transition function of a Petri net was presented in [17]. The idea was that the transition functions were associated with transitions instead of places.

4 Synthesis Methodology

The synthesis methodology starts from the behavioral specification of a controller, which is given in terms of a synchronous interpreted Petri net[2]. The Petri net specification of a controller is implemented by using a state register to represent places and

---

[2] A translator of a behavioral VHDL into a Petri net, based on the method presented in [9], is available as an alternative entry interface to the system.

combinational logic to realize transitions. To optimize state assignment, flip–flops in the state register can be reused to implement different places under the constraint that any two places holding a token simultaneously and using the same flip–flops never try to set and reset any of them at the same time. Symbolic traversal of the Petri net is used to establish all relations between places (i.e. local states of the controller). Next, the behavioral properties of the Petri net model are verified. Finally, a state assignment is generated and a logic–level description of the controller is created in a standard format that can be accepted by logic synthesis tools. The description is first optimized and then transformed into a form suitable for final implementation, e.g. FPGA–dedicated packages. The following sections describe the above methodology in detail.

4.1 Symbolic Traversal of a Petri Net

To generate all reachable markings of a Petri net a method similar to a symbolic breadth–first traversal of an FSM [8] is used. Several markings are simultaneously calculated using their characteristic function and the transition functions. The traversal algorithm is defined as follows:

$$\textbf{symbolic\_traversal\_of\_Petri\_net}(initial\_marking, \chi_{[M_0]}, X) \ \{$$

[1] $\quad \chi_{[M_0]} = current\_marking = initial\_marking;$
$\quad \textbf{while} \ (\ current\_marking \ != \ \emptyset \ ) \ \{$
[2] $\quad\quad next\_marking = \textbf{image\_computation}(current\_marking, X);$
[3] $\quad\quad current\_marking = next\_marking * \overline{\chi_{[M_0]}};$
[4] $\quad\quad \chi_{[M_0]} = current\_marking + \chi_{[M_0]};$
$\quad \}$
$\}$

In the above, all variables in italics represent characteristic functions of corresponding sets of markings, e.g. given the Petri net in Fig. 1 (a) for the first iteration the appropriate functions are defined as follows:

[1] $\chi_{[M_0]} = current\_marking = initial\_marking = \ p_1 \, \overline{p_2} \, \overline{p_3} \, \overline{p_4} \, \overline{p_5} \, \overline{p_6} \, p_7 \, \overline{p_8} \, \overline{p_9}$

[2] $next\_marking = \ \overline{p_1} \, p_2 \, \overline{p_3} \, \overline{p_4} \, \overline{p_5} \, \overline{p_6} \, p_7 \, \overline{p_8} \, \overline{p_9} + \overline{p_1} \, p_2 \, \overline{p_3} \, \overline{p_4} \, \overline{p_5} \, p_6 \, \overline{p_7} \, \overline{p_8} \, p_9$
$\quad\quad + p_1 \, \overline{p_2} \, \overline{p_3} \, \overline{p_4} \, \overline{p_5} \, p_6 \, \overline{p_7} \, \overline{p_8} \, p_9 + p_1 \, \overline{p_2} \, \overline{p_3} \, \overline{p_4} \, \overline{p_5} \, p_6 \, \overline{p_7} \, \overline{p_8} \, p_9$

[3] $current\_marking = \ \overline{p_1} \, p_2 \, \overline{p_3} \, \overline{p_4} \, \overline{p_5} \, \overline{p_6} \, p_7 \, \overline{p_8} \, \overline{p_9} + \overline{p_1} \, p_2 \, \overline{p_3} \, \overline{p_4} \, \overline{p_5} \, p_6 \, \overline{p_7} \, \overline{p_8} \, p_9$
$\quad\quad + p_1 \, \overline{p_2} \, \overline{p_3} \, \overline{p_4} \, \overline{p_5} \, p_6 \, \overline{p_7} \, \overline{p_8} \, p_9$

[4] $\chi_{[M_0]} = \ \overline{p_1} \, p_2 \, \overline{p_3} \, \overline{p_4} \, \overline{p_5} \, \overline{p_6} \, p_7 \, \overline{p_8} \, \overline{p_9} + \overline{p_1} \, p_2 \, \overline{p_3} \, \overline{p_4} \, \overline{p_5} \, p_6 \, \overline{p_7} \, \overline{p_8} \, p_9$
$\quad\quad + p_1 \, \overline{p_2} \, \overline{p_3} \, \overline{p_4} \, \overline{p_5} \, p_6 \, \overline{p_7} \, \overline{p_8} \, p_9 + p_1 \, \overline{p_2} \, \overline{p_3} \, \overline{p_4} \, \overline{p_5} \, p_6 \, \overline{p_7} \, \overline{p_8} \, p_9$

The **image_computation** function performs a computation of a set of markings which can be reached from *current_marking* in one iteration according to the following equations:

$$next\_marking = \exists_p \exists_x \left(current\_marking * \underbrace{\underbrace{\prod_{i=1}^{n} [p_i' \odot \delta_i(p, x)]}_{a}}_{b}\right) \qquad (8)$$

$$next\_marking = next\_marking \langle p' \leftarrow p \rangle \qquad (9)$$

where: p, p', x denote the present state, the next state and the input signal variables; $\exists_p$ and $\exists_x$ represent existential quantifications of the present state and the input signal variables; symbols $\odot$ and $*$ represent logic operators XNOR and AND respectively.

The bottleneck of the presented approach is the size of BDDs being quantified i.e. part "b" of the equation (8), since the quantification operation has exponential computation complexity. To optimize the computation process a method applying an "early quantification of variables" [6] is used. The method is based on the following observations:

- although, existential quantification does not distribute over conjunction, formulas can be moved out of the scope of existential quantification if they do not depend on any of the variables being quantified;
- many nets (circuits) exhibit locality, so many of the transition functions δ_i depends only on a small number of the places and input signals.

Thus, the following equation:

$$A = \exists_{p_1} \exists_{p_2} \exists_{p_3} \exists_{x_1} \exists_{x_2} [(\overline{p_1}\, p_2\, \overline{p_3}\, \overline{p_4})(p_1' \odot (p_2\, p_3 + p_1\, x_2))(p_2' \odot p_1 x_1)(p_3' \odot p_1 p_3)] \quad (10)$$

can be rewritten as follows, since the conjunction is commutative and associative:

$$A = \exists_{p_1} \exists_{x_1} [\exists_{p_3} [\exists_{p_2} \exists_{x_2} [(\overline{p_1}\, p_2\, \overline{p_3}\, \overline{p_4})(p_1' \odot (p_2\, p_3 + p_1\, x_2))](p_3' \odot p_1 p_3)](p_2' \odot p_1 x_1)] \quad (11)$$

In [10] a method of automated ordering of partitions (i.e. parts "a" of equation (8)) was reported. The method is especially efficient for designs which are described using a set of quasi–independent nets, each of which specifies a single process. An example of such a design is presented in Sect. 5. Having a set of all reachable markings computed, the Petri net behavior is verified.

4.2 Verification of Petri Net Properties

For the Petri net model of a controller, three Petri net properties are essential: safeness, liveness and determinism. They can be easily verified using Boolean manipulations on the characteristic functions of sets of markings and places. Most of the verification methods presented here have been extracted from [17], and adopted to synchronous interpreted Petri nets.

Safeness. In a parallel controller any local state can be active or inactive at any time. This means that an appropriate flip–flop (or a state of a state register) in a hardware implementation is set or reset. In the Petri net model of the controller this situation is equivalent to marking any place with at most one token at any time. Thus, the Petri net specifying the controller has to be safe. A place of the safe Petri net can be implemented by a single flip–flop. For any marking there are two situations when a place "p_i" can become unsafe:

- Two or more input transitions fire simultaneously:

$$\underset{p_i \in P}{\exists} \underset{t_k, t_l \in \, ^\bullet p_i}{\exists} \quad t_k \neq t_l, \quad current\_marking * \, ^\circ t_l * \, ^\circ t_k \neq 0 \quad (12)$$

- The place already has a token and any of its input transitions is enabled while all its output transitions are disabled:

$$\underset{p_i \in P}{\exists} \underset{t_k \in \, ^\bullet p_i}{\exists} \left(\begin{array}{l} p_i * current\_marking \neq 0 \ and \\ current\_marking * \, ^\circ t_k \neq 0 \ and \\ current\_marking * (\sum_{t_j \in \, p_i^\bullet} \, ^\circ t_j) = 0 \end{array} \right) \quad (13)$$

The symbols $^\bullet t$, $t^\bullet$, $^\bullet p$, and $p^\bullet$ define characteristic functions of the pre–set and post–set of every transition t or place p respectively. For example, given the Petri net depicted in Fig. 1 (a), $^\bullet t_7 = p_3 * p_5$ and $^\bullet p_1 = t_1$. The symbol $^\circ t$ represents a firing condition of the transition t, i.e. $^\circ t_5 = p_6 * al\_sh$ while $^\bullet t_5 = p_6$.

Liveness. Liveness is a very strong property of Petri nets and a complete liveness test of the Petri net specification of parallel controller can be very time-consuming. In the presented approach the test for liveness has been restricted to the test of the main features which have to be satisfied if the controller is expected to be deadlock–free (i.e. every designed operation is activated at least once) and each reachable global state of the controller has at least one successor global state. Thus the test is looking for deadlock transitions and terminal markings:

- A transition is called a deadlock transition, if it will never fire during the net execution:

$$\underset{t_i \in T}{\exists} \quad \chi_{[M_0]} * {}^\bullet t_i = 0 \quad \Longrightarrow \quad t_i \text{ is deadlock transition} \tag{14}$$

- A terminal marking of a Petri net is one in which no transitions are enabled:

$$\chi_{[M_0]} * \prod_{t_i \in T} \overline{{}^\bullet t_i} \neq 0 \quad \Longrightarrow \quad \text{there is terminal marking in Petri net} \tag{15}$$

Determinism. As the specification of a controller should describe its operation in an entirely deterministic way, the Petri net model of the controller must be conflict–free. If the firing of one transition disables another transition, the situation is referred to as a conflict. The appearance of a conflict in the specification means the specification is erroneous or incomplete, formally t_k and t_l are in conflict if:

$$\underset{t_k, t_l \in T}{\exists} t_k \neq t_l, \underset{(p_i \in P,\, p_i \in {}^\bullet t_k,\, p_i \in {}^\bullet t_l)}{\exists} current\_marking * {}^\circ t_k * {}^\circ t_l \neq 0 \tag{16}$$

Safeness and determinism are verified after every *current_marking* is computed, while liveness is checked after symbolic traversal of the Petri net has finished. Having verified the net properties the Petri net state assignment is performed.

4.3 Petri Net State Assignment

State assignment assigns a state code to each place in the Petri net. The number of flip-flops needed to implement a parallel controller depends on the way they are assigned to the places. The silicon area occupied by the controller is usually a trade–off between the number of flip–flops comprising the state register and the number of gates comprising the combinational logic blocks. An optimal solution is obtained if places share the same multiple–flip–flop state register. Since different local states can be active simultaneously, constraints on the use of the register are necessary. The *state assignment constraints* can be formulated as follows [16]:

- The local states which can be active simultaneously (i.e. places which hold tokens simultaneously) must have non-orthogonal codes, whilst local states that are non-concurrent can have orthogonal codes.
- Moreover, if places which hold tokens simultaneously share some state variables, then their codes for these variables must be the same.

Two codes are said to be non–orthogonal if they differ by at least one state variable.

The Petri net state assignment algorithm operates on the BDD representation of the characteristic function $\chi_{[M_0]}$. To simplify the state assignment process, the characteristic function $\chi_{[M_0]}$ is simplified prior to state assignment.

Characteristic function simplification. The simplification algorithm consists of the following steps. First, a pair p_i and p_j is identified, such that both p_i and p_j have the same marking relationship with all other places, i.e:

$$\underset{p_k \in P}{\forall} \; k \neq i, j, \; \begin{cases} \exists_{p_i, p_j}(\chi_{[M_0]} * p_i) * p_k \neq 0 \; \Leftrightarrow \; \exists_{p_i, p_j}(\chi_{[M_0]} * p_j) * p_k \neq 0 \\ \exists_{p_i, p_j}(\chi_{[M_0]} * p_i) * p_k = 0 \; \Leftrightarrow \; \exists_{p_i, p_j}(\chi_{[M_0]} * p_j) * p_k = 0 \end{cases} \quad (17)$$

Next, p_i and p_j are merged by forming their Boolean product or sum and replacing them by a new variable p_i:

$$p_i = \begin{cases} p_i.p_j & if \; \chi_{[M_0]} * p_i * p_j = 0 \quad i.e. \; p_i, p_j \; are \; sequentially \; related \\ p_i + p_j & if \; \chi_{[M_0]} * p_i * p_j \neq 0 \quad i.e. \; p_i, p_j \; are \; concurrently \; related \end{cases} \quad (18)$$

Finally, p_j is removed from the characteristic function. The algorithm is reiterated until no further simplification can be achieved. Figure 4 shows the BDD representation of the simplified characteristic function χ_R of the function $\chi_{[M_0]}$ shown in Fig. 3. For simplicity, the expressions representing Boolean product or sum will be explicitly used, instead of using the variables by which they are replaced, e.g. $p_5.p_6$ instead of p_5.

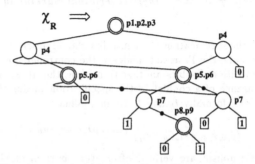

Fig. 4. The characteristic function from Fig. 3 after simplification.

The simplification of the characteristic function $\chi_{[M_0]}$ is equivalent to the reduction of the Petri net. The sum of variables corresponds to fusion of parallel places, while the product of variables corresponds to fusion of serial places [14] (e.g. the variable $p_1.p_2.p_3$ represents the fusion of serial places p_1, p_2 and p_3). In [14] a set of transformations which operate on the net's structure were proposed. We believe that our reduction method is more efficient, since even places which are not structurally related can be merged together, e.g. places which belong to different sub-nets. The available empirical results support this conclusion, although more experimentation is necessary before a definitive statement can be made.

Having constructed the χ_R, its attribute is evaluated. The χ_R attribute, which can be sequential or parallel, determines the performance of the state assignment algorithm. The evaluation algorithm is defined as follows:

function_attribute(χ_R) {
    $par = 0$;
    foreach_pair_of_variables($p_i, p_j \in supp(\chi_R)$, $p_j \neq p_j$)
        if ($\chi_R * p_i * p_j \neq 0$) $par = par + 1$;
    if ($\frac{|supp(\chi_R)|^2 - par)}{|supp(\chi_R)|^2} > 0.5$)   **return** *parallel*;
    else   **return** *sequential*;
}

where: $supp(\chi_R)$ is a support of χ_R, i.e. a set of all variables in χ_R.

State assignment algorithm. If the number of variables in the BDD representation of the characteristic function χ_R is greater than one, a parallel state assignment is performed as follows:

1. Identify BDD variable p_i satisfying the following criterion according to χ_R attribute:

$$parallel \quad \longrightarrow \quad \underset{p_k \in supp(\chi_R)}{\forall} \quad k \neq i, \; gcn(p_i * \chi_R) \geq gcn(p_k * \chi_R) \qquad (19)$$

$$sequential \quad \longrightarrow \quad \underset{p_k \in supp(\chi_R)}{\forall} \quad k \neq i, \; gcn(p_i * \chi_R) \leq gcn(p_k * \chi_R) \qquad (20)$$

where: $gcn(bdd)$ returns the number of product terms represented by bdd (i.e. number of markings represented by bdd).

2. Assign $Q_l = 1$ to the variable p_i, so that p_i is uniquely coded.

3. Assign $Q_l = 0$ to each variable p_k satisfying the criterion:

$$\underset{p_k \in supp(\chi_R)}{\forall} \quad k \neq i, \; p_i * p_k * \chi_R = 0 \qquad (21)$$

Applying these steps to the reduced characteristic function in Fig. 4, variable p_7 is selected (the χ_R has parallel attribute), thus p_7 is assigned $Q_l = 1$. Then, since variables p_4, $p_5.p_6$, and $p_8.p_9$ satisfy the criterion (21), p_4, $p_5.p_6$, and $p_8.p_9$ are all assigned $Q_l = 0$.

4. For $l > 1$ allocate $Q_l = {}'-'$ (meaning the flip–flop can be reused during sequential state assignment) to variables that are already uniquely coded, provided these variables are mutually sequential, and sequential to the variables identified in Steps 2 and 3.

5. Remove variable p_i from the characteristic function χ_R.

6. Reiterate the characteristic function simplification algorithm.

The reduced characteristic function after removing p_7 and simplification is shown in Fig. 5. Note, that after removing variable p_7 from the BDD of Fig. 4, the sub–tree starting at the right node $p_5.p_6$ expresses a function which equals 1, so that the entire sub–tree is replaced by the sink–node 1.

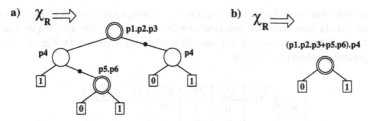

Fig. 5. Characteristic function χ_R after: (a) removing of the variable p_7, and simplification (b).

7. If the number of variables in the BDD representation of χ_R is greater then one go to Step 1.

The speed of the χ_R reduction and the probability of the reuse of flip–flops which were already allocated depends on the χ_R attribute and is shown in Tab. 1.

When the characteristic function χ_R is reduced to one variable, a sequential state assignment of the places represented by each BDD node is performed as follows:

χ_R attribute	reduction speed	probability of the flip-flops reuse
sequential	low	high
parallel	high	low

Table 1. The χ_R attribute impact on the χ_R reduction and the flip–flops reusing.

8. Allocate extra flip–flops to all variables to satisfy the *state assignment constraints*. The number of flip–flops needed to encode a product of other variables is calculated as follows:

$$p_1.p_2. \dots .p_n \;:\; min(ff) \ge \log_2 n, \;\; ff \in N^+ \tag{22}$$

where: ff is an integer representing the number of flip–flops. The number of flip–flops needed to encode a sum of variables is calculated as sum of flip–flops needed to implement each of its arguments, e.g. given the expression depicted in Fig. 5(b): $(p_1.p_2.p_3 + p_5.p_6).p_4$ the number of flip–flops needed is calculated as follows:

$$\underbrace{\underbrace{(p_1.p_2.p_3}_{2} + \underbrace{p_5.p_6)}_{1}.p_4}_{1} \;\;\Longrightarrow\;\; 4 \text{ flip–flops needed}$$

9. Using only allocated flip-flops, assign a unique code to places represented by each variable.

The state assignment for the current example prior to Step 9 (i.e. flip–flops allocation) is shown in Tab. 2.

	Q_1	Q_2	Q_5	Q_4	Q_5	Q_6
$p_1.p_2.p_3$	—	—	—	—		
p_4	0	—	—	—	—	
$p_5.p_6$	0	—	—	—	—	
p_7	1					
$p_8.p_9$	0					—

Table 2. Allocation of flip–flops.

One flip–flop was used during the parallel state assignment, and five flip–flops were allocated to satisfy the *state assignment constraints*. Note, that the sequential state assignment algorithm operates on all variables of the χ_R shown in Fig. 4. The final state assignment is shown in Tab. 3.

	Q_1	Q_2	Q_3	Q_4	Q_5	Q_6
p_1		0	0	0		
p_2		0	0	1		
p_3		0	1	0		
p_4	0	1				
p_5	0	0			0	
p_6	0	0			1	
p_7	1					
p_8	0					0
p_9	0					1

Table 3. Final state assignment.

5 Design Example

In this section, an application of the algorithms presented in the preceding section is demonstrated. The design of MAXbus port controller for a high performance Transputer Framestore with parallel image processing and display capabilities was taken from [22]. First, a brief description of the design is given. Next, a Petri net model of the control circuit is presented, and finally the design implementation is discussed.

5.1 Design Specification

The Transputer Framestore consists of two identical MAXbus port controllers, two 1 MByte banks of video memory (VRAM) and a 20 MHz T800 transputer with 1 to 4 MBytes of program memory, which is dynamic RAM accessed using 150ns read/write cycle, the fastest this processor offers. Each MAXbus port controller organizes transfers of 16 bit data between the one of the banks of VRAM and external video data paths. The MAXbus port controller is build around a Petri net designed control circuit. The major parts of the controller are shown in Fig. 6.

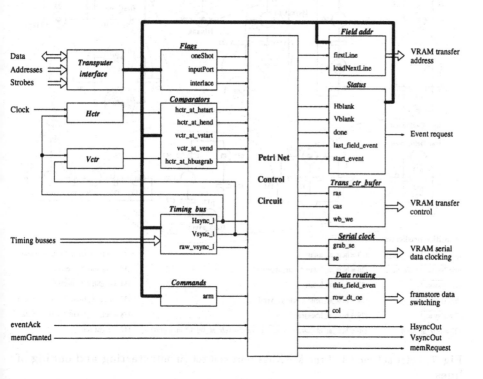

Fig. 6. Simplified block diagram of the MAXbus port controller.

The basic task of the controller is to clock data in and out of the appropriate VRAM chips as and when required. The controller may operate as a timing master, (i.e. generates *Hsync* and *Vsync* signals, c.f. TV) or as a slave (i.e. receives *Hsync* and *Vsync*). When operated as a slave, the controller can accept irregular (i.e. of variable period) *Hsync* and *Vsync* signals. The four timing buses are used to define the *sync* signals for region–of–interest (ROI) transfers, which allow any small image patch to be moved around the system at a rate much faster than the full frame rate. Images and

ROI may be rectangles of any size and may be located anywhere within a bank of video memory. The controller may operate in continuous or in single shot mode. Interlaced and non–interlaced transfers are supported.

5.2 Petri Net Model of the Control Circuit

The Petri net specification of the MAXbus port control circuit is constituted of eighteen quasi–independent nets which communicate with each other using inhibitor and enabling arcs. The two parts of that specification are shown in Fig. 7 and 8. Shaded circles represent either places or entire sub–nets, which are not shown on the diagram. The detail description of the Petri net model would exceed the scope of this paper. However, to show how a Petri net can be used as a hardware description language, a small part of the specification is discussed.

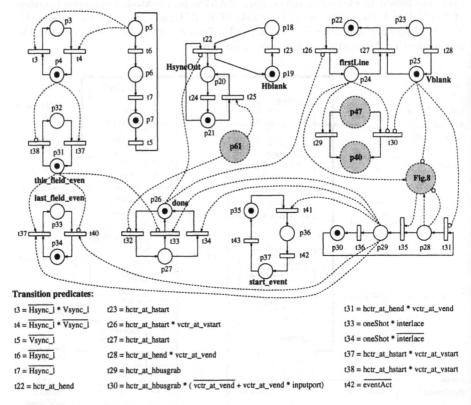

Transition predicates:

t3 = $\overline{\text{Hsync\_l}}$ * Vsync_l	t23 = hctr_at_hstart	t31 = hctr_at_hend * vctr_at_vend
t4 = $\overline{\text{Hsync\_l}}$ * $\overline{\text{Vsync\_l}}$	t26 = hctr_at_hstart * vctr_at_vstart	t33 = oneShot * interlace
t5 = $\overline{\text{Vsync\_l}}$	t27 = hctr_at_hstart	t34 = oneShot * $\overline{\text{interlace}}$
t6 = $\overline{\text{Hsync\_l}}$	t28 = hctr_at_hend * vctr_at_vend	t37 = hctr_at_hstart * vctr_at_vstart
t7 = $\overline{\text{Hsync\_l}}$	t29 = hctr_at_hbusgrab	t38 = hctr_at_hstart * vctr_at_vstart
t22 = hctr_at_hend	t30 = hctr_at_hbusgrab * ($\overline{\text{vctr\_at\_vend}}$ + vctr_at_vend * inputport)	t42 = eventAct

Fig. 7. Petri net model of the MAXbus port control circuit: **starting and ending of lines.**

The Petri net shown in Fig. 7 specifies the starting and ending of lines of transferred images. In the middle of the top line of the diagram, the *HsyncOut* signal is generated, together with the horizontal blanking signal *Hblank*. A one clock pulse wide *HsyncOut* is generated when the arming command occurs (place p_{61} has a token), and this starts the whole transfer sequence off. The *HsyncOut* is also generated when the *hctr_at_hend* signal is received, which corresponds to the end of active data on the line. *Hblank* is also asserted at this time to indicate the start of the horizontal blanking period to the outside world, and remains asserted until the *hctr_at_hstart* signal is asserted. The arming command also causes the *done* flag to be released at the end of a field, when the *wrap_up_field* process is active (place p_{29} has a token).

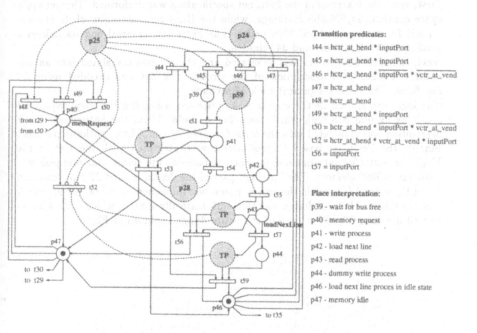

Fig. 8. Petri net model of the MAXbus port control circuit: **load next line process**.

A full Petri net model of the MAXbus port control circuit is composed of 62 places and 64 transitions. The net behavior is controlled using 14 input signals and 18 control outputs are driven by the net. An example of detailed place interpretation is shown in Fig. 8.

5.3 Design Implementation

The presented design was first implemented on a Logic Cell Array (LCA) architecture[3] in 1992 at the University of Bristol. The Petri net specification was manually encoded using a similar approach to that presented in [18]. In the approach each transition is mapped onto an AND gate, while each place is implemented with a D–type flip–flop and an OR gate collecting its input transitions. The design was entered in schematic form to the ViewLogic CAD environment, and then mapped into Xilinx xc3090 element[4]. The Petri net based control circuit needed 57 flip–flops and 74 Configurable Logic Blocks (CLBs) in the LCA, while the entire MAXbus controller used 242 flip–flops and 269 CLBs. That design could not have been routed automatically within one xc3090 element using Xilinx automatic placement and routing tool APR. The entire implementation process (including manual routing) took over four months of continuous effort. The behavioral properties of the Petri net representation of the control circuit had not been formally verified due to complexity of the net state space explosion. All computations were done using a 386DX25 PC.

---

[3] The LCA is the Xilinx version of the Field Programmable Gate Array (FPGA) architecture.

[4] The xc3090 was the largest LCA element available at the time of implementation.

Using the methods presented in the preceding sections, the design has been reimplemented. The new implementation process was divided into the following stages:

- First, symbolic traversal of the Petri net specification was performed. The net state space contains 21,575,805 markings, while the BDD representation of its characteristic function consists of 8086 nodes. The Petri net traversal and its behavior verification took 2 hours and 34 minutes.
- Next, the state assignment was generated and the net was translated into an unoptimized logic level description of the circuit. The net was encoded using 37 flip–flops. The state-assignment took 17 seconds.
- The logic description of the circuit was optimized using the Berkeley sequential synthesis system SIS [20], which took 243 seconds. Then, the optimized circuit description was translated into the form accepted by the Xilinx synthesis package.
- Finally, the Petri net based control circuit was integrated with the rest of the MAXbus controller into one design. For comparison, the design was mapped into a similar xc3090 element. The Petri net part of the design needed 37 flip–flops and 61 CLBs in the LCA, while the entire design was implemented using 222 flip–flops and 243 CLBs. The new design was then automatically routed, which took 4 hours and 40 minutes. The LCA map of the entire design is shown in Fig. 9

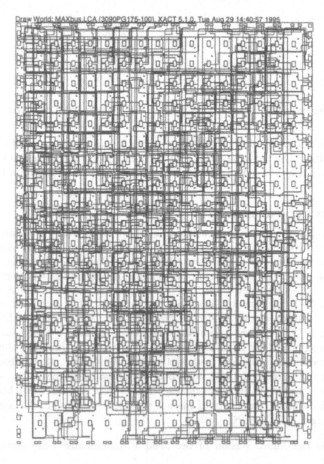

Fig. 9. LCA map of the MAXbus controller.

All implementation steps except the last were done using SPARC station2 computer with 64 MByte of memory, while the routing was performed using a 486DX33 PC. Applying the new encoding method to the Petri net based control circuit we achieved 35% improvement in the number of flip–flops and 17.5% improvement in the number of CLBs. Considering the entire design the improvements are 8.3% and 9.8% respectively. The obtained improvements enable the APR to perform fully automatic routing of the design. As a result, the implementation time was reduced from four months to less than eight hours[5].

6 Experimental Results

A CAD system for high level synthesis of synchronous parallel controllers based on the algorithms presented in the preceding sections has been developed. In this section the synthesis results are described. All benchmarks were run on a Sun SPARC–Station2 computer with 64 MByte of memory.

Statistics of the benchmarks are given in Tab. 4. The benchmarks were taken from published papers and various other sources. "Frcntr" and "minctr" represent 16 and 8 bit versions of the MAXbus port controller for parallel image processing system respectively [22]. The "gf2" example describes the main process of the above controllers. "Armstr" is a multifunctional counter, "rs232" describes the control circuit of RS232 transmitter/receiver system and "am2901" represents the four–bit am2901 microprocessor; these were taken from a set of high level synthesis benchmarks *HLSynth92*. The examples "sm7" and "bar" taken from [21], are based on the *cigarette smokers* and *sleeping barber* problems respectively. The example "rascas", a video memory transfer controller used in a pipeline dedicated VME–based frame grabber, coming from [23]. A parallel controller of communication printer–spooler protocol "spool" is presented in [2]. The examples "zigzag" and "zscan" coming from [15], are parallel controllers for zigzag scanner.

| name | #in | #out | Petri net | | | | Equivalent STG | |
			#places	#trans.	$\|[M_0]\|$	$\|\chi_{[M_0]}\|$	#vertices	#edges
frcntr	14	18	62	64	21575805	8086	-	-†
minctr	13	17	46	48	93242	2006	-	-†
armstr	12	8	91	102	54933	2603	54933	117870
sm7	6	18	56	42	28516	879	28516	72545
bar	9	12	30	31	18797	308	18797	460968
rascas	8	4	28	28	1214	487	1214	10673
gf2	10	4	20	20	485	106	485	8954
rs232	10	24	44	45	165	248	165	246
am2901	34	90	130	153	130	257	130	302
spool	4	19	28	24	77	64	77	214
zscan	5	5	14	20	44	37	44	178
zigzag	6	18	20	18	40	38	40	165

†The computation cannot be completed within 24 hour time limit.

Table 4. Statistics of examples: parallel and sequential controller representations.

---

[5] During reimplementation a new more efficient version of Xilinx APR tool was used. The same program was also applied to the old design. However, the old design still cannot be automatically routed.

The first four columns show the number of input and output signals and the size of the Petri net model of each controller. $|[M_0]|$ represents the number of all reachable markings of the Petri net (i.e. the number of all global states of the controller), and $|\chi_{[M_0]}|$ represents the size of the BDD representation of the characteristic function $\chi_{[M_0]}$. The last two columns show the statistics of the functionally equivalent state transition graphs (STGs) with which the sequential controllers can be synthesized.

Each of the examples was encoded using the algorithms presented in the preceding sections (*BDD encoding*). To evaluate the synthesis results an alternative parallel implementation of each of the examples was created using the method presented in [16] (*RG encoding*). The *RG encoding* method operates on a concurrency matrix in order to produce a state assignment. The method uses a reachability graph to verify the net properties and to generate the concurrency matrix. In addition, functionally equivalent sequential controllers were also generated [13] (*FSM encoding*). The method starts from converting the Petri net description into a single state transition graph, from which a sequential controller is next synthesized. This is the main existing method to synthesize sequential controllers from the Petri net specification, however this method does not produce optimum designs. Every example, after state assignment, was optimized using SIS and synthesized using the Cadence Design Framework II with the CMOS $1\mu m$ ES2 cell libraries.

Table 5 summarizes the synthesis results. For each of the examples the number of flip-flops #ff used to implement the controller, the *area*, the *delay*[6] of the final design, and the synthesis cpu run-time T_{cpu} were considered. The "am2901" example represents a purely sequential design, for which all methods produced similar results, since the parallel state assignment cannot be applied (i.e. the χ_R for this example is simplified to one node, thus only Steps 8 and 9 of the state assignment algorithm are applied). The rest of the benchmarks represent parallel designs.

	BDD encoding				RG encoding				FSM encoding			
	#ff	area	delay	T_{cpu}	#ff	area	delay	T_{cpu}	#ff	area	delay	T_{cpu}
name		$[\mu m^2]$	[ns]	[s]		$[\mu m^2]$	[ns]	[s]		$[\mu m^2]$	[ns]	[s]
frcntr	37	565001	7.82	9720	-	-	-	-†	-	-	-	-†
minctr	24	395930	7.22	658	-	-	-	-†	-	-	-	-†
armstr	46	487864	6.59	432	47	541360	9.23	5012	16	2113463	39.41	11196
sm7	38	329805	5.00	147	38	486784	6.13	2627	15	1269147	34.12	7832
bar	18	274921	4.29	135	18	283651	4.88	1351	15	945712	32.43	7345
rascas	19	362627	4.10	171	19	379220	3.80	240	11	1065913	30.92	6687
gf2	11	154944	4.39	65	12	157783	4.61	95	9	805482	23.35	3291
rs232	24	188699	7.06	132	23	215711	11.95	203	8	836723	24.42	2467
am2901	8	787315	18.71	2138	8	787315	18.71	2138	8	787315	18.71	2138
spool	16	341075	5.86	95	19	356621	7.57	91	7	677176	22.15	1157
zscan	8	196938	6.27	60	8	217433	7.22	70	6	598271	17.28	421
zigzag	13	247779	6.76	61	14	249251	6.83	69	6	858190	24.48	585

†The computation cannot be completed within 24 hour time limit.

Table 5. Synthesis results.

Summarizing the experimental results we can observe, that for all the examples the parallel implementations produce significantly better results when compared to their sequential counterparts. Although, the *FSM encoding* needs the smallest number of flip-

---

[6] The delay denotes the longest feedback time in the circuit.

flops, the method produces the worst results due to the very complex combinational part of these designs. The best results are produced using the *BDD encoding* approach — mainly because of the better grouping of flip–flops during the state assignment in comparison to the *RG encoding* approach. The proposed method is also able to handle the synthesis of much more complex controllers than the alternative approaches.

7 Conclusion

A new methodology for synchronous parallel controller synthesis together with its application to complex controller design have been presented. The approach is based on the symbolic manipulation of the Petri net model of the controller. The Petri net state explosion problem was successfully managed by using BDDs. An efficient parallel state assignment algorithm is introduced, which operates on the BDD representation of the controller's state space. The proposed synthesis methodology proved to be capable of dealing with large designs, due to the modeling power of Petri nets to specify systems with concurrency and the efficient data representation provided by BDDs. The experimental results clearly demonstrate the advantages of the method. Further, a significant increase of the applicability of the approach has also been achieved.

Acknowledgment

The authors want to thank Jordi Cortadella, Tomasz Kozłowski and Alex Yakovlev for many helpful comments on an earlier version of this paper, and David Milford for his help during the reimplementation of the MAXbus port controller.

References

1. P. Azema, R. Valette, and M. Diaz. Petri nets as a common tool for design verification and hardware simulation. In *Proceedings of the 12th ACM/IEEE Design Automation Conference*, pages 109 – 116. IEEE Computer Society Press, 1976.

2. G. Balbo. Performance Issues inParallel Programming. In K. Jensen, editor, *Proceedings of the 13th International Conference: Application and Theory of Petri Nets*, volume 616 of *Lecture Notes in Computer Science*, pages 1 – 23. Springer-Verlag, June 1992.

3. K. Bilinski, M. Adamski, J. Saul, and E. Dagless. Parallel Controller Synthesis From A Petri Net Specification. In *Proceedings of the European Design Automation Conference EURO-DAC'94*, pages 96 – 101, Grenoble, September 19–23, 1994. IEEE Computer Society Press.

4. K.S. Brace, R.L. Rudell, and R.E. Bryant. An Efficient Implamentation of a BDD package. In *Proceedings of the 27th ACM/IEEE Design Automation Conference*, pages 40 – 45. IEEE Computer Society Press, 1990.

5. R.E. Bryant. Graph–Based Algorithms for Boolean Function Manipulation. *IEEE Transaction on Computers*, C-35(12):1035 – 1043, December 1986.

6. J.R. Burch, E.M. Clarke, D.E. Long, K.L. McMillan, and D.L. Dill. Symbolic Model Checking for Sequential Circuit Verification. *IEEE Transactions on Computer Aided Design of Integrated Circuits and Systems*, 13(4):401–424, April 1994.

7. J.R. Burch, E.M. Clarke, K.L. McMillan, and D.L. Dill. Sequential Circuit Verification Using Symbolic Model Checking. In *27th ACM/IEEE Design Automation Conference*, pages 46 – 51. IEEE Computer Society Press, 1990.

8. O. Coudert, J.C. Madre, and C. Berthet. Verifying Temporal Properties of Sequential Machines without Building their State Diagrams. In E.M. Clarke and

R.P. Kurshan, editors, *Proceedings of Computer-Aided Verification 2nd International Conference CAV'90*, volume 531 of *Lecture Notes in Computer Science*, pages 23 – 32. Springer-Verlag, June 1990.

9. P. Eles, K. Kuchcinski, Z. Peng, and M. Minea. Synthesis of VHDL Concurrent Processes. In *Proceedings of the European Design Automation Conference EURO-DAC'94*, pages 540 – 545, Grenoble, September 19–23, 1994.

10. D. Geist and I. Beer. Efficient Model Checking by Automated Ordering of transition Relation Partitions. In D. L. Dill, editor, *Proceedings of Computer-Aided Verification 6th International Conference CAV'94*, volume 818 of *Lecture Notes in Computer Science*, pages 299 – 310. Springer-Verlag, June 1994.

11. D. C. Hendry. Heterogeneous Petri Net Methodology for the Design of Complex Controllers. *IEE Proceedings – Computers and Digital Techniques*, 141(5):293 – 297, September, 1994.

12. A. Kondratyev, J. Cortadella, M. Kishinevski, E. Pastor, O. Roig, A. Yakovlev. Checking Signal Transition Graph Implementability by Symbolic BDD Traversal. In *Proceedings of the European Design and Test Conference ED&TC'95*, pages 325 – 332. IEEE Computer Society Press, 1995.

13. T. Kozlowski, E. Dagless, J. Saul, M. Adamski, and J. Szajna. Parallel Controller Synthesis using Petri Nets. *IEE Proceedings – Computers and Digital Techniques*, 142(4):263 – 271, July, 1995.

14. T. Murata. Petri Nets: Properties, Analysis and Applications. *Proceedings of the IEEE*, 77(4):548 – 580, 1989.

15. J. Pardey and M Bolton. Logic Synthesis of Synchronous Parallel Controlers. In *Proceedings of the IEEE International Conference on Computer Design*, pages 454–457. IEEE Computer Society Press, 1991.

16. J. Pardey, T. Kozlowski, J. Saul, and M. Bolton. State Assignment Algorithms for Parallel Controller Synthesis. In *Proceedings of the IEEE International Conference on Computer Design*, pages 316 – 319. IEEE Computer Society Press, 1992.

17. E. Pastor, O. Roig, J. Cortadella, and R. Badia. Petri Net Analysis Using Boolean Manipulation. In R. Valette, editor, *Proceedings of 15th International Conference: Application and Theory of Petri Nets*, volume 815 of *Lecture Notes in Computer Science*, pages 416 – 435. Springer-Verlag, June 1994.

18. M.R.K. Patel. Random Logic Implementation of Extended Timed Petri Nets. In *Microprocessing and Microprogramming*, volume 30, pages 313 – 320. North-Holland, 1990.

19. O. Roig, J. Cortadella, and E. Pastor. Verification of Asynchronous Circuits by BDD–based Madel Checking of Petri nets. In G. DeMichelis and M. Diaz, editors, *Proceedings of 16th International Conference: Application and Theory of Petri Nets*, volume 935 of *Lecture Notes in Computer Science*, pages 374 – 391. Springer-Verlag, June 1995.

20. E.M. Sentovich, K.J. Singh, L. Lavagno, C. Moon, R. Murgai, A. Saldanha, H. Savoj, P.R. Stephan, R.K. Brayton, and A. Sangiovanni-Vincentelli. *SIS: A System for Sequential Circuit Synthesis*. University of California, Berkelay, May 1992. Memorandum No. UCB/ERL M92/41.

21. A. Silberschatz and J.L. Peterson. *Operating Systems Concepts*. Addison-Wesley, 1988.

22. J. Stewart, E. Dagless, D. Milford, and O. Miles. A Petri Net Based Framstore. In W. Moore and W. Luk, editors, *International Workshop on Field Programmable Logic and Applications*, pages 332 – 342, 1991.

23. K. Wiatr, J. Kasperek, and P.J. Rajda. Xilinx FPGA–based Frame–grabber for Image Pre–processing. In *Proceedings of the 7th School – VLSI and ASIC Design*, pages 273 – 279. Format Publisher, 1995. ISBN 83-900859-3-3.

Non Sequential Semantics for Contextual P/T Nets*

Nadia Busi, G. Michele Pinna

Dipartimento di Matematica, Università di Siena,
Via del Capitano 15, I-53100 Siena, Italy
e-mail: busi@cs.unibo.it, pinna@di.unipi.it

Abstract. The problem of finding a true concurrent semantics for contextual P/T nets has not been deeply studied yet. The interest for such
a semantics has been renewed by some recent proposals to equip mobile
process algebras with a net semantics based on this model. In [3] we
proposed a causal semantics for such nets, in this paper we study the
non sequential semantics of contextual P/T nets. The semantics is based
on a suitable notion of occurrence net and, equipped with a notion of
history preserving bisimulation, is more discriminating of the contextual
one in general, whereas both are equivalent if we restrict our attention
to positive context only.

1 Introduction

Place/Transition nets [13] are a well known model used to describe concurrent
systems. Recently the interest in contextual nets, i.e. nets that are able to test
for presence and absence of tokens, besides the usual flow relation, has grown,
mainly due to the use of contextual nets to give a net semantics to the π-calculus
([8]) studied in [2], or to model the concurrent access to shared data [14, 4], or to
their relation with constraint programming [9], or to model priorities [7]. Quite
clearly contextual Petri nets give a faithful and simpler representation of real
systems, for instance timeouts can be modeled just requiring that the various
activities (that may be independent to each other) test for the presence of a
token in a certain place that is removed after a certain time: the possibility of
testing for absence or presence of tokens simplify a lot the representation of the
system. The contextual nets considered in mentioned approaches are contextual
Condition/Event nets, or P/T nets testing only for presence of tokens, and true
concurrent semantics are provided. Here we study how to give a true concurrent
semantics for contextual P/T nets.

Usually a true concurrent semantics for P/T nets is provided by means of processes [6]; then, to take into account also the branching structure of systems, the

---

* Research partially supported by EC BRA n. 9102 COORDINATION, and MURST,
  quota 60%.

history preserving bisimulation (based on processes) has been proposed [1]. However the semantics of P/T nets in term of processes can not so easily adjusted to keep into account contextual and inhibitor arcs, still keeping the same flavor of [6], as a partial order models *faithfully* causality and concurrency only, whereas contextual and inhibitor arcs act as *contexts* conditions in the firing of transitions, hence they represent other types of dependencies. In [3] we introduced a notion of causal semantics and we proved that this semantics is a conservative extension of the non sequential one, i.e. two nets with empty contexts that are history preserving bisimilar are causal bisimilar (and vice versa). The idea being this semantics is to map a behaviour of the contextual net in a so called *causal firing sequence* recording the dependencies between the occurrence of transitions. This is achieved enriching the information about the current state, decorating each token with its history, recording how many times a transition has fired and recording, for each place, the occurrence of transitions that have removed tokens from it. This enriched "state" is called configuration. We obtain information about the dependencies between occurrences of transitions just looking at the tokens consumed by one and produced by the other. Tokens in the initial marking are decorated with '*' and consumption of them doesn't add causal dependencies. The major drawback to this approach is that it consider on the same level *causal dependencies, positive context* and *negative context*, reducing everything to causality. This is acceptable when considering positive context: the place in the context must contain a token (and possibly a transition has filled that place with some); but more questionable when considering negative context: the place in the context must be unmarked, we can say that a dependency exists between the transition and those removing the tokens, but if the place is unmarked at the initial marking and the transition may fire *before* one filling that place with tokens we can't say anything regarding the dependencies between the occurrences of the transitions. Consider the following net:

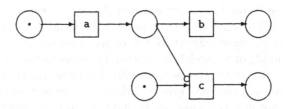

there are two possible executions: either *a* followed by *b* and then *c* or *c* followed by *a* and then *b*. In the former we can say that *c depends* on *b* because *b* removes the inhibiting token from the postset of *a*, whereas in the latter *c* is completely independent from *a* and *b*, but it cannot happen *between* the two. This phenomenon happens as negative contexts are used to forbid certain executions (as in the case of priorities).

In this paper we propose a non sequential semantics for contextual P/T net that keeps apart causal dependencies, positive context dependencies and negative context dependencies. We define a contextual occurrence net following [9], i.e. representing positive contexts as contextual arcs, as they state a *positive* dependency, but where negative contexts are represented via a labeling function. To such a net it is still possible to associate a partial order representing the causal dependencies and the positive contexts dependencies at once, and where the labeling is used to find the computations associated to the net. As the previous example shows, inhibitor arcs are used to *reduce* the number of computations of a net, and then what we really need is a way of selecting these computations. The proposed notion of occurrence net coincides with the one proposed in [9] (but not with the one proposed in [7][2], except that the associated partial order is different: we never say that the occurrence of the transition removing a token from a positive context *depends* on the occurrence of the transition using that token as positive context, but we say that an asymmetric conflict between these two occurrences arises, as defined in [12]. Indeed, if occurrence nets with an empty labeling are considered, it is possible to associate an event structure with possible event, as defined in [12].

A notion of history preserving bisimulation, based on non sequential processes, can be defined in this setting, following the schema in [1]. This bisimulation is more discriminating of the causal one, defined in [3], as the latter cannot distinguish between positive and negative context whereas the former does.

The paper is organized as follows: in the next section we recall some basic definitions about (contextual) P/T nets then, in section 3 we introduce the non sequential semantics and we prove that it agrees with the sequential one. In section 4 we recall the formal definition of causal semantics introduced in [3] and in section 5 we compare the two, showing that the one based on non sequential process is more discriminating than the causal one in general, and that they coincide in the case of nets without inhibitor arcs. Finally we discuss some relations with other works and we point out possible further development of this work.

2 Basic Definitions

We recall some basic notions on P/T nets [15]. Let ω be the set of natural numbers and $\omega^+ = \omega \setminus \{0\}$.

Definition 1. Given a set S, a *finite multiset* over S is a function $m : S \to \omega$ such that the set $dom(m) = \{s \in S \mid m(s) \neq 0\}$ is finite. The *multiplicity* of an element s in m is given by the natural number $m(s)$. The set of all finite multisets over S, denoted by $\mathcal{M}_{fin}(S)$, is ranged over by m. A multiset m such that $dom(m) = \emptyset$ is called *empty*. The set of all finite sets over S is denoted by $\wp_{fin}(S)$. If A is a finite subset of S, with abuse of notation we use A to denote the multiset m_A defined as follows: $m_A(s) = $ if $s \in A$ then 1 else 0. We write $m \subseteq m'$ if $m(s) \leq m'(s)$ for all $s \in S$. The operator $\oplus$ denotes *multiset*

---

[2] This because as they deal with inhibitor arcs on C/E nets, they may represent *positively* both the holding and the not holding of a condition, which is impossible, in general, in P/T net.

union: $(m \oplus m')(s) = m(s) + m'(s)$. The operator $\setminus$ denotes *multiset difference*: $(m \setminus m')(s) = m(s) \dotminus m'(s)$ where $i \dotminus j = i - j$ if $i > j$, $i \dotminus j = 0$ otherwise. Finally, the *scalar product*, $j \cdot m$, of a number j with a multiset m is $(j \cdot m)(s) = j \cdot (m(s))$. $\square$

Let $f : A \to B$ be a function. It is extended to a function from $\mathcal{M}_{fin}(A)$ to $\mathcal{M}_{fin}(B)$ in the following way: if $m \in \mathcal{M}_{fin}(A)$ then, for all $a \in A$, $(f(m))(a) = \sum_{f(a')=a} m(a')$.

Definition 2. A *net* is a tuple (S, T, F) such that $S \cap T = \emptyset$ and $F \subseteq (S \times T) \cup (T \times S)$. $\square$

The elements of S are called *places* and the elements of T are called *transitions*. For $x \in S \cup T$, we define ${}^\circ x = \{y \in S \cup T \mid (y, x) \in F\}$ and $x^\circ = \{y \in s \cup T \mid (x, y) \in F\}$. A net N is T-restricted if for all transitions t we have that ${}^\circ t \neq \emptyset \neq t^\circ$.

Definition 3. A *place/transition net* is a tuple $N = (S, T, F, W, m_0)$ where

- (S, T, F) is a net;
- $W : F \to \omega$ is the *weight function* such that $\forall (x, y) \in F, W(x, y) > 0$;
- $m_0 : S \to \omega$ is the *initial marking* $\square$

A finite multiset over the set S of places is called a *marking*. Given a marking m and a place s, we say that the place s contains $m(s)$ *tokens*.

The *preset* of a transition t is the multiset ${}^\bullet t(s) = W(s, t)$ and the *postset* of t is $t^\bullet(s) = W(t, s)$.

A transition t is *enabled* at m if ${}^\bullet t \subseteq m$. The execution of a transition t enabled at m produces the marking $m' = (m \setminus {}^\bullet t) \oplus t^\bullet$. This is written as $m[t\rangle m'$.

A *firing sequence* is defined inductively as follows:

- m_0 is a firing sequence;
- if $m_0[t_1\rangle m_1 \ldots [t_{n-1}\rangle m_{n-1}$ is a firing sequence and $m_{n-1}[t_n\rangle m_n$ then $m_0[t_1\rangle m_1 \ldots [t_{n-1}\rangle m_{n-1}[t_n\rangle m_n$ is a firing sequence.

The *interleaving marking graph* of N is $IMG(N) = (\mathcal{M}_{fin}(S), \to, T)$, where $\to \subseteq \mathcal{M}_{fin}(S) \times T \times \mathcal{M}_{fin}(S)$ is defined by $m \xrightarrow{t} m'$ iff there exists a transition $t \in T$ such that $m[t\rangle m'$.

Given two nets N_1 and N_2, a binary relation $R \subseteq \mathcal{M}_{fin}(S_1) \times \mathcal{M}_{fin}(S_2)$ is a *bisimulation* if:

- If $(m_1, m_2) \in R$ and $m_1 \xrightarrow{a} m_1'$ there exists then m_2' such that $m_2 \xrightarrow{a} m_2'$ and $(m_1', m_2') \in R$.
- If $(m_1, m_2) \in R$ and $m_2 \xrightarrow{a} m_2'$ there exists then m_1' such that $m_1 \xrightarrow{a} m_1'$ and $(m_1', m_2') \in R$.

Let m_{0i} be the initial marking of N_i ($i = 1, 2$). The nets N_1 and N_2 are *bisimilar* ($N_1 \sim N_2$) iff there exists a bisimulation R such that $(m_{01}, m_{02}) \in R$.

A finite, non empty multiset over the set T is called a *step*.

A step G is enabled at m if $m_1 \subseteq m$, where $m_1 = \bigoplus_t G(t) \cdot {}^\bullet t$. The execution of a step G enabled at m produces the marking $m' = (m \setminus m_1) \oplus m_2$, where $m_2 = \bigoplus_t G(t) \cdot t^\bullet$. This is written as $m[G\rangle m'$.

A *step firing sequence* is defined inductively as follows:

- m_0 is a step firing sequence;
- if $m_0[G_1\rangle m_1 \ldots [G_{n-1}\rangle m_{n-1}$ is a step firing sequence and $m_{n-1}[G_n\rangle m_n$ then $m_0[G_1\rangle m_1 \ldots [G_{n-1}\rangle m_{n-1} [G_n\rangle m_n$ is a step firing sequence.

Our definition of contextual P/T net is an extension of the ones in [9, 14].

Definition 4. A *contextual place/transition net* is a tuple (S, T, F, I, C, W, m_0) where

- (S, T, F, W, m_0) is a place/transition net;
- $C \subseteq S \times T$ is the *contextual relation*;
- $I \subseteq S \times T$ is the *inhibiting relation*;

such that $(C \cup I) \cap F = \emptyset$ and $I \cap C = \emptyset$. $\qquad\qquad\qquad\square$

The structural restriction that contextual arcs, flow relation and inhibitor arcs are totally separated is imposed because otherwise the intended meaning of contexts (positive and negative) is lost.

The *inhibitor set* of a transition t is the set ${}^\circ t = \{s \in S \mid (s, t) \in I\}$ and the *contextual set* of t is the set $\hat{t} = \{s \in S \mid (s, t) \in C\}$.

A transition t is enabled at m if ${}^\bullet t \subseteq m$, $\hat{t} \subseteq dom(m)$ and ${}^\circ t \cap dom(m) = \emptyset$. The execution of a transition t enabled at m producing the marking m', written $m[t\rangle m'$, is defined as before, i.e. $m' = (m \setminus {}^\bullet t) \oplus t^\bullet$.

According to [9], two transitions can happen in the same step iff they can happen in either order. We have to check that not all tokens in a place tested for presence by (an occurrence of) a transition are consumed by the others and that (an occurrence of) a transition does not produce tokens in a place tested for absence by another.

A step G is enabled at m iff

- $m_1 \oplus m_3 \subseteq m$, where $m_1 = \bigoplus_t G(t) \cdot {}^\bullet t$ and $m_3 = \bigcup_{t \in dom(G)} \hat{t}$
- for all $t \in dom(G)$ ${}^\circ(t) \cap dom(m) = \emptyset$
- for all $t_1, t_2 \in dom(G)$ such that $t_1 = t_2 \Rightarrow G(t_1) > 1$ we have that $dom(t_1^\bullet) \cap {}^\circ t_2 = \emptyset$

The third condition ensures that, for each pair of instances of transitions, it is never the case that one of them puts a token in an inhibiting place of the other.

The execution of a step G enabled at m producing the marking m', written $m[G\rangle m'$, is defined as before.

We briefly illustrate how contextual nets work. A contextual arc is depicted with a line, whereas an inhibitor arc is represented as a special arrow – with a final, white circle – from the inhibiting place to the transition.

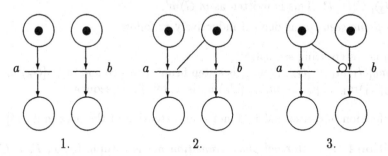

The first net can perform a and b in any order or together, the second one can do a followed by b or only b (in this case a isn't any longer enabled) and finally the third one can do only a followed by b. Though positive context may enlarge the amount of concurrency in the net (two transitions, both enabled at the same marking, that consume and produce a token in a common place, need to be sequentialized in usual nets, whereas if the arcs are contextual ones, they may happen in the same step [9]), in general they reduce the possible different run of the net, as they work as contexts.

3 Non sequential semantics

Occurrence nets have been introduced by Petri in [10]. Starting from a partial order, where the partial order is intended to model the causal dependencies and the absence of order (disorder) reflects the concurrency, it is possible to define a net with some characteristics: places are unbranched (i.e. there is just one outgoing and one ingoing arc) and the transitive closure of the flow relation is the partial order we started with. A sequence of ordered elements (i.e. such that they are pairwise comparable) is a subprocess.

Definition 5. A (finite) *occurrence net* is a net (B, E, F), whose places and transitions are called *conditions* and *events* respectively, such that:

- B and E are finite;
- the conditions are not branched, i.e. for all $b \in B$, $|^\circ b| \le 1$ and $|b^\circ| \le 1$;
- the net is acyclic, i.e. for all $x, y \in B \cup E$ $(x, y) \in F^+$ implies $(y, x) \notin F^+$ (where F^+ is the transitive closure of F). $\qquad\square$

As we discussed in the introduction, the reality is more complicated: there are situations, like timeouts in systems representing industrial processes, where events happen if certain side conditions are satisfied. Though it is possible to reduce everything to a rather strong relation of dependency (modeling the possible temporal executions), it seems more reasonable to keep apart as possible the notion of causality from the one of context. In this section we develop a notion of non sequential process, based on a notion of contextual occurrence net, where positive contexts, flow relation and inhibitor conditions are kept apart as much as it is possible.

We recall some notions we will use in the following. With $(X, \leq)$ we denote a partially ordered set (poset), i.e. $\leq \subseteq X \times X$ is transitive, reflexive and antisymmetric relation. We will write often X instead of $(X, \leq)$. Starting from $\leq$ we define $x \text{ li } y$ iff $x \leq y$ or $y \leq x$ and $x \text{ co } y$ iff $x = y$ or $\neg x \text{ li } y$. $l \subseteq X$ is a *li-set* (chain) iff $\forall x, y \in l, \ x \text{ li } y$; a *line* l is a li-set such that $\forall x \in X \setminus l, \exists y \in l$ $\neg x \text{ li } y$, i.e. it isn't possible to add elements to a line because it is a maximal li-set. $c \subseteq X$ such that $\forall x, y \in L, \ x \text{ co } y$ is called a *co-set*, and a maximal co-set c is called a *cut*, i.e. $\forall x \in X \setminus c, \exists y \in c \ \neg x \text{ co } y$. A $(X, \leq)$ is well founded iff $\forall x \in X \ |\downarrow x| < \infty$, where $\downarrow x = \{y \in X \mid y \leq x\}$. As we are considering well founded partial orders, we can define $Min(X, \leq)$ as the cut c such that $\forall x \in c$, $\forall y \in X \setminus c, \neg(y \leq x)$, and $Max(X, \leq)$ as the cut c such that $\forall x \in c, \forall y \in X \setminus c$, $\neg(x \leq y)$. As usual we will write simply $Min(X)$ or $Max(X)$. A *labeled partial order* (over the set A) $(X, \leq, l)$ consists of a partial order $(X, \leq)$ and a labeling function $l : X \to A$. Two labeled partial orders $(X_1, \leq_1, l_1)$ and $(X_2, \leq_2, l_2)$ are isomorphic if there exists an isomorphism $f : X_1 \to X_2$, i.e. a bijection which is order preserving $(x \leq_1 y \Leftrightarrow f(x) \leq_2 f(y))$ and label preserving $(l_2(f(x)) = l_1(x)$ for all $x \in X_1)$.

Definition 6. A *finite contextual occurrence net* is a labeled net with contextual arcs $(B, E, F, C, \lambda : E \to \wp_{fin}(B))$ where

1. (B, E, F) is a finite occurrence net,
2. $\lambda : E \to \wp_{fin}(B)$ is a labeling function that associate to each event the (set of) conditions inhibiting such event,
3. $\forall e \in E \ \forall b \in B \ (b, e) \in C \Rightarrow ((b, e) \notin F^* \wedge (e, b) \notin F^*)$,
4. $\forall e \in E \ \forall b, b' \in \hat{e} \ (b, b') \in F^* \Rightarrow b = b'$,
5. $\forall e, e' \in E \ (e, e') \in F^* \Rightarrow \forall b \in \hat{e}, \forall b' \in \hat{e'} \ (b', b) \notin F^+$. $\qquad \square$

In this definition an occurrence net is enriched with contextual arcs (C) and a labeling function λ. For each event e, $\lambda(e)$ is the set of conditions inhibiting e. Positive contexts are treated quite differently from negative ones: the reason is that an event needs that a positive context is present (a condition holds) and hence we can say that, among the causes of the event itself, there are also the ones that caused the holding of the context condition, whereas the only dependencies between the happening of an event and the conditions inhibiting is a temporal one, as the event can happen at states where no inhibiting condition is present. Moreover a negative context can be faithfully represented only when we have a mean to say that a condition doesn't hold, mean that exists only when dealing with condition/event Petri nets (C/E nets), e.g. [7]. The last three conditions are structural requirements for the contextual arcs: $\forall e \in E \; \forall b \in B \; (b,e) \in C \Rightarrow ((b,e) \notin F^* \wedge (e,b) \notin F^*)$ says that a positive context for an event e cannot belong to the same subprocess of e, $\forall e \in E \; \forall b, b' \in \hat{e} \; (b,b') \in F^* \Rightarrow b = b'$ asserts that two positive contexts of the same event must be unrelated and the last condition guarantees that two events belonging to the same subprocess have contexts that agree with the dependency between them.

To each contextual occurrence net we associate a relation which reflexive and transitive closure turns out to be a partial order. Positive contexts (namely holdings of conditions) are *removed* by the happening of events: with $e \prec_c e'$ we state that the context used by e is removed by e'. As we said in the introduction we don't see any reason to say that $e \prec e'$, this one is a coding of a temporal execution where the event e happens before e' because if e' happens first, e cannot happen any longer. The right interpretation of this situation, from our point of view, is that e and e' are in asymmetric conflict. We will discuss the matter more extensively in the last section.

Definition 7. Let $(B, E, F, C, \lambda : E \to \wp_{fin}(B))$ be a finite contextual occurrence net. We associate to it two relations $\prec$ and $\prec_c$ defined on $(B \cup E) \times (B \cup E)$ defined as follows:

1. $\prec$ is the least relation such that $(x,y) \in F$ implies $x \prec y$ and $(e',b) \in F \wedge (b,e) \in C$ implies $e' \prec e$,
2. $\prec_c$ is the least relation such that $(b,e) \in C$ and $(b,e') \in F$ implies $e \prec_c e'$. $\square$

Proposition 8. Let $(B, E, F, C, \lambda : E \to \wp_{fin}(B))$ be a contextual occurrence net. Then $(B \cup E, \prec^*)$ is a partial order, where $\prec$ is the relation defined in 7. $\square$

The partial order $\prec^*$ represents the causal dependencies between events in an occurrence net. As we have also negative context, it isn't any longer true that a subset of unrelated conditions (a cut of the poset associated to an occurrence net) represent a state of the system: consider the following occurrence net where $\lambda(e) = \{b\}$.

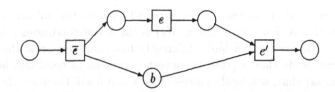

Here e' can happen only when e has happened but e can happen only at states where no inhibiting condition is present, and then only if $\bar{e}$ hasn't happened yet or e' has happened, but this is clearly impossible. To overcome this problem we require that a linearization of the events of an occurrence net compatible with the causal dependencies and with the inhibitions exists. Formally:

Definition 9. Let $K = (B, E, F, C, \lambda : E \to \wp_{fin}(B))$ be an occurrence net and $\prec^*$ be the associated partial order. We say that K is *linearizable* if there exists a total ordering $\leq$ on E such that (a) $e_i \prec^* e_j \Rightarrow e_i \leq e_j$, (b) $e_i \prec_c e_j \Rightarrow e_i \leq e_j$ and (c) $\forall e \in E$ such that $\lambda(e) \neq \emptyset$ then $\forall b \in \lambda(e)$ either $b \prec e_j$ and $e_j \leq e$ or $e_j \prec b$ and $e \leq e_j$. □

Proposition 10. Let $K = (B, E, F, C, \lambda : E \to \wp_{fin}(B))$ be an occurrence net with $C = \emptyset$ and $\lambda(e) = \emptyset$ for each $e \in E$, and let $\prec^*$ be the associated partial order. Then d is a linearization of K iff is a linearization of (B, E, F). □

We restate in this framework the definition of B-cut.

Definition 11. Let $K = (B, E, F, C, \lambda : E \to \wp_{fin}(B))$ be an occurrence net and $\prec^*$ be the associated partial order. A B-cut D is a subset of B such that:

- D is a cut of $(B \cup E, \prec^*)$, and
- $(B', E', F', C', \lambda' : E' \to \wp_{fin}(B'))$ is linearizable, where $B' = B \cap \downarrow D$, $E' = E \cap \downarrow D$, $F' = F \cap ((B' \times E') \cup (E' \times B'))$, $C' = C \cap (B' \times E')$ and $\lambda'(e) = \lambda(e) \cap B'$. □

B-cuts are meant to represent states (distribute ones) of the system. the second condition assure that the state is a reachable one. As a consequence of the proposition 10 we have that in an occurrence net without contexts (positive and negative), B-cuts are the slices as defined, for instance, in [6], i.e. cuts of the poset associated to an occurrence net composed by conditions only.

Proposition 12. Let $K = (B, E, F, C, \lambda : E \to \wp_{fin}(B))$ be an occurrence net with $C = \emptyset$ and $\lambda(e) = \emptyset$ for each $e \in E$, and let $\prec^*$ be the associated partial order. Then D is a B-cut of K iff is a slice of (B, E, F). □

As occurrence nets are E restricted, the minima and the maxima are always subsets of conditions.

Proposition 13. Let $K = (B, E, F, C, \lambda : E \to \wp_{fin}(B))$ be a finite occurrence net and $\prec^*$ be the associated partial order. Then $Min(B \cup E, \prec^*)$ and $Max(B \cup E, \prec^*)$ are B-cuts. □

Proposition 14. Let $K = (B, E, F, C, \lambda : E \to \wp_{fin}(B))$ be a finite and linearizable occurrence net, and let $e_1 \ldots e_{n-1} e_n$ be a linearization. Then:

- $e_n \in °(Max(B \cup E, \prec^*))$, and
- $e_1 \ldots e_{n-1}$ is a linearization of $(B \setminus e_n°, E \setminus \{e_n\}, F', C', \lambda')$ where F', C' and λ' are the restriction of F, C and λ to the elements in $B \setminus e_n°$ and $E \setminus \{e_n\}$. □

The non sequential behaviour of P/T nets is described as a non sequential process, i.e. an occurrence net and labeling mapping relating the occurrence net to the P/T net itself.

Definition 15. Given a P/T net with contextual and inhibitor arcs $N = (S, T, F, I, C, W, m)$, then a *non sequential process* of N is the pair $\pi = (K, \phi)$ where $K = (B_\pi, E_\pi, F_\pi, C_\pi, \lambda : E_\pi \to \wp_{fin}(B_\pi))$ is a linearizable occurrence net and $\phi : B_\pi \cup E_\pi \to S_N \cup T_N$ such that

1. $\phi(B_\pi) \subseteq S_N$ and $\phi(E_\pi) \subseteq T_N$,
2. $\forall s \in S_N \ m(s) = |\{b \in B_\pi \mid \phi(b) = s \wedge {}^\circ b = \emptyset\}|$,
3. $\forall e \in E_\pi, \forall s \in S_N \ {}^\bullet\phi(e)(s) = |{}^\circ e \cap \phi^{-1}(s)|$ and $\phi(e)^\bullet(s) = |e^\circ \cap \phi^{-1}(s)|$,
4. $\forall s \in \phi(\hat{e}) \ \exists b \in \phi^{-1}(s)$ such that $(b, e) \in C_\pi$,
5. $b \in \lambda(e) \Leftrightarrow (\phi(b), \phi(e)) \in I_N$. □

The definition is standard: conditions are mapped onto places, events onto transitions, the minima of the occurrence net represent the initial marking and the neighborhood of transitions is respected; moreover contextual arcs are preserved (but not reflected, as each event *happens* in a context). Finally $b \in \lambda(e)$ iff an inhibitor arc exists in the underlying P/T net.

This notion coincide with the usual one when $C_\pi = \emptyset$ and $\lambda(e) = \emptyset$.

We prove now that this semantics agrees with the sequential semantics.

Theorem 16. *Let $N = (S, T, F, I, C, W, m)$ be a P/T net m' be a reachable marking. Then there exists a non sequential process $\pi = ((B_\pi, E_\pi, F_\pi, C_\pi, \lambda : E_\pi \to \wp_{fin}(B_\pi)), \phi)$ such that $\phi(Max((B_\pi \cup E_\pi, \prec^*))) = m'$.*

Proof. The proof is by induction on the length of the derivation yielding to the marking m'.

If $m' = m$ then $((B_0, \emptyset, \emptyset, \emptyset, \emptyset), \phi_0)$ with $B_0 = \bigcup B_s$ where $B_s = \{b \mid \phi_0(b) = s\}$ and $|B_s| = m(s)$. This defines trivially a non sequential process, as the net is clearly a linearizable occurrence net.

Assume that this is true for a derivation of length n, and let $((B'_{\pi'}, E'_{\pi'}, F'_{\pi'}, C'_{\pi'}, \lambda' : E_{\pi'} \to \wp_{fin}(B'_{\pi'})), \phi')$ be the corresponding non sequential process, with $\phi'(Max((B'_{\pi'} \cup E'_{\pi'}, (\prec')^*))) = m''$ and such that $(B'_{\pi'}, E'_{\pi'}, F'_{\pi'}, C'_{\pi'}, \lambda' : E_{\pi'} \to \wp_{fin}(B'_{\pi'}))$ a linearizable occurrence net. Let us consider $m''[t\rangle m'$, where t is a transition enabled at m''. For t to be enabled it is necessary that ${}^\bullet t \subseteq m''$, $\hat{t} \subseteq dom(m'')$ and ${}^\circ t \cap dom(m'') = \emptyset$. Define $((B_\pi, E_\pi, F_\pi, C_\pi, \lambda : E_\pi \to \wp_{fin}(B_\pi)), \phi)$ as follows:

$B_\pi = B'_{\pi'} \cup B_{\{s \mid t^\bullet(s) \neq 0\}}$ where $B_{\{s \mid t^\bullet(s) \neq 0\}} = \bigcup_{\{s \mid t^\bullet(s) \neq 0\}} \{b_s \notin B'_{\pi'} \mid \phi(b_s) = s\}$ such that $t^\bullet(s) = |\{b_s \notin B'_{\pi'} \mid \phi(b_s) = s\}|$ (just add as many conditions as the tokens produced by the firing of the transition, and these are labeled, via ϕ, as the places in the underlying net),

$E_\pi = E'_{\pi'} \cup \{e\}$, with $e \notin E'_{\pi'}$ and $\phi(e) = t$ (we add an event, labeled via ϕ with t),

$F_\pi = F_{\pi'} \cup (B_{\{s| {}^\bullet t(s) \neq 0\}} \times \{e\}) \cup (\{e\} \times B_{\{s| t^\bullet(s) \neq 0\}})$ where $B_{\{s| {}^\bullet t(s) \neq 0\}} = \bigcup_{\{s| {}^\bullet t(s) \neq 0\}} \{b_s \in Max((B'_{\pi'} \cup E'_{\pi'}, (\prec')^*)) \mid \phi'(b_s) = s\}$ such that ${}^\bullet t(s) = |\{b_s \in Max((B'_{\pi'} \cup E'_{\pi'}, (\prec')^*)) \mid \phi'(b_s) = s\}|$ (we extend the flow relation connecting maxima of the previous occurrence net with the new event e and connecting it with the new conditions),

$C_\pi = C_{\pi'} \cup (B_{\{s \in \hat{t}\}} \times \{e\})$ where $B_{\{s \in \hat{t}\}} = \bigcup_{\{s \in \hat{t}\}} \{b_s \in Max((B'_{\pi'} \cup E'_{\pi'}, (\prec')^*)) \mid \phi'(b_s) = s\}$ and $|\{b_s \in Max((B'_{\pi'} \cup E'_{\pi'}, (\prec')^*)) \mid \phi'(b_s) = s\}| = 1$ (contextual arcs are added, when necessary), and

$\lambda(e) = \phi^{-1}({}^\circ \phi(e))$ and ϕ behaves as ϕ' on $B'_{\pi'}$, $E'_{\pi'}$, $F'_{\pi'}$ and $C'_{\pi'}$.

We first must show that this construction is indeed an occurrence net. The only conditions to be verified are:

- $\forall e \in E \ \forall b \in B \ (b, e) \in C \Rightarrow ((b, e) \notin F^* \wedge (e, b) \notin F^*)$,

  we consider only the added event, and the the only possibility is that we add between the same condition in $Max((B'_{\pi'} \cup E'_{\pi'}, (\prec')^*))$ and the new even e a contextual arc and a causal arc, but this is possible only if in the underlying net $F \cap C \neq \emptyset$, which is absurd,

- $\forall e \in E \ |\hat{e}| > 1 \Rightarrow \forall b, b' \in \hat{e} \ (b, b') \notin F^*$ and $(b', b) \notin F^*$,

  which is trivially true as we add contextual arcs only between the new event and the conditions belonging to the maximal B-cut of $Max((B'_{\pi'} \cup E'_{\pi'}, (\prec')^*))$, and

- $\forall e, e' \in E \ (e, e') \in F^* \Rightarrow \forall b \in \hat{e}, \forall b' \in \hat{e'} \ (b', b) \notin F^+$.

  also this condition is trivially true, as we are adding contextual arcs between the added event and the conditions of $Max((B'_{\pi'} \cup E'_{\pi'}, (\prec')^*))$

Now we have to prove that $(B_\pi, E_\pi, F_\pi, C_\pi, \lambda)$ is linearizable. By induction we know that $(B'_{\pi'}, E'_{\pi'}, F'_{\pi'}, C'_{\pi'}, \lambda')$ is linearizable and let $e_1 \ldots e_n$ be a linearization of it. Then $e_1 \ldots e_n e$ is a linearization of $(B_\pi, E_\pi, F_\pi, C_\pi, \lambda)$. First we observe that certainly it is never be the case that $e \prec^* e_i$, and then the only condition violating the definition of linearization is $\exists b \in \lambda e$ and $\forall j \ e_j \prec b$ but $e \not\leq e_j$, which is false because it means that $b \in Max((B'_{\pi'} \cup E'_{\pi'}, (\prec')^*))$ and then, as $m'' = \phi'(Max((B'_{\pi'} \cup E'_{\pi'}, (\prec')^*)))$ and by the definition of ϕ, we have that t is not enabled at m''.

Observing that the new relation $\prec$ is obtained by adding to $\prec'$ all the new pairs induced by the extension of the flow relations and the ones using the added contextual arcs, as $t^\bullet(s) = |\{b_s \notin B'_{\pi'} \mid \phi(b_s) = s\}|$ and ${}^\bullet t(s) = |\{b_s \in Max(B'_{\pi'} \cup E'_{\pi'}, (\prec')^*) \mid \phi'(b_s) = s\}|$ we have that $m' = (m'' \setminus {}^\bullet t) \oplus t^\bullet = \phi(Max(B'_{\pi'} \cup E'_{\pi'}, (\prec')^*))$. $\qquad \square$

Theorem 17. *Let* $N = (S, T, F, I, C, W, m)$ *be a P/T net and* $\pi = ((B_\pi, E_\pi, F_\pi, C_\pi, \lambda : E_\pi \to \wp_{fin}(B_\pi)), \phi)$ *be a non sequential process. Then for each B-cut D we have that $\forall s \in S \ m'(s) = |D \cap \phi^{-1}(s)|$, and $m' \in [m\rangle$.*

Proof. Following [6] we prove the theorem by induction on the number of elements of E_π.

If $|E_\pi| = 0$ then, by definition of non sequential process, we have that B_π is the unique B-cut and $\forall s \in S \ m(s) = |D \cap \phi^{-1}(s)|$.

Assume the thesis for $|E_\pi| = n$ and consider $|E_\pi| = n + 1$. The occurrence net $(B_\pi, E_\pi, F_\pi, C_\pi, \lambda : E_\pi \to \wp_{fin}(B_\pi))$ is linearizable by definition of occurrence net, and let $e_1 \ldots e_n e_{n+1}$ be such a linearization. From the proposition 14 we know that $e_{n+1} \in {}^\circ(Max(B_\pi \cup E_\pi, \prec^*))$ and that $e_1 \ldots e_n$ is a linearization of $(B_\pi \setminus e_n^\circ, E_\pi \setminus \{e_n\}, F_\pi', C_\pi', \lambda')$ with the elements F_π', C_π' and λ' being the restriction of F_π, C_π and λ to $B_\pi' = B_\pi \setminus e_n^\circ$ and $E_\pi' = E_\pi \setminus \{e_n\}$, hence we know that $m''(s) = |Max(B_\pi' \cup E_\pi', \prec'^*) \cap \phi^{-1}(s)|$ is a reachable marking. Consider now the transition $\phi(e_n)$. This transition is clearly enabled at m'', and then we have that $m''[\phi(e_n)\rangle m' = \phi(Max(B_\pi \cup E_\pi, \prec^*))$. □

Remark. If we consider nets with empty positive and negative contexts, then the non sequential semantics introduced so far is the same introduced in [6], that means that our semantics is a conservative extension of it. □

We end this section introducing the notion of history preserving bisimulation and showing that it finer than the bisimulation based on the firing sequences.

Two processes π_1 and π_2 are isomorphic ($\pi_1 \cong \pi_2$) if there exists a bijection $f : B_1 \cup E_1 \to B_2 \cup E_2$ such that $f(B_1) = B_2$, $f(E_1) = E_2$, for all $x, y \in B_1 \cup E_1$ $(x, y) \in F_1 \Leftrightarrow (f(x), f(y)) \in F_2$, $x, y \in B_1 \cup E_1 \ (x, y) \in C_1 \Leftrightarrow (f(x), f(y)) \in C_2$, $\forall b \in B_1 \ \forall e \in E_1 \ b \in \lambda_1(e) \Leftrightarrow f(b) \in \lambda_2(f(e))$ and for all $x \in B_1 \cup E_1$ $\phi_1(x) = \phi_2(f(x))$.

The *initial process* π_0 of a net N is the unique (up to isomorphism) process of N with empty set of events.

By definition of a process $\pi = ((B, E, F, C, \lambda), \phi)$, we have that $lpo(\pi) = (B \cup E, \prec^*, \phi)$, where $\prec^*$ is the reflexive and transitive closure of relation of definition 7, is a labeled partial order. We define $Max(\pi) = Max(lpo(\pi))$ and $Min(\pi) = Min(lpo(\pi))$.

The *action structure* $ac(\pi)$ of the process π of the net N is the labeled partial order $(E, \prec^* |_{E \times E}, \phi|_E)$.

If some $t \in T$ is enabled under the marking $\phi(Max(\pi))$ we can extend π to a process π' obtained from π by adding a new event e labeled with t, for all $s \in dom(t^\bullet)$, a set of conditions B_s, with label s, such that $|B_s| = t^\bullet(s)$, arcs (causal ones) such that $e^\circ = \bigcup_{s \in dom(t^\bullet)} B_s$ and ${}^\circ e = \bigcup_{s \in dom(^\bullet t)} B_s'$, where $B_s' \subseteq Max(\pi)$, $|B_s'| = {}^\bullet t(s)$ and the conditions in B_s' are labeled with s, contextual arcs (b, e) with $b \in Max(\pi)$ such that $\phi(b) \in \phi(\hat{e})$ and $\forall b, b' \in \hat{e} \ b \neq b' \Rightarrow \phi(b) \neq \phi(b')$ and $\lambda'(e) = \phi^{-1}({}^\circ t)$.

Definition 18. Let N_1 and N_2 be nets. A set R of triples (π_1, π_2, f) is a *history preserving bisimulation* if

- $(\pi_{0,1}, \pi_{0,2}, \emptyset) \in R$, where π_{0i} is the initial process of N_i, for $i = 1, 2$;

- if $(\pi_1, \pi_2, f) \in R$ then π_i is a process of N_i for $i = 1, 2$ and f is an isomorphism from $ac(\pi_1)$ to $ac(\pi_2)$;

- if $(\pi_1, \pi_2, f) \in R$ and $\pi_1 \xrightarrow{t_1} \pi_1'$ for some $t_1 \in T_1$, then there exists t_2, π_2', f' such that $\pi_2 \xrightarrow{t_2} \pi_2'$, $(\pi_1', \pi_2', f') \in R$ and $f'|_{ac(\pi_1)} = f$;

- if $(\pi_1, \pi_2, f) \in R$ and $\pi_2 \xrightarrow{t_2} \pi_2'$ for some $t_2 \in T_2$, then there exists t_1, π_1', f' such that $\pi_1 \xrightarrow{t_1} \pi_1'$, $(\pi_1', \pi_2', f') \in R$ and $f'|_{ac(\pi_1)} = f$.

The nets N_1 and N_2 are called *history preserving bisimilar* $(N_1 \sim_{hp} N_2)$ if there exists a history preserving bisimulation between them. □

The semantics defined here is finer than the sequential one.

Proposition 19. *Let $CtoM(\pi, D)$ be the marking corresponding to the B-cut D of π. Let R be an history preserving bisimulation between N_1 and N_2. Then $\{(CtoM(\pi_1, Max(\pi_1)), CtoM(\pi_2, Max(\pi_2))) \mid (\pi_1, \pi_2, f) \in R\}$ is a bisimulation.* □

The viceversa doesn't hold: just consider two nets without contextual and inhibitor arcs, in this case our definition of history preserving bisimulation coincide with the one of [1] which is finer then the one defined on firing sequences.

Theorem 20. *Let N_1 and N_2 be two nets, then $N_1 \sim_{hpb} N_2 \Rightarrow N_1 \sim N_2$* □

4 Causal Semantics

In [3] we have introduced a notion of causal execution of a net. We recall here some definitions and results. To obtain information about the causal dependencies of occurrences of transitions, the state of the net is enriched by adding information on the history of tokens and on the number of occurrences of each transition already happened. The construction proceeds first by defining token types: they are essentially a decoration that records some information about the way the token has been generated, namely the occurrence of the transition which produced it; now a place does not contain simply a set of tokens, but rather a multiset of token types, because each individual token remembers its origin. Then, we introduce a notion of configuration of a marked net which essentially defines three pieces of information: the first is the present marking of the net, where the tokens are decorated by their type/history, the second records, for each place, the (occurrences of) transitions that have removed a token from it, and the last one is a counter of occurrences of transitions. Formally:

Definition 21. *Let $N = (S, T, F, I, C, W, m_0)$ be a net.*
The set of *token types* is $\Theta = (T \times \omega^+) \cup \{*\}$, ranged over by θ, where (t, i) is the type of tokens produced by the i-th occurrence of transition t and $*$ is the type of tokens in the initial marking.

A *configuration* γ of a net is a triple $(\mathbf{p}, \mathbf{e}, \mathbf{o})$, where

- $\mathbf{p} : S \to \Theta \to \omega$ describes for each place the number of tokens of each type it contains,
- $\mathbf{o} : T \to \omega$ defines the number of occurrences of each transition,
- $\mathbf{e} : S \to \wp(T \times \omega^+)$ describes for each place the set of (occurrences of) transitions which have consumed tokens from that place.

The *initial configuration* of the net is $\gamma_0 = (\mathbf{p}_0, \mathbf{e}_0, \mathbf{o}_0)$, where

$$\mathbf{p}_0(s)(\theta) = \begin{cases} m_0(s) & \text{if } \theta = * \\ 0 & \text{otherwise} \end{cases} \qquad \mathbf{e}_0(s) = \emptyset \qquad \mathbf{o}_0(t) = 0 \qquad \square$$

Definition 22. The rule for the i-causal firing rule is as follows:[3]
$(\mathbf{p}, \mathbf{e}, \mathbf{o})[(t, i), C\rangle(\mathbf{p}', \mathbf{e}', \mathbf{o}')$ if and only if

- $\exists \mathbf{p}_1 \subseteq \mathbf{p}$ such that for all $s \in S$ ${}^\bullet t(s) = \sum_\theta \mathbf{p}_1(s)(\theta)$
- $\exists \mathbf{p}_2 \subseteq \mathbf{p}$ such that for all $s \in S$ $\sum_\theta \mathbf{p}_2(s)(\theta) = \begin{cases} 1 & \text{if } s \in \hat{t} \\ 0 & \text{otherwise} \end{cases}$
- $\forall s \in {}^\circ t$ $\sum_\theta \mathbf{p}(s)(\theta) = 0$
- $\mathbf{o}(t) = i - 1$
- $C = C_1 \cup C_2 \cup C_3$ where
  $C_1 = \{(t, i) \mid \exists s : \mathbf{p}_1(s)(t, i) > 0\}$,
  $C_2 = \{(t, i) \mid \exists s : \mathbf{p}_2(s)(t, i) > 0\}$ and
  $C_3 = \bigcup_{s \in {}^\circ t} \mathbf{e}(s)$
- $\mathbf{p}' = (\mathbf{p} \setminus \mathbf{p}_1) \oplus \mathbf{p}_3$, where $\mathbf{p}_3(s)(\theta) = \begin{cases} t^\bullet(s) & \text{if } \theta = (t, i) \\ 0 & \text{otherwise} \end{cases}$
- $\mathbf{e}'(s) = \begin{cases} \mathbf{e}(s) \cup \{(t, i)\} & \text{if } s \in {}^\bullet t \setminus t^\bullet \\ \mathbf{e}(s) & \text{otherwise} \end{cases}$
- $\mathbf{o}'(u) = \begin{cases} i & \text{if } u = t \\ \mathbf{o}(u) & \text{otherwise} \end{cases}$ $\qquad \square$

With $\gamma[(t, i), C\rangle\gamma'$ we denote the firing of transition t from configuration γ to configuration γ'. Actually, it is the i-th time that transition t is fired. The set C records the immediate causes for the firing of this occurrence; its elements are occurrences of transitions.

An *i-causal firing sequence* (CFS) is defined inductively as follows:

- γ_0 is a CFS;
- if $\gamma_0[(\tau_1, C_1\rangle\gamma_1 \ldots [\tau_{n-1}, C_{n-1}\rangle\gamma_{n-1}$ is a CFS and $\gamma_{n-1}[\tau_n, C_n\rangle\gamma_n$ then $\gamma_0[\tau_1, C_1\rangle\gamma_1 \ldots [\tau_{n-1}, C_{n-1}\rangle\gamma_{n-1} [\tau_n, C_n\rangle\gamma_n$ is a CFS.

---

[3] Note that $S \to \Theta \to \omega$ is isomorphic to $(S \times \Theta) \to \omega$, so we can treat its elements as multisets.

The set of i-causal firing sequences of a net N is denoted by $CFS(N)$.

The causal dependencies produce a relation between events whose transitive closure is a partial order.

Definition 23. Let $\gamma_0[\tau_1, \mathcal{C}_1\rangle\gamma_1 \ldots [\tau_n, \mathcal{C}_n\rangle\gamma_n$ be an i-causal firing sequence. We define the relation $\prec$ as $\tau_i \prec \tau_j$ iff $\tau_i \in \mathcal{C}_j$. □

Proposition 24. $\prec^+$ is a (strict) partial order. □

Causal trees [5] are trees labeled with pairs $(a, \mathcal{I})$, where a is an action and $\mathcal{I}$ is a set of relative pointers to all the predecessors which caused the present action a.

Definition 25. A *causal tree* over Act is a tree (V, A, f), where

- V is a set of nodes;
- $A \subseteq V \times V$ is a set of arcs;
- $f : A \to Act \times \wp_{fin}(\omega)$ is a labeling function. □

We associate a causal tree to the set of the i-causal firing sequences $CFS(N)$. Here Act will be the set of (names of) transitions in the net N, and $l(\tau_i) = l(t, i) = t$.

Definition 26. The causal tree of a net N is the tree $CT(N) = (V, A, f)$ defined as follows:

- $V = CFS(N)$
- $A = \{(\sigma, \sigma[\tau, \mathcal{C}\rangle\gamma) \mid \sigma, \sigma[\tau, \mathcal{C}\rangle\gamma \in CFS(N)\}$
- Let $\sigma = \gamma_0[\tau_1, \mathcal{C}_1\rangle\gamma_1 \ldots [\tau_{n-1}, \mathcal{C}_{n-1}\rangle\gamma_{n-1}$; then $f(\sigma, \sigma[\tau_n, \mathcal{C}_n\rangle\gamma_n) = (l(\tau_n), \{n - i \mid \tau_i \prec^+ \tau_n\})$. □

With $\sigma \xrightarrow{a, \mathcal{I}} \sigma'$ we mean that there exists an arc labeled with $(a, \mathcal{I})$ between the nodes σ and σ'.

Definition 27. Given two nets N_1 and N_2, a binary relation $R \subseteq CFS(N_1) \times CFS(N_2)$ is a *causal bisimulation* if:

- $(\gamma_{1,0}, \gamma_{2,0}) \in R$, where $\gamma_{i,0}$ is the initial configuration of N_i for $i = 1, 2$
- If $(\sigma_1, \sigma_2) \in R$ and $\sigma_1 \xrightarrow{a, \mathcal{I}} \sigma_1'$ there exists then σ_2' such that $\sigma_2 \xrightarrow{a, \mathcal{I}} \sigma_2'$ and $(\sigma_1', \sigma_2') \in R$.
- If $(\sigma_1, \sigma_2) \in R$ and $\sigma_2 \xrightarrow{a, \mathcal{I}} \sigma_2'$ there exists then σ_1' such that $\sigma_1 \xrightarrow{a, \mathcal{I}} \sigma_1'$ and $(\sigma_1', \sigma_2') \in R$.

The nets N_1 and N_2 are *causal bisimilar* $(N_1 \sim_c N_2)$ iff there exists a causal bisimulation R between them. □

5 Causal versus non sequential semantics

We compare the two semantics formally. For the case of nets without contextual and inhibitor arcs the two semantics are equivalent ([3]). We first show how to associate to a causal firing sequence a linearizable occurrence net. The construction is almost the same of proposition 16, with the difference that now we know which "conditions" have to be used.

Definition 28. Let $N = (S, T, F, I, C, W, m)$ be a contextual place/transition net. Let ϕ be the following mapping: $\phi(s, \theta, i) = s$ and $\phi(t, i) = t$.

Then it is possible to associate to a casual firing sequence σ of $CFS(N)$ a contextual occurrence net and a labeling mapping, denoted with $\mathcal{F}(\sigma)$, defined inductively in the following way:

Let $\gamma_0 = (\mathbf{p}_0, \mathbf{e}_0, \mathbf{o}_0)$ be the initial configuration of N. $\mathcal{F}(\gamma_0) = ((B, \emptyset, \emptyset, \emptyset, \emptyset), \phi|_B)$, where

$$B = \bigcup_{s \in S} \{(s, *, i) \mid 1 \le i \le \mathbf{p}_0(s)(*)\}$$

Let $\gamma_0 \ldots \gamma_{n-1}$ be a causal firing sequence and $\mathcal{F}(\gamma_0 \ldots \gamma_{n-1}) = ((B', E', F', C', \lambda' : E \to \wp_{fin}(B')), \phi|_{B' \cup E'})$. Let $\gamma_n = (\mathbf{p}_n, \mathbf{e}_n, \mathbf{o}_n)$. Then
$\mathcal{F}(\gamma_0 \ldots \gamma_{n-1}[\tau_n, \mathcal{C}_n)\gamma_n) = ((B, E, F, C, \lambda : E \to \wp_{fin}(B)), \phi|_{B \cup E})$, where

$B = B' \cup \bigcup_{s \in S} \{(s, \tau_n, i) \mid 1 \le i \le \mathbf{p}_n(s)(\tau_n)\}$

$E = E' \cup \{\tau_n\}$

$F = F' \cup \bigcup_{s \in S} \{(\tau_n, (s, \tau_n, i)) \mid 1 \le i \le \mathbf{p}_n(s)(\tau_n)\}$
$\cup \bigcup_{s \in S, \theta \in E' \cup \{*\}} \{((s, \theta, i), \tau_n) \mid m \le i \le m + \mathbf{p}_{n-1}(s)(\theta) - \mathbf{p}_n(s)(\theta) \wedge m = min\{i \mid (s, \theta, i)^\circ = \emptyset\}\}$

$C = C' \cup \{((s, \theta, i), \tau_n) \mid (s, \theta, i)^\circ = \emptyset \wedge s \in \phi(\hat{\tau}_n)\}$ such that $\forall b, b' \in \hat{\tau}_n, b \ne b' \Rightarrow \phi|_B(b) \ne \phi|_B(b')$

$\lambda(e) = \phi|_B^{-1}({}^\circ\phi|_E(e))$ $\qquad\qquad\square$

The next proposition states the correspondence between the final configuration of a CFS σ of N and the places of $\mathcal{F}(\sigma)$ that do not have outgoing arcs. It is used to prove that $\mathcal{F}(\sigma)$ is a non sequential process of N.

Proposition 29. Let $\sigma = \gamma_0[\tau_1, \mathcal{C}_1\rangle \ldots [\tau_n, \mathcal{C}_n\rangle\gamma_n$ be a CFS and $\mathcal{F}(\sigma) = ((B, E, F, C, \lambda : E \to \wp_{fin}(B)), \phi|_{B \cup E})$. Then

$\forall s \in S : p_n(s)(*) = |\{b \in B \mid b^\circ = \emptyset \wedge \phi(b) = s \wedge {}^\circ b = \emptyset\}|$

$\forall s \in S, \forall \tau_i (1 \le i \le n) : p_n(s)(\tau_i) = |\{b \in B \mid b^\circ = \emptyset \wedge \phi(b) = s \wedge {}^\circ b = \{\tau_i\}\}|.$ $\square$

This account to say that, given a CFS σ of N, $\mathcal{F}(\sigma)$ is a non sequential process of N.

We show now how to associate to a non sequential process a causal firing sequence corresponding to a given linearization of it.

Definition 30. Let $\pi = ((B, E, F, C, \lambda), f)$ be a (finite) process of the net N and $e_1 \ldots e_n$ a linearization of π. Then

$$\mathcal{G}(\pi) = (\mathbf{p}_0, \mathbf{e}_0, \mathbf{o}_0)[\tau_1, \mathcal{C}_1\rangle(\mathbf{p}_1, \mathbf{e}_1, \mathbf{o}_1) \ldots [\tau_n, \mathcal{C}_n\rangle(\mathbf{p}_n, \mathbf{e}_n, \mathbf{o}_n)$$

where:

$\tau_i = (f(e_i), |\{j \mid j < i \wedge f(e_j) = f(e_i)\}|)$

$\mathcal{C}_i = \{\tau_j \mid (e_j F^2 e_i) \vee ((e_j, b) \in F \wedge (b, e_i) \in C)\} \cup \{\tau_j \mid \exists b \in \lambda(e_i) \ (b, e_j) \in F \text{ and } j \leq i\}$

$$\mathbf{p}_i(s)(\theta) = \begin{cases} |\{b \in B \mid f(b) = s \wedge {}^\circ b = \emptyset \wedge (e_k \in b^\circ \Rightarrow j > i)\}| & \text{if } \theta = * \\ |\{b \in B \mid f(b) = s \wedge e_j \in {}^\circ b \wedge (e_k \in b^\circ \Rightarrow j > i)\}| & \text{if } \theta = \tau_j \\ & \text{and } 1 \leq j \leq i \\ 0 & \text{otherwise} \end{cases}$$

$\mathbf{e}_i(s) = \{\tau_j \mid \text{ for some } b \in f^{-1}(s) \ (b, e_j) \in F \text{ and } j \leq i\}$

$\mathbf{o}_i(t) = |\{j \mid j \leq i \wedge f(e_j) = t\}|$

for $i = 1, \ldots, n$. □

As every prefix of a linearization is a linearization of the (sub)occurrence net with the events in the prefix, the following proposition is obvious.

Proposition 31. *Let π be a process of the net N and $\bar{e}$ a linearization of π. Let $\mathcal{G}(\pi, \bar{e}) = \gamma_0[\tau_1, \mathcal{C}_1\rangle\gamma_1 \ldots [\tau_n, \mathcal{C}_n\rangle\gamma_n$. Then for $0 \leq i \leq n$ $\gamma_0 \ldots [\tau_i, \mathcal{C}_i\rangle\gamma_i$ is a CFS of N.* □

Proposition 32. *Let π be a (finite) process and $\bar{e}$ a linearization of π. Then $\mathcal{F}(\mathcal{G}(\pi, \bar{e})) \cong \pi$.* □

This proposition shows that no information is lost in associating to a non sequential process and a linearization the corresponding causal firing sequence.

Proposition 33. *If $\pi \overset{t}{\longrightarrow} \pi'$, $\bar{d}$ is a linearization of π, $E' \setminus E = \{e\}$ and $\mathcal{G}(\pi, \bar{d}) = \gamma_0[\tau_1, \mathcal{C}_1\rangle\gamma_1 \ldots [\tau_n, \mathcal{C}_n\rangle\gamma_n$ then $\mathcal{G}(\pi', \bar{d}e) = \mathcal{G}(\pi, \bar{d})[(t, i), \mathcal{C}\rangle\gamma$, where $\mathcal{C} = \{\tau_i \mid d_i F^2 e \text{ or } (d_i, b) \in F \wedge (b, e) \in C\} \cup \{\tau_i \mid \text{ for some } b \in \lambda'(e) \ (b, d_i) \in F\}$.* □

If $\bar{e} = e_1 \ldots e_n$ then $f(\bar{e}) = f(e_1) \ldots f(e_n)$. We show that to an history preserving bisimulation a causal one corresponds.

Proposition 34. *Let R be a history preserving bisimulation between N_1 and N_2. Then $R_c = \{(\mathcal{G}(\pi_1, \bar{e}), \mathcal{G}(\pi_2, f(\bar{e}))) \mid (\pi_1, \pi_2, f) \in R \wedge \bar{e} \text{ is a linearisation of } \pi_1\}$ is a causal bisimulation.* □

Theorem 35. *Two history preserving bisimilar nets (N_1 and N_2) are causal bisimilar, i.e. $N_1 \sim_{hbp} N_2 \Rightarrow N_1 \sim_c N_2$.* □

The history preserving bisimulation is more discriminating than the causal one, as it can distinguish between the two kind of contexts, as the following example shows:

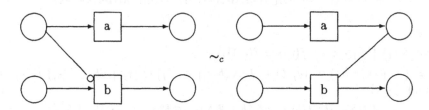

The two nets are causal bisimilar but not history preserving bisimilar as in the action structure associated to the second one we have a dependency between a and b that is absent in the one associated to the first one. However the two semantics are equivalent if we consider nets without inhibitor arcs but only contextual arcs. Formally:

Proposition 36. *Let R be a causal bisimulation between two nets N_1 and N_2 such that $I_1 = \emptyset = I_2$. Then $R_h = \{(\mathcal{F}(\sigma_1), \mathcal{F}(\sigma_2)) \mid (\sigma_1, \sigma_2) \in R \wedge \sigma_i = \gamma_{i0}[\tau_{i1}, C_{i1}] \dots [\tau_{in}, C_{in}]\gamma_{in}, i = 1, 2 \wedge f = \{(\tau_{1i}, \tau_{2i}) \mid i = 1, \dots, n\}$ is an history preserving bisimulation.* □

Theorem 37. *Let N_1 and N_2 be two nets without inhibitor arcs. Then $N_1 \sim_{hbp} N_2$ iff $N_1 \sim_c N_2$.* □

6 Relation with other works and conclusions

To the best of our knowledge, there are two other different definitions of contextual occurrence net. One is proposed by Janicki and Koutny in [7] and the other by Montanari and Rossi in [9]. The first one is hardly comparable with ours as occurrence nets are enriched with places that represent the non holding of the corresponding condition in the net, and inhibitor arcs are special arcs connecting these new places to the event, which is possible as a condition in the C/E net either holds or not; in a P/T net we cannot say anything analogous. Regarding the second one, they don't consider inhibitor arcs, that can be removed via an operation of complementation we cannot define as we are considering P/T nets. However a comparison is possible when we consider contextual P/T nets without inhibitor arcs and safe, i.e. such that for every reachable marking, the marked places have at most one token (this means that the marking is always a set). As we already pointed out, the difference is mainly in the different definition of partial order associated to the occurrence net.

We briefly show how our definition leads to an event structure with possible event and that the configurations of this one correspond to the possible linearization of our definition of occurrence net.

We recall the definition of prime event structure with possible event given in [12].

Definition 38. – A prime event event structure with possible events is the 4-tuple $\mathbf{E} = (E, E_p, \leq, \#)$ where $(E, \leq, \#)$ is a prime event structure and $E_p \subseteq E$ is a set of possible events.

- A configuration X is any subset of events satisfying the following conditions:
  1. X is conflict free;
  2. $\forall e \in X \ \forall e' \leq e$ either $e' \in X$ or $e' \in E_p$. □

The domain of configurations of a such an event structure is still a prime algebraic domain, as proved in [11], but the order isn't any longer the set theoretic inclusion: for all the configurations X and Y, $X \sqsubseteq Y$ iff $X \subseteq Y$ and, for all $e \in E_p$, $e \in X$ if $e \in Y$ and if $e \leq e'$ for some $e' \in X$.

Given a contextual occurrence net (B, E, F, C) (we omit the λ as we are considering nets without inhibitor arcs), and the associated relations $\prec$ and $\prec_c$, we associate the prime event structure with possible events $(E, E_p, \prec^*, \emptyset)$ where $E_p = \{e \in E \mid \text{for some } e' \in E \ e \prec_c e'\}$. Recalling that a linearization of a configuration corresponds to say that it is secured ([16]) it is easy to see that the linearizations of the configurations (compatible with the partial order and taking into account the fact that we have now *possible* events) and the one of the occurrence net correspond. We can conclude that our definition, in the case of safe nets without inhibitor arcs, does capture the causality more precisely than the one proposed by Montanari and Rossi in [9]. When we add inhibitor arcs we face the problem that events must be duplicated, as events come along with their past and in the past of some events there may be some condition inhibiting it that have been removed. The definition is then rather complicated.

Summarizing the results of the paper, we have developed a notion of contextual occurrence nets and defined a non sequential semantics for contextual P/T nets. The semantics is finer than the other ones presented in literature, as we can define a notion of history preserving bisimulation that distinguishes positive contexts from negative ones. We aren't yet completely satisfied with the non sequential semantics we have introduced as we would like to give simpler conditions in order to say that a "B-cut" represents a state of the net (a marking), for instance using a suitable relation instead of the notion of linearization. We are already pursuing this research direction considering another weak ordering modeling mainly the linearizations.

References

1. E. Best, R. Devillers, A. Kiehn, L. Pomello, "Concurrent bisimulations in Petri nets", *Acta Informatica* 28, 231-264, 1991.
2. N. Busi, R. Gorrieri, "A Petri Net Semantics for π- calculus", in Proc. *Concur'95*, LNCS 962, Springer, 145-159, 1995.

3. N. Busi, G. M. Pinna, "A Causal Semantics for Contextual P/T Nets", To appear in Proc. *ICTCS'95* , 1995.
4. N. De Francesco, U. Montanari, G. Ristori, "Modelling Concurrent Access to Shared Data via Petri Nets", in Proc. *Procomet'94*, North-Holland, 337 - 396, 1994.
5. Ph. Darondeau, P. Degano, "Causal Trees", in Proc. *ICALP'89*, LNCS 372, Springer, 234-248, 1989.
6. U. Goltz, W. Reisig, "The Nonsequential Behaviour of Petri Nets", *Information and Computation* 57, 125-147, 1983.
7. R. Janicki, M. Koutny, "Invariant Semantics of Nets with Inhibitor Arcs", in Proc. *CONCUR'91*, LNCS 527, Springer, 317-331, 1991.
8. R. Milner, J. Parrow, D. Walker, "A Calculus of Mobile Processes", *Information and Computation* 100, 1-77, 1992.
9. U. Montanari, F. Rossi "Contextual Nets", *Acta Informatica* 36, 545-596, 1995.
10. C. A. Petri, "Non-Sequential Processes", Gesellschaft für Mathematik und Datenverarbeitung Bonn, Interner Bericht ISF-77-5, 1977.
11. G. M. Pinna, A. Poigné "On the Nature of Events", in Proc. *MFCS'92*, LNCS 629, Springer, 430-441, 1992.
12. G. M. Pinna, A. Poigné "On the Nature of Events: another Perspective in Concurrency", *Theoretical Computer Sciences* 138, 425-454, 1995.
13. W. Reisig, "Petri Nets: An Introduction", EATCS Monographs in Computer Science, Springer, 1985.
14. G. Ristori, "Modelling Systems with Shared Resources via Petri Nets", Ph.D. Thesis, Università di Pisa, TD-5/94, 1994.
15. W. Vogler, "Modular Construction and Partial Order Semantics of Petri Nets" LNCS 625, Springer, 1992.
16. G. Winskel, "Event Structures", in *Petri Nets: Applications and Relationships to other Models of Concurrency*, LNCS 255, Springer, 325-392, 1986.

The PSR Methodology: Integrating Hardware and Software Models

Susanna Donatelli and Giuliana Franceschinis *

Dipartimento di Informatica, Università di Torino
Corso Svizzera 185, 10149 Torino, Italy

Abstract. This paper proposes a new methodology for the construction of integrated hardware-software GSPN models. The PSR methodology is an extension of the Process/Resource Box methodology defined in [5]: it consists of defining three submodels, the $\mathcal{P}$ level (processes), the $\mathcal{S}$ level (services) and the $\mathcal{R}$ level (resources), and the composition rules to combine them into a complete integrated GSPN model of the whole system.

This work has been motivated by the need of a systematic approach to the construction of (parallel) hardware-software models. The adequacy of the methodology is discussed through a running example of a relatively complex system.

1   Introduction

In the theory and application of Petri nets, and of stochastic Petri nets in special mode, there is a lack of precise guidelines that can assist a non expert of Petri nets, in the process of modelling complex systems. We feel that a general solution to this problem is not feasible, and we have therefore concentrated on a specific field: indeed this paper proposes a methodology for the construction of GSPN models of (parallel) hardware and software.

This methodology is the result of our experience in a joint project with ENEL-CRA (the research center of the major Italian electricity company) to model a real hardware and software system called "complex-node"[6], that will be used as a running example.

The final goal of this research effort is to make the construction of integrated hardware and software systems easier by defining precise guidelines on how to organize the modeling process. The methodology should produce models for which it is relatively easy to define the timing information (hence the timed transitions should model hardware operations whose duration could be easily derived from the performance characteristics of the various components) and that facilitate reusability.

There has been quite a bit of work aimed at simplifying the task of building models of complex systems. In recent years the DEMON project has investigated

---
* This work has been partially supported by the Italian MURST "40%" project and by the ENEL-CRA contract n. 73/93

the issue of compositionality in untimed Petri nets, indeed the Box calculus [3] provides the modeller with an algebra of boxes that includes operations like synchronous and asynchronous communication. For our application domain the box calculus lacks a notion of time, and of synchronization over immediate transitions, that is instead crucial when dealing with resource acquisition, as it will be shown in the paper.

Stochastic timing associated with actions is instead the basic idea behind the timed extension of process algebra called TIPP [15],MPA [2, 8], and PEPA [14, 17], but again there are no immediate transitions in MPA and PEPA, while this extension is under development in TIPP.

Composition is also a central issue in the work by Buchholz on hierarchies for colored GSPN [7]. Unfortunately the hierarchical framework may not be an adequate one to model systems that are usually defined in a layered fashion.

The previous works that are more strictly related to ours are the Process Resource Mapping (PRM) net approach by Ferscha [12, 13] and the process and resource boxes (PR) methodology by Botti and DeCindio [10, 5].

The work by Ferscha introduces a new class of timed nets in which models are built as a mapping of a process model on a simple resource model, unfortunately too simple to describe complex resource behaviour (as explained in the example of Fig. 6 in [13], PRM may pose problems in the simultaneous acquisition of resources).

Our work actually started from the paper by Botti and DeCindio [5], that proposes a methodology (called PR) to build GSPN models of software mapped onto hardware, using as basic building blocks a timed flavour of Petri net boxes. We tried to apply this approach to our case study application, and observed some problems due to the limitations imposed by the availability of a single, general-purpose, compositional operator on transitions, by the resource model that is too simple, and by the uncomplete formalization.

All the papers mentioned so far propose formalisms and operators that are simple and general, which is an important requirement for a general theory of modularity, but we consider as a strict requirement also the availability of precise guidelines. Our proposal would therefore lose in "elegancy" and generality, by providing specific operators at different modelling levels, but we hope to have reached the goal of *providing precise guidelines in a formalized setting*. A similar approach has been followed also for other application fields, as for example the work on GRAMAN [18] for the modelling of manufacturing systems.

The basic idea behind the methodology described in this paper consists of clearly separating the model of the process ($\mathcal{P}$ level) from the pattern of resource usage ($\mathcal{R}$ level). An intermediate model acting as an interface between the process and resource models (the $\mathcal{S}$ level) allow to map the complex operations required by the processes to the simple operations allowed by the resources. The timing information is defined at the resource level, expressed as average duration of the basic services offered by the resources. By composing the levels, in a "stack" like fashion, the "time" will emerge from bottom to the top, to finally get a GSPN model.

The reasons for keeping the $\mathcal{P}$ level separated are *simplicity* and *reusability*: simplicity since the model does not have to take into consideration the physical implementation of the process operations, and reusability of the process model in conjunction with different hardware models

The separation between the $\mathcal{R}$ and $\mathcal{S}$ levels reflects the layered structure of modern computing systems (it is natural to extend the method to allow several intermediate ($\mathcal{S}$) levels, however this possibility won't be discussed in this paper), and allows the timing of complex operations in $\mathcal{S}$ to be derived from the simpler operations of $\mathcal{R}$ to which it is easier to associate a time specification, being the basic operations of the considered hardware.

The operators for the combination of the three levels are based on different flavours of the operator of transition superposition. To the best of our knowledge, in the GSPN literature has not appeared so far any organic study on the several problems that arise when composing GSPNs by superposition over timed or immediate transitions. In this paper we shall discuss some peculiar problems of transition superposition in GSPN, especially the problem of the interference between the race policy and the superposition of timed transitions, the confusion that can be introduced in the model by the superposition of immediate transitions, and the problems of fork and join structures with multiple tokens.

Section 2, 3, and 4 define the construction of the $\mathcal{R}, \mathcal{S}$ and $\mathcal{P}$ levels, respectively. Each section presents first an informal explanation of the concepts that are then shown on the running example and finally described in a more formal setting. Section 5 concludes the paper.

In this paper we shall adhere to the notation defined in [1], therefore A GSPN system is a 8-tuple $\mathcal{G} = (P, T, \Pi, I, O, H, W, M_0)$ where P is the set of places; T is the set of transitions, $T \cap P = \emptyset$; $\Pi : T \to I\!N$, is the priority function that maps transitions onto natural numbers representing their priority level. $I, O, H : T \to Bag(P)$, are the input, output and inhibition functions, respectively, where $Bag(P)$ represents all multisets over P. $W : T \to I\!R$, is a weight function (defining rates of timed transitions and probabilities of immediate transitions). $M_0 : P \to I\!N$, is the initial marking.

A labelled GSPN system (LGSPN) is defined as $\mathcal{LG} = (\lambda, \mathcal{G})$ where $\mathcal{G}$ is a GSPN system, and $\lambda : T \to L \cup \tau$, assigns a label from a set L to each transition. τ labelled transitions are considered to be internal, non-τ are considered external (i.e., interface transitions). If λ is injective (no two transitions are assigned the same label) we term the LGSPN as *injective*.

A multilabelled GSPN (MLGSPN) is defined as $\mathcal{MLG} = (\lambda, \mathcal{G})$ where $\mathcal{G}$ is a GSPN system, and $\lambda : T \to 2^L \cup \tau$, where 2^L is the power set of L.

Our running example is the model of the hardware and software of a node of a parallel architecture proposed by ENEL-CRA as the support system for the control of the automation of their plants. A previous model of the node developed following the PR methodology (when possible), has been presented at the "case study" minitrack of the 1996 Petri Net conference [6].

The system considered is a processing unit made up by connecting two Transputers through a dual port shared memory, depicted in Fig. 1. Each Transputer has a CPU, a local memory and four links, therefore a complex node has eight links available for connection with the rest of the world (so that, for example, an hypercube of up to eight dimensions could be constructed).

The behavior of the complex node has been studied under a parametric synthetic workload: by properly setting the parameters of the workload the full range of applications from CPU-bound to I/O bound can be modelled. This workload consists of four cyclic processes (names A, B, C, D), each performing a sequence of computation and communication, plus a number (N_e) of processes representing those running on other nodes in the architecture (these nodes are simply modelled as "external resources"). The processes are connected on a ring through synchronous directed (virtual) channels (named AB, BC, CD and DA), moreover process X (X $\in$ {A, B, C, D}) is connected with the external world (processes running on the external resources) through the synchronous channel Xext (X $\in$ {A, B, C, D}). Each time a process decides to perform a communication it can choose to synchronize with any of the three different processes it is connected to. In Fig. 1 we assume that processes A and B are allocated on Transputer 1, and B and C on Transputer 2.

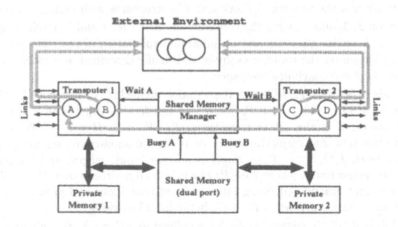

Fig. 1. System architecture and mapping of processes

2 Representing the hardware: the $\mathcal{R}$ level

The resource level is the bottom level in the pile: it describes the basic operations offered by the physical resources, that is to say the low level operations that can be performed by the hardware. Example of basic services of the $\mathcal{R}$ level are a cpu basic instruction –like an integer addition– (or a "burst" of given length of basic instructions), a dual port memory write, a transmission of a given amount of data over the link, etc.

It is the only level comprising both immediate and timed transitions, while the other two levels only contain immediate transitions: by composing the models of the three levels, the "time" emerges from bottom to the top, to finally get a GSPN model.

Since the hardware system can be quite complex, composed of different basic (interacting) components, we propose that level $\mathcal{R}$ is obtained as the composition of the models of the single components: to this aim we introduce a generic model of a component and a compositional rule.

Figure 2(a) shows the LGSPN model of a generic component $\mathcal{R}_i$ with n basic operations $OP1, \ldots, OPn$: this model will be referred to as LGSPN of "resource type". The timed transition that represents operation OP on a resource is preceded by an immediate transition of label S_op that models the start of the activity and is followed by an immediate transition of label E_op that models the end of that activity. Note in the model the presence of two transitions labelled *lockRes* and *unlockRes*, that enforce the need for operations $OP1, \ldots, OPk$ to acquire a lock on the resource before the operation can be performed.

We require each $\mathcal{R}_i$ to be safe (indeed place IdleRes in Fig. 2(a) is initialized with exactly one token), since we want to keep a one to one mapping between the hardware elements and their models (as we shall see in Section 3 safeness is also a necessary condition for a correct stochastic behaviour of the composition).

Labels are partitioned in two sets: the timed transition labels, representing basic hardware operations ($L_{loc}(\mathcal{R}_i)$), and the immediate transition labels ($L_{off}(\mathcal{R}_i)$), representing the start and end of the operations that are offered to the upper level. We assign the same label to timed transitions that represent basic operations of different hardware components that need to *cooperate*, for example a copy of data into memory requires a joint activity of memory and cpu.

The complete model of level $\mathcal{R}$ is then obtained as superposition over timed transitions of equal label of the components $\mathcal{R}_i$. Since the cooperation between two resources can be interpreted as two partners working in parallel, then the duration of the combined operation is taken as the maximum of the time (minimum rate) taken by the two superposed transitions (see [16] for a complete discussion on the possible alternatives for the rate).

The timed transition labels are used only while constructing the hardware model, and will therefore be "hidden" before composing the $\mathcal{R}$ level model with the upper level.

2.1 The $\mathcal{R}$ level model of the "complex node"

We identify 5 types of hardware components in our node: CPU, memory, link, dual port for the shared memory, and a fictitious external resource. The node comprises 2 CPUs, two memories, eight serial links, two ports to the shared memory, and N_e fictitious external resources, for a total of $14 + N_e$ elements.

Fig. 2(d) shows the model of the memory of a generic CPU (the two memory submodels can be obtained by suffixing each label with the CPU identifier). The memory operations considered are: the instruction fetch, the copy of data from a location into another in memory, the read of data from memory, the write of data into memory, and the transfer of data from memory to a link buffer in DMA (direct memory access) mode. Consequently we have $L_{\mathrm{loc}} = \{$ *FETCH, MEMCOPY, MEMRD, MEMWR*$\}$, and $L_{\mathrm{off}} = \{$ *S_fetch_mem, E_fetch_mem, S_dma_mem, E_dma_mem, S_copy_mem, E_copy_mem, S_rd_mem, E_rd_mem, S_wr_cpu, E_wr_cpu* $\}$, where label *S_op* (*E_op*) stays for start (end) of operation *op*. Note that we have explicitly assigned a label only to timed transitions that models operation that require a cooperation. Indeed the operation *DMA_MEM* (the transfer of data from memory to links in DMA mode) requires only the memory resource, and therefore the corresponding timed transition is labelled with τ. As a consequence the operation *DMA_MEM* is not in L_{loc}, while, of course, the labels of the two immediate transitions that model the start and end of a *DMA_MEM* operation, *S_dma_mem* and *E_dma_mem*, are in L_{off}.

Fig.s 2(e), (b), (c), (f) show the LGSPN models of a generic CPU, of the i link for a generic CPU, of the shared memory port connected to a generic CPU, and of a generic fictitious external resource. The actual models for each hardware element can be derived by adding the CPU identifier (number 1 or 2), and when needed the link identifier $(0, \ldots, 3)$, to the labels of Fig. 2. As in the case of memory, we can define for each model the sets of operations and the corresponding L_{loc} and L_{off} labels. Since the labels shown in the figures are quite evocative, we do not list them here for the sake of brevity.

Observe that some labels are common to more than one model meaning that a certain operation require the cooperative use of more than one resource. For example a memory copy operation between memory and CPU, corresponds to a physical activity in both CPU and memory: these activities have indeed the same label *MEMCOPY*. Other operations that require a cooperation are *FETCH*, *MEMRD* and *MEMWR* (for cooperation between CPU and memory), *DMRD*, and *DMWR* (between CPU and port of the shared memory), *SETUP* and *RESET* (between CPU and a link).

The model of level $\mathcal{R}$ is then obtained by simply composing the $14 + N_e$ resource models over the set of labels of timed transitions (that is to say over $L_{\mathrm{loc}}(\mathcal{R}) = \bigcup L_{\mathrm{loc}}(\mathcal{R}_i))$.

2.2 Formal definition

We now give a formal definition of the composition operator, of the hiding operator and of level $\mathcal{R}$. The composition is realized by superposing transitions of equal label.

Definition 1. Given two LGSPN systems LG_1 and LG_2 we define the LGSPN system LG as the *cooperation of* LG_1 and LG_2 over the set L of labels,

$$LG = LG_1 \underset{L}{\bowtie} LG_2$$

iff $LG = (\lambda, (P, T, \Pi, I, O, H, W, M_0))$ is defined as follows. Let $E = L \cap \lambda_1(T_1) \cap \lambda_2(T_2)$ be the subset of L comprising labels that are common to

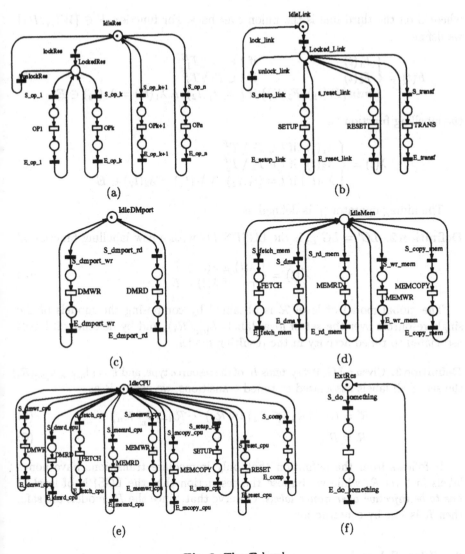

Fig. 2. The $\mathcal{R}$ level

the two LGSPNs, T_1^l be the set of transitions of LG_1 that are labelled l, and T_1^E be the set of all transitions in LG_1 that are labelled with a label in E. Same definitions apply to LG_2.

Then $P = P_1 \cup P_2$, $T = T_1 \setminus T_1^E \cup T_2 \setminus T_2^E \cup_{l \in E} \{T_1^l \times T_2^l\}$; for function $F \in \{I(), O(), H()\}$ we define

$$F(t) = \begin{cases} F_1(t) & \text{if } t \in T_1 \setminus T_1^E \\ F_2(t) & \text{if } t \in T_2 \setminus T_2^E \\ F_1(t_1) \cup F_2(t_2) & \text{if } t = (t_1, t_2) \wedge \lambda_1(t_1) = \lambda_2(t_2) \in E \end{cases} \tag{1}$$

where $\cup$ on the third line is the union over bags. For function $F \in \{W(), \Pi()\}$ we define

$$
F(t) = \begin{cases} F_1(t) & \text{if } t \in T_1 \setminus T_1^E \\ F_2(t) & \text{if } t \in T_2 \setminus T_2^E \\ min\{F_1(t_1), F_2(t_2)\} & \text{if } t = (t_1, t_2) \wedge \lambda_1(t_1) = \lambda_2(t_2) \in E \end{cases} \tag{2}
$$

the labelling function is:

$$
\lambda(t) = \begin{cases} \lambda_1(t) & \text{if } t \in T_1 \setminus T_1^E \\ \lambda_2(t) & \text{if } t \in T_2 \setminus T_2^E \\ \lambda_1(t_1) & \text{if } t = (t_1, t_2) \wedge \lambda_1(t_1) = \lambda_2(t_2) \in E \end{cases} \tag{3}
$$

The hiding operator $|_L$ is defined as

Definition 2. $LG' = LG_{|L}$, is the LGSPN LG with a new labelling function λ'

$$
\lambda'(t) = \begin{cases} \lambda(t) & \text{if } \lambda(t) \notin L \\ \tau & \text{if } \lambda(t) \in L \end{cases} \tag{4}
$$

The global model of level $\mathcal{R}$ is obtained by composing the models of the single resources over the set of local labels $L_{loc}(\mathcal{R}_i)$, and by deleting all labels associated to timed activity in the resulting model.

Definition 3. Given n LGSPN systems R_i of the resource type, and $L = \bigcup_{i \in 1..n} L_{loc}(\mathcal{R}_i)$ the set of all labels associated to timed transitions, we define $\mathcal{R}$ as

$$
R' = ((\cdots(\mathcal{R}_1 \bowtie_L \mathcal{R}_2) \bowtie_L \mathcal{R}_3) \cdots \mathcal{R}_{n-1} \bowtie_L \mathcal{R}_n) \tag{5}
$$

$$
\mathcal{R} = R'_{|L} \tag{6}
$$

It follows from the definition that only immediate transitions have non-τ labels in level $\mathcal{R}$, while we impose the restriction that the LGSPN of level $\mathcal{R}$ *has to be injective over non-τ labels*. Observe that, since the $\mathcal{R}_i$ models are safe, then $\mathcal{R}$ is safe by construction.

3 The $\mathcal{S}$ level

The service level $\mathcal{S}$ is used to implement the services requested at the top level with the one offered by the resources at the bottom one. The interface of the level is a set of services offered to the process level, and a set of services requested from the bottom level.

The service level is intended to hide from level $\mathcal{P}$ all the implementation details. For example, a service offered by level $\mathcal{S}$ to level $\mathcal{P}$ can be made up of a sequence of operations offered by level $\mathcal{R}$, or it can be made as a non-deterministic/probabilistic choice among different implementations of the same service.

The service part is motivated by the fact that some operations may use more than one physical resource in a complex pattern, while at the process level, we want to be independent of the implementation of the operations on resources. Example of intermediate services are: communication over a link, communication through a dual port memory.

The S level model should thus provide the upper level with *macro* operations obtained by composition of low-level hardware operations. It is obtained by parallel composition (without any kind of synchronization at this level) of the submodels implementing the single services.

A service model S_i for service SRV is an MLGSPN composed by an idleSRV place with a single output transition (labelled S_SRV) and a single input transition (labelled E_SRV), both immediate. The two transitions are connected through an arbitrary subnet that models the implementation of the service in terms of the hardware operations offered by level $\mathcal{R}$. We impose that there are no timed transitions in S_i, since the time is associated to operations implemented in the $\mathcal{R}$ level. We indicate with $L_{req}(S_i)$ the set of labels of operations of level $\mathcal{R}$ required by model S_i, and with $L_{off}(S_i)$ the pair of labels of the service offered by S_i.

To ensure that the S_i acquire each lower level operation OP through a pair of transitions labelled S_op – E_op, we impose the constraint that for any possible (labelled) firing sequence of S_i starting with S_SRV and ending with E_SRV, each firing of a transition labelled E_op is preceded by the firing of a transition labelled S_op.

We want to allow different implementation of the same service, and we can therefore have two service models S_i and S_j that offer the same pair of S_SRV – E_SRV labels, therefore with $L_{off}(S_i) = L_{off}(S_j)$, but possibly $L_{req}(S_i) \neq L_{req}(S_j)$.

Since it may be necessary to start several (low level) operations at the same instant, transitions can be labelled with multiple distinct labels. In particular, since it may be useful to have the possibility of selecting a given service only upon the availability of some hardware operation, transitions representing the start (end) of a service can have a multilabel composed by the S_SRV (E_SRV) label and any subset of $L_{req}(S_i)$. Therefore λ is defined as

$$\lambda : T \rightarrow (L_{off}(S) \cup \tau) \times (2^{L_{req}(S)} \cup \tau) \cup \tau.$$

3.1 The S model of the "complex node"

In our application we can define the following services: computation burst ($COMP$), communication inside the node (CON), communication inside a CPU (COC) communication with the external world (COL), and computation in the external resource (EXT_ACT).

The MLGSPN model of the computation burst is shown in Fig. 3(a), it consists of a sequence of instruction fetch (requiring the simultaneous acquisition of both memory and CPU), and a pure computation phase (requiring only a service by the CPU).

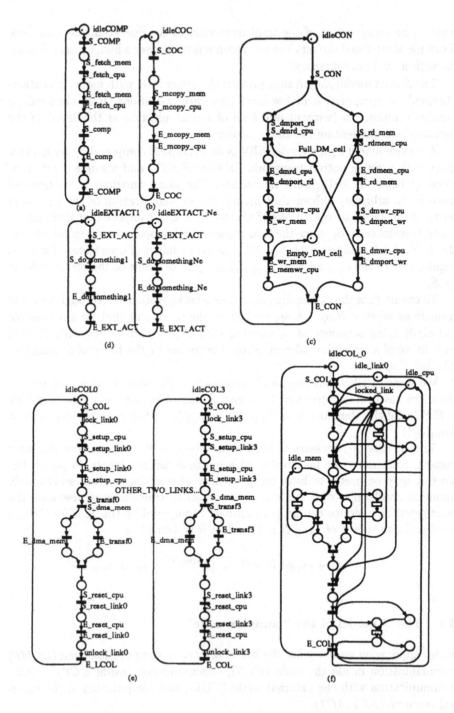

Fig. 3. The $\mathcal{S}$ level and its composition with the $\mathcal{R}$ level

The two MLGSPN models shown in Fig.s 3(b) and (c), correspond to a communication between two processes: using local memory, if the processes are allocated on the same CPU, or using the the dual port shared memory, if the processes are allocated on different CPUs. Indeed the subnet that implements the service in the dual port memory case is rather complex: it comprises two branches that are activated in parallel. The first one consists of a read from local memory (using CPU plus memory) followed by a write onto the dual port memory; the second one is a read from shared memory, followed by a write into the local memory. The two branches are actually not executed in parallel since the protocol used to implement a virtual communication channel using the dual port shared memory requires the sender request to be completed before the corresponding receiver request can be satisfied (producer-consumer model).

Fig. 3(e) shows two of the four MLGSPNs that provide different implementations of the same COL service on the four links. Communication through a serial link is realized by a sequence of three phases: link setup, data transfer, link reset. Observe that a communication over the link is offered only if it is possible to lock a link, indeed the two labels S_COL and *lock_link0* are assigned to the same transition, so that the begin-end service transitions are labelled with a pair of labels (S_COL, *lock_link* and E_COL, *unlock_link* respectively). The communication through a link comprises the setup requiring both the link and the CPU, the DMA data transfer between the memory and link controller in pipeline with the actual transfer of data through the serial link and finally the link reset involving both the CPU and the link.

The composition of level S with level R is performed by superposition over immediate transitions with matching requested-offered operation label; we require that the set of request labels at level S, is a subset of the offered labels at level R. Observe that since the S level is multilabelled, the superposition operation involves sets of transitions rather than pairs of transitions. The transition obtained by superposition of matching request-offer transitions, loses the corresponding pair of labels, while the R level transitions that offer an operation that is not used by any service are removed from the net.

Fig. 3(f) is an example of composition of the model of service COL over link of index 0 with the R level model; note that the transition offering service S_COL is enabled only when place idle_link0 is marked, and that timed activities have "emerged" from level R.

3.2 Problems

Before proceeding to the formal definition of the S level model and of its composition with the R level, let us discuss some problems related with composition of GSPN models, that motivate the constraints imposed on the structure of the first two levels.

The first issue is the synchronization over immediate vs. timed transitions. when synchronization is mixed with choice. Using the model in Fig. 4(b) (a non free choice conflict among two processes using a common resource, obtained by composing the three submodels in Fig. 4(a)), we can enlight two problems:

servers can become faster, and a later arriving process can preempt a process that is already using the resource. Indeed when a race policy is chosen to solve conflicts between timed activities (as in GSPN) the rate at which the resource moves from a state with place $p13$ marked to a state with place $p14$ marked is $W(t')$ if P is the only process ready to synchronize (place $p11$ marked and $p15$ not marked), while it is $W(t')+W(t'')$ is both processes are ready to synchronize (also $p15$ is marked): an obviously undesirable behaviour.

Another undesirable behaviour is that, while process P is already using the resource, process Q may also require the resource (due to the firing of some timed transition not shown in the picture that puts a token in $p15$): again due to the race policy there is absolutely no guarantee that the transition that will actually fire is t', since process P may be "preempted" by Q.

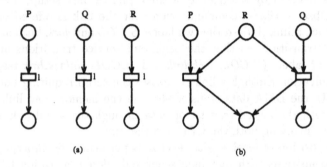

Fig. 4. Resource acquisition in a "all-timed" context

These arguments justify the modelling of a resource through a "double" synchronization, as it is done in the proposed methodology.

The second problem is called *min-match*. Fig. 5(a) shows a process (P) composed with the models of two resources ($R1$ and $R2$) through a double synchronization. If we allow an initial marking $n > 1$ for the P subnet, meaning that several concurrent instances of the same type of process (implementing a given service) are available, and an initial marking $m1 > 1$ and $m2 > 1$ for resources $R1$ and $R2$, meaning that several resources of type Ri are available, then an odd situation may arise because transition t_2 can match tokens in p_3 and p_4 that were not put in p_1 and p_2 by the same firing of t_1 as t_1 matches the tokens that have traversed the two subnets with the smallest delay.

Note that a solution to this problem requires that both processes and resources are safe, as indeed the same problem is present in the models of Fig. 5(b) (two processes represented by two safe subnets composed with a non-safe resource submodels) and of Fig. 5(c) (three safe resource submodels composed with a non safe process model).

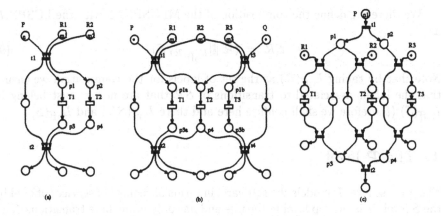

Fig. 5. The min-match delay problem

3.3 Formal definition

The operator required to build level S from the S_i is the parallel composition with no synchronization (of trivial definition). For its composition with R we need instead an operator for the synchronization over a subset E of labels of an MLGSPN with a safe, injective, LGSPN, that we call $\|\|\|_E$.

Definition 4. Given a LGSPN system LG_1, injective over the set of labels $E \subseteq L_1$, and an MLGSPN system MLG_2 we define the MLGSPN system MLG as the *superposition of LG_1 and MLG_2* over the set E of labels,

$$MLG = LG_1 \|\|\|_E MLG_2$$

iff $MLG = (\lambda, (P, T, \Pi, I, O, H, W, M_0))$ is defined as follows. Let T_1^l be the set of transitions in LG_1 that are labelled l and T_1^E be the set of all transitions in LG_1 that are labelled with a label belonging to E. Same definitions apply to MLG_2. Then $P = P_1 \cup P_2$, and $T = T_1 \setminus T_1^E \cup T_2$. The labelling function is

$$\lambda(t) = \begin{cases} \lambda_1(t) & \text{if } t \in T_1 \setminus T_1^E \\ \lambda_2(t) \setminus E & \text{if } t \in T_2 \end{cases} \qquad (7)$$

For functions $F \in \{I(), O(), H()\}$ we define

$$F(t) = \begin{cases} F_1(t) & \text{if } t \in T_1 \setminus T_1^E \\ F_2(t) \cup \bigcup_{t' \in T_1 : \lambda_1(t') \in \{\lambda_2(t) \cap E\}} F_1(t') & \text{if } t \in T_2 \end{cases} \qquad (8)$$

We leave open the problem of assigning weights and priorities to transitions, by simply saying that $W(t) = f_W(t)$ and $\Pi(t) = f_\Pi(t)$.

The composition of level S over R requires that all operations used by S are actually offered by R: $L_{\text{req}}(S) \subseteq L_{\text{off}}(R)$.

We therefore define the composition of the MLGSPNs S with the LGSPN $\mathcal{R}$ as

$$SR = (\mathcal{R} \; ||||_{L_{\text{off}}(\mathcal{R})} \; S) \tag{9}$$

Note that by taking $L_{\text{off}}(\mathcal{R})$ as the set of labels for the composition we ensure that the labels associated to transitions of SR that are non τ must belong to $L_{\text{off}}(S)$ (therefore we shall confuse here and there $L_{\text{off}}(SR)$ and $L_{\text{off}}(S)$.

4 The $\mathcal{P}$ level

The process level $\mathcal{P}$ models the software, in terms of request of services offered by the S level. It is the top level in the pile and has only immediate transitions. This level can be built "from-scratch" or with any tool for the automatic generation of nets[4, 11]. Since we want to allow the greatest flexibility at this level a request of a service is simply a transition labelled with the name of the requested service, with the requirement that each transition is assigned a single label, and that a level $\mathcal{P}$ can be composed with SR only if the service requested are a subset of the offered ones.

The $\mathcal{P}$ level for our application is depicted in Fig. 6. The labels associated with the transitions already contain the mapping information (A,B $\to$ Transputer 1, C,D $\to$ Transputer 2).

The requested services are $COMP\_i$, $i \in \{1,2\}$ (request of a computation burst mapped on Transputer i), COC_i, $i \in \{1,2\}$ (request of a communication between processes placed on CPU i), CON_ij, $i,j \in \{1,2\} : i \neq j$ (request of a communication from a process in CPU i to one in CPU j), COL_i, $i \in \{1,2\}$ (request of a communication between a process placed on Transputer i and a process in the external world), EXT_ACT (request of a generic activity by an external process).

The labels of $\mathcal{P}$ are simply the name of service, say SRV, while at level S there are pairs of label for the acquisition and release of the service (S_SRV and E_SRV), therefore the composition of the $\mathcal{P}$ level with the SR model requires a matching function α between each label of $\mathcal{P}$ (the request of a service SRV) and the pair of labels that offer that service at level SR. With our naming convention, function α is simply $\alpha(SRV) = (S\_SRV, E\_SRV)$.

To define also this composition in terms of synchronization over transitions, we need to "expand" net $\mathcal{P}$, that is to say to substitute each transition that requires service SRV into the sequence transition t' - place - transition t'', where t' and t'' are immediate transitions labelled with the first and second component of the α function. The composition can then be defined as the parallel composition with synchronization over transitions of equal label.

Unfortunately this simple procedure does not work if level S offers different implementations of the same service, (in our application this is true for the communication over the links of a given CPU), as the following problem can arise. The two processes P and Q in Fig. 7 can obtain a service from two equivalent servers SR1 and SR2 (that may have different rates associated with their

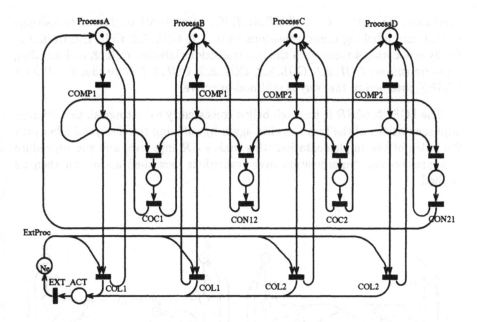

Fig. 6. GSPN model of AllProc

timed transitions). By simply superposing each start-end request transition in the processes with the two start-end service transitions in the resources we get the model in Fig. 7(a). Since the synchronization transitions t_2 and t_6 share the same input place P_2 and transitions t_4 and t_8 share the same input place P5, it might happen that process P starts a service offered by $SR1$, Q starts a service offered by $SR2$, but then P synchronizes on the end-service with SR2 while Q synchronizes on the end-service with SR1.

This undesirable exchange of servers can be avoided by setting up an appropriate composition rule that composes level $\mathcal{P}$ with a non injective $\mathcal{SR}$ model as follows: (1) make $\mathcal{SR}$ injective with an appropriate relabelling that maintains the association between pairs of transitions representing the start and end activity of the same implementation of the service; (2) replicate n times each transition of $\mathcal{P}$ that is labelled with a request for a service for which there are n implementations available in $\mathcal{SR}$; (3) expand each replica into the sequence transition - place - transition (as for the injective case); (4) label the newly created transition in a consistent way with respect to the relabelling done in $\mathcal{SR}$. Fig. 7(b) shows the solution for the net of Fig. 7(a).

In the complex node application there are two services for which we have more than one implementation: N_e implementations of EXT_ACT (activity in the fictitious external resource) and 4 implementations of COL_i, for each of the two CPUs ($i \in \{1, 2\}$) (communication over a link for a process in CPU i). The composition process would therefore modify the LGSPN of $\mathcal{P}$ by replicating N_e times the single transition labelled EXT_ACT, and labelling these transitions EXT_ACT_1, ..., $EXT\_ACT\_N_e$, 4 times each of the two transitions COL_1, and labelling these transitions COL_1_1, COL_1_2, COL_1_3, COL_1_4, finally replicating 4 times each the two transitions labelled COL_2, and labelling these transitions COL_2_1, COL_2_2, COL_2_3, COL_2_4. Note that the $\mathcal{P}$ level LGSPN obtained in this way is still non-injective.

The LGSPN of $\mathcal{SR}$ is then relabelled consistently to distinguish the different implementations of the same service, again by suffixing the start/end label with the index of the implementation: this makes $\mathcal{SR}$ injective, and the procedure of expansion and superposition over transitions described above can then be applied.

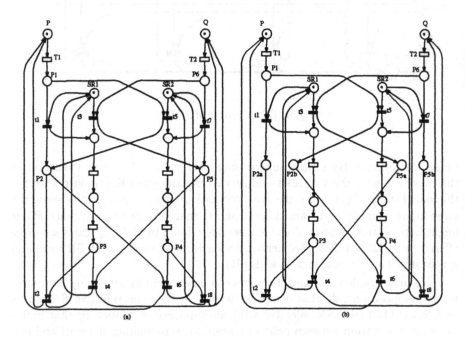

Fig. 7. Problems with non injective labels for offered services

The $\mathcal{P}$ level of our application includes already a mapping of the required services over the offered one: for example the computation in process A requires service $COMP1$, meaning that process A is mapped on CPU 1. From a methodological point of view, it is very important that the modeller can easily change the mapping: a possible solution is that the modeller should be allowed to define a process model $\mathcal{P}$ that is labelled with virtual services and the composition of $\mathcal{P}$ with $\mathcal{SR}$ should then be enriched with a mapping of the labels requested by $\mathcal{P}$ (virtual services) to the labels offered by $\mathcal{SR}$ (implemented services).

4.1 Formal definition

To compose level $\mathcal{P}$ with model $\mathcal{SR}$ we need a special type of operator that performs a sort of client-server composition, called $\diamondsuit_\alpha^L \mathrm{off}^{(S)}$, that performs the connection between a service requested and all its possible implementations at the lower level. For sake of brevity we define here only the composition that assumes an injective $\mathcal{SR}$ model, since the non injective case, already described informally, requires a long, although conceptually simple, formal definition.

The operator for the client-server composition is defined in terms of a parallel composition and of a vertical expansion. The parallel composition is performed using the superposition operator $\|\|_E$ already defined for the composition of $\mathcal{S}$ and $\mathcal{R}$: this is possible because we are assuming an injective $\mathcal{SR}$ model, and because a LGSPN is a special case of MLGSPN.

The vertical expansion of an LGSPN LG with respect to transition t and the pair of labels (l', l'') is the LGSPN $LG' = \mathrm{VEXP}_t^{(l',l'')}(LG)$ obtained by substituting transition t of LG with the sequence transition t' - place p - transition t'', with $I'(t') = I(t), O'(t') = p, H'(t') = H(t), I'(t'') = p, O'(t'') = O(t), H'(t'') = \emptyset, \lambda'(t') = l'$, and $\lambda'(t'') = l''$; the operator can be easily extended to deal with a set of transitions and a set of pairs of labels.

We now define the client server cooperation between two LGSPN, LG_1 and LG_2, for the case of injective LG_1.

Definition 5. Let LG_1 be an injective LGSPN, LG_2 a LGSPN, L_i the set of labels of LG_i, and $\alpha : L_2 \to L_1 \times L_1$. The *client-server cooperation* between LG_1 (the server) and LG_2 (the client) over the set of labels L_1, with respect to function α is the LGSPN $LG' = LG_1 \diamondsuit_\alpha^{L_1} LG_2$ defined as

$$LG_1 \|\|_{L_1} \mathrm{VEXP}_{T_2^{L_2}}^\alpha (LG_2)$$

where $T_2^{L_2}$ is the set of labelled transitions of T_2.

To obtain the $\mathcal{PSR}$ model we apply the client-server composition between $\mathcal{SR}$ (the server) and $\mathcal{P}$ (the client) over the set of offered labels of $\mathcal{S}$: this requires the definition of α, (simply $\alpha(SRV) = (S\_SRV, E\_SRV)$, with our naming convention).

The global system $\mathcal{PSR}$ is therefore:

$$(\mathcal{PSR} = \mathcal{SR} \diamondsuit_\alpha^L \mathrm{off}^{(S)} \mathcal{P}) \tag{10}$$

5 Conclusion

In this paper we have proposed a new methodology for constructing GSPN models of (parallel) systems, including both the hardware and the software. The method extends the original idea proposed in [5]. The hardware and software are described in the $\mathcal{R}$(esource) and $\mathcal{P}$(rocess) levels respectively. The $\mathcal{R}$ level describes the basic components of the hardware and the (basic) service they offer. The $\mathcal{P}$ level describes the software behaviour, independently of the mapping onto the hardware. Since the type of operations represented by the transitions in the $\mathcal{P}$ level model (usually) doesn't correspond to the simple basic operations offered by the hardware, an intermediate level $\mathcal{S}$(ervice) is introduced to describe how the complex operations required by the $\mathcal{P}$ level can be implemented by composing the basic services provided by the $\mathcal{R}$ level.

The three levels are then composed using different flavours of labelled transitions superposition operations. For this methodology to be applicable on a wide range of test cases an implementation inside a Petri net tool is indeed a must: the formal definition of the composition process provided in Sections 2.2, 3.3, and 4.1 can be seen as a specification of such a task.

To augment the usability of the methodology we plan to study the "confusion problem" and an extension to high level nets (specifically to Stochastic Well Formed Nets[9]).

Indeed additional rules should be defined to deal with the composition of priority levels and of weights of immediate transitions explicitly, to avoid the possibility that the final model be "confused", a problem that is often encountered when superposing subnets over immediate transitions, and whose solution is a careful definition of priorities [1]. Up to now the check for confusion is possible only on the final model, but it may be possible to define rules on the single components, or at least guidelines, to limit the problem. Also the assignment of weights to immediate transitions, that in GSPN requires the computation of extended conflict sets [1], can, for the time being, be done only on the global $\mathcal{PSR}$ model.

The extension to high level models can help in many ways, for example (1) make the submodels representing each level simpler (e.g., when several resources or processes of the same type are available only one (colored) subnet could be included, representing their common behavior), (2) use colors to distinguish processes asking the same type of service to solve the problems arising when multiple servers (resources) offer the same service, (3) use colors to represent mapping information (thus allowing a definition of $\mathcal{P}$ independent of the mapping), (4) use colors to allow "parametric services", for example to associate with a service request at level $\mathcal{P}$ a "scaling factor" that may modify the duration of the activities implementing the required service.

Acknowledgements. We would like to thank the anonymous referees for the many valuable suggestions and Oliver Botti of ENEL-CRA for the numerous discussions on the "complex node" application and on the PR definition.

References

1. M. Ajmone Marsan, G. Balbo, G. Conte, S. Donatelli, and G. Franceschinis. *Modelling with Generalized Stochastic Petri Nets.* J. Wiley, 1995.

2. M. Bernardo, L. Donatiello, and R. Gorrieri. Integrated analysis of concurrent distributed systems using Markovian Process Algebra. In *Proc. Seventh International Conference on Formal Description Techniques*, Berna,Switzerland, October 1994.

3. E. Best, R. Devillers, and J. Hall. The Petri box calculus: a new causal algebra with multilabel communication. In G. Rozenberg, editor, *Advances in Petri Nets*, volume 609 of *LNCS*, pages 21–69. Springer Verlag, 1992.

4. E. Best, H. Fleischack, W. Fraczak, R.P. Hopkins, H. Klaudel, and E. Pelz. A class of composable high level Petri nets with an application to the semantics of $B(PN)^2$. In *Proc. of 16th International Conference on Application and Theory of Petri Nets*, Turin, Italy, June 1995. Springer Verlag. LNCS 935.

5. O. Botti and F. De Cindio. Process and resource boxes: an integrated pn performance model for applications and architectures. In *IEEE Proc. of the Int. Conf. on Systems, Man and Cybernetics*, Le Toquet, France, 1993.

6. O. Botti, S. Donatelli, and G. Franceschinis. Assessing the performance of multiprocessor architectures through swn models simulation: a case study in the field of plant automation systems. Submitted to the 29th Annual Simulation Symposium (SCS), New Orleans, Louisiana, U.S.A april 1996, 1995.

7. P. Buchholz. Hierarchies in colored GSPNs. In *Proc. 14th Intern. Conference on Application and Theory of Petri Nets*, volume 691 of *LNCS*, Chicago, Illinois, June 1993. Springer Verlag.

8. P. Buchholz. Markovian process algebra: Composition and equivalence. In U. Herzog and M. Rettelbach, editors, *Proc. 2nd Workshop on Process Algebra and Performance Modelling*, Erlangen, 1994.

9. G. Chiola, C. Dutheillet, G. Franceschinis, and S. Haddad. Stochastic well-formed coloured nets for symmetric modelling applications. *IEEE Transactions on Computers*, 42(11), November 1993.

10. F. De Cindio and O. Botti. Comparing Occam2 program placements by a GSPN model. In *Proc. 4th Intern. Workshop on Petri Nets and Performance Models*, pages 216–221, Melbourne, Australia, December 1991. IEEE-CS Press.

11. S. Donatelli, G. Franceschinis, M. Ribaudo, and S. Russo. Use of GSPNs for concurrent software validation in EPOCA. *Information and Software Technology*, 36(7):443–448, 1994.

12. A. Ferscha. Modelling mappings of parallel computations onto parallel architectures with PRM-net model. In *Proc. IFIP-WG 10.3 Working Conference on Decentralized Systems*, Lyon, December 1989.

13. A. Ferscha. A Petri net approach for performance oriented parallel program design. *Journal of Parallel and Distributed Computing*, 15(3):188–206, July 1992. Special Issue on Petri Net Modelling of Parallel Computers.

14. Stephen Gilmore and Jane Hillston. The PEPA workbench: A tool to support a Process Algebra based approach to performance modelling. In *Proc. Seventh International Conference on Modelling Techniques and Tools for Computer Performance Evaluation*, Vienna, 1994.

15. N. Gotz, U. Herzog, and M. Rettelbach. Multiprocessor and distributed system design: The integration of functional specification and performance analysis using stochastic process algebra. In *Tutorial Proc. Performance 1993*, volume 729 of *LNCS*, Rome, 1994.

16. Jane Hillston. The nature of synchronization. In U. Herzog and M. Rettelbach, editors, *Proc. 2nd Workshop on Process Algebra and Performance Modelling*, Erlangen, 1994.

17. Jane Hillston. Compositional Markovian modelling using a process algebra. In *Proc. 2nd International Workshop on the Numerical Solution of Markov Chains*, Raleigh, North Carolina, January 1995.
18. J.L. Villaroel, J. Martinez, and M. Silva. Graman: A graphic system for manufacturing system design. In *Proc. of the IMACS Symposium on System Modelling and Simulation*, Cetraro, Italy, 18-21 September 1988, 1989. North Holland - Elsevier.

Designing and Verifying a Communications Gateway Using Coloured Petri Nets and Design/CPN™

D. J. Floreani †‡, J. Billington †, A. Dadej †,

† *Telecommunications System Engineering Centre (TSEC),*
Institute for Telecommunications Research (ITR),
University of South Australia (U–SA).

‡ *Communications Division,*
Defence Science and Technology Organisation (DSTO),
Adelaide, Australia.

Abstract: A gateway between a packet radio network and B–ISDN is being designed as part of a larger project that aims to bring modern telecommunications services to the Australian Defence Force. The modelling procedure employs Coloured Petri Nets to investigate the gateway architecture and behaviour prior to implementation. Part of the modelling involves the specification of the gateway call control using Coloured Petri Nets and the Design/CPN™ tool. The specification is then checked for correctness by simulation and observation of the Occurrence Graph generated by the Design/CPN™ tool. The form of the refined specification is discussed and future verification tests using the PROTEAN tool outlined

Key words: Coloured Petri Nets, Gateway Architecture, Protocol Modelling.

1   Introduction

There are precious few papers in the current literature on the design of gateways using Petri Nets and Formal Techniques [1][2]. In particular, the authors are unaware of any other attempts to model and design gateways between Radio Networks and Broadband Integrated Services Digital Networks (B–ISDN). This paper describes our attempt to specify and design a gateway between a narrowband packet radio network and a B–ISDN using Coloured Petri Nets and Design/CPN™ .

The Australian Department of Defence has chosen Asynchronous Transfer Mode (ATM)[3] technology as the target communications architecture for its command and control in the 21st century[4]. This will align defence communications with the evolving standards for the commercial "information super–highways" of the future. The use of the commercial communications infrastructure and equipment is a major thrust of planning to provide communications to the soldier in the field. The Tactical Packet Radio Network (TPRN) [5][6] is just one of the proposed solutions, and is a leading theme of a research program for both the Defence Science and Technology Organisation in Adelaide (DSTO) and the Defence Communications Research Centre (DCRC) at the University of South Australia. A TPRN is a group of mobile radio nodes with, on average, one user terminal per node. These nodes pass messages in a store and forward manner between themselves, allowing information to propagate across the network. A more complete description appears in [15]. It is based on the utilisation of modern adaptive high frequency (HF) modems to form a fully distributed packet radio network for the provision of versatile communication services. The desire to utilise B–ISDN infrastructure as a backbone between TPRNs, leads to the requirement of gateways to be present in the TPRN to allow inter–working between the two different net-working environments (radio vs optical fibre). ATM is media independent as it is a method of sharing access to underlying media.

The gateway between a TPRN and a B–ISDN will have to act as a normal node within the TPRN as well as carry out its gateway responsibilities. These responsibilities in-volve the following tasks:

- **service conversion**, an example of this is the conversion of standard 64kb/sec digital voice encoding to a low bit rate voice coding that only requires 2.4kb/sec [7].

- **protocol conversion**, the gateway will access the B–ISDN with the Internation-al Telecommunications Union (ITU–T) Q.2931 access protocol [8], and will need to convert the call control information into the TPRN's Packet Radio Signalling protocol (PRS) and visa versa.

- **handling intra–net and inter–net calls**, the gateway must handle two types of calls: calls between a TPRN user and a B–ISDN user, and calls between TPRN users in different subnets (via the broadband backbone).

- **handling node mobility**, the gateway must be able to monitor node locations in order to update its own node location tables. This information is then distributed to other gateways via the B–ISDN.

- **providing supplementary services**, the gateway will provide its own supplementary services to the user, as well as provide access to the commercial services, such as call waiting and call redirection.

- **media independence of a gateway**, the gateway must be capable of utilising any available connection to the B–ISDN in the case of bearer faults (increasing survivability). This means that a gateway must be able to interface to copper, satellite, radio and fibre links whenever necessary.

2 Methodology

This paper will detail the procedures involved in specifying a communications gateway using Formal Description Techniques, and then lead onto the refinement of the speci-fication. These techniques have previously been devised for protocol design by many authors and will just be referenced here [9][10][11]. In our case, however, the process is slightly different as we are designing a gateway and must take into account that a sys-tem is being designed instead of a protocol. The methodology is based on [9] and is sum-marised here.

1 The user requirements of the gateway must be defined and documented. These requirements are based on the features available in each of the networks that the gateway must interconnect. The list above is a brief form of our user requirements.

2 A protocol architecture of the gateway must be devised, based on the de-signer's experience in communication protocol architectures Figure 1. At this stage the type of gateway must be chosen, and hence the location of the mapping routines within the protocol stack of the gateway. Standard protocol implementa-tions may also be chosen to serve most purposes within the protocol stacks, how-ever a few may need to be adapted to suit particular needs (the radio environment in our case). A detailed description of the gateway architecture is given in [15].

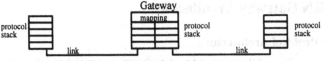

Fig. 1. Protocol Architecture

3   A service specification is then created by defining the network services to be provided by the gateway. The services that are required are dependent on the layer at which the mapping occurs within the gateway and whether all services are mapped between networks, or just a subset. At this stage, only the interfaces to the gateway are modelled and not the internal structure of the gateway (Figure 2). The specification procedure is carried out using a specification language that describes the interaction of the interfaces with the gateway as a whole. The model of the service specification is then validated against its definition, normally text based.

Fig. 2. Service Specification

4   State machines must then be designed and specified to carry out the internal mapping of the services between the interfaces to the gateway, Figure 3. The interfaces, shown in Figure 2, may also be refined with the addition of Protocol Entity state machines. Both of these tasks are carried out using a specification language that describes their behaviour. This can be seen as a refinement of the service specification.

Fig. 3. Refinement of Service Specification

5   The refined specification is then verified for correct behaviour with reference to the service specification, before implementation. In our case, the verification techniques include the simulation of the specification, reachability analysis, and language comparison. Verifying the gateway implementation has two main benefits: discovering and correcting protocol specification errors, and allowing a better understanding of the operation of the gateway before the next stage of development.

6   The final stage is to implement the gateway in software and hardware, and then validate the implementation. This last step is not discussed in this paper.

The paper is arranged along the same lines as the methodology stated above, with the gateway architecture design being briefly covered in section 3, the service specification covered in section 4, and the refinement of the specification outlined in section 5. The verification techniques are discussed in section 6. Coloured Petri Nets (CPN) [12] will be used as the specification language and Design/CPN™ as the simulation tool for editing and verifying the models. The PROTEAN tool [26] will also be used for language analysis. The design work will ultimately result in a verified specification for the call control architecture and protocols used within a TPRN gateway.

3 The TPRN Gateway Architecture

3.1 Gateway Protocol Architecture

A basic understanding of the gateway architecture is essential to present a full picture of the modelling process, and to understand the design decisions made during the creation of the gateway service specification. The following papers present methodologies for designing communication gateway architectures, including the use of formal design methods: [16][17][18].

The gateway has been designed to act as a standard TPRN node with user access to the narrowband radio network, and to contain the Networking and Radio Protocol entities necessary for distributed radio communications. A standard B–ISDN User–Network Interface ((C) Figure 4) for access to the broadband network must also be present. The user access protocol used within the TPRN is called Q.93R ((B) Figure 4), and is planned to be a reduced version of the B–ISDN user access protocol entity Q.2931 [8]. The TPRN Protocol Entity ((A) Figure 4) is a network protocol used for call control throughout the TPRN. We call this the Packet Radio Signalling Protocol (PRS).

The role of the Call Control Application (CCA) is to handle the transfer of call control intent between the TPRN and BISDN. This is achieved by the occurrence of service primitives. Service primitives define information to be transferred across a boundary or interface. The occurrence of service primitives could be realised in many ways such as a function call, message passing, or the pushing of a button. The action of the underlying Access Protocol Entities is to gain a connection within their corresponding network. Thus each protocol entity is providing a service, ie connectivity to another user, to the CCA layer above.

We used the stratified model [14] to represent the gateway and also proposed extensions to this model to cope with the extra complexity due to the inclusion of complex radio protocols [15]. Only the control plane (or signalling part) of the gateway architecture is shown below. Interested readers are advised to consult [15] for a more complete description of the gateway architecture.

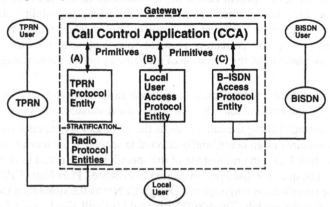

Fig. 4. Partial Gateway Protocol Stack

The CCA receives primitives from the underlying Protocol Entities of one network, and then issues the primitives to request the corresponding service from the other network's service provider. This corresponds to a mapping of service primitives and not protocols.

In many cases of inter–working, the services provided by one network are not compatible with the other network. In this case there are two approaches to implementing the service mapping within the CCA. The least common denominator approach only allows the smaller set of services that are compatible on each side to be transferred through the CCA. The other approach is to build a sub–service on top of the weaker service to provide a compatible service to the other network. In our case the services provided by the call control part of each network are similar. This is due to the design of the TPRN Network and User Access protocols being based on Q.2931 [8].

4 The Service Specification

4.1 Specification Techniques and Tools

The specification of protocols and process behaviour can be carried out using many different languages: Specification and Description Language (SDL) [13], LOTOS [19] and Coloured Petri Nets [12] just to name a few. The behaviour of the gateway was originally specified using SDL to model the process hierarchy and process behaviour. This method was satisfactory until it came to testing the specification. The SDL package used did not have any verification tools associated with it. The package contained a set of tools for converting graphical SDL to C code, which must then be compiled. This would have caused long cycle times for protocol development as specification testing must be carried out at the software implementation level. After looking around, the use of the Design/CPN^{TM} Coloured Petri Net tool became more appealing for two main reasons; verification tools existed with fairly standardised verification techniques, as well as the ability to simulate the specification with instant graphical feedback of the results. Coloured Petri Nets are also flexible and abstract enough to be able to describe messages, primitives, and function calls in the same manner, releasing the creator of the specification from the implementation issues.

Due to the flexibility of CPNs, it is up to the user to form a structure to suit the specification purpose. This flexibility includes the ability to draw places and transitions in any shape and layout that suits. This is different from SDL where the symbols have predefined shapes, and the top to bottom flow of control is standard. SDL requires that inter–process communication be via FIFO queues, whereas Petri Nets allow inter–process communication mechanisms to be specified, increasing the designer's choices. CPNs do, however, have construction rules, but they are simple, easy to understand, and very powerful [12]. Design/CPN™ uses Standard Meta Language (Standard ML) [25] for its inscriptions as well as code segments. This language has a simple syntax, which can still describe complex data structures and functions.

4.2 Specifying the Service Primitives

The first step towards designing the service mapping gateway is defining the service primitives at the interfaces to the CCA. In our case we have 3 Protocol Entities (Figure 4), and a Service Handler (Figure 5a). The Service Handler is a process that manages the specific service transfer through the gateway, here service means voice, data, messaging, etc.

The TPRN and Local User Access protocol entities' service primitives were originally based on the primitives defined in the ISDN access protocol Q.931 [21], and modified to suit the new Q.2931 protocol [8]. The old CCITT specifications [21] define primitives that relate closely to the messages actually sent between protocol entities. This is not necessary. The set of primitives may be reduced to a set that closely resembles those of the Open Systems Interconnection (OSI) Network Layer service [22]. Now the A, B, and C interfaces (Figure5a) have identical service primitives. The service handler primitives were designed to allow the CCA to initialise, re–configure, and terminate the service handler. The service primitives that are defined for the interfaces to the CCA are listed in Figure 6.

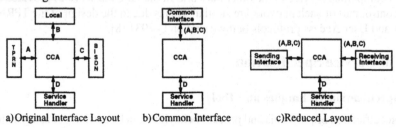

a) Original Interface Layout   b) Common Interface   c) Reduced Layout

Fig. 5. CCA Interfaces

The CCA has 4 interfaces including the Service Handler as shown in Figure 5a. However, as we have aligned the primitives, it is possible to describe each call control entity with the same interface model, see Figure 5b. To simplify the model even further we have separated this common interface model into primitives associated with call origination and call accepting. These interfaces are then called the Sending Interface and Receiving Interface respectively in Figure 5c. This modelling simplification does not result in a loss of detail as the sending and receiving processes are independent. Note that the terminology can become confusing at this stage, as the gateway's sending interface is actually the receiver of the call between the sending user and the gateway.

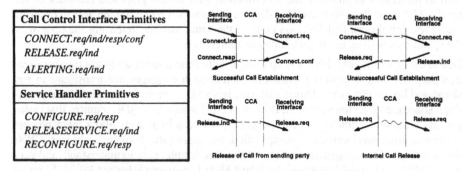

Fig. 6. List of Primitives and some indicative time sequence diagrams at the sending and receiving interfaces to the CCA

Time sequence diagrams are then created to illustrate the primitive interactions at the CCA interfaces. Figure 6 shows some of the CCA's interactions with the sending and receiving interfaces. The Service Handler Interface is not shown here, but equivalent diagrams can be made for this interface. These diagrams are used for gaining an understanding of the CCA and the primitive nomenclature before the modelling process is undertaken in a more detailed manner.

4.3 Creating the CCA Specification using Coloured Petri Nets

The first task when specifying the behaviour of the CCA is to create specifications of the interfaces to it. In this modelling the interfaces modelled will be based in the simplified interface model shown in Figure 5b above.

An interface is a point where interaction occurs between two regions or systems. In our case the interface is a point in which we can define the service primitives and the orders in which they occur. To model this (Figure 7), we have separate Petri Net places to represent the state of each interface, and use transitions to model the possible primitives. The occurrence of a primitive is normally associated with the transfer of information. At this stage we did not model this information, it is included in the next stage of the specification.

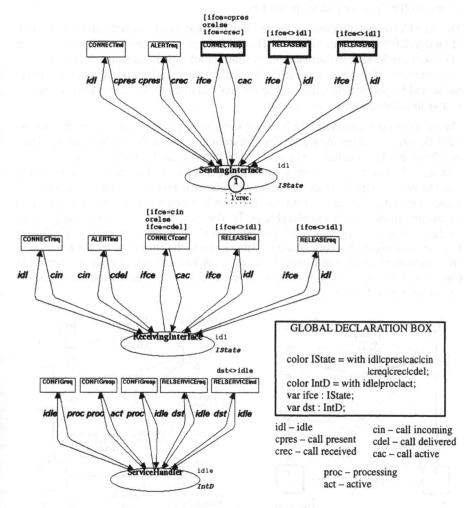

Fig. 7. Sending Interface Specification

The *IState* (Interface State) colour set represents the possible states that the Sending and Receiving interfaces may be in, and is defined in the global declaration box (Figure 7). The *IState* colour set is defined as an enumerated type. The *IntD* (Interface D) colour set represents the possible states the Service Handler interface may be in, and again is an enumerated type. The Service Handler is different from the call control interfaces, and hence has its own colour set. The *var* statement in the global declaration box defines *ifce* as a variable of the *IState* colour set.

The places representing the interfaces initially contain the *idl* or *idle* token to represent that they are in the idle state. This is represented by the text on the top right of each place. Every place must have an associated colour set (eg *IntD*), this is inscribed on the lower right hand side of each place. Next to some transitions are conditions enclosed by square brackets (eg *[ifce<>idl]*). These are called guards, and allow extra conditions to be placed on the enabling of the transition. In our case we are putting restrictions on which values of the variable *ifce* can enable the associated transition. The <> symbol represents the "is not equal to" predicate.

The Petri Net model representing the *Sending Interface* is shown in the simulation mode of Design/CPN (see top of Figure 7). The bolded transitions are those that are enabled. The token can be seen in the place and the current token value is shown below (*1'crec*). The behaviour of the interfaces can be checked by running the models in the simulation mode, and by performing a full state space analysis of each model checking the behaviour against the designer's expectations.

The behavioural properties of the interface specifications are checked using the Design/CPN Occurrence Graph Analyser. The occurrence graph of the sending and receiving interfaces are shown below in Figure 8. Node 1 represents the initial state in both cases with the *idl* token being present in the Sending and Receiving Interfaces respectively. The Occurrence Graph Analyser has an 8 character maximum for transition and page names, hence the shorter inscriptions. The arcs between nodes represent the firing of a transition in the Petri Net model (Figure 7). The current state of the Sending Interface model at the top of Figure 7 is represented by node 4 in the occurrence graph (Figure 8a). We have highlighted one of the possible paths to the current state. Note how there are 3 transitions enabled in Figure 7 and there are two arcs out of node 4 (Figure 8a). One arc has two inscriptions representing two transitions, and hence the total is the expected three possibilities.

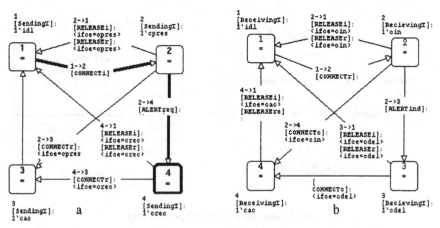

Fig. 8. Sending and Receiving Interface Occurrence Graphs

The Occurrence Graphs and the Petri Net models both contain the same information about the behaviour of the interface specification, however the graphs portray more information to the reader at first glance. However as the number of states in a model increases the Occurrence Graphs become more confusing and the Petri Net model normally conveys more information at first glance.

The Design/CPN tool also supports the use of hierarchical Coloured Petri Nets [12]. This allows fusion places and substitution transitions. Fusion places are useful for depicting the same place on different pages of the Petri Net. Substitution Transitions can reduce the complexity of petri nets by allowing one transition to represent complex behaviour that is modelled on another page. In Design/CPN a fusion place is identified by a box with *FG* next to the place, and a substitution transition has a box with *HS* next to it, Figure 9. Substitution transitions require the allocation of ports, which mark which places are common between the super page and the sub page. These port places are marked with a box with a *P*, see Figure 10.

Once the Interface Specifications are complete, the next task is to create a Coloured Petri Net to model the gateway specification. The nets we created to represent the gateway specification take advantage of both substitution transitions and fusion places to enhance the readability of the model. The Super page of the complete gateway specification (Figure 9) shows the 3 interfaces and the CCA connected with substitution transitions. There is also an internal page (*clean*) for handling housekeeping during call termination. The rounded boxes are places and the complete diagram is arranged to conform to Figure 5b.

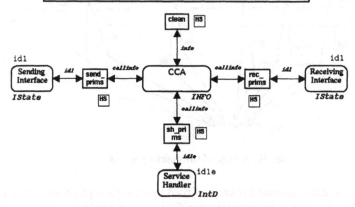

```
GLOBAL DECLARATION BOX
color IState = with idl|cpres|creq|crec|cin|cdel|cac
color IntD = with idle|proc|act;
color SR = with s|r;
color INFO = union callrelease:SR + callinfo + callaccept +
callreject + callresp + reject + alert + servrelease + clean;
color COUNT = int with 0..5;
color GLOB_INT_STATE = with init|refuse|allow;
color COMP = product COUNT*GLOB_INT_STATE;
var ifce:IState;
var dst:IntD;
var info:INFO;
var x:COUNT;
var m:GLOBAL_INT_STATE;
```

Fig. 9. Gateway Specification Super Page

A global declaration box is present to describe the data types. The *IState* colour describes the call control interface states, the *IntD* colour describes the Service Handler interface states. The *INFO* colour contains the types of information passed to the *CCA* on the occurrence of a primitive. It is defined as a union type, where the elements are identifiers associated with different colour sets. A union type is more flexible than an enumerated type, and was chosen to ease future additions to the service specification due to the addition of new services. In our case we only use one non trivial colour set, *SR*, which is used to determine whether the Sending (s) or Receiving (r) interface originated the information. The *COUNT* colour set is defined as an integer from 0 to 5, and is used to count the number of pieces of information present within the CCA. This is used to assist in removing this information when the call is terminated (*clean* transition in Figure 9). The *GLOB_INT_STATE* colour set is used to change the behaviour of the *CCA* on the receipt of certain primitives. This behaviour is classified into three states: initial state (*init*), normal operation (*allow*), and the releasing process (*refuse*) in which certain primitives are not allowed or have different emphasis. For example a release primitive that occurs after others have occurred is redundant and may be ignored. Note that interface specifications (Figure 7) do not allow the inhibiting of these primitives and hence they must be handled by the *CCA*. The *COMP* colour set is a product of the *COUNT* and *GLOB_INT_STATE* sets. Below the declarations in the global declaration box, the variables used within the model are defined.

The *send_prims* substitution transition in Figure 9 is represented in full in Figure 10. Here you can see that when the *CONNECTind* primitive occurs, the Sending Interface changes state from *idl* to *cpres*, and the *CCA* is passed a token that represents the information required to set up the call.

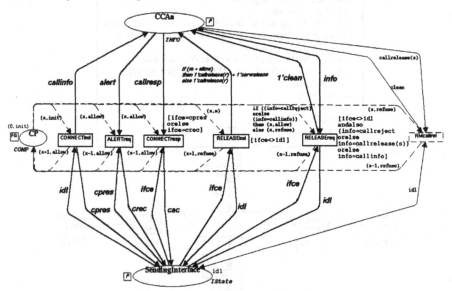

Fig. 10. A page of the gateway specification

It can be seen that whenever a token is placed into the *CCA* place, the first element of the *COMP* tuple (x,m), in the counter place *CP* is incremented $(x+1)$. Conversely, when a token is removed the value is decremented $(x-1)$. The second element of the *COMP* tuple (x,m) is modified when the behaviour of the *CCA* is to change, namely after *CONNECT* or *RELEASE* transitions. The *CP* place initially has a $(0,init)$ token in it. The *init* element is never replaced after its initial consumption, and this forces the model to only simulate one call entity. The *RMcallrel* transition is used as a clearing function to remove extra *callrelease()* information tokens, and is only enabled once another *RELEA-SEind* or *RELEASESERVICE* primitive has been issued.

The Occurrence Graph Analyser is used to obtain occurrence graph of the Specification. Initially it is used to test for deadlocks and to see whether the gateway ever established a call. Once this basic behaviour is shown, then the behaviour can be analysed more closely to refine the call establishment and clearing behaviour. The removal of unwanted tokens during call clearing was the most challenging task of the modelling process and required the creation of the counter place (*CP*). The Occurrence Graph at this stage still contains the states and arcs related to the *clear* substitution transition. It contains 365 nodes and 860 arcs. As we are interested in the interaction of the interfaces with the CCA, we don't want added complexity of the transitions represented by the *clear* substitution transition as well. It is here that we now turn to the PROTEAN [26] tool to reduce the language of the specification to its essential elements, namely the interface primitives. Here the word language means the set of all possible sequences of primitives. To achieve this, code was written to extract this information from the Occurrence Graph Analyser. The text output from this step is then massaged into a form that is acceptable to the PROTEAN tool. The new, reduced language now contained 17 states which is much more manageable and is shown in Figure 11 below. This graph can be plotted by the tool or saved in text form for later analysis. Code has also been written to reorganise the original occurrence graph in accordance to the reduced one, allowing the drawing tools of the Occurrence Graph Analyser to be used. This language definition is the behavioural specification of the gateway that we will use to test further refinements against.

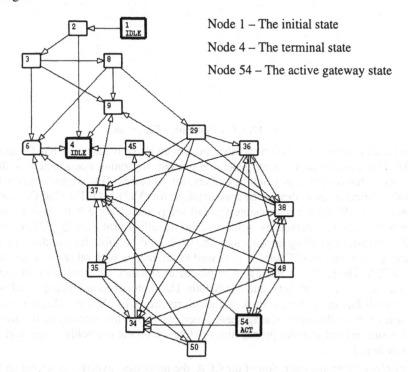

Node 1 – The initial state

Node 4 – The terminal state

Node 54 – The active gateway state

Fig. 11. Occurrence Graph of Gateway Specification

The first attempt to model the CCA specification allowed the occurrence of a primitive at one interface to instantly effect other interfaces. While this specification behaved correctly with a state space of 102 nodes and a reduced state space of 33 nodes, it was later realised that it was physically impossible to refine or implement this specification. This event reinforces the importance of using formal methods to validate designs and specifications for complex systems.

5 Refinement of the Service Specification

5.1 Call Control Application Functional Architecture

Once the service specification is completed it is possible to consider refinement of the specification. This involves proposing an architecture that will provide the functionality of the specification. It is also possible to add in extra functionality not mentioned in the specification and check that the refinement still complies with the specification. An example of this is the addition of supplementary services and checking if it interferes with the basic call setup and clearing functions.

The proposed functional architecture of the Call Control Application (CCA) is shown in Figure 12. The CCA, described in section 4, is now segmented into 3 functional blocks, comprising the sending, receiving and gateway Call Control Applications (SCCA,RCCA,GCCA).

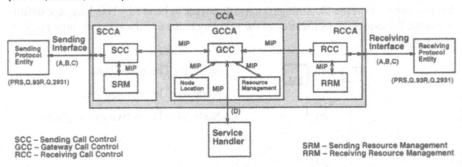

Fig. 12. CCA Functional Diagrams

These Call Control Applications are then broken up into processes that handle different tasks. This classification is very general at present, separating management of the resources within the gateway from the call control processes. This is similar to the functional architecture described in the specifications for Q.931 [21]. The Call Control processes (SCC, RCC) are responsible for call setup and clearing with their respective interfaces. This is achieved by utilising their signalling protocols Q.2931, Q.93R or PRS to transfer signalling information. The GCC is responsible for handling the interworking between the SCC and RCC as well as interacting with other processes within the GCCA. The Resource Management processes simulate the monitoring of computing resources, call identifiers, and bandwidth. These processes are then queried during call establishment to check if there is enough resources for the call. This is especially complex for the Resource Management processes within the gateway as it must cope with issues related to radio propagation. (Note we do not explicitly model this complexity here)

To briefly explain the operation of the CCA, the processes involved in setting up a call through the CCA will be explained. Call Setup messages from the call initiator are received by the Sending Protocol Entity (either PRS,Q.93R or Q.2931 see Figures 4 and 12). This entity then passes primitives to its call control (SCC) located within the Sending Call Control Application. If the call is accepted (depending on interaction with the SRM), the SCC process requests a gateway connection from the Gateway Call Control process (GCC) via an interface. The GCC then initiates the Node Location process and the Gateway Resource Management processes to look up routing information and evaluate the available gateway resources respectively. If the destination address can be found and there are enough resources available within the gateway then the GCC passes on the call setup information to the RCC for transfer to the end user. If the end user accepts the call, a connect message is passed back through its Network, the gateway receiving protocol entity and RCC to the GCC. The GCC then initiates the Service Handler (SH) to handle the setup, negotiation of mapping parameters, and the actual user service mapping within the gateway. The connect message is then passed back via the SCC to the calling user, and the user service, such as a file transfer, can commence. The protocol for passing information within the gateway is called the Media Independent Protocol (MIP) and is being developed in conjunction with the gateway design.

Additions to the GLOBAL DECLARTION BOX

```
color INTERFACES = with glslrlp;
color ADDRESS = product INTERFACES * INTERFACES;
color RESPONSE = with succlfaillredir;
color REPLY = with oklnok;
color RESULT = product INTERFACES * INTERFACES * RESPONSE;
color IOcnt = int with 0..10;
/*Process States*/
color Process = with idlelproclactlhalt;
/*Call Control States*/
color GST = with G0lG1lG2lG3lG3alG4lG4alG5lG5alG6lG7lG8lG10lG11lG12lG14;
color SST = with S0lS1lS2lS3lS4lS5lS6lS7lS10lS11lS12lS14lS15;
color RST = with R0lR1lR2lR3lR4lR5lR6lR7lR8lR10lR11lR12;
/*Message Declarations */
color MIP = union CallSetup:ADDRESS + CallConnect:ADDRESS + ReleaseCall:ADDRESS + Relea-
seCallComplete:ADDRESS + Redirection:ADDRESS + Alert:ADDRESS + AddCheckReq:ADDRESS
+ AddCheckResp:RESULT + ResQueryReq:ADDRESS + ResQueryResp:RESULT + ResReleaseR-
eq:ADDRESS + Configure:ADDRESS + ConfigureResp:RESULT + ServRelease:ADDRESS + Stop-
Serv:ADDRESS + RQreq + RQresp:REPLY + NL:REPLY + NLS:REPLY + CR:REPLY;

var st : Process;
var mip : MIP;
var Gx : GST;
var Sx,sst : SST;
var Rx,rst :RST;
var x:IOcnt;
```

Fig. 13. Global Declaration Box

At this stage of the design the interfaces between the CCAs have not been specified, and it is now that we apply the same formal approach to the refined specification as was used for the specification. The CCAs pass MIP messages between each other. The message structure was defined using ASN.1 [24], and time sequence diagrams were created to describe the purpose of each message. The MIP messages are defined by the MIP colour set in the global declaration box, Figure 13. These messages have fields associated with them, in the same way a real protocol message contains information fields. Most messages have an address field to identify the source and destination process of the message (ie *CallSetup:ADD*). Other messages contain a result field to indicate the success or failure of certain events (ie *ResQueryResp:RESULT*).

The super page for the refined specification is shown below in Figure 14. Again every effort has been made to align the petri net model with the functional diagram. The current work models the interactions of the MIP, ie the substitution transitions labelled with the MIP prefix, leaving the sending, receiving and service handler interfaces alone. The result will provide a protocol that will meet the specification that is defined by the behaviour of the interfaces to the CCA.

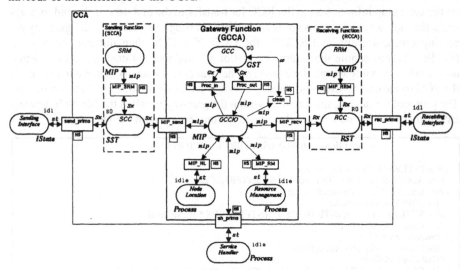

Fig. 14. The refined specification super page

The *GCCIO* place does not have an equivalent in the functional diagram. It represents the communication mechanism within the gateway. At present it is modelled as a place with an unlimited capacity that contains MIP messages. There is no queuing, any MIP message present may be consumed by its relevant process at any time. This could be replaced by FIFO queues to each process, but this is an implementation issue to be made at a later refinement stage. As in the specification in Figure 9, there is a *clean* substitution transition for cleaning up the remaining messages in the *GCCIO* during call clearing. Definitions for the inscriptions appear in the global declarations box in Figure 12. These are additional definitions, the colour sets and variables defined in the earlier specification still hold, and are part of this specification. There is one change however, the Service Handler interface is now covered by the *Process* colour set as it is identical to the *IntD* colour set defined in the specification.

The SCC, RCC, and GCC hold state information with the possible states defined in the colour sets SST, RST, and GST respectively in Figure 13. Each State place is initially in its idle state (S0,R0,G0), as shown in the super page.

The *send_prims* substitution transition of Figure 14, models the interaction between the sending interface and the Sending Call Control (SCC) process. This page is shown in Figure 15 and it can be seen that the Interface specification part of the model (see top of Figure 7) is unchanged. The additional conditions imposed by its interaction with the SCC may change the behaviour of the interface. It is the purpose of formal analysis to discover this and modify the behaviour of the SCC process so as to meet the original specification. The place named *one* is present to force only one instance of a call being originated at any time, this is because we have decided to analyse the properties of one call at a time, to reduce complexity.

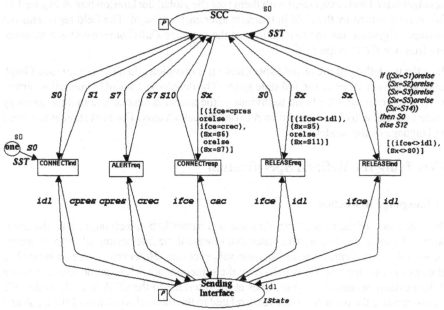

Fig. 15. The send_prims substitution page of the refined gateway specification

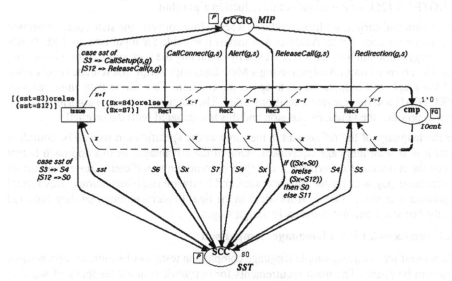

Fig. 16. The MIP_send substitution page of the refined gateway specification

The *MIP_send* page, Figure 16, is responsible for the sending and receipt of MIP messages from the *GCCIO* place. The message sent depends on the state of the *SCC*, this is modelled by the case statement on the outgoing arcs of the *Issue* transition. The receipt of a message forces the *SCC* to enter a new state depending on the incoming message. This is modelled by the *Rec* messages. The *cmp* place is used to count MIP messages in the *GCCIO* place, this information is used when call clearing occurs. The MIP messages have fields associated with them (see the global declaration box of Figure 13). This is represented by the field in brackets after each message. The field represents the message originator and the destination, so the message *CallConnect(g,s)* is a message sent from the *GCC* to the *SCC*.

Once the modelling of the refined specification is complete, then the Occurrence Graph Analyser is used to create the full state space. The deadlocks caused by modelling errors are then removed and the basic behaviour of the model is checked, ie that the gateway reaches the call active state, and then releases the call cleanly. The next stage is to record the language of the model.

6 Verifying the Refined Specification

6.1 Language Reduction

The state space of the refined specification will naturally be much larger than the original specification due to the extra states that represent the interactions of the MIP protocol with the call control entities and the subprocesses. However we are interested in whether the two specifications behave in the same manner. This can be tested in terms of the possible sequences of primitives at the interface to the CCA in each model. We need to remove the transitions or events related to the internal workings of the CCA and leave the skeleton of the interface primitives. To do this, we use a standard automata reduction algorithm [27]. Telstra Research Laboratories developed a package called PROTEAN [26], which includes this reduction algorithm.

As mentioned earlier, we have produced a script that converts the state space generated by the Occurrence Graph Analyser and makes it suitable for input into the PROTEAN language reduction tool. This is achieved by extracting the state space information from the Occurrence Graph Analyser using a Meta Language script that searches every node of the graph and outputs the state space information. This file is then imported into PROTEAN and the relevant transitions listed. The language reduction algorithm is then run to produce the language skeleton and this is compared with the original.

If the language of the refinement is the same as the specification we can be confident that it is at least language equivalent. One of the advantages of this approach is that once the refinement is verified, it can then become the specification for either further refinement stages or for the final software and hardware implementations. This formal approach guarantees that if errors occur in the final implementation that they occurred in the last stage and not at the early design stages.

6.2 Results so far from Language Comparison

At present we are using simple language comparison tests, until a more comprehensive test can be found. The main requirements for our work is quick feedback of whether the languages are the same, and if not where the language graph deviates from the specification. To achieve this quick feedback a file with a list of each node and its outgoing arcs is created at the same time as the information for plotting the graph is created. The ascii total of the arc names is also given at the end of each line, see the list for the specification in Figure 17 below. The node numbers in the specification and refinement are not the same, and in order to identify the equivalence of the nodes we compare ascii totals. This is made simpler by rearranging the list in order of increasing ascii totals. Nodes that do not meet the spec are then found. The erroneous behaviour is then discovered by studying the text and graphical forms of the occurrence graph in conjunction with the petri net of the refinement. Note that the transitions names in the table below are different to those in the primitive listing (Figure 6). This is due to the Occurrence Graph Analyser requiring 8 character names, and so different names are actually used to represent the primitives in this stage. The primitive names identify the interface it which they occur to make it easier for debugging. Hence SENDRELi is the RELEASEind primitive at the sending interface, and RECRELre is the RELEASEreq primitive at the receiving interface.

```
Node 1:CONind = 539
Node 2:SENDRELi SENDRELr CONFIGre = 1922
Node 3:SENDRELi CFIG1res CFIG2res RELSind = 2576
Node 4: = 0
Node 6:SENDRELi SENDRELr = 1269
Node 8:CONreq SENDRELi RELSind = 1807
Node 9:RELSreq = 638
Node 29:CONconf ALERTind SENDRELi RECRELin RELSind = 3252
Node 34:SENDRELi SENDRELr RECRELin RECRELre = 2589
Node 35:CONconf ALERTreq SENDRELi RECRELin RELSind = 3265
Node 36:CONresp SENDRELi RECRELin RELSind = 2581
Node 37:SENDRELi SENDRELr RELSreq = 1907
Node 38:RECRELin RECRELre RELSreq = 1958
Node 45:RECRELin RECRELre = 1320
Node 48:CONresp ALERTreq SENDRELi RECRELin RELSind = 3285
Node 50:CONconf SENDRELi RECRELin RELSind = 2561
Node 54:SENDRELi RECRELin RELSind = 1915
```

Fig. 17. Simple language comparison table

As mentioned at the end of section 4, the original specification contained 17 states after being through the process of language reduction. When this technique was applied to the refined specification the result was 35 states. So far the main type of error found in the modelling of the refinement was the difference in possible release methods for each state.

7 Further Work

At the time of writing this paper the specification had been created and tested using the Design/CPN and Occurrence Graph Analysis tools. The refined specification has been created and is currently being verified and modified to meet the specification. Once the modelling of the refined gateway specification has been completed, it will be possible to proceed in a number of ways:

- The modelling detail can be increased for the subprocesses within the gateway.

- The sending and receiving protocol entities can be modelled. These would interact with the interface to the CCA via primitives, but amongst peer protocol entities via messages.

- The addition of extra services to the gateway such as call modification, multiparty calls, and supplementary services.

This last step may involve removing some of the symmetry within the model, as the protocols at each end of the gateway will be different. However the modelling technique allows this to occur relatively seamlessly. This is due to the fact that main changes will be the number of CCAs and the addressing information within the MIP messages, which will not be too complex.

8 Conclusion

The formal techniques for the design of a communications gateway described in this paper are being used on a real project and, so far, have provided good results. The approach provides a structure to the design process that results in a designed product that should meet the user's requirements.

The use of Coloured Petri Nets and Design/CPN™ within the specification stage has been successful so far. The fact that the first attempt to refine the specification did not meet the specification shows how easy it is to make mistakes when specifying the behaviour of systems.

The readability of the specification is up to the individual to decide, but as long as efforts are made to present the Petri Nets in a simple and readable form, the method should gain wide acceptance.

9 Acknowledgments

The authors would like to express their appreciation for the efforts of John Gilmour, Telstra Research Laboratories, in providing assistance with PROTEAN and the Language Reduction code. The work reported in this paper has been sponsored by the Communications Division, Defence Science and Technology Organisation (DSTO), Adelaide, South Australia. The permission of the sponsors to publish this paper is hereby gratefully acknowledged.

10 References

[1] G. Juanole, C. Faure, "On Gateway for Interworking through ISDN: Architecture and Formal Modelling with Petri Nets", IEEE INFOCOM'89, Proceedings of the 8th Annual Joint Conference of the IEEE Computer and Communications Societies, 1989, pg 458–467.

[2] G. Juanole, and A. Onodi, "On Gateway Architecture, Formal Modelling and Verification", Protocol Specification, Testing, and Verification, VI, Elsevier, IFIP 1987.

[3] Rainer Handel, Manfred N.Huber, "Integrated Broadband Networks, An introduction to ATM–based networks", ADDISON–WESLEY, 1991.

[4] F. B. Andrews and G. I. Kollar, "The Australian Defence Communications Corporate Plan and its Underpinning Research Program", NATO Symposium on Military Communication Networks Interoperability and Standards, The Hague, The Netherlands, June'93.

[5] D. J. Floreani and A. J. Dadej, "An Architecture for a Distributed Narrowband Packet Radio Network", Australian Telecommunications Research Journal (ATR), Vol.27 No.2, 1993.

[6] A. J. Dadej and D. J. Floreani, "Interconnected Mobile Radio Networks – A Step Towards Integrated Multimedia Military Communications", Proc. IEEE International Conference on Networks, SICON'93, 6–11 Sept. 1993.

[7] Federal Standard, "LPC–10", US Dept of Defence, MIL–STD–188–133.

[8] ITU–TS Q.2931, "B–ISDN Capability Set 1 User–Network Interface Layer 3 Specification", Dec 1993, Geneva.

[9] J. Billington, "Formal Specification of Protocols: Protocol Engineering", Encyclopedia of Microcomputers, Vol. 7, pg. 299–314, Marcel Dekker, NY 1991.

[10] J. Billington and M. Wilbur–Ham, "Automated Protocol Verification", Protocol Specification, Testing, and Verification V, Elsevier, IFIP 1986.

[11] G. Holzmann, "Protocol Design: Redefining the State of the Art", IEEE Software, Vol 9, No. 1, Jan 1992.

[12] K. Jensen ,"Coloured Petri Nets – Basic Concepts, Analysis Methods and Practical Use", EATCS Monographs on Theoretical Computer Science, Springer–Verlag, 1992.

[13] CCITT/SGX/WP3–1, "Specification and Description Language SDL", CCITT Recommendations Z.100–Z.104, 1988.

[14] CCITT/COMXVIII–R/WP18–4, "ISDN Protocol Reference Model", CCITT Revised Recommendation I.320, 1992.

[15] D. J. Floreani and A. J. Dadej, "Application of the Stratification Concept to Radio Networks and their Gateways", to appear in Computer Networks and ISDN Systems, Elseiver.

[16] M.T. Rose, "The Open Book : A Practical Perspective on OSI", Prentice Hall, 1990.

[17] G.V. Bochmann "Deriving Protocol Converters for Communications Gateways", IEEE Transactions on Communications, Vol. 38, No. 9, Sept 1990.

[18] G.V Bochmann "Design Principles for Communication Gateways", IEEE Journal on Selected Areas in Communications, Vol. 8, No. 1, Jan 1990.

[19] IS 8807, "Information Processing Systems, Open Systems Interconnection, LOTOS, A Formal Description Technique Based on the Temporal Ordering of Observational Behaviour ", ISO, 1989.

[20] CCITT/SGX1/WP4–3,"Baseline Text for the Harmonised Signalling Requirements ", Geneva, March 1992.

[21] CCITT Q.931, "ISDN User–Network Interface Layer 3 Specification for Basic Call Control", CCITT Recommendation Q.931, Fascicle, Blue Book, Melbourne 1988.

[22] CCITT Recommendation X.213, Blue Book, Melbourne 1988.

[23] Design/CPN™ Manuals, Meta Software Corporation.

[24] Information processing systems – Open Systems Interconnection – Specification of Abstract Syntax Notation One (ASN.1), International Standard ISO–8824.

[25] A. Wikstrom, "Functional Programming using Standard ML", Prentice Hall International Series in Computer Science, 1987.

[26] J. Billington, G. Wheeler, M. Wilbur–Ham, "PROTEAN: A High–level Petri Net Tool for the Specification and Verification of Communication Protocols", IEEE Transactions on Software Engineering, Vol 14, No. 3, March 1988, pp 301..316.

[27] W. Barrett, R. Bates, D. Gustafson, J. Couch "Compiler Construction – theory and practice", second edition, Science Research Associates, 1986.

Expected Impulse Rewards in Markov Regenerative Stochastic Petri Nets *

Reinhard German[1], Aad P. A. van Moorsel[2], Muhammad A. Qureshi[2], and William H. Sanders[2]

[1] Prozeßdatenverarbeitung und Robotik, Technische Universität Berlin, Franklinstr. 28/29, 10587 Berlin, Germany, rge@cs.tu-berlin.de
[2] Center for Reliable and High-Performance Computing, Coordinated Science Laboratory, University of Illinois at Urbana-Champaign, 1308 W. Main St., Urbana, IL 61801 USA, {moorsel,qureshi,whs}@crhc.uiuc.edu

Abstract. Reward structures provide a versatile tool for the definition of performance and dependability measures in stochastic Petri nets. In this paper we derive formulas for the computation of expected reward measures in Markov regenerative stochastic Petri nets, which allow for transitions with non-exponentially distributed firing times. The reward measures may be composed of rate rewards which are obtained in certain markings and of impulse rewards which are obtained when transitions fire. The main result of the paper is the derivation of formulas for the expected impulse reward of transitions with non-exponentially distributed firing times. The analysis is based on the method of supplementary variables. Numerical examples are given for an M/D/1/K queueing system with service breakdowns.

Key words: Markov regenerative stochastic Petri nets, method of supplementary variables, rate and impulse reward measures.

1 Introduction

The inclusion of reward structures in stochastic Petri nets (SPNs) facilitates the specification of a variety of interesting performance and dependability measures. This fact has motivated the introduction of reward structures as an integrated part of modeling in two SPN variants: in *stochastic activity networks* [19, 20] and in *stochastic reward nets* [6]. In both cases, rewards are specified at the net level, so that the modeling and the specification of measures is done at the same level. Two types of rewards are typically considered: *rate rewards* and *impulse rewards*. Rate rewards are associated with markings of the SPN and are collected

---

* This work was initiated while Reinhard German visited the University of Illinois at Urbana-Champaign. His visit was supported by Siemens Corporate, Research and Development.

during the time the SPN resides in the marking. Impulse rewards, on the other hand, are associated with transition firings and are collected when a transition fires. In a recently proposed specification of reward structures [18], impulse rewards may depend on the state in which a transition fires and on the untimed events which take place immediately after the firing.

If all transition firing times are exponentially distributed, the underlying stochastic process is a Markov chain and formulas are known for the computation of the expected values of the reward measures (e.g., [6]). The computation of the distribution of the reward measure is computationally more expensive, but results are also available (e.g., [16, 17]). If the firing times of the transitions are generally distributed the underlying stochastic process is not a Markov chain. Under the constraint that in each marking at most one transition with a generally distributed firing time is enabled, the underlying process is a Markov regenerative stochastic process [5, 7]. Therefore this class of SPNs is referred to as *Markov regenerative stochastic Petri nets* (MRSPNs) [3].

Two approaches have been suggested for the analysis of MRSPNs. First, an embedded Markov chain can be considered at a suitable selected subset of time points, an approach taken by most authors. The approach is suitable for stationary [2, 5, 7, 21] as well as transient analysis [4, 5, 14, 3, 15, 12]. Secondly, the process can be made Markovian by the inclusion of supplementary variables [8, 9]. This approach has been used for the stationary analysis of MRSPNs in [11] and has been extended for the transient analysis in [10, 12]. A comparison of both approaches can be found in [12]. The aim of the analysis in the reported references is to compute the vector of state probabilities, either in the transient or stationary case, and to derive the measures of interest from these state probabilities. However, if the reward measure contains impulse rewards of transitions with general firing time distributions, it is no longer possible to derive the expected reward directly from the state probabilities.

The aim of this paper is therefore to derive formulas for the computation of expected reward measures in MRSPNs, given that the reward measures are composed of both rate and impulse rewards. For this purpose we propose a mathematical framework for the representation of combined rate and impulse reward measures. This extends the formalisms suggested in [6, 20]. The analysis is then conducted by the method of supplemen-

---

[3] This constraint can easily be formulated and checked on a computer. However, the class of SPNs with an underlying Markov regenerative process is actually larger. It is also important to consider the firing policy of the transitions. In this paper we always assume a firing policy known as *race with enabling memory*. See [3, 21] for a more detailed discussion.

tary variables. State equations are derived and analyzed which describe the dynamics of the underlying stochastic process. The analysis combines results from [11, 10, 12], and the derivation of this general class of state equations has not yet been published. The general form of the state equations allows the formulation of the main result of this paper: computational formulas for the expected impulse reward of transitions with generally distributed firing times. The result is given as an integral over the value of the supplementary variable. The formula appears natural in the approach of supplementary variables, since the equations are valid for the full set of time points. A comparable result is not known for the approach of the embedded Markov chain.

The paper is organized as follows. In Sec. 2 the considered class of SPNs is introduced. The mathematical formalism for the representation of reward structures is given in Sec. 3. The analysis of the stochastic process is presented in Sec. 4 and formulas for the expected reward are given in Sec. 5. A numerical example is shown in Sec. 6 and conclusions are given in Sec. 7.

2 Markov Regenerative Stochastic Petri Nets

We assume that the reader is familiar with SPNs, and will therefore only briefly discuss the specific class of SPNs considered in this paper. The primitives of the considered of SPNs are: *places*, *transitions*, *arcs*, and indistinguishable *tokens*. The arcs are divided into *input*, *output*, and *inhibitor* arcs, and the transitions are divided into *immediate* and *timed* ones. Furthermore, the timed transitions are divided into those which fire after an exponentially distributed time (referred to as *exponential transitions*) and those which fire after a generally distributed time (referred to as *general transitions*).

The set of all transitions is denoted by T. T can be partitioned into the set of exponential transitions T^E and the set of general transitions T^G. Single transitions are denoted by letters $a, g, h \in T$. The firing time of a transition $g \in T^G$ is specified by the *probability distribution function* (PDF) $F^g(x)$. We assume that the PDF has no mass at zero [4], (i.e., $F^g(0) = 0$), and use x^g_{max} to denote the range over which the firing time is defined (i.e., $x^g_{max} = \min \{x \geq 0 : F^g(x) = 1\}$). If the distribution has infinite support, x^g_{max} will be interpreted as $x^g_{max} = \infty$. The firing time can also be represented by the *probability density function* (pdf) $f^g(x)$ or

---

[4] This does not affect the use of immediate transitions.

by the *instantaneous rate function* (irf) $\lambda^g(x)$, defined by:

$$\lambda^g(x) = \frac{f^g(x)}{1 - F^g(x)}, \quad \text{for } x < x_{max}^g. \tag{1}$$

$\lambda^g(x)$ is undefined for $x \geq x_{max}^g$. The firing time may be a mixed random variable containing both a continuous and a discrete part. The discrete part corresponds to discontinuities of $F^g(x)$. These discontinuities can be represented as Dirac impulses in the pdf and irf (see [13], pp. 341–343 and 372–373 for an informal discussion of generalized functions in the context of random variables). Using this formalism of generalized functions, $f^g(x)$ and $\lambda^g(x)$ are always existing. In the special case of a deterministic transition with delay τ, the PDF and pdf are given by

$$F^g(x) = \begin{cases} 0 & x \leq \tau \\ 1 & x > \tau \end{cases}, \quad f^g(x) = \delta(x - \tau), \tag{2}$$

where $\delta(x - \tau)$ denotes the Dirac unit impulse located at $x = \tau$. The irf is then given by

$$\lambda^g(x) = \begin{cases} \delta(x - \tau) & x \leq \tau \\ \text{undefined} & x > \tau \end{cases}. \tag{3}$$

In this paper we consider SPNs in which at most one general transition is enabled in each marking and in which the firing policy of each transition is *race with enabling memory* (as defined in [1]). Under this policy, a new delay has to be sampled if a disabled transition becomes enabled again. Since a Markov regenerative stochastic process is underlying an SPN of this class, we refer to these SPNs as *Markov regenerative stochastic Petri nets* (MRSPNs).

In order to illustrate the formalism, an example is given and used throughout the paper. Figure 1 shows an SPN model of an M/D/1/K-queueing system with service breakdowns. The exponential transition $a1$ models arrival of customers and the deterministic transition $a2$ service of customers. Tokens in $P2$ represent customers inside the system. The service facility is operating if a token is in $P3$ and failed if a token is in $P4$. The exponential transitions $a3$ and $a4$ model the failure and repair. The rates λ, ρ, and σ are associated with the exponential transitions $a1$, $a3$, and $a4$, respectively. $a2$ has a constant firing delay τ. The arcs between $P2$ and $a3$ model that a failure is only possible, if the system is utilized. Note that due to the race-enabling memory policy the service time is lost if it is interrupted by a failure.

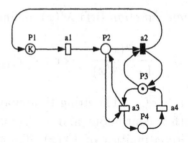

Fig. 1. SPN of an M/D/1/K queueing system with service breakdowns.

The *tangible markings* of an MRSPN constitute the states of the underlying stochastic process. We denote the discrete state space by $\mathcal{S}$ and refer to the single states with integers $i, j \in \mathcal{S}$. It is assumed that the discrete state space is finite. The stochastic process is a tuple of random variables:

$$\left\{ (N(t), X(t)), t \in \mathbb{R}_0^+ \right\}, \tag{4}$$

where $N(t)$ gives the tangible marking at time t. If in $N(t)$ a general transition g is enabled, $X(t)$ gives the times since the enabling of g. Otherwise $X(t)$ is undefined. $N(t)$ is discrete and $X(t)$ is continuous. $X(t)$ is referred to as the *supplementary variable*.

The example SPN shown in Fig. 1 contains only tangible markings. The discrete state space $\mathcal{S}$ is given by the tuples (n, m), where n gives the number of customers inside the system and $m \in \{o, f\}$ denotes whether the server is in mode operating (o) or failed (f). The state number is given by $i = n + K \cdot 1_{m=f}$. Figure 2 shows the stochastic process for $K = 5$. Each state transition is labeled with the rate, which is constant (λ) in case of state transitions caused by exponential transitions and which depends on the value of $X(t)$ in case of state transitions caused by the deterministic transition $(\mu(x))$.

3 Reward Measures

For a formal introduction of reward measures we use a framework similar to that suggested in [20]. This framework has been extended in [18] to allow for more general impulse reward definitions, but to keep the notation concise we will only allow a single impulse reward for each transition on the state-space level. The extension of the results obtained in this paper can easily be extended to include the reward structure in [18]. A *reward*

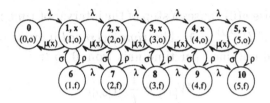

Fig. 2. Underlying stochastic process.

structure is given by the vector $\mathbf{r}$ and the matrices $\mathbf{C}^a$ for each timed transition $a \in T$:

- $\mathbf{r}$, where $r_i \in \mathbb{R}$ is a rate reward which is obtained when the SPN is in state i,

- $\mathbf{C}^a$, where $c_{i,j}^a \in \mathbb{R}$ is an impulse reward which is obtained when the transition a fires in state i and causes a state transition to state j.

Furthermore, let I_t^i and $I_t^{(i,a,j)}$ be *indicator variables*:

$$I_t^i = \begin{cases} 1 & N(t) = i \\ 0 & \text{otherwise} \end{cases}, \tag{5}$$

$$I_t^{(i,a,j)} = \begin{cases} 1 & a \text{ causes a state transition from } i \text{ to } j \text{ at time } t \\ 0 & \text{otherwise} \end{cases}. \tag{6}$$

These indicator variables can be used to express the reward which is collected in a certain interval of time. Note that the variable I_t^i equals one over time intervals and that the variable $I_t^{(i,a,j)}$ equals one only at single instants of time. In other words, I_t^i represents continuous quantities on each interval and $I_t^{(i,a,j)}$ represents discrete quantities located at the instants of transition firings. One can deal with both quantities by integrating I_t^i over a time interval and summing over $I_t^{(i,a,j)}$ at the instants of transition firings. The integration and summation can be explicitly expressed. This approach is taken in [6]. Motivated by the use of generalized functions for mixed random variables (as discussed at the beginning of the last section), we represent here the discrete quantities by Dirac impulses. This leads to a compact formalism. The *interval-of-time* or accumulated reward $Y_{[0,t]}$ gives the rate and impulse reward obtained from 0 to t and is defined as:

$$Y_{[0,t]} = \sum_{i \in S} \int_0^t r_i \cdot I_x^i \, dx + \sum_{a \in T} \sum_{i \in S} \sum_{j \in S} \int_0^t c_{i,j}^a \cdot I_x^{(i,a,j)} \cdot \delta(0) \, dx. \tag{7}$$

The Dirac impulse $\delta(0)$ causes a step of the interval-of-time reward at time t of height $c_{i,j}^a$ if $I_t^{(i,a,j)} = 1$. The *time-averaged* interval-of-time reward gives the reward obtained per time unit from 0 to t:

$$W_{[0,t]} = \frac{1}{t} \cdot Y_{[0,t]}. \tag{8}$$

Both the interval-of-time and the time-averaged interval-of-time reward are random variables. It is possible to define the distribution and expectation of the interval-of-time reward:

$$F_{Y_{[0,t]}}(y) = \Pr\left\{Y_{[0,t]} \leq y\right\}, \quad E\left[Y_{[0,t]}\right] = \int_0^\infty y\, dF_{Y_{[0,t]}}(y). \tag{9}$$

The expectation exists only in case of absolute convergence of the integral. The distribution and expectation of the time-averaged interval-of-time reward can be defined similarly.

Another interesting quantity is the *instant-of-time* reward measure. For rate rewards the instant-of-time measure can be defined straightforwardly, but in case of impulse rewards defining a useful interpretation is more involved. Different approaches have been taken to overcome this problem (e.g., [20, 6]). Here we consider the proposal in [6], where the instant-of-time measure is defined as the change of the interval-of-time measure. For rate rewards the resulting measure corresponds to the natural definition of the instant-of-time measure.

We first consider the random variable $W_{[t,t+\Delta t]}$, which is the difference quotient of the interval-of-time reward:

$$W_{[t,t+\Delta t]} = \frac{Y_{[0,t+\Delta t]} - Y_{[0,t]}}{\Delta t}. \tag{10}$$

The instant-of-time reward V_t could then be defined as the differential quotient $\lim_{\Delta t \to 0} W_{[t,t+\Delta t]}$. Using the derivation rules for generalized functions the steps of the interval-of-time reward correspond to Dirac impulses in the instant-of-time reward. The value of V_t is infinite at these points and the area is equal to the step height of $Y_{[0,t]}$. Intuitively, the Dirac impulse describes the change in case of a step. Since V_t contains Dirac impulses it is not a proper random variable. However, we can define its "expectation" as the following limit:

$$E[V_t] = \lim_{\Delta t \to 0} \int_0^\infty v\, dF_{W_{[t,t+\Delta t]}}(v) = \lim_{\Delta t \to 0} E\left[W_{[t,t+\Delta t]}\right]. \tag{11}$$

For the interval-of-time reward and the instant-of-time reward one can define the stationary limits:

$$E[W] = \lim_{t \to \infty} E\left[W_{[0,t]}\right], \quad E[V] = \lim_{t \to \infty} E[V_t] \tag{12}$$

These limits not necessarily both exist. Below, an example will be shown where the limit $E[V]$ does not exist. If $E[V]$ is existing, it is equal to $E[W]$. A proof of existence of $E[W]$ is beyond the scope of this paper.

Example An example is investigated in order to illustrate the framework of reward measures. Figure 3 shows an SPN with two deterministic transitions which let the token circulate between the two places. The firing delay of $a1$ is equal to 3 and the delay of $a2$ equal to 2. Let the state

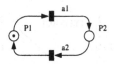

Fig. 3. SPN with two states.

space be $S = \{0, 1\}$, the state number denotes the number of tokens in $P2$. The following reward structure is defined: a rate reward equal to -1 is obtained when $N(t) = 1$ and an impulse reward equal to 3 is obtained when $a2$ fires. Figure 4 shows the evolution of the instant-of-time reward V_t. V_t consists of a Dirac impulse when $a2$ fires (at multiples of 5). A Dirac impulse is graphically represented by an arrow, the area of the impulse is shown in the circle at the destination of the arrow. V_t describes the change of the interval-of-time reward. The limit $E[V]$ is not existing. Figure 5 shows the interval-of-time and time-averaged interval-of-time reward. The time-averaged interval-of-time reward converges to a limit. Note that due to the deterministic nature of the process all reward variables are equal to their expected values. In Fig. 6 the corresponding expected rewards are shown if both transitions have exponentially distributed firing times with unchanged mean firing times.

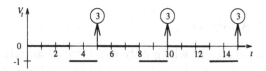

Fig. 4. Evolution of the instant-of-time reward.

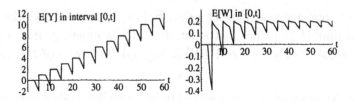

Fig. 5. Expected values of the reward variables in the deterministic case.

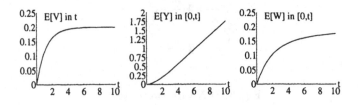

Fig. 6. Expected values of the reward variables in the exponential case.

4 Analysis of the Stochastic Process

For the computation of the reward measures of an MRSPN the underlying stochastic process has to be analyzed. We use the *method of supplementary variables* [8, 9]: based on the state space representation given in Eq. (4) the forward Kolmogorov state equations are formulated. These equations express the state occupancy distribution as function of the time t and the value of the supplementary variable x. In Sec. 5 it will be shown, how from these quantities the values of the expected rewards can be derived. In this section we derive the state equations and discuss their analysis. The derivation in this section is a generalization of the material published in [11, 10, 12].

4.1 Transient Analysis

Let S^E contain all states in which only exponential activities are enabled and let S^g contain all states in which general transition $g \in T^G$ is enabled (for each $g \in T^G$ exists a separate set S^g). The transient probabilities are denoted as $\pi_i(t) = \Pr\{N(t) = i\}$. For states with a supplementary variable (i.e., $i \in \mathcal{S}^g, g \in T^g$), the *age density functions* are defined as:

$$\pi_i(t, x) = \frac{d}{dx}\Pr\{N(t) = i, X(t) \le x\},\qquad(13)$$

$\pi_i(t, x)$ is a probability with respect to t and a (defective) pdf with respect to x. $\pi_i(t, x)dx$ represents the probability of being in state i in an infinitesimal environment of x at time t. In order to derive the state equations in a generic vector-matrix form the following vectors are defined:

$$- \boldsymbol{\pi}^E(t),\ \boldsymbol{\pi}^g(t),\ \text{and}\ \boldsymbol{\pi}^g(t, x),\quad g \in T^G.$$

All vectors are of dimension $|\mathcal{S}|$. $\boldsymbol{\pi}^E(t)$ contains all single probabilities in states of $\mathcal{S}^E$, $\boldsymbol{\pi}^g(t)$ and $\boldsymbol{\pi}^g(t, x)$ contain all single probabilities and age densities in states of $\mathcal{S}^g$, respectively. The elements that do not correspond to the states indicated by the superscripts are equal to 0. Additionally, the following matrices are defined:

$$- \mathbf{Q}^{E,E},\ \mathbf{Q}^{E,g},\ \mathbf{Q}^{g,E},\ \mathbf{Q}^{g,h},\ \mathbf{Q}^g,\ \text{and}\ \Delta^{g,E},\ \Delta^{g,h},\quad g, h \in T^G.$$

All matrices are of dimension $|\mathcal{S}| \times |\mathcal{S}|$. The $\mathbf{Q}$-matrices contain the rates of state changes caused by exponential transitions and the Δ-matrices describe branching probabilities after a general transition has fired. The superscripts indicate the subsets of the state space between which the state transitions take place (e.g., $\mathbf{Q}^{E,g}$ contains all rates of transitions starting in states $i \in \mathcal{S}^E$ and leading to states $j \in \mathcal{S}^g$). The elements that do not correspond to the states indicated by the superscripts are equal to 0. $\mathbf{Q}^g$ is a special case: it contains all rates of exponential state transitions which can take place during g is enabled and which do not disable g, it also contains the negative sum of all outgoing rates in states $i \in \mathcal{S}^g$.

In states with a supplementary variable, (i.e., $i \in \mathcal{S}^g, g \in T^G$) the forward equations are given by:

$$\boldsymbol{\pi}^g(t + \Delta t, x + \Delta t) = \boldsymbol{\pi}^g(t, x) \cdot (\mathbf{I} + \mathbf{Q}^g \Delta t - \lambda^g(x)\Delta t) + \mathrm{o}(\Delta t),\quad(14)$$

for $0 \le x < x_{max}^g$. $\mathbf{I}$ denotes the identity matrix and $\mathrm{o}(\Delta t)$ is a function tending to zero more rapidly than Δt. For states without a supplementary variable (i.e., $i \in \mathcal{S}^E$) the forward equations are given by:

$$\boldsymbol{\pi}^E(t + \Delta t) = \boldsymbol{\pi}^E(t) \cdot \left(\mathbf{I} + \mathbf{Q}^{E,E}\Delta t\right)\qquad(15)$$

$$+ \sum_{g \in T^G} \int_0^{x_{max}^g} \boldsymbol{\pi}^g(t, x) \cdot \lambda^g(x) \cdot \Delta^{g,E}\, dx\, \Delta t$$

$$+ \sum_{g \in T^G} \int_0^{x_{max}^g} \boldsymbol{\pi}^g(t, x) \cdot \mathbf{Q}^{g,E}\, dx\, \Delta t + \mathrm{o}(\Delta t).$$

Let the vectors $\boldsymbol{\pi}_0^E$, $\boldsymbol{\pi}_0^g$, $g \in T^G$, represent the initial state occupancy distribution. Assuming that no general transition was enabled before $t = 0$ we obtain as initial conditions:

$$\boldsymbol{\pi}^E(0) = \boldsymbol{\pi}_0^E, \quad \boldsymbol{\pi}^g(0) = \boldsymbol{\pi}_0^g, \quad \boldsymbol{\pi}^g(0, x) = \boldsymbol{\pi}_0^g \cdot \delta(0). \tag{16}$$

The supplementary variable is set to zero in the instant of enabling of a general transition g:

$$\boldsymbol{\pi}^g(t, 0) = \boldsymbol{\pi}^E(t) \cdot \mathbf{Q}^{E,g} + \sum_{h \in T^G} \int_0^{x_{max}^h} \boldsymbol{\pi}^h(t, x) \cdot \lambda^h(x) \cdot \Delta^{h,g} \, dx +$$

$$\sum_{h \in T^G} \int_0^{x_{max}^h} \boldsymbol{\pi}^h(t, x) \cdot \mathbf{Q}^{h,g} \, dx. \tag{17}$$

Integrating the age densities over x yields the state probabilities:

$$\boldsymbol{\pi}^g(t) = \int_0^\infty \boldsymbol{\pi}^g(t, x) \, dx \tag{18}$$

Equations (14–18) describe the dynamic behavior of the process. Following the line presented in [8] these equations can be transformed into a more convenient form. Subtracting $\boldsymbol{\pi}^g(t, x)$ from both sides of Eq. (14), dividing both sides by Δt, and taking the limit $\Delta t \to 0$ leads to the following system of partial differential equations (PDEs):

$$\frac{\partial}{\partial t} \boldsymbol{\pi}^g(t, x) + \frac{\partial}{\partial x} \boldsymbol{\pi}^g(t, x) = \boldsymbol{\pi}^g(t, x) \cdot \mathbf{Q}^g - \boldsymbol{\pi}^g(t, x) \cdot \lambda^g(x), \tag{19}$$

for $0 < x < x_{max}^g$. In order to simplify the PDE system further, the following quantities are defined:

$$p_i(t, x) = \frac{\pi_i(t, x)}{1 - F^g(x)}, \quad x < x_{max}^g, i \in S^g, g \in T^G. \tag{20}$$

Since $p_i(t, x)dx$ can be interpreted as the probability of being in state i in an infinitesimal environment of x, given that the firing time has not yet elapsed, $p_i(t, x)$ is referred to as the *instantaneous age rate*. $\mathbf{p}^g(t, x)$ is defined as the vector of all single age rates of g. Application of the chain rule for differentiation to the left side of Eq. (19) gives:

$$\frac{\partial}{\partial t} \boldsymbol{\pi}^g(t, x) + \frac{\partial}{\partial x} \boldsymbol{\pi}^g(t, x) = \frac{\partial}{\partial t} \mathbf{p}^g(t, x) \cdot (1 - F^g(x)) + \tag{21}$$

$$\frac{\partial}{\partial x} \mathbf{p}^g(t, x) \cdot (1 - F^g(x)) - \mathbf{p}^g(t, x) \cdot f^g(x).$$

Moreover, since $\pi^g(t, x) \cdot \lambda^g(x) = \mathbf{p}^g(t, x) \cdot f^g(x)$, we obtain:

$$\frac{\partial}{\partial t}\mathbf{p}^g(t, x) + \frac{\partial}{\partial x}\mathbf{p}^g(t, x) = \mathbf{p}^g(t, x) \cdot \mathbf{Q}^g. \tag{22}$$

Using the method of characteristics, Eq. (22) can be simplified to a system of ordinary differential equations (ODEs) with solution [10]:

$$\mathbf{p}^g(t_0 + h, x_0 + h) = \mathbf{p}^g(t_0, x_0) \cdot e^{\mathbf{Q}^g \cdot h}, \quad t_0, x_0, h \in \mathbb{R}_0^+. \tag{23}$$

Equation (15) can be transformed into a system of ODEs by subtracting $\pi^E(t)$ from both sides, dividing both sides by Δt, taking the limit $\Delta t \to 0$, and substituting the age densities by the age rates:

$$\frac{d}{dt}\pi^E(t) = \pi^E(t) \cdot \mathbf{Q}^{E,E} + \sum_{g \in T^G} \int_0^\infty \mathbf{p}^g(t, x)\, dF^g(x) \cdot \Delta^{g,E} + \sum_{g \in T^G} \pi^g(t) \cdot \mathbf{Q}^{g,E}. \tag{24}$$

Substituting the age densities by the age rates in the initial conditions (16), boundary conditions (17), and integral equations (18) yields:

$$\pi^E(0) = \pi_0^E, \quad \pi^g(0) = \pi_0^g, \quad \mathbf{p}^g(0, x) = \pi_0^g \cdot \delta(x), \tag{25}$$

$$\mathbf{p}^g(t, 0) = \pi^E(t) \cdot \mathbf{Q}^{E,g} + \sum_{h \in T^G} \int_0^\infty \mathbf{p}^h(t, x)\, dF^h(x) \cdot \Delta^{h,g} + \sum_{h \in T^G} \pi^h(t) \cdot \mathbf{Q}^{h,g}, \tag{26}$$

$$\pi^g(t) = \int_0^\infty \mathbf{p}^g(t, x) \cdot (1 - F^g(x))\, dx. \tag{27}$$

Equations (23–27) uniquely describe the time-dependent behavior of the stochastic process underlying an MRSPN and constitute transient state equations. The numerical analysis of the transient equations is discussed in [10, 12]: the continuous variables t and x can be discretized, the age rates and the state probabilities can then be computed at the grid points by an iterative scheme.

Simplifications for Deterministic Firing Times

If a transition g has a deterministic firing time τ^g, the integrals in Eqs. (24, 26, 27) simplify to:

$$\int_0^\infty \mathbf{p}^g(t, x)\, dF^g(x) = \mathbf{p}^g(t, \tau^g), \tag{28}$$

$$\int_0^\infty \mathbf{p}^g(t, x) \cdot (1 - F^g(x))\, dx = \int_0^{\tau^g} \mathbf{p}^g(t, x)\, dx. \tag{29}$$

4.2 Stationary Analysis

The time-averaged limits of the transient state probabilities and age rates are defined as:

$$\pi_i = \lim_{y \to \infty} \frac{1}{y} \int_0^y \pi_i(t)dt, \quad p_i(x) = \lim_{y \to \infty} \frac{1}{y} \int_0^y p_i(t,x)dt. \tag{30}$$

$\boldsymbol{\pi}^E$, $\boldsymbol{\pi}^g$, and $\mathbf{p}^g(x)$ are vectors containing the single state probabilities and age rates of states of $\mathcal{S}^E$ and $\mathcal{S}^g$, respectively. Taking the time-averaged limits on both sides of the transient state equations (22, 24 – 27) eliminates the variable t in these equations. The system of PDEs in Eq. (22) reduces to a system of ODEs:

$$\frac{d}{dx} \mathbf{p}^g(x) = \mathbf{p}^g(x) \cdot \mathbf{Q}^g. \tag{31}$$

The system of ODEs (24) reduces to the following system of balance equations:

$$0 = \boldsymbol{\pi}^E \cdot \mathbf{Q}^{E,E} + \sum_{g \in T^G} \int_0^\infty \mathbf{p}^g(x) \, dF^g(x) \cdot \Delta^{g,E} + \sum_{g \in T^G} \boldsymbol{\pi}^g \cdot \mathbf{Q}^{g,E}. \tag{32}$$

The boundary conditions (26) simplify to the following equations:

$$\mathbf{p}^g(0) = \boldsymbol{\pi}^E \cdot \mathbf{Q}^{E,g} + \sum_{h \in T^G} \int_0^\infty \mathbf{p}^h(x) \, dF^h(x) \cdot \Delta^{h,g} + \sum_{h \in T^G} \boldsymbol{\pi}^h \cdot \mathbf{Q}^{h,g}. \tag{33}$$

Note that the initial conditions (25) are not relevant for the stationary case (a formal characterization of the conditions is beyond the scope of this paper). The integral equations (27) reduce to:

$$\boldsymbol{\pi}^g = \int_0^\infty \mathbf{p}^g(x) \cdot (1 - F^g(x)) \, dx. \tag{34}$$

Finally, an additional normalization condition is given by:

$$\sum_{i \in \mathcal{S}} \pi_i = 1. \tag{35}$$

The numerical analysis of the stationary equations is discussed in [11]. The solution of the ODE system (32) is given by the matrix exponential $\mathbf{p}^g(x) = \mathbf{p}^g(0) \cdot e^{\mathbf{Q}^g \cdot x}$. The integrals occurring in Eqs. (31, 33) can then iteratively be computed by a generalized version of Jensen's method, given that the PDFs $F^g(x)$ of the firing times can piecewise be represented by exponential polynomials [11, 7]. The remaining equations constitute a linear system of equations in $|\mathcal{S}|$ unknowns (the unknowns are the entries of the vectors $\boldsymbol{\pi}^E$ and $\mathbf{p}^g(0)$, $g \in T^G$). All state probabilities can be obtained from this solution.

Simplifications for Deterministic Firing Times If a transition g has a deterministic firing time τ^g, the integrals in Eqs. (32, 33, 34) simplify to:

$$\int_0^\infty \mathbf{p}^g(x)\, dF^g(x) = \mathbf{p}^g(\tau^g), \tag{36}$$

$$\int_0^\infty \mathbf{p}^g(x) \cdot (1 - F^g(x))\, dx = \int_0^{\tau^g} \mathbf{p}^g(x)\, dx. \tag{37}$$

5 Computation of Expected Rewards

As discussed in Sec. 3, a reward structure is given by the vector $\mathbf{r}$ of rate rewards and by the matrices $\mathbf{C}^a$, $a \in T$, of impulse rewards. Let $\boldsymbol{\pi}(t)$ denote the vector of transient state probabilities and let $\boldsymbol{\pi}$ denote the vector of stationary state probabilities. The expected rate reward can be derived from the state probabilities, as can the expected impulse reward corresponding to exponential transitions. However, for the expected impulse reward of general transitions the age densities are required: it can be obtained by integrating the product of the age densities and of the instantaneous rate functions of the firing times over the supplementary variable x.

A vector $\mathbf{s}$ is defined to contain all constants which have to be multiplied with the state probabilities. These constants are the rate rewards and the impulse rewards of the exponential transitions. The ith entry of $\mathbf{s}$ is given by:

$$s_i = r_i + \sum_{a \in T^E} \sum_{j \in S} c_{i,j}^a \cdot \lambda_i^a \cdot \delta_{i,j}^a. \tag{38}$$

$\delta_{i,j}^a$ denotes a branching probability (probability that a state transition to state j takes place, given that a fires in state i). A vector $\mathbf{d}^g$ is defined with all constants needed to compute the expected impulse rewards of the general transition g. The ith entry of $\mathbf{d}^g$ is defined as:

$$d_i^g = \sum_{j \in S} c_{i,j}^g \cdot \delta_{i,j}^g. \tag{39}$$

The expected instant-of-time reward V_t can now be expressed as:

$$\mathrm{E}\left[V_t\right] = \boldsymbol{\pi}(t) \cdot \mathbf{s} + \sum_{g \in T^G} \int_0^{x_{max}^g} \boldsymbol{\pi}^g(t, x) \cdot \lambda^g(x) \cdot \mathbf{d}^g\, dx. \tag{40}$$

The expected impulse reward of a general transition g is obtained by integrating over the product of age densities in which g is enabled and the instantaneous rate function associated with g (the impulse reward is thus conditioned on the value of the supplementary variable). Substituting the age density by the age rate according to Eq. (20) modifies the expression:

$$E\left[V_t\right] = \boldsymbol{\pi}(t) \cdot \mathbf{s} + \sum_{g \in T^G} \int_0^{x_{max}^g} \mathbf{p}^g(t, x)\, dF^g(x) \cdot \mathbf{d}^g. \tag{41}$$

Note that the integrals in Eq. (41) have to be computed as an intermediate step during the numerical analysis of the transient state equations (even if one only solves for the state probabilities). Therefore, only minor changes to the solution algorithm are required to compute the expected impulse rewards according to Eq. (41). The expected accumulated reward is given by the integral of Eq. (41):

$$E\left[Y_{[0,t]}\right] = \int_0^t \boldsymbol{\pi}(y)\, dy \cdot \mathbf{s} + \sum_{g \in T^G} \int_0^t \int_0^{x_{max}^g} \mathbf{p}^g(y, x)\, dF^g(x)\, dy \cdot \mathbf{d}^g, \tag{42}$$

which can be computed by numerical integration over the values at discrete points.

From Eq. (42) the computation of the expected time-averaged reward $E\left[W_{[0,t]}\right]$ is straightforward. The stationary limit $E\left[W\right] = \lim_{t \to \infty} E\left[W_{[0,t]}\right]$ is given by:

$$E\left[W\right] = \boldsymbol{\pi} \cdot \mathbf{s} + \sum_{g \in T^G} \int_0^{\infty} \mathbf{p}^g(x)\, dF^g(x) \cdot \mathbf{d}^g. \tag{43}$$

The integrals in this equation also have to be computed during the numerical analysis of the stationary state equations. Hence, as for the transient case, no significant additional computational costs is involved in the computation of expected rewards containing impulse rewards.

Simplifications for Deterministic Firing Times If all transitions $g \in T^G$ have deterministic firing times τ^g, Eqs. (41, 43) simplify to:

$$E\left[V_t\right] = \boldsymbol{\pi}(t) \cdot \mathbf{s} + \sum_{g \in T^G} \mathbf{p}^g(t, \tau^g) \cdot \mathbf{d}^g, \tag{44}$$

$$E\left[W\right] = \boldsymbol{\pi} \cdot \mathbf{s} + \sum_{g \in T^G} \mathbf{p}^g(\tau^g) \cdot \mathbf{d}^g. \tag{45}$$

Special Case: Throughput of a Transition The throughput of a transition is defined as the expected number of firings per time unit and is normally considered in the stationary limit. The throughput is a measure of interest in many performance studies. Since it is a special case of an impulse reward, it is discussed here.

For the throughput of a transition a each firing of a is counted: an impulse reward equal to one is obtained for each firing of a. This can be represented by setting all entries of $\mathbf{C}^a$ equal to one. The stationary throughput is then given by $S(a) = \mathrm{E}[W]$. We define the transient throughput as $S_t(a) = \mathrm{E}[V_t]$; this quantity describes the rate of firing of the transition varying over time t. In case of a general transition $g \in T^G$ the transient throughput is thus given by:

$$S_t(g) = \int_0^\infty \boldsymbol{\pi}^g(t,x) \cdot \boldsymbol{\lambda}^g(x) \cdot \mathbf{e}\, dx \tag{46}$$

$$= \int_0^\infty \mathbf{p}^g(t,x)\, dF^g(x) \cdot \mathbf{e}, \tag{47}$$

where $\mathbf{e}$ denotes a vector containing values equal to one. In the stationary case this simplifies to:

$$S(g) = \mathbf{p}^g(0) \cdot \int_0^\infty e^{\mathbf{Q}^g \cdot x}\, dF^g(x) \cdot \mathbf{e}. \tag{48}$$

Common knowledge tells that the stationary throughput of a transition a can be computed by summing the probabilities of states in which a is enabled and dividing this sum by the mean firing time $\overline{X^a}$ of a. In case of a general transition g (and firing policy race with enabling memory) this result is only valid, if g cannot be disabled before firing: in the following we show that in this case Eq. (48) can be transformed to the simple expression described above. Since g cannot be preempted, the rows of the matrix exponential sum to one, and Eq. (48) can be simplified:

$$S(g) = \mathbf{p}^g(0) \cdot \int_0^{x_{max}^g} e^{\mathbf{Q}^g \cdot x}\, dF^g(x) \cdot \mathbf{e} = \mathbf{p}^g(0) \cdot \mathbf{e} \tag{49}$$

$$= \mathbf{p}^g(0) \cdot \frac{\overline{X^g}}{\overline{X^g}} \cdot \mathbf{e} = \mathbf{p}^g(0) \cdot \int_0^{x_{max}^g} e^{\mathbf{Q}^g \cdot x} \cdot \mathbf{e} \cdot (1 - F^g(x))\, dx \cdot \frac{1}{\overline{X^g}}$$

$$= \boldsymbol{\pi}^g \cdot \mathbf{e} \cdot \frac{1}{\overline{X^g}}.$$

Note that this result is not valid, if g can be disabled before firing. In case the counting of transition firings depends on the state change, however, the expression in (43) can be used in order to to take the branching probabilities into account.

6 Numerical Examples

We have performed numerical experiments for the SPN of the M/D/1/K queueing system with service breakdowns shown in Fig. 1. The parameters of the model have been chosen as follows: $K = 5$, $\lambda = 0.5$, $\tau = 1$, $\rho = 0.1$, and $\sigma = 0.2$. The initial state of the SPN is as shown in Fig. 1, which corresponds on the state space level to state 0, see Fig. 2. A Mathematica proto-type implementation has been used for the numerical experiments. For all transient results a fixed discretization step-size $h = 0.01$ is used (the deterministic service time is thus discretized into 100 steps).

In a first set of experiments the throughput of the exponential transition $a1$ (modeling arrivals) and of the deterministic transition $a2$ (modeling services) was investigated. Figure 7 shows three curves: the transient throughput $S_t(a1)$ and $S_t(a2)$ of $a1$ and $a2$, respectively, and the stationary throughput of both transitions (the horizontal straight line). It can be seen that the transient throughput of the deterministic transition behaves very irregularly. The transient throughput peaks at multiples of the service time $\tau = 1$, but smooths out when t increases. In the stationary limit the values for $S_t(a1)$ and $S_t(a2)$ go to the same limiting value, showing an example of the law of marking flow.

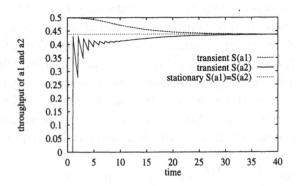

Fig. 7. Transient and stationary throughput of transitions $a1$ and $a2$.

A more complicated reward structure is used to compute what can be interpreted as the "gain" of the system. It combines rate rewards and impulse rewards, as follows. A rate reward -0.1 is obtained if at least one token is waiting in the system, which can be interpreted as the cost of an occupied server. Impulse rewards are collected whenever a service completes (value 1) or when a repair completes (value -5). The negative impulse reward indicates costs while the positive impulse reward indicates benefits. Figure 8 shows the expected instant-of-time gain. The stationary gain is shown by the horizontal straight line in Fig. 8. The instant-of-time gain can be interpreted as time-dependent change of the interval-of-time reward, as explained in Sec. 3. If the actually accumulated reward over a time interval is considered, the curve looks considerably smoother, as can be seen in Fig. 9.

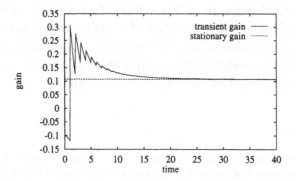

Fig. 8. Expected transient and stationary reward "gain".

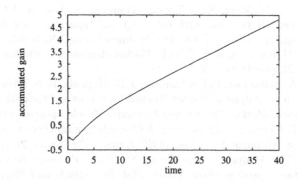

Fig. 9. Expected accumulated reward "gain".

7 Conclusion

In this paper the analysis of reward measures in Markov regenerative stochastic Petri nets has been investigated. The considered reward measures incorporate both rate and impulse rewards, and results have been derived for the expectation of transient as well as stationary measures. If impulse rewards are considered, the definition and solution of measures becomes more involved. Therefore, a mathematical formalism has been proposed that facilitates convenient definition of reward measures based on rate and impulse rewards, and allows for straightforward incorporation of reward measures in the solution formulas. For the analysis we have successfully adopted the method of supplementary variables. The expected impulse rewards can be expressed by quantities which are required in intermediate steps during the computation of the state probabilities. Therefore no significant additional costs are required.

References

1. M. Ajmone Marsan, G. Balbo, A. Bobbio, G. Chiola, G. Conte, A. Cumani. The Effect of Execution Policies on the Semantics of Stochastic Petri Nets. *IEEE Trans. Softw. Engin.* **15** (1989) 832-846.
2. M. Ajmone Marsan, G. Chiola. On Petri Nets with Deterministic and Exponentially Distributed Firing Times. *Advances on Petri Nets '87*, pp. 132–145, Springer-Verlag LNCS 266, 1987.
3. A. Bobbio, M. Telek. Markov Regenerative SPN with Non-Overlapping Activity Cycles. *Proc. IEEE Int. Performance and Dependability Symp.*, Erlangen, Germany, pp. 124–133, 1995.
4. H. Choi, V. G. Kulkarni, K. S. Trivedi. Transient Analysis of Deterministic and Stochastic Petri Nets. *Proc. 14th Int. Conf. on Application and Theory of Petri Nets*, Chicago, IL , USA, pp. 166–185, Springer-Verlag LNCS 691, 1993.
5. H. Choi, V. G. Kulkarni, K. S. Trivedi. Markov Regenerative Stochastic Petri Nets. *Perf. Eval.* **20** (1994) 337–357.
6. G. Ciardo, A. Blakemore, P. F. Chimento, J. K. Muppala, K. S. Trivedi. Automated Generation and Analysis of Markov Reward Models using Stochastic Reward Nets. *Linear Algebra, Markov Chains, and Queueing Models*, Springer-Verlag, 1993.
7. G. Ciardo, R. German, C. Lindemann. A Characterization of the Stochastic Process Underlying a Stochastic Petri Net. *IEEE Trans. Softw. Engin.* **20** (1994) 506–515.
8. D. R. Cox. The Analysis of Non-Markov Stochastic Processes by the Inclusion of Supplementary Variables. *Proc. Camb. Phil. Soc. (Math. and Phys. Sciences)* **51** (1955) 433–441.
9. D. R. Cox, H. D. Miller. *The Theory of Stochastic Processes.* Chapman and Hall, 1965.
10. R. German. New Results for the Analysis of Deterministic and Stochastic Petri Nets. *Proc. IEEE Int. Performance and Dependability Symp.*, Erlangen, Germany, pp. 114–123, 1995.

11. R. German, C. Lindemann. Analysis of Stochastic Petri Nets by the Method of Supplementary Variables. *Perf. Eval.* **20** (1994) 317–335.

12. R. German, D. Logothetis, K. S. Trivedi. Transient Analysis of Markov Regenerative Stochastic Petri Nets: A Comparison of Approaches. *Proc. Petri Nets and Performance Models '95*, Durham, North Carolina, USA, pp. 103–112, 1995.

13. L. Kleinrock. *Queueing Systems*, Volume 1: Theory, John Wiley, 1975.

14. D. Logothetis, K. S. Trivedi. Time-Dependent Behavior of Redundant Systems with Deterministic Repair. *Proc. 2nd Int. Workshop Numerical Solution of Markov Chains*, Raleigh, NC, USA, 1995.

15. D. Logothetis, K. S. Trivedi, A. Puliafito. Markov Regenerative Models. *Proc. IEEE Int. Performance and Dependability Symp.*, Erlangen, Germany, pp. 134–142, 1995.

16. M. A. Qureshi, W. H. Sanders. Reward Model Solution Methods with Impulse and Rate Rewards: An Algorithm and Numerical Results. *Perf. Eval.* **20** (1994) 413–436.

17. M. A. Qureshi, W. H. Sanders. A New Methodology for Calculating Distributions of Reward Accumulated During a Finite Interval, *Proc. 26th Fault-Tolerant Computing Symposium*, Sendai, Japan, 1996.

18. M. A. Qureshi, W. H. Sanders, A. P. A. van Moorsel, R. German. Algorithms for the Generation of State-Level Representations of Stochastic Activity Networks with General Reward Structures. *Proc. Petri Nets and Performance Models '95*, Durham, North Carolina, USA, pp. 180–190.

19. W. H. Sanders, J. F. Meyer. Performability Modeling of Distributed Systems Using Stochastic Activity Networks, *Proc. Int. Workshop on Petri Nets and Performance Models*, 1987, 111-120.

20. W. H. Sanders, J. F. Meyer. A Unified Approach for Specifying Measures of Performance, Dependability, and Performability. *Dependable Computing for Critical Applications*, Vol. 4, pp. 215–238, Springer Verlag, 1991.

21. M. Telek, A. Bobbio, L. Jereb, A. Puliafito, K.S. Trivedi. Steady State Analysis of Markov Regenerative SPN with Age Memory Policy. H. Beilner, F. Bause (Eds.): *Quantitative Evaluation of Computing and Communication Systems*, pp. 165–179, Springer LNCS 977, 1995.

Asynchronous Composition of High Level Petri Nets: A Quantitative Approach

Serge Haddad   Patrice Moreaux

LAMSADE – URA CNRS 825, Université Paris Dauphine, Place du Maréchal de Lattre de Tassigny, 75775 PARIS Cedex 16, FRANCE

Abstract. Stochastic Well Formed Nets (SWNs) are a powerful Petri Net model which allows the computation of performance indices with an aggregation method. Decomposition methods initiated by B. Plateau are another way to reduce the complexity of such a computation. We have shown in a previous work, how to combine these two approaches for systems with synchronous composition. Despite similarities between the asynchronous and synchronous cases, it turns out that the former presents specificities that need theoretical foundations. We undertake this task in the present paper. We derive necessary conditions on the modeled systems that allow for the two methods to be combined. For parallel systems satisfying these necessary conditions we develop a model with the corresponding algorithm. This model, based upon synchronization of "global" tokens moving across submodels, covers a large range of real life systems. An example shows the intuitive ideas behind these developments.

Introduction

It is well known that complex systems with synchronization forbid the use of analytical results to find the steady state probabilities of the corresponding stochastic models. We have then to work at the Markov chain level or to use approximate methods.

Our work deals with Markov chains. In this case, because of the huge size of the state space, one is lead to find efficient methods that avoid the building of the whole chain. The most effective methods are based upon the structure of the models generating the Markov chain. Two main methods follow this approach, aggregation and decomposition using tensor products.

The aggregation method builds a partition of the state space compatible with stochastic behaviour to get a new Markov chain on the classes of the partition with much fewer states. Then this chain is solved providing steady state probabilities of the aggregates and the results are possibly used to find the steady state probabilities of the original chain. This method is very effective since, for instance, with the Stochastic Well formed Net (SWN) model ([3]), we can build the partition a priori, and furthermore it has an interpretation at the system level (a class is a set of states for which a condition is satisfied, . . .).

The decomposition method describes the state space as a product of smaller state spaces and from this, gives an expression of the infinitesimal generator of the Markov chain using only generators on these smaller spaces and operators from tensor algebra. The chain is then solved using directly this expression which contains only "small" matrices. This method, introduced by Plateau [11], with the Stochastic Automata Network (SAN) model was extended to Generalized Stochastic Petri Nets (GSPN) by Donatelli [6] and Buchholz [2].

The purpose of our work is to merge these two methods keeping the benefits of both. We have shown in [9] how to combine the SWN model and the Plateau approach for synchronous composition of subsystems. In this paper, we show how to develop the same combination in the asynchronous case already tackled by Buchholz for GSPNs. As shown by Buchholz in several papers, the asynchronous composition needs the definition of the environment of each subnet: we propose the formally defined notion of *abstract view* of a subnet, based on qualitative criteria, which allows application of the decomposition method.

However, as in the synchronous case, we first point out the two main difficulties of the combination approach, that is to say the specification problem of external/internal synchronization and the resolution problem of synchronization memory. If we have a system with several objects of the same kind (processes for instance), which are synchronized together (through common resources sharing for instance), we may design a model of this system as a synchronized product of models of the activities of these objects, but this product does not allow any aggregation; the synchronization occurs between objects and there are no object classes. We say in this case that the system exhibits an internal synchronization. Therefore we focus on systems with external synchronization, that is to say with synchronization between objects of different kinds. In this context, we observe the synchronization memory phenomenon: the successive firings of synchronization transitions change the state space in such a way that, as a general rule, the original chain cannot be lumped as composition of smaller aggregated chains, preventing the combination of the two methods.

Fortunately, for models involving synchronization memory, which is the general case, we show that a control of the synchronization memory may be managed for restricted asynchronous composition of SWNs. We derive the steady state probabilities computation algorithm which is composed of four stages: definition of an abstract view of each component (subnet), definition of enlarged SWNs, derivation of a tensor expression of the generator of the model, computation of steady state probabilities via iterative methods with that expression.

The outline of the paper is as follows: in Sect. 1 we remind the reader of the methods we want to merge and review related works. We then summarize the basic problems and our proposed framework in Sect. 2 before giving theoretical results together with an illustrative example in Sect. 3. For concision of the paper, proofs of results are carried over to appendix A and definitions about flows and implicit places may be found in App. B.

1 General framework and related works

In this presentation, we restrict ourselves for the sake of simplicity to Continous Time Markov Chain (CTMC).

1.1 Aggregation and decomposition

Let us recall that aggregation methods may be summarized in the following steps: given a CTMC with state space E and infinitesimal generator $Q = [q_{ij}]$:

- find *a partition* of the state space E, say $(E^{(k)})_{k=1,...,K}$, so that the behaviour of states belonging to the same $E^{(k)}$ are stochastically equivalent, that is to say:

$$\forall k, h \in \{1, \ldots, K\}, \quad \forall e, e' \in E^{(k)} \quad \sum_{e_h \in E^{(h)}} q_{e,e_h} = \sum_{e_h \in E^{(h)}} q_{e',e_h} = \widetilde{q}_{k,h} \quad (1)$$

- construct a new CTMC with state space $\widetilde{E} = \{E^{(k)} \mid k = 1, \ldots, K\}$, and infinitesimal generator $\widetilde{Q} = [\widetilde{q}_{k,h}]$,
- solve this CTMC,
- possibly compute the probabilities of the states $e_k \in E^{(k)}$ using the previous solution (note that this step requires additional information about the e_k).

Kemeny and Snell ([10]) showed (first for Discret Time Markov Chain (DTMC), the result was later on extended to CTMC) that *the strong lumpability condition* (1) is necessary and sufficient for the aggregated process to be markovian hence a CTMC. Usually we have $K \ll \sum_k |E^{(k)}|$ so the computation of the steady state probabilities is easier for $\widetilde{Q}$ than for Q. SWN ([3]) is a Petri Net model which supports such a method.

Now, the basic steps in decomposition methods are:
- describe the CTMC state space E as a subset of *a cartesian product* of smaller spaces, say $E \subseteq \prod_{k=1}^{K} E_k$,
- use this decomposition to get an expression of Q as function $f(Q_1, \ldots, Q_K)$, where Q_k is the infinitesimal generator of the CTMC restricted to E_k,
- compute the solution π with $\pi.f(Q_1, \ldots, Q_K) = 0$.

In our context, the functions f are sums of tensor products of the Q_k (see [5] and [11] for details about tensor algebra and its use in the area of stochastic transition systems).

The main interest of this method is to enable the steady state probabilities computation *without computing the Q matrix* but instead, *directly using the tensor expression of Q*.

Trying to merge the two methods above means: from a CTMC C with state space E and infinitesimal generator Q, get an aggregated CTMC of C as a "tensor composition" of smaller aggregated CTMCs. Hence, the successive steps for such a method are:

- build a state space decomposition of E, getting $E \subseteq E' = \prod_{k=1}^{K} E_k$,
- use an aggregation method verifying the strong lumpability condition (1) on each of the CTMCs (E_k, Q_k) leading to $\widetilde{E_k} = \{E_k^{(j)} \mid j = 1, \ldots, n_k\}$ and infinitesimal generators $\widetilde{Q_k}$,
- build the product $(\widetilde{E}' = \prod_{k=1}^{K} \widetilde{E_k}, \widetilde{Q} = f(\widetilde{Q_1}, \ldots, \widetilde{Q_K}))$ of the aggregated CTMCs and define the aggregated image $\widetilde{E} \subseteq \widetilde{E}'$ of E.

Unfortunately, as a general rule, $(\widetilde{E}, \widetilde{Q})$ *is not an aggregation of* (E, Q) *verifying* (1). *So the problem is to find additional conditions on the initial CTMC* C which ensure that the combination satisfies the condition (1). We give in Sect. 3 a solution for SWNs models, using tensor expression for the function f.

1.2 GSPNs asynchronous composition

Asynchronous composition of Petri nets models communicating subsystems (subnets), with entities (tokens) moving from one subsystem to another one. The communication links between subnets are common places which are not input places of transitions of the source net (the PO_k places in the source, the input places in the destination) and transitions from the source subnet with at least one output place in the destination subnet. The formal expression of asynchronous composition for a given class of Petri nets (GSPN, SWN, ...) is summarized in the following

Definition 1. The Petri net $N = (P, T, \ldots)$ is the asynchronous composition of the nets $(N_k = (P_k, T_k, \ldots))_{k \in K}$ iff[1]
- For all k there is a subset PO_k of P_k s.t. $\forall t \in T_k \ {}^\bullet t \cap PO_k = \emptyset$ (the set of output places); we denote $TO_k = \{t \in T_k \text{ s.t. } t^\bullet \cap PO_k \neq \emptyset\}$ the set of output transitions of N_k, and $TO = \bigcup_{k \in K} TO_k$ the set of all output transitions,
- $P = \bigcup_{k \in K} P_k$ and $P_k \cap P_{k'} \subseteq [PO_k \cap (P_{k'} \backslash PO_{k'})] \bigcup [(P_k \backslash PO_k) \cap PO_{k'}]$ for $k \neq k'$ (the set of places and the input/output places),
- $T = \bigcup_{k \in K} T_k$ and $T_k \cap T_{k'} = \emptyset$ for $k \neq k'$ (the set of transitions); $T \backslash TO$ is the set of local transitions,
- Each additional parameter of N for the class of nets involved (designated by ...) is such that its restriction to N_k is the corresponding parameter of N_k.

In the rest of the paper, for any marking M of N, we denote $M_k = M(P_k \backslash PO_k)$, so that $M = (M_k)_{k \in K}$.

P. Buchholz studied the asynchronous composition of GSPNs and other kinds of Petri nets in several papers. His approach may be summarized as follows.

The global net N is decomposed in K subnets N_k as defined above. For each subnet, one define an *aggregated view*, discarding local behaviour of the subnet (in his papers he proposed several definitions of such aggregated views and we refer here to the one proposed in [1, 2]: a *virtual place* p_k and a *virtual transition* t_k summarize the global behaviour of the subnet).

---

[1] from now on, we shall use for ease of writing, K as set of k indexes or as maximal k index when no confusion can arise.

The global behaviour of the net is summarized in the net N_0 composed with aggregated views of all the subnets and is studied for itself giving a Reachability Set (RS) RS_0.

The RS RS_k of the subnet k is computed using N_k and the behaviour of all other subnets, also summarized with a single virtual place transition pair. RS_k is decomposed in partition $(RS_k(m_k))$, all markings of $RS_k(m_k)$ providing the same marking of the virtual place p_k.

Buchholz has proved that the generator of the CTMC of the tangible states of N may be expressed as linear combinations of tensor products of three kinds of matrices: $Q_k(m_k)$ giving local transition firings of N_k, $U_k(n, m_k)$ for marking changes due to incoming bags n in N_k and $S_k(m_k, c)$ for those due to firings of output transitions of N_k for a colour c. So, to compute the steady state probabilities, it is sufficient to use these "small" matrices.

The main advantages of this method are:
- reduced data structures allow the study of large nets
- elimination of vanishing states may be done at the subnet level which reduces both state space sizes and time computation w.r.t to global elimination
- aggregated views may be defined at different levels leading to hierarchical decomposition from coarsest views to more detailed ones.

These works also point out the very important fact that the study of a subnet in isolation requires to define its *environment*. We propose in Sect. 3.1 such a definition for SWNs in a formal way.

However the following points must be highlighted:
- although [2] deals with SWNs, only Tangible Reachability Graphs (TRG) – not Symbolic Reachability Graph (SRG) – are used to compute the solution, the Well Formed aspects of the net being used only to compute ordinary markings and firings: the Q matrix relates to the unfolded net, and *no aggregation –in the stochastic meaning– is exploited*.
- *no automatic method to build aggregated views* based upon the net description is provided, which may lead, as pointed out by the author ([1]), to consistency problems.

The present work provides solutions for these two important problems.

2   Theoretical Context

In this section we first set the framework of our research in order to extend the results that we have reviewed above, then point out the key problems about such extensions. Let us recall that we want to develop an aggregation method based upon the SWN formalism while keeping the advantages of the decomposition methods for asynchronous composition of subsystems. Because of space constraint, we refer the reader to [3] for a detailed presentation of SWN and SRG, and to [7] for a complete study.

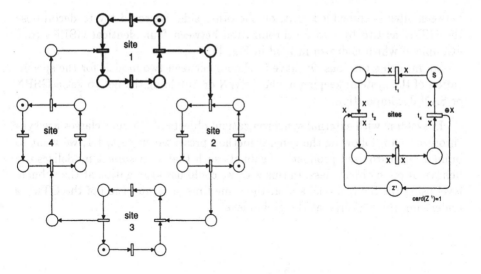

Fig. 1. GSPN and SWN of a logical token ring

2.1 The specification problem

The first problem with this approach relates to the kind of synchronization found at the system specification level: if several objects of the same "type" (processes for instance) are synchronized together, then we may build a model with a product structure , each of the terms of this synchronized product being a model of one object behaviour, leading to a product of CTMCs; but then, as we modelize *one* object behaviour with *one* subnet, there is no object class at all, and therefore we cannot use aggregation. In these situations, we say that the system exhibits *internal synchronization*, "internal" meaning "between objects of the same kind". Alternatively, we may also build SWN models of such systems, and we get an aggregated CTMC of the whole model, but no decomposition may be used.

We say that the system exhibits *external synchronization* if there is synchronization between objects of *different classes*.

A simple example of internal synchronization is a system of sites executing sequential code with a section in mutual exclusion, the enabling of critical section execution being allocated in a cyclic manner to each site (logical token ring). The GSPN and SWN of this system (with 4 sites) is shown in Fig. 1: starting from the idle state, each site does a first job (first transition) and then waits for the mutual exclusion token to continue its work (second transition). When the critical section work is done, it releases the mutual exclusion token (fourth transition) and returns to idle state. In the SWN model, we have only one basic class C_s for the Sites. The S marking indicates that all sites are idle in the initial state and the Z^1 dynamic marking means that this place holds *any single token* of the colour class C_s.

As we see, the SWN is the "folding" of the GSPN and the synchronization between sites is embedded in it; on the other side, we could try to decompose the GSPN as two by two synchronization between four identical GSPNs (one example of which is drawn in bold in Fig. 1).

So, in such situations, we have to choose between two models for the specification of the system: keeping a whole SWN or unfolding the net to get a GSPN or SAN decomposition.

In systems with external synchronization, objects of different classes are synchronized and, following the general method presented in Sect. 1.1, we want to build a "synchronized product" of subnets such that each subnet modelizes the behaviour of an object class. In this way, we could use aggregation at the subnets level using the SWN model and at the same time, a composition of the CTMCs underlying these SWNs, at the global level.

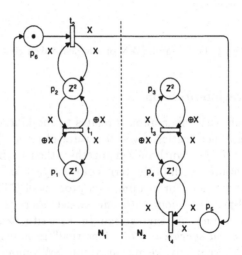

Fig. 2. Asynchronous composition of SWNs without combined aggregation and decomposition

2.2 The resolution problem

Given a system with external synchronization only, we can modelize it as a SWN N, *asynchronous composition* of SWNs $(N_k)_{k \in K}$ each N_k being a SWN model of an object class. Unfortunately, as a general rule, a direct extension of the GSPNs composition cannot be used *to solve* the initial CTMC because composition (i.e. cartesian product) of aggregates provided by the SRGs of the N_k *does not provide an aggregation of the whole CTMC of the model* verifying the strong lumpability condition (1). The main reason for this is that synchronization of *coloured* tokens is not preserved in such a direct composition.

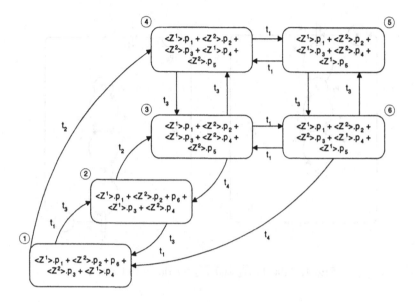

Fig. 3. SRG of the net of Fig. 2

Let us give an example of this problem with the SWN of Fig. 2. N is the composition of 2 SWNs N_1 and N_2. The colour domain of all places except p_6 and all transitions is a single ordered colour class C with $|C| = 2$. Firing of t_1 (t_3) exchanges the colour of places p_2 and p_1 (p_4 and p_3). Firing of t_2 provides the colour of p_2 to p_5 and firing of t_4 returns an non coloured token in p_6 when p_4 and p_5 have the same token colour. The SRG of N is given in Fig. 3: symbolic markings 1 and 2 are the only ones with one token in p_6 and they differ by the colour of tokens in p_2 and p_4 (same – Z^2 – or different – Z^1 and Z^2 – colours). Let us emphasis that *these two markings cannot be aggregated because the firing of t_2 produces the markings (3 and 4) from which t_4 is enabled (in 3) or nor (in 4)*. Figure 4 shows two subnets $\overline{N}_1$ and $\overline{N}_2$ which are extensions of N_1 and N_2 with an abstract view of the complementary net reduced to one place and one transition and Fig. 5 gives the SRGs of $\overline{N}_1$, $\overline{N}_2$ and their "product". The marking 12 corresponds to markings 1 *and* 2 in the original SRG. The key point is that *the abstract view does not catch the colour synchronization which will append in t_4*, and it is easy to see that no other abstract view could do it.

Furthermore, even if we can define a "synchronized product" of SRGs, we have to find how to express the label associated with an output transition, that is to say its rate, with information provided by these SRGs.

All of the following work consists of building nets in which the colour synchronization are never conditioned by earlier firings in more than one subnet. This is done through:

– the definition of an abstract view of each component (subnet),

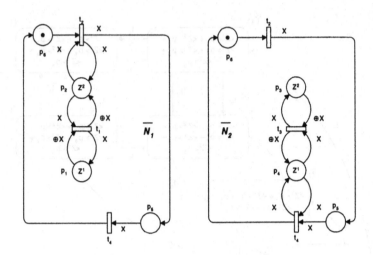

Fig. 4. Subnets $\overline{N}_1$ and $\overline{N}_2$ for the net of Fig. 2

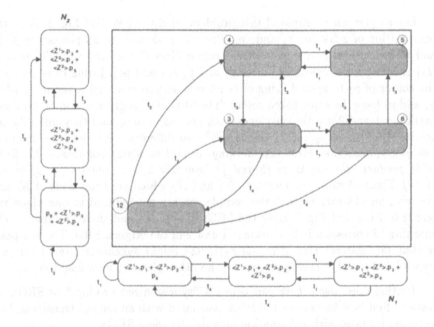

Fig. 5. SRGs of $\overline{N}_1$, $\overline{N}_2$ (Fig. 4) and their "product"

– the definition of *modified* – that is to say enlarged – SWNs (denoted $\overline{N_k}$) of the subsystems including *a representation of their environment* via abstract views of all other subnets.

From the SRGs of these SWNs we derive of a tensor expression of the generator of the model, and we compute performance measures with iterative methods.

3 Asynchronous composition of SWNs

In the present work we assume that each transition has exponential firing time, so that the stochastic process defined by N is a CTMC, and that the transition rate is marking independent (future work will partially relax these conditions).

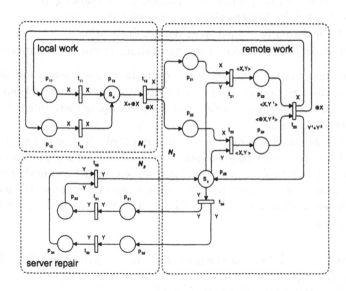

Fig. 6. Example of asynchronous composition of SWNs

We give in Fig. 6 an example of asynchronous composition of SWNs to which we shall refer throughout this section to explain our definitions and results. A brief explanation of the system is as follows: we have a client-server system in which clients are initially located in N_1 (place p_{13}) and servers in N_2 (place p_{25}). Clients emit a server request by pairs of neighbour (variables X and $\oplus X$, the client class C_c is ordered). The requests are treated in N_2 where servers (class C_s) execute the requests (transitions t_{21}, t_{22} and t_{23}). A server may fails: in this case it must be repaired with two jobs located in N_3 (transitions t_{31}, t_{32} and t_{33}).

The basic colour classes are hence C_c and C_s and the colour domains of each node (not shown in the figure for ease of reading) are: C_c for p_{11}, p_{12}, p_{13}, p_{21},

p_{22}, t_{11}, t_{12} and t_{13}, C_s for p_{25}, p_{31}, p_{32}, p_{33}, p_{34}, t_{24}, t_{31}, t_{32} and t_{33}, $C_c \times C_s$ for p_{23}, p_{24}, t_{21} and t_{22}, $C_c \times C_s^2$ for t_{23}.

3.1 The abstraction process

As we have seen in previous sections, the first problem is to be able to define subnets $\overline{N}_k$ embedding N_k and *an aggregated view of its environment*. This means that the modelizer has to define an abstract view of each N_k which will be used in the description of the environment of other subnets.

An abstract view has to enforce a set of constraints:

- it allows to hide details of behaviour of the subnet.
- it is consistent, that is to say: if $M[\delta\rangle M'$ with a sequence of firings $\delta = \tau_1^* \sigma \tau_2^*$ composed of local firings τ_1^* and τ_2^* to be hidden, then we must have: $a(M)[\sigma\rangle a(M')$ with $a(M)$ the abstraction of the marking M.
- it must let "visible" interactions between global entities.
- it has to be formally defined from the net description.
- it is compatible with a combined aggregation/decomposition method.

To take into account the previous constraints, this abstraction should be *guided by qualitative considerations*, especially by observing interactions between global entities (colour classes) inside each subnet keeping these interactions "visible" in the abstraction.

Furthermore, more generally and unlike Buchholz, it does not seem possible to abstract a subnet with *only one place*: we need at least one place to modelize entities of each basic colour class moving from one subnet to another one.

We propose to define each place through a partial semiflow[2] to ensure consistency of the abstraction and deal with formal definitions deduced from the net description.

Definition 2. An *abstraction semiflow* f of N_k is a partial semiflow of N with respect to $T_k \backslash TO_k$ s.t.:
- there is a colour class C_i s.t. $C(f) = C_i$.
- $\forall p \in P_k \backslash PO_k$, f_p is 0 or b times a projection (with b a positive constant).

Let F_k be a set of abstraction semiflows of N_k. The *abstract view* of N_k w.r.t F_k is the set of places $PA_k = \{p_f \; ; \; f \in F_k\}$ with:
- $C(p_f) = C(f)$ (the colour domain of p_f).
- $M_0(p_f) = \sum_{p \in P_k \backslash PO_k} f(M_0(p))$ (the initial marking of p_f).

For any marking $M = (M_k)_{k \in K}$ of N, the *abstract marking* of M_k is $am(M_k) = (\sum_{p \in P_k \backslash PO_k} f(M(p)))_{f \in F_k}$, also denoted $am_k(M)$ or $M(PA_k)$ (the values of the semiflows of F_k in the marking M_k)

The *global colours* of N_k are $\{C(f) \; ; \; f \in F_k\}$.

---

[2] f is a partial semiflow on $N = (P, T, \ldots)$ w.r.t a set $T' \subseteq T$ of transitions iff f is a semiflow on the net $N' = (P, T', \ldots)$

The objects of the global classes of the (N_k) are the only ones which move between subnets. Let us note that a global class for N_k may be a non global one for another $N_{k'}$. Such classes may be renamed with different names in each N_k where it is non global. From now on, *we assume that such a renaming has been done.*

In our example SWN, we have two partial semiflows in N_2 ($f_{2c} = X.p_{21} + X.p_{23} + X.p_{22} + X.p_{24}$ and $f_{2s} = X.p_{25} + Y.p_{23} + Y.p_{24}$) and one partial semiflow in N_3 ($f_{3s} = X.p_{31} + X.p_{33}$).

We can now define the modified subnets allowing decomposition under appropriate conditions, set additional notations and define the aggregation function we shall use. In the sequel, the abstractions of the (N_k) are given.

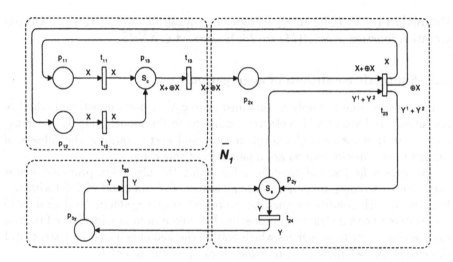

Fig. 7. Subnet $\overline{N}_1$ for the example SWN

Definition 3. Let N be an asynchronous composition of the $(N_k)_{k \in K}$. The *extension* $\overline{N}_k$ of N_k is the SWN $(\overline{P}_k, \overline{T}_k, C, J, \overline{W_k^-}, \overline{W_k^+}, \Phi, \omega, \overline{M}_{0,k}, \theta_k)$ with:

- $\overline{P}_k = (P_k \backslash PO_k) \bigcup_{k' \neq k} PA_{k'}$ (For each marking M of N, the corresponding marking of $\overline{N}_k$ is $\overline{M}_k = (M_k, (M(PA_{k'}))_{k' \neq k})$).
- $\overline{T}_k = T_k \bigcup_{k' \neq k} TO_{k'}$
- $\forall p \in P_k \backslash PO_k, \forall t \in \overline{T}_k, \overline{W_k^-}(p,t) = W_k^-(p,t)$ and $\overline{W_k^+}(p,t) = W_k^+(p,t)$
- $\forall p_f \in PA_{k'}, \forall t \in \overline{T}_k$:
  $$\overline{W}^-(p_f,t) = \sum_{p' \in {}^\bullet t} f_{p'} \circ W^-(p',t) \text{ and } \overline{W}^+(p_f,t) = \sum_{p' \in t^\bullet} f_{p'} \circ W^+(p',t)$$

$\overline{M}_k(PA_{k'})$ is still named the *abstract marking* of $N_{k'}$ (in $\overline{M}_k$).

Fig. 7 shows the extension $\overline{N}_1$ of N_1. We see that in PA_2 we have only two places p_{2x} and p_{2y} for the colours C_c and C_s, and in PA_3 one place for C_s. The transitions t_{13}, t_{23}, t_{24} and t_{33} are also modified accordingly with our definition.

We could also define a full abstract view $\overline{N}$ of N with abstract views only of all subnets, useful in a hierarchical design process, however our method does not use $\overline{N}$.

Notations
- $\overline{SRS}_k$ (resp. $\overline{SRG}_k$) is the SRS (resp. the SRG) of $\overline{N}_k$,
- $\overline{XSRS} = \prod_{k \in K} \overline{SRS}_k$.
- $\overline{M}_k$ is the symbolic marking of $\overline{M}_k$ in $\overline{SRS}_k$ and $\overline{M} = (\overline{M}_k)_{k \in K} \in \overline{XSRS}$.
- $\mathcal{D}(\overline{XSRS}) = \{M \mid \exists \overline{M} \in \overline{XSRS} \text{ s.t. } \forall k \in K,\ \overline{M}_k \in \mathcal{M}_k\}$

Definition 4. Let N be an asynchronous composition of the $(N_k)_{k \in K}$. The aggregation function $\mathcal{A}$ is: $\mathcal{A}(M) = (\overline{M}_k)_{k \in K} = \overline{M} \in \overline{XSRS}$

3.2 Syntactic conditions of aggregation

In order to be able to apply a combined aggregation/decomposition method of resolution, we have to add syntactic conditions to the model: roughly speaking, they mean that for each global colour class and every marking, the subsets of colours in the abstract views are a partition of this class.

We give a first set of conditions for which the algorithm proposed below may be used to compute performance measures of N. As usual with (stochastic) Petri nets, such conditions must be expressed at the *syntactic* level that is to say relative to either structural properties like colour domains, incidence function expressions, ... or to properties which may be checked only using these structural properties like semiflows, ..., to avoid checking of the RG of N.

Definition 5. We say that N fulfills the syntactic conditions of aggregation iff $\forall k \in K$ we have the following properties:

1. $\forall p \in {}^{\bullet}TO_k$ with $C(p) = \prod_{i \in I} C_i^{e_i}$, $\forall i \in I$ s.t. C_i is a global colour of N_k and $e_i > 0$ we have: $\forall 1 \leq j \leq e_i$, X_i^j (the jth projection on C_i) is in one abstraction semiflow $f \in F_k$.

2. $\forall t \in TO_k \ \forall X \in Var(t)$ corresponding to a global colour class of N_k, X is in a positive term of some $W^-(p, t)$.

3. $\forall p \in PA_k$ with $C(p) = C_i$, $\forall k' \neq k$, $\forall p' \in P_{k'}$ with $C(p') = \prod_{i \in I} C_i^{e'_i}$:
   $\forall 1 \leq j' \leq e'_i$ there is a semiflow $g = (g_q)_{q \in \overline{P}_{k'}}$ on C_i in $\overline{N}_{k'}$, s.t. $\forall M\ g(M) = S_i$ with:
   $$g_q = \begin{cases} 0 \text{ or a projection} & \text{if } q \neq p, q \neq p' \\ Identity & \text{if } q = p \\ j'\text{th projection} & \text{if } q = p' \end{cases}$$

Condition 1 ensures that the abstract view of N_k is not too coarse and memorizes colours moving between components: the basic colour classes involved in the firings of $t \in TO_k$ have to be in the colour domain of PA_k. Condition 2 means that the firing colours of $t \in TO_k$ must be in the input places of t. At last, condition 3 implies that in each marking, we have a partition of each occurrence of each global colour C_i, between colours usable to fire t and colours in other subnets $N_{k'}$.

It is easy to show that our example SWN fulfills these conditions.

Let us notice these conditions are fulfilled for many nets which are models of systems with several components each having some kind of autonomy.

3.3 Performance measures computation algorithm

As in the synchronous case, the basis of the algorithm is a combination of the tensor expression of the generator Q of the CTMC of the synchronized product of the $\overline{SRG}_k$ and of the regular computation for SWNs.

The CTMC transitions come from the firing of a transition $t \in T \backslash TO$ hence changing *only the kth component of the global state*, or from the firing of an output transition $t \in TO$ changing several components $k_1, \ldots, k_l$ of the state.

As transition firing rates depend only on the static subclasses of the chosen colors, the rate $r(t)$ from $\mathcal{M}$ to another $\mathcal{M}'$ sums the rates $r(t, d, \langle \lambda, \mu \rangle)$ of all symbolic firings $\langle \lambda, \mu \rangle$ of t which fit the choice d, for all given color static subclasses choices d. Moreover $r(t, d, \langle \lambda, \mu \rangle)$ is card($\mathcal{F}$) (also denoted $|\mathcal{F}|$) times the mean firing time $\theta(t, d)$ of t for d where $\mathcal{F}$ is the set of ordinary firings of t from any ordinary marking of $\mathcal{M}$ to ordinary markings of the symbolic marking $\mathcal{M}'$.

It can be shown then, that the matrix Q of the CTMC can be written as sub-matrix of

$$Q' = \bigoplus_{k=1}^{K} Q'_k + \sum_{t \in TO} \sum_{d} \theta(t, d) \left[\bigotimes_{k=1}^{K} C_k(t, d) - \bigotimes_{k=1}^{K} A_k(t, d) \right] \quad (2)$$

where $d = (d_i^j)_n^{e_i(t)}$ (with n_i the number of static subclasses of C_i and $1 \leq d_{ij} \leq n_i$) is a choice of static subclasses for the symbolic firings of t ($((a_i^j)_n^{b_i}$ denotes the tuple $(a_1^1, \ldots, a_1^{b_1}, \ldots, a_i^1, \ldots, a_i^{b_i}, \ldots, a_n^1, \ldots, a_n^{b_n})$).

The Q'_k matrices come from the generator of the CTMC of the $\overline{N}_k$ nets in isolation using only local transitions: they can be built from classical SWN technics, discarding any output transition effect.

The $\bigoplus$ operator means that these CTMCs are independent stochastic processes. The A_k and C_k matrices are obtained as consequence of output transition firings: any such firing produces a state change in each component $\overline{N}_k$ involved in the marking change.

By sub-matrix we mean that non zero terms (for same pair of states) of Q equal those of Q' and that if $\overline{\mathcal{M}}$ is reachable and $\overline{\mathcal{M}}'$ is unreachable then $q'_{\overline{\mathcal{M}}, \overline{\mathcal{M}}'} = 0$.

The $C_k(t, d)$ and $A_k(t, d)$ matrices are computed by the algorithm below.

Let us denote:

− $\forall \langle \lambda, \mu \rangle$ (instantiation functions in $\overline{N}_k$ for t), $\overline{\mathcal{M}}_k$ and $\overline{\mathcal{M}}'_k$ in $\overline{SRS}_k$:

$$1_{(t,d,\langle \lambda,\mu \rangle),\overline{\mathcal{M}}_k,\overline{\mathcal{M}}'_k} = \begin{cases} 1 & \text{if} \quad d = (d(Z_i^{\lambda_i(j)}))_n^{e_i(t)} \\ & \text{and} \quad \overline{\mathcal{M}}_k[t(\langle \lambda,\mu \rangle)) \rangle \overline{\mathcal{M}}'_k \\ 0 & \text{else} \end{cases}$$

$$1_{(t,d,\overline{\mathcal{M}}_k,\overline{\mathcal{M}}'_k)} = \bigvee_{(\lambda,\mu)} 1_{(t,d,\langle \lambda,\mu \rangle),\overline{\mathcal{M}}_k,\overline{\mathcal{M}}'_k}$$

with $\bigvee$ denoting the Boolean addition (logical or).

− for $t \in TO_k$

$$F_{(\langle \lambda,\mu \rangle),\overline{\mathcal{M}}_k,\overline{\mathcal{M}}'_k} = \prod_{i=1}^{h} \prod_{j=1}^{m_i} \frac{\mathrm{card}(Z_i^j)!}{(\mathrm{card}(Z_i^j) - \mu_i^j)!}$$

with h the highest index of non ordered basic colour classes of $C(t)$. Then we have the following algorithm.

Algorithm:

1. for each $k \in K$ compute $\overline{SRG}_k$ (hence $\overline{SRS}_k$) ($r_k = |\overline{SRS}_k|$)
2. for each $k \in K$ compute the Q'_k matrices from $\overline{SRG}_k$, using only local transitions
3. for each $t \in TO$ (say $t \in TO_k$)
    for each $h \in K$ compute $C_h(t,d)$ and $A_h(t,d)$ from $\overline{SRG}_h$:
    if $h \neq k$ and $t^\bullet \cap P_h = \emptyset$ then $C_h(t,d) = A_h(t,d) = I_{r_h}$
    if $h = k$ then $c_h(t,d)_{\overline{\mathcal{M}}_h,\overline{\mathcal{M}}'_h} = \sum_{(\lambda,\mu)} 1_{(t,d,\langle \lambda,\mu \rangle),\overline{\mathcal{M}}_h,\overline{\mathcal{M}}'_h} F_{(\langle \lambda,\mu \rangle),\overline{\mathcal{M}}_h,\overline{\mathcal{M}}'_h}$
    if $h \neq k$ and $t^\bullet \cap P_h \neq \emptyset$ then $c_h(t,d)_{\overline{\mathcal{M}}_h,\overline{\mathcal{M}}'_h} = 1_{(t,d,\overline{\mathcal{M}}_h,\overline{\mathcal{M}}'_h)}$

$$a_h(t,d)_{\overline{\mathcal{M}}_h,\overline{\mathcal{M}}'_h} = \begin{cases} \sum_{\overline{\mathcal{M}}''_h} c_h(t,d)_{\overline{\mathcal{M}}_h,\overline{\mathcal{M}}''_h} & \text{if} \quad \overline{\mathcal{M}}_h = \overline{\mathcal{M}}'_h \\ 0 & \text{else} \end{cases}$$

4. compute the performance measure using the tensor expression of Q'.

To use the tensor expression of Q', the numerical method computing a given measure has to verify the conditions:

− only linear functions of products $V.Q^m$ are used, with V a vector
− no unreachable state is involved in the computation

Let us emphasize that such computations never use Q' directly but instead the Q'_k, C_k and A_k matrices.

An important example of performance measure is the steady state probability distribution vector of the aggregated CTMC. We can then use iterative methods like the power method or the faster GMRES method ([12]) to compute these probabilities with the proposed algorithm. However, to ensure the above conditions, the initial vector must have non zero components for reachable states only: we can for instance, choose the vector $v^{(0)}$ with $v_{\mathcal{M}_0}^{(0)} = 1$ and $v_{\mathcal{M}}^{(0)} = 0$ if $\mathcal{M} \neq \mathcal{M}_0$.

We state in the following theorem the results giving the correctness of the algorithm.

Theorem 6. *Let N be an asynchronous composition of the $(N_k)_{k \in K}$, then:*
1. *$RS \subseteq \mathcal{D}(\overline{XSRS})$*
2. *the function $\mathcal{A}$ defined by $\mathcal{A}(M) = (\overline{M_k})_{k \in K} = \overline{M} \in \overline{XSRS}$ is an aggregation function verifying the strong lumpability condition (1)*
3. *the transition rate from any marking $\overline{M}$ of $\overline{XSRS}$ is given in the previous algorithm.*

The proof is given in App.A.

Conclusion

In this paper we have shown how to combine two methods, aggregation and decomposition, to deal with the increasing complexity of parallel systems. The studied systems are composed of subsystems communicating via entities moving between them. Decomposition expresses the state space of the system as a cartesian product of smaller spaces. Aggregation reduces the state space by grouping states and solving the Markov chain on the set of state classes; the SWN model moreover enables the reduction directly from the net description.

We have shown that in the case of internal synchronization, we have to choose between either decomposition or aggregation to specify the studied system. Now, in the case of external synchronization, we have shown that it is the synchronization memory phenomenon which allows or disallows the merging of the two methods to solve its Markov chain.

We have given a new method, allowing a combined aggregation–composition approach: we construct new subnets, built from original ones and abstract views of the others to deal with synchronization memory – capturing the synchronization memory – and we apply the decomposition approach on their SRGs. We have established a set of syntactic conditions under which such a method can be used.

Future work will extend the results to SWNs with immediate transitions and will experiment the method for large nets with a Petri net tool ([4]).

A   Proof of the algorithm

The proof[3] of theorem 6 is established in several steps:
- definition of a set of semantic conditions, that is to say at the marking level (definition 8) for which
- we prove that they verify the strong lumpability condition (1) (theorem 10 based upon lemma 9).
- proof that the syntactic conditions of aggregation imply those semantic conditions (theorem 11).
- proof that the generator of the aggregated CTMC is a submatrix of Q' given in the algorithm.

---

[3] This section may be skipped by readers how have no interest in theoretical developments

We use a semantic intermediate level in this proof. The reason of which is threefold:

- the semantic conditions allow a structured definition of syntactic conditions: for each of the first one, we try to find a syntactic translation.
- given the same set of semantic conditions, we shall be able to find other sets of syntactic conditions, for special classes of nets, reducing the new proof to the fact that syntactic conditions imply semantic conditions.
- the correctness of the expression of Q' depends on syntactic, not semantic, conditions.

A.1 Semantic conditions of aggregation

We now exhibit the marking level conditions under which we can define a combined aggregation and decomposition. From the above aggregation function definition, two markings M and M' will be in the same aggregate if there is a K-tuple of admissible permutations s_k such that $\forall k \in K$, $s_k(\overline{M}_k) = M'_k$. However, for any marking, the (s_k) must be equal on the markings of the common abstract places of the $(\overline{N}_k)$ so we introduce the definition:

Definition 7. Two admissible permutations s and r are *compatible w.r.t.* M_k (or $\overline{M}_k$) iff $s(\text{am}(M_k)) = r(\text{am}(M_k))$.
A *compatible family of permutations* $s = (s_k)_{k \in K}$ w.r.t M is a family of admissible permutations s_k s.t. for any k, k', k'', s_k and $s_{k'}$ are compatible w.r.t. $M_{k''}$; for such a family we denote $s(M) = (s_k(M_k))_{k \in K}$.

Definition 8. We say that N fulfills the semantic conditions of aggregation iff for every k and every reachable marking $\overline{M}_k$ we can find a subset $C_{i,k}$ of every $C_i \in C$ with the following properties:

1. for every admissible permutation s:
   $s(\text{am}(\overline{M}_k)) = \text{am}(\overline{M}_k) \Rightarrow (\forall i \in I \ s(C_{i,k}) = C_{i,k})$.
2. $\forall t \in TO_k :$ if $\overline{M}_k[t(c)\rangle$ then c is composed of colours of $(C_{i,k})_{i \in I}$.
3. for every compatible permutations r and s w.r.t $\overline{M}_k$ s.t. $\forall i \in I \ r_{|C_i \backslash C_{i,k}} = s_{|C_i \backslash C_{i,k}}$, we have:
   $\forall k', \forall M'_{k'}: \overline{M}'_{k'}(PA_k) = \text{am}(\overline{M}_k) \Rightarrow r(\overline{M}'_{k'}) = s(\overline{M}'_{k'})$.

Intuitively, condition 2 means that the $C_{i,k}$ are the set of basic colours enabling the firings of $t \in TO_k$. Condition 1 ensures that the abstract view of $\overline{M}_k$ is consistent with respect to firings of $t \in TO_k$. Condition 3 expresses the fact that only colours not in the $C_{i,k}$ are relevant outside $\overline{M}_k$.

Lemma 9. *Let N fulfill the semantic conditions above and M s.t. $\forall k \in K \ \overline{M}_k$ is reachable in $\overline{N}_k$. Let $s = (s_k)_{k \in K}$ be a compatible family of permutations w.r.t M.*

If $M[t(c)\rangle M'$ then there is a permutation r and a compatible family $q = (q_k)_{k \in K}$ w.r.t M' s.t. $s(M)[t(r(c))\rangle q(M')$, with r depending only upon t and s.

Proof. We have to find r and q such that $\forall k \in K$, $s_k(M_k)[t(r(c))\rangle q_k(M'_k)$. From the definition of the $(\overline{M}_k)_{k \in K}$ and the compatibility of the $(s_k)_{k \in K}$, it is equivalent to show: $\forall k \in K$, $s_k(\overline{M}_k)[t(r(c))\rangle q_k(\overline{M}'_k)$.

• *First case:* $t \in T_{k_0} \backslash TO_{k_0}$ is a local transition.

We have then $\forall k \neq k_0$ $\overline{M}'_k = \overline{M}_k$; and $t \notin T_k \Rightarrow s_k(\overline{M}_k)[t(s_{k_0}(c))\rangle s_k(\overline{M}_k) = s_k(\overline{M}'_k)$.

As s_{k_0} is admissible $s_{k_0}(\overline{M}_{k_0})[t(s_{k_0}(c))\rangle s_{k_0}(\overline{M}'_{k_0})$.

The result follows with $r = s_{k_0}$ and $\forall k \in K$ $q_k = s_k$ i.e. $q = s$.

• *Second case:* $t \in TO_{k_0}$ is an output transition.

We apply the semantic conditions with $s_{k_0}(\overline{M}_{k_0})$ and show that we can choose $r = s_{k_0}$.

Since s_{k_0} is admissible, $s_{k_0}(\overline{M}_{k_0})[t(s_{k_0}(c))\rangle s_{k_0}(\overline{M}'_{k_0})$

We now prove that for each $k \neq k_0$ we have $s_k(\overline{M}_k)[t(s_{k_0}(c))\rangle q_k(\overline{M}'_k)$ with adapted q_k.

Let us denote $M'' = s_{k_0}(M)$ and $(C_{i,k_0})_{i \in I}$ the subsets from the semantic conditions w.r.t. M''_{k_0}.

For each $k \neq k_0$, let us denote $u_k = s_k \circ s_{k_0}^{-1}$. As s is a compatible family and by definition of $\overline{M}''_{k_0}$ and u_k, we have[4]: $u_k(am(\overline{M}''_{k_0})) = am(u_k(\overline{M}''_{k_0})) = am(s_k(\overline{M}_{k_0})) = am(s_{k_0}(\overline{M}_{k_0})) = am(\overline{M}''_{k_0})$. So we have, from semantic condition 1, $u_k(C_{i,k_0}) = C_{i,k_0}$ for every i.

Let us now define the permutation v_k:

$\forall i \in I$, $v_{k|C_{i,k_0}} = Id_{C_{i,k_0}}$ and $v_{k|C_i \backslash C_{i,k_0}} = u_{k|C_i \backslash C_{i,k_0}}$. It is clear that v_k is admissible, hence so is $v_k \circ s_{k_0}$ and $v_k \circ s_{k_0}(\overline{M}_k)[\, t(v_k \circ s_{k_0}(c))\rangle \, v_k \circ s_{k_0}(\overline{M}'_k)$

From semantic condition 2, s_{k_0} is composed of colours from the C_{i,k_0}, hence, by definition of v_k, $v_k \circ s_{k_0}(c) = s_{k_0}(c)$.

We have for every $i \in I$, $v_{k|C_i \backslash C_{i,k_0}} = u_{k|C_i \backslash C_{i,k_0}}$. Applying the semantic condition 3, we get $v_k(\overline{M}'_k) = u_k(\overline{M}''_k)$, that is to say $v_k(\overline{M}'_k) = s_k(\overline{M}_k)$, and finally $s_k(\overline{M}_k)[t(s_{k_0}(c))\rangle q_k(\overline{M}'_k)$ with $q_k = v_k \circ s_{k_0}$ ($k \neq k_0$) and $q_{k_0} = s_{k_0}$ ($q = (q_k)_{k \in K}$ is clearly a compatible family w.r.t $\overline{M}'$).

From this lemma we deduce the following theorem:

Theorem 10. *Let N fulfill the semantic conditions above. Then the aggregation function $\mathcal{A}$ fulfills the strong lumpability condition (1) on $\mathcal{D}(\overline{XSRS})$.*

Proof is omitted (see [8]).

A.2 From syntactic conditions to semantic conditions

Theorem 11. *Let N fulfill the syntactic conditions of aggregation. Then N fulfills the semantic conditions of aggregation.*

---

[4] from the definition of the flows $f \in F_k$, $s(am(M_k)) = am(s(M_k))$ for any admissible permutation s.

Sketch of proof (see [8] for a detailed proof): Let k and $\overline{M}_k$ be given. We define the partition of the semantic conditions as: $C_{i,k} = \{$colours of C_i in am$(\overline{M}_k)\}$ (let us remark that $C_{i,k} = \emptyset$ for any non global colour of N_k).

We then prove successively the semantic condition 1 from the definition above, the second semantic condition from syntactic conditions 1 and 2 and the third semantic condition from syntactic condition 3.

A.3 Generator of the aggregated CTMC

The sketch of the proof is the following (see [8] for a detailed proof):

Let $\theta(\overline{M}, \overline{M'})$ be the rate from the reachable state $\overline{M}$ to the reachable state $\overline{M'}$ in the aggregated CTMC. We have $\theta(\overline{M}, \overline{M'}) = \sum_t \sum_d \theta(t, d)|U_{t,d}|$ with $U_{t,d} = \{\overline{M} \xrightarrow{t(c)} \overline{M'} \mid \overline{M'} \in \overline{\mathcal{M}'}, \; d(c) = d\}$ and $\theta(t, d)$ the rate of the transition t for any colour with static partition d and $\overline{M}$ fixed in $\overline{\mathcal{M}}$.

For $t \notin TO$ it is clear that $\bigoplus_{k=1}^{K} Q'_k$ gives the correct rates.

For $t \in TO_h$ we rewrite $U_{t,d}$ with firing sets of the $(\overline{N}_k)_{k \in K}$ which leads to the expression of the algorithm.

B Flows, semiflows and implicit places in WNs

Flows are structural invariant of Petri nets: to each flow is associated a constant sum of weighed markings of places which gives information about the behaviour of the net. For coloured PN, the definition of flows uses *place colour functions* instead of constants hence the name *symbolic* flows.

Definition 12. Let N be a WN with places P and incidence matrix W (of linear functions from $\text{Bag}_{\mathbb{Q}}(C(t))$ to $\text{Bag}_{\mathbb{Q}}(C(p))$). Let A be a set. A (symbolic) flow of N on A is a vector $f = (f_p)_{p \in P} \neq 0$ of linear functions from $\text{Bag}_{\mathbb{Q}}(C(p))$ to $\text{Bag}_{\mathbb{Q}}(A)$ s.t.:

$$\forall t \in T, \; \sum_{p \in P} f_p \circ W(p, t) = 0$$

For any reachable marking M, we have then:

$$\sum_{p \in P} f_p(M(p)) = \sum_{p \in P} f_p(M_0(p)) = \sum_{a \in A} \alpha(a).a \quad \text{(a constant)}$$

A semiflow is a flow with positive functions f_p: $\forall a \in A, \; \forall c \in \text{Bag}(C(p))$, $f_p(c)(a) \geq 0$

For WNs, following linear functions play an important role (for a place p with colour domain $\prod_{i \in I} C_i^{e_i}$ and a colour c of p):
- Identity: $Id(c) = c$ (and also $n.Id$ with n a constant number)
- Projections: for a colour class C_i the jth projection is $\text{Proj}_i^j(c) = c_i^j$

An implicit place with respect to a set P' of places, does not disable the firing of any transition for which preconditions are satisfied in P'.

Definition 13. A place p of a coloured Petri net $N = (P, T, \ldots)$ is implicit w.r.t. $P' \subseteq P$ $(p \notin P')$ iff:

1. there is a symbolic flow f with domain $C(p)$ s.t.: $f_p = Id$ and $\forall q \in P'$, $f_q < 0$

2. $\forall t \in T$, we have $\forall c \in C(t)$:

$$f_p(M_0(p)) + \sum_{q \in P'} f_q(M_0(q)) \geq f_p(W^-(p,t)(c)) + \sum_{q \in P'} f_q(W^-(q,t)(c))$$

An *implicit place* is an implicit place w.r.t some P'.

References

1. P. Buchholz. Hierarchies in colored GSPNs. In *Proc. of the 14th International Conference on Application and Theory of Petri Nets*, number 691 in LNCS, pages 106–125, Chicago, Illinois, USA, June 1993. Springer–Verlag.

2. P. Buchholz. Aggregation and reduction techniques for hierarchical GCSPN. In *Proc. of the 5th International Workshop on Petri Nets and Performance Models*, pages 216–225, Toulouse, France, October 19–22 1993. IEEE Computer Society Press.

3. G. Chiola, C. Dutheillet, G. Franceschinis, and S. Haddad. Stochastic well-formed colored nets and symetric modeling applications. *IEEE Transactions on Computers*, 42(11):1343–1360, November 1993.

4. G. Chiola, G. Franceschinis, R. Gaeta, and M. Ribaudo. Greatspn 1.7: graphical editor and analyzer for timed and stochastic Petri nets. *Performance Evaluation*, 24(1&2):47–68, 1995.

5. M. Davio. Kronecker products and shuffle algebra. *IEEE Transactions on Computers*, 30(2):116–125, 1981.

6. S. Donatelli. Superposed generalized stochastic Petri nets: definition and efficient solution. In Robert Valette, editor, *Proc. of the 15th International Conference on Application and Theory of Petri Nets*, number 815 in LNCS, pages 258–277, Zaragoza, Spain, June 20–24 1994. Springer–Verlag.

7. C. Dutheillet. *Symétries dans les réseaux colorés. Définition, analyse et application à l'évaluation de performance.* Thèse, Université Paris VI, Paris, France, January 28 1991.

8. S. Haddad and P. Moreaux. Asynchronous composition of stochastic high level Petri nets. Research report, LAMSADE, Université Paris Dauphine (to appear).

9. S. Haddad and P. Moreaux. Evaluation of high level Petri nets by means of aggregation and decomposition. In *Proc. of the 6th International Workshop on Petri Nets and Performance Models*, pages 11–20, Durham, NC, USA, October 3–6 1995. IEEE Computer Society Press.

10. J. G. Kemeny and J. L. Snell. *Finite Markov Chains*. V. Nostrand, Princeton, NJ, 1960.

11. B. Plateau and J.M. Fourneau. A methodology for solving Markov models of parallel systems. *Journal of parallel and distributed computing*, 12:370–387, 1991.

12. Y. Saad and M. H. Schultz. Gmres: A generalized minimal residual algorithm for solving nonsymetric linear systems. *SIAM J. Sci. Stat. Comput.*, 7:856–869, 1986.

A Formal Definition of Hierarchical Predicate Transition Nets

Xudong He

Department of Computer Science
North Dakota State University
Fargo, ND 58105, U.S.A.

Abstract. Hierarchical predicate transition nets have recently been introduced as a visual formalism for specifying complex reactive systems. They extend predicate transition nets with hierarchical structures so that large systems can be specified and understood stepwisely, and thus are more suitable for real-world applications. In this paper, we provide a formal syntax and an algebraic semantics for hierarchical predicate transition nets, which establish the theory of hierarchical predicate transition nets for precise specification and formal reasoning.

1 Introduction

Petri nets are an excellent model for studying concurrent and distributed systems due to their modeling power and simple graphical notation. Petri nets, unlike many other graphical modeling techniques, have a well-defined algebraic semantics supporting formal analysis of system properties as well as an operational semantics for exhibiting dynamic system behaviors. However traditional Petri nets have the distinct drawback of producing very large and unstructured system specifications for even small systems, which are normally very difficult to understand due to their low-levelness and primitive structures. In the past decade, various types of high-level Petri nets ([1], [5], [13], [19]) have been developed to partially overcome the above drawback; which represents a major progress similar to that from low-level programming languages to high-level programming languages. As the next natural step, modular and hierarchical mechanisms have recently been incorporated into high-level Petri nets ([10], [12], [14], [15], [20]) to make them more suitable for real-world applications.

Heuristic principles and rules for composing and refining large predicate transition net specifications (one type of high-level Petri nets) were first proposed in [20] and then further investigated and formulated in [10]. Although the above research effort has provided a systematic way to incorporate modularity and hierarchies into predicate transition nets, there is a lack of formal definition of both syntax and semantics of the underlying net model.

This paper presents both a formal syntax and an algebraic semantics of hierarchical predicate transition nets, which establish the foundation of hierarchical predicate transition nets for precise specification and formal reasoning. With the formal syntax and semantics, the validity of a net specification can be automatically checked during its construction process and many properties of the net specification can be statically derived without its execution. The formal syntax and semantics also provide a basis for rigorous comparison among different types of hierarchical high-level Petri nets and between hierarchical high-level Petri nets and other formal specification methods. This paper also presents a result for deriving a behavioral equivalent non-hierarchical predicate transition net from a given hierarchical predicate transition net, which provides an alternative semantic definition of hierarchical predicate transition nets and opens the possibility of adapting existing analysis techniques of flat predicate transition nets to hierarchical predicate transition nets.

2 Brief History of Hierarchical Predicate Transition Nets

The development of hierarchical predicate transition nets (HPrT nets in the sequel) was motivated by the need to construct specifications for large systems using Petri nets and inspired by the development of modern high-level programming languages and other hierarchical and graphical specification methods such as data flow diagrams [22] and statecharts [6]. With the introduction of hierarchical structures into predicate transition nets (PrT nets in the sequel), not only the resulting net specifications are more understandable but also the specification construction process becomes more manageable.

During the development of HPrT nets, the following rationals and criteria were followed:
(1) Simplicity: the introduction of new concepts and notations should be kept minimal;
(2) Understandability: the new concepts and notations should resemble and be closely related to those in PrT nets;
(3) Hierarchy and information hiding: different levels of abstraction should be supported;
(4) Executability: HPrT nets should be direct executable without the need being translated into behavioral equivalent PrT nets. The semantics concepts of HPrT nets should be closely related to those of PrT nets;
(5) Compositionality: the new concepts and notations should facilitate the compositional development of large HPrT nets from small existing HPrT nets;
(6) Stepwise abstraction and refinement: the new concepts and notations should support top-down as well as bottom-up development approaches including stepwise abstraction and refinement of predicates, transitions, constraints, labels, and tokens; and
(7) Maintainability: the new concepts and notations should facilitate simple specification modification and extension.

The basic ideas of HPrT nets were first proposed in [20], which were refined in [7] with the introduction of super nodes (dotted predicates and transitions) and non-terminating arcs adapted from statecharts [6]. The above notations along with the four transformational development rules (abstraction, refinement, decomposition, and synthesis) were formalized in [10]. In [10], two label construction operators, + (non-determinancy) and × (concurrency), and the associated rules were also developed based on the semantics of PrT nets and inspired by the data flow balancing concept in data flow diagrams [22]. HPrT nets have been applied to specifying several small systems including an elevator system [10] and a library system [11]. Heuristics and strategies adapted from modern structured analysis [22] were proposed for constructing HPrT net specifications in [11]. An hybrid analysis technique for HPrT nets was presented in [8], which adapts two temporal induction rules ([17], [18]) and combines structural, behavioral, and logic reasoning.

3 Syntax and Static Semantics of Hierarchical Predicate Transition Nets

The syntax and static semantics (typing information) of traditional Petri nets can be represented graphically as well as defined algebraically. In this section, we generalize the traditional graphical representation and algebraic definition to HPrT nets.

3.1 Graphical Notation

The traditional graphical notation of Petri nets include circles denoting predicates, boxes denoting transitions, and directed arcs denoting flow relation, which are naturally retained in HPrT nets. To represent the hierarchical structure and behavior of HPrT nets, we further introduce the following symbols and conventions:

(1) dotted circles / boxes (super nodes) are used to stand for either abstractions or refinements of existing HPrT nets. Predicate p_1 and transition t_2 in Fig. 1 are such examples. Hierarchies are introduced through a sequence of dotted nodes where a lower level one reveals more details than its ancestors. The idea is similar to those of data flow diagrams [22].

(2) non-terminating arcs (one end being connected to the boundary of a dotted node) are introduced to keep track of data flow relationships between a child net with its external environment. If the tail of a non-terminating arc connects to a boundary, it is called an incoming non-terminating arc and is referenced as ($\bullet n$, n) (where n is the node, and $\bullet n$ is the traditional notation for the pre-set of n and denotes the external environment related to n); otherwise it is called a outgoing non-terminating arc (n, $n\bullet$) ($n\bullet$ is the post-set of n). Arcs (p_3, $p_3\bullet$) and (p_4, $p_4\bullet$) in Fig. 1(2) are such examples. The use of non-terminating arcs has been motivated by statecharts [6].

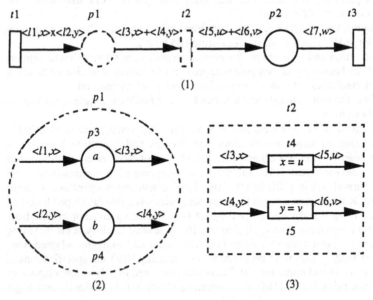

Fig. 1 - An HPrT net

3.2 Arc Labels

In an HPrT net, an arc may corresponds to many different arcs in a PrT net, which have been merged during the construction of the HPrT net. Thus the role of a label defines not only the type and number of tokens to be moved along an arc, but also its component arcs. In order to distinguish component arcs, we need to associate an identification with an arc, i.e. make an identification part of its label. In HPrT nets, a compound arc has a label expression having label constructor + or × (+ and × express non-determinancy and concurrency respectively, and specify proper control flow relations between different abstraction levels [10]. For example, label $<l_1, x> \times <l_2, y>$ of arc $(t_1, p_1$) in Fig.1(1) defines that two tokens instantiated with x and y flow to p_1 concurrently; and label $<l_1, x> + <l_2, y>$ of arc $(p_1, t_2$) in Fig.1(1) expresses that either a token instantiated with x or a token instantiated with y can enable transition t_2); therefore, it is only necessary to assign identifications to simple labels (labels without + or $\times$). To avoid ambiguity, the identifications of all simple labels in a compound label must be distinct, and the identifications of all simple labels in arcs connected to a node must be distinct. Arcs are related through the same identification of their simple labels. For example, arc $(p_1, t_2$) in Fig.1(1), arc $(p_3, p_3\bullet)$ in Fig.1(2), and arc $(\bullet t_4, t_4$) in Fig.1(3) are related through label identification l_1.

3.3 Algebraic Definition

There are a few possible ways to define HPrT nets algebraically. One obvious way is to define an entire HPrT net as a collection of related individual PrT nets, which, we found, does not provide a satisfactory unified algebraic framework to represent and analyze the distinct features of HPrT nets. Thus we choose to give a single algebraic definition for an entire HPrT net. An HPrT net N consists of
(1) a finite hierarchical net structure (P, T, F, ρ),
(2) an algebraic specification *SPEC* , and
(3) a net inscription (φ, L, R, M_0)
The definitions of the above components are given and discussed in the following sections.

3.3.1 Hierarchical Net Structure

An HPrT net N has a finite hierarchical net structure (P, T, F, ρ). (P, T, F) is the essential net structure, where $P \cup T$ is the set of nodes satisfying the condition $P \cap T = \varnothing$. P is called the set of *predicates* and T is called the set of *transitions*. In particular, we identify two subsets $IN \subseteq P \cup T$ and $OUT \subseteq P \cup T$ such that IN contains the heads of all incoming non-terminating arcs and OUT contains the tails of all outgoing non-terminating arcs. We use $\bullet IN$ to denote the set of the pre-sets of all elements in IN, i.e. $\bullet IN = \{\bullet n \mid n \in IN\}$; and $OUT\bullet$ to denote the set of post-sets of all elements in OUT. F is the set of arcs and is called the *flow relation* satisfying the condition: $F \subseteq (\bullet IN \times IN \cup P \times T \cup T \times P \cup OUT \times OUT\bullet)$. Notice that we use the pre-set and post-set notations for two differnt purposes: (1) traditional use to denote a set, and (2) as an implicit node of a non-terminating arc, i.e. treating the pre-set or post-set as a single entity. The above uses not only do not cause any ambiguity, but also reflect the true meaning of the pre-set and post-set of any node in an HPrT net.

$\rho: P \cup T \to \wp(P \cup T)$ is a hierarchical function that defines the hierarchical relationships among the nodes in P and T. For any node n, $\rho(n)$ defines the immediate descendant nodes of n. Let ρ^{-1}, ρ^+, and ρ^* denote the inverse, irreflexive transitive closure, and reflexive transitive closure of ρ respectively, the ancester and descendants of any node can be easily expressed using the above notations. To avoid complexity and ensure the correctness of refinement, the following constraints are imposed: (1) a node cannot be its own descendant (no recursive definition), (2) the refinement hierarchies have a tree structure (no structure sharing), (3) consistent interfaces among different hierarchies (the interface nodes $\in IN \cup OUT$ be all predicates if their parent node is a predicate or all transitions if their parent node is a transition) [20], and (4) completeness (a refined net must have a parent node). The above constraints are formulated as following rules:

- **Rule 1:** $\forall n \in P \cup T.(n \notin \rho^+(n))$,
- **Rule 2:** $\forall n_1, n_2 \in P \cup T.(\rho(n_1) \cap \rho(n_2) \neq \emptyset \Rightarrow n_1 = n_2)$,
- **Rule 3:** $\forall p \in P.(\rho(p) \cap (IN \cup OUT) \subseteq P) \wedge$

  $\forall t \in T.(\rho(t) \cap (IN \cup OUT) \subseteq T)$

- **Rule 4:** $\forall n \in P \cup T.(n \in IN \cup OUT \Rightarrow \rho^{-1}(n) \neq \emptyset)$.

The hierarchical net structure of Fig.1 is algebraically defined as follows:

$P = \{p_1, p_2, p_3, p_4\}$,

$T = \{t_1, t_2, t_3, t_4, t_5\}$,

$F = \{(t_1, p_1), (p_1, t_2), (t_2, p_2), (p_2, t_3), (\bullet p_3, p_3), (p_3, p_3 \bullet),$

$(\bullet p_4, p_4), (p_4, p_4 \bullet), (p_5, t_4), (t_4, t_4 \bullet), (\bullet t_5, t_5), (t_5, t_5 \bullet)\}$,

$\rho = \{t_1 \mapsto \emptyset, \ t_2 \mapsto \{t_4, t_5\}, \ t_3 \mapsto \emptyset, \ t_4 \mapsto \emptyset, \ t_5 \mapsto \emptyset,$

$p_1 \mapsto \{p_3, p_4\}, \ p_2 \mapsto \emptyset, \ p_3 \mapsto \emptyset, \ p_4 \mapsto \emptyset\}$.

3.3.2 Underlying Specification

An HPrT net N contains a underlying specification $SPEC = (S, OP, Eq)$ consisting of a signature $\Sigma = (S, OP)$ and a set Eq of Σ-equations. Signature $\Sigma = (S, OP)$ includes a set of sorts and a family $OP = (OP_{s_1, \ldots, s_n}, s)$ of sorted operations for $s_1, \ldots, s_n, s \in S$. For each $s \in S$, we use CON_s to denote $OP_{,s}$ (the 0-ry operation of sort s), i.e. the set of constant symbols of sort s. The Σ-equations in Eq define the maenings and properties of operations in OP, for example, the associativity and commutativity of addition operation on real numbers. Algebraic specifications for many familiar sorts can be found in [2] and [3]. Several integrations of Petri nets with algebraic specifications are given in [21] and [16]. We often simply use familiar operations and their properties without explicitly listing the relevant equations. Based on $SPEC$, tokens, labels, and constraints of an HPrT net are defined .

Tokens of an HPrT net are essentially constant symbols of the family OP. The tokens of sort s are elements in CON_s .

To express the number and type of (identical and / or different) tokens to be moved along an arc, the following multi-set expression $\{k_1 v_1, \ldots, k_n v_n\}$ is adequate, in which each k_i is a natural number, and each v_i is either a member of X_s (the set of variables of sort s disjoint to OP). Often we drop the set notation $\{ \ \}$ when there is only one distinctive element in the multi-set. We define the set of all such expressions of sort s as:

$Exp_s (X_s) = \{ \ \{k_1 v_1, \ldots, k_n v_n\} \mid 1 \leq i \leq n \wedge k_i \in CON_s \wedge v_i \in X_s\}$.

The set of simple labels is defined as follows:

$Slabel_S (X) = \{ \ <l, e> \mid l \in CON_{id} \wedge e \in \bigcup_{s \in S} Exp_s (X_s)\}$

where id is a distinctive sort of label identifications and CON_{id} denotes the set of constants of sort id. $X = (X_s)_{s \in S}$ is a family of variables disjoint to OP.

The set of compound labels (or labels) are recursively defined as follows:

$Label_S(X) = Slabel_S \cup \{ l_1 + l_2 \mid l_1, l_2 \in Label_S(X) \} \cup$

$$\{ l_1 \times l_2 \mid l_1, l_2 \in Label_S(X) \}$$

Let l be any label, we use $slab(l)$ to denote the multi-set of simple labels in l. The set of labels defines syntactically valid labels, and the net inscription (to be defined in the next section) assures semantically valid labels for an HPrT net.

Constraints of an HPrT net are logic formulas, which are Σ_terms of sort *bool* over X, denoted as $Term_{OP,bool}(X)$. The set $Term_{OP,bool}(X)$ is formally defined as follows:

(1) if $v \in CON_{bool} \cup X_{bool}$, then $v \in Term_{OP,bool}(X)$,

(2) if $v_1 \in Term_{OP,s1}(X), ..., v_k \in Term_{OP,sk}(X), op \in OP_{s1...sk,bool}$, then

$$op(v_1,...,v_k) \in Term_{OP,bool}(X).$$

By treating logical connectives and quantifiers as algebraic operations, the resulting definition is simpler than traditional definitions of logic formulas.

The underlying *SPEC* for the net shown in Fig. 1 is as follows:

$S = \{alpha, bool\}$, $OP_{,alpha} = \{a, b\}$, $OP_{,bool} = \{true, false\}$,

$OP_{alpha,alpha,bool} = \{ = \}$,

and the family of sorted variables is: $X_{alpha} = \{x, y, u, v, w\}$. *Eq* includes conventional equations about equality [2].

3.3.3 Net Inscription

An HPrT net N contains a net inscription (φ, L, R, M_0), which associates each graphical symbol of the net structure (P, T, F, ρ) with an entity in the underlying *SPEC*, and thus defines the static semantics of an HPrT net.

Each predicate in an HPrT net is a data structure and a component of the overall system state. The sort of each predicate defines its valid values, i.e. proper tokens. The sorts of elementary predicates are members of S. The sort of a super predicate is defined as the union of sorts of its descendant predicates. Therefore, we associate each predicate p in P with a subset of sorts in S, and give the following sort assignment: $\varphi : P \to \wp(S)$ ($\wp$ is the power set operation).

$L: F \to Labels(X)$ is a sort-respecting labeling of N. As discussed in Arc Label section, the identifications of simple labels must be distinct when they are related through (1) a compound label (denoting a merged arc), and (2) a node (possiblly involving non-terminating arcs). The constraints for the above two situations are formulated as the following four identification uniqueness rules:

• **Rule 5:** $\quad \forall a \in F.(l \in slab(L(a)) \Rightarrow 2l \notin slab(L(a)))$

Rule 5 specifies that all simple labels of any compound label are distinct, i.e. *slab* returns a set instead of a multi-set.

• **Rule 6:** $\quad \forall a \in F.(l_1, l_2 \in slab(L(a)) \wedge l_1 \neq l_2 \Rightarrow l_1[1] \neq l_2[1])$

$l[1]$ represents the projection on the 1st component of l. Rule 6 defines the uniqueness of the identifications of all simple labels in any compound label.

• **Rule 7:** $\forall n \in P \cup T.(\forall n_1, n_2 \in \bullet n.(n_1 \neq n_2 \Rightarrow slab(L(n_1,n)) \cap slab(L(n_2,n)) = \varnothing)$

$\quad \wedge \; \forall n_1, n_2 \in n\bullet.(n_1 \neq n_2 \Rightarrow slab(L(n,n_1)) \cap slab(L(n,n_2)) = \varnothing)$

$\quad \wedge \; \forall n_1 \in \bullet n, n_2 \in n\bullet.(slab(L(n_1,n)) \cap slab(L(n,n_2)) = \varnothing) \;)$

Rule 7 specifies that all simple labels of arcs connected to a node are distinct.

- **Rule 8:** $\forall n \in P \cup T.((l_1, l_2 \in \bigcup_{n' \in \bullet n} slab(L(n', n)) \cup$

$$\bigcup_{n' \in n\bullet} slab(L(n, n'))) \wedge l_1 \neq l_2 \Rightarrow l_1[1] \neq l_2[1])$$

Rule 8 defines the uniqueness of the identifications of all simple labels of all arcs connected to a node.

Since compound labels defines data flows as well as control flows. The following basic control flow patterns [10] must be correctly labeled:
(1) data flows into and out of an elementary transition must take place concurrently, and
(2) data flows into and out of an elementary predicate can occur at different times.
Thus the following control flow preserving rules are needed:

- **Rule 9:** $\forall t \in T.(\rho(t) = \emptyset \wedge ((t', t) \in F \Rightarrow l_1, l_2 \in Label_S(X)$

$\wedge (L(t', t) \neq l_1 + l_2)) \wedge ((t, t') \in F \Rightarrow l_1, l_2 \in Label_S(X)$

$\wedge (L(t, t') \neq l_1 + l_2)))$

- **Rule 10:** $\forall p \in P.(\rho(p) = \emptyset \wedge ((p', p) \in F \Rightarrow l_1, l_2 \in Label_S(X)$

$\wedge (L(p', p) \neq l_1 \times l_2)) \wedge ((p, p') \in F \Rightarrow l_1, l_2 \in Label_S(X)$

$\wedge (L(p, p') \neq l_1 \times l_2)))$

Further the data flows between different levels of hierarchies must be balanced, i.e. a simple label occurs in a non-terminating arc if and only if it also appears in an arc with the same direction connected to the enclosing super node. This constraint is defined as the following rule:

- **Rule 11:** $\forall n \in P \cup T.(\rho(n) \neq \emptyset \Rightarrow (l \in \bigcup_{n' \in \bullet n} slab(L(n', n)) \Leftrightarrow$

$l \in \bigcup_{\bar{n} \in \rho(n) \cap IN} slab(L(\bullet\bar{n}, \bar{n})))$

$\wedge (l \in \bigcup_{n' \in n\bullet} slab(L(n, n')) \Leftrightarrow l \in \bigcup_{\bar{n} \in \rho(n) \cap OUT} slab(L(\bar{n}, \bar{n}\bullet))))$

With Rules 5 to 11, we can precisely determine the external environment of an interface node using the following rules:

- **Rule 12:** $\forall n \in IN.(\bullet n = \{n' \mid n' \in \bullet\rho^{-1}(n) \wedge (slab(L(\bullet n, n))$

$\cap slab(L(n', \rho^{-1}(n))) \neq \emptyset)\})$

- **Rule 13:** $\forall n \in OUT.(n\bullet = \{n' \mid n' \in \rho^{-1}(n)\bullet \wedge (slab(L(n, n\bullet))$

$\cap slab(L(\rho^{-1}(n), n')) \neq \emptyset)\})$

For example, in Fig.1(2), $\rho^{-1}(p_3) = p_1$, $t_2 \in p_1\bullet$, $slab(L(p_3, p_3\bullet)) = \{< l_3, x >\}$, and $slab(L(p_1, t_2)) = \{< l_3, x >, < l_4, y >\}$; thus $p_3\bullet = \{ t_2 \}$.

$R : T \to Term_{OP,bool}(X)$ is a well-defined constraining mapping of N, which associates each transition t in T with a logic formula. Since a super transition is an abstraction of lower level transitions, it should not have any constraint, i.e. its constraint is always true as defined by the following rule:

- **Rule 14:** $\forall t \in T.(\rho(t) \neq \emptyset \Rightarrow R(t) = true).$

By convention, the logical constant symbol *true* is not explicitly represented in a transition.

$M_0: P \to MCON_S$ is a sort-respecting initial marking of N, which assigns a multi-set of tokens to each predicate p in P with the same sort (where $MCON_S = \{ kc \mid k \in CON_{nat} \wedge c \in CON_S \}$). The above requirement is defined by the following rule:

- **Rule 15:** $\quad \forall p \in P.(kc \in M_0(p) \Rightarrow c \in CON_{\varphi(p)})$

Since a super predicate's state is defined by its lower level predicates, we define the tokens in a super predicate as the union of tokens of its next level predicates as follows:

- **Rule 16:** $\quad \forall p \in P.(\rho(p) \neq \varnothing \Rightarrow M_0(p) = \bigcup_{p' \in \rho(p)} M_0(p')).$

The net inscription of Fig. 1 is as follows:

$\varphi = \{ p_1 \mapsto alpha, \; p_2 \mapsto alpha, \; p_3 \mapsto alpha, \; p_4 \mapsto alpha \};$

$L = \{(t_1, p_1) \mapsto < l_1, x > \times < l_2, y >, \; (p_1, t_2) \mapsto < l_3, x > + < l_4, y >,$

$\quad (t_2, p_2) \mapsto < l_5, u > + < l_6, v >, \; (p_2, t_3) \mapsto < l_7, w >, (\bullet p_3, p_3) \mapsto < l_1, x >,$

$\quad (p_3, p_3\bullet) \mapsto < l_3, x >, \; (\bullet p_4, p_4) \mapsto < l_2, y >, \; (p_4, p_4\bullet) \mapsto < l_4, y >,$

$\quad (\bullet t_4, t_4) \mapsto < l_3, x >, \; (t_4, t_4\bullet) \mapsto < l_5, u >, \; (\bullet t_5, t_5) \mapsto < l_4, y >, \; (t_5, t_5\bullet) \mapsto < l_6, v >\};$

$R = \{ t_1 \mapsto true, \; t_2 \mapsto true, \; t_3 \mapsto true, \; t_4 \mapsto x = u, \; t_5 \mapsto y = v \};$

$M_0 = \{ p_1 \mapsto \{a, b\}, \; p_2 \mapsto \{ \}, \; p_3 \mapsto \{a\}, \; p_4 \mapsto \{b\} \}.$

Theorem 1. Each simple label creates a unique data flow link between one elementary predicate and one elementary transition.
Proof. the existence of a data flow link between an elementary predicate and an elementary transition identified by a simple label is ensured by data flow balance Rule 11, and the uniqueness of such a data flow link is ensured by Rules 5 to 8. []

4 Dynamic Semantics of Hierarchical Predicate Transition Nets

In this section, we define the dynamic semantics of HPrT nets including markings, transition enabling conditions, and transition firing rules.

4.1 Markings

A marking M of an HPrT net is a mapping $P \to MCON_S$ from the set of predicates to multi-sets of tokens. Since a super predicate denotes an abstraction of a lower level net; thus its state is defined its next level predicates. We define the tokens of a super predicate as the union of the tokens in its next level predicates. Therefore the following rules similar to the initial marking are required:

- **Rule 17:** $\quad \forall p \in P.(kc \in M(p) \Rightarrow c \in CON_{\varphi(p)}) \quad$ and

- **Rule 18:** $\quad \forall p \in P.(\rho(p) \neq \varnothing \Rightarrow M(p) = \bigcup_{p' \in \rho(p)} M(p')).$

Further new markings resulted from transition firings also satisfy the above rules, which is ensured by the transition firing rule to be given in a following section.

4.2 Transition Enabling Conditions

An elementary transition in an HPrT net defines some concrete token processing as in a PrT net, thus it is natural to adapt the transition enabling conditions in PrT nets to HPrT nets. Let t be an elementary transition, α be an occurrence mode instantiating all the variables related to t, and M be a marking, t enabled with α under M, written as $enabled(M[t/\alpha>)$, would be defined as follows if the definition of transition enabling condition for PrT nets were adapted:

$$\forall p \in {}^{\bullet}t.(\bigcup_{l \in slab(L(p,t))} \alpha(l[2]) \subseteq M(p)) \wedge \alpha(R(t))$$

But the above formula is incorrect since the tokens in a super predicate defined by Rule 18 are not necessarily in relevant predicates (related through label identification). Fig. 2 shows a simple example, in which t_1 is not enabled, but would be enabled according to the above formula ($p_1 \in {}^{\bullet}t_1$ and $M(p_1) = \{a,b\}$).

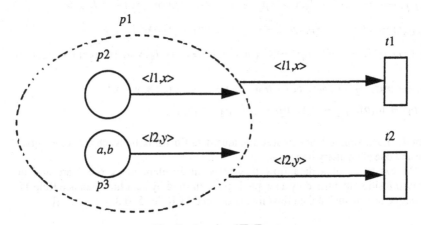

Fig. 2 - Another HPrT net

In order to define the correct enabling condition of a transition, we must use the tokens in relevant predicates. According to Theorem 1, we only need to consider the tokens in relevant elementary predicates or simply the exterior interface elementary predicates of any predicate in the pre-set of a transition. To determine the exterior interface elementary predicates, the following two functions are defined for any $p \in P$:

$$in(p) = \rho(p) \cap IN \quad \text{and} \quad out(p) = \rho(p) \cap OUT.$$

Further we define $in^0(p) = out^0(p) = \{p\}$, $in^{i+1}(p) = in(in^i(p))$, and $out^{i+1}(p) = out(out^i(p))$. Let m and n be the smallest numbers such that $in^m(p) = \emptyset$, and $out^n(p) = \emptyset$ respectively, we define the finite irreflexive transitive closures for in and out as follows respectively:

$$in^+(p) = \bigcup_{1 \leq i \leq m} in^i(p) \quad \text{and} \quad out^+(p) = \bigcup_{1 \leq i \leq n} out^i(p)$$

$enabled(M[t/\alpha\!\!>\!)$ is thus defined by the following rule:

- **Rule 19:**

$$\forall p' \in \bullet t. ((\rho(p') = \varnothing \Rightarrow \alpha(L(p',t)[2]) \subseteq M(p')) \wedge$$

$$(\forall p \in out^+(p').(\rho(p)) = \varnothing \Rightarrow$$
$$\alpha((slab(L(p',t)) \cap slab(L(p,p\bullet)))[2]) \subseteq M(p)))) \wedge \alpha(R(t))$$

A super transition is an abstraction of some low-level action. As long as some lower level transition is enabled, we consider the super transition is enabled since local changes and extenal visible changes can take place. Therefore we define the enabling condition for a super transition t with occurrence mode α under marking M as follows:

- **Rule 20:** $\quad \exists t' \in \rho(t).(enabled(M[t'/\alpha >)).$

As an example, the transition t_2 in Fig.1 (1) is enabled with occurrence mode $\alpha = \{x \mapsto a, u \mapsto a\}$ under the initial marking M_0 since elementary transition t_4 in Fig. 1 (3) is enabled with the same occurrence mode under the initial marking.

4.3 Transition Firing Rule

The firing of an enabled transition results in a new marking. Since the enabling condition of a super transition is defined in terms of some of its lower level transition(s), the firing of a super transition is thus the firing of its lower level transition(s). Therefore it is adequate to only define the transition firing rule for elementary transitions. Since the tokens in a super predicate are defined in terms of the union of tokens in its next level predicates specified by Rule 17 and eventually defined by the tokens in all its descendant elementary predicates; but a transition firing involving a super predicate can only affect the tokens in its interface predicates. Therefore we only need to define token changes in its exterior interface elementary predicates caused by the transition firing.

To simplify the definition of the transition firing rule, we define the extended version of labeling function:

$$\bar{L}(f) = \begin{cases} L(f) & \text{if } f \in F \\ \varnothing & \text{if } f \notin F \end{cases}$$

Let t be an elementary transition enabled with occurrence mode α under marking M; the firing of t results in a following marking M', written as $M[t/\alpha\!\!>\!M'$, defined by:

- **Rule 21:**

$$\forall p' \in \bullet t \cup t\bullet.((\rho(p') = \varnothing \Rightarrow$$

$$M'(p') = M(p') \cup \alpha(slab(\bar{L}(t,p')[2]) - \alpha(slab(\bar{L}(p',t)[2])) \wedge$$

$$\forall p \in (in^+(p') \cup out^+(p')).(\rho(p) = \varnothing \Rightarrow M'(p) = M(p) \cup Gain(t,p) - Loss(p,t)) \wedge$$

$$\forall p \in (\rho^+(p') \cap P - (in^+(p') \cup out^+(p'))).(\rho(p) = \varnothing \Rightarrow M'(p) = M(p))) \wedge$$

$$\forall p' \notin \bullet t \cup t\bullet.(\forall p \in \rho^\bullet(p') \cap (P \cup T).(M'(p) = M(p)))$$

where $\quad Gain(t, p) = \alpha((slab(\overline{L}(t, p')) \cap slab(\overline{L}(\bullet p, p))))[2])$, and

$\quad\quad Loss(p, t) = \alpha((slab(\overline{L}(p', t)) \cap slab(\overline{L}(p, p\bullet))))[2])$.

In Rule 21, (1) the first line defines token changes in p' when predicate p' connected to t is an elementary predicate, (2) the second line defines token changes in any exterior interface elementary predicate p of p', (3) the third line specifes that no changes to any interior elementary predicate of p', and (4) the fourth line specifies that no changes to any elementary predicate unrelated to t.

4.4 Behavior (Dynamic Semantics) of An HPrT Net

Two enabled transitions are said to be in conflict when the firing of one of them disables the other. Conflicts are resolved nondeterministically. Non-conflicting enabled transitions can fire at the same time, which also includes the situation where the same transition is enabled with two different occurrence modes. The firing of a super transition is always considered to be concurrent with some of its lower level transition(s). Therefore three types of concurrency can occur: (1) two completely unrelated enabled transitions fire simultaneously, (2) two versions of the same transition enabled with two occurrence modes fire at the same time, and (3) the firing of a super transition concurrently with some of its lower level transition(s).

The *behavior* of an HPrT net is defined as the set of *maximal execution sequences*, in which each execution step consists of a set of firing transitions. We call an execution sequence an *essential execution sequence* if we drop all super transitions in all steps.

The following is an example of an execution step of the HPrT net in Fig. 1:
The marking M_1 resulted in from firing transition t_1 and t_4 (thus t_2) at the same time with occurrence modes $\alpha_1 = \{x \mapsto a, y \mapsto b\}$ and $\alpha_2 = \{x \mapsto a, y \mapsto b\}$ respectively under marking M_0 (notice the x and y associated with arc (t_1, p_1) are different from x and y associated with arc (p_1, t_2)): $M_1 = \{p_1 \mapsto \{a, 2b\}, p_2 \mapsto \{a\}, p_3 \mapsto \{a\}, p_4 \mapsto \{2b\}\}$.

5 Equivalent PrT Net for An HPrT Net

In this section, we show that there is a unique behavioral equivalent PrT net for any HPrT net. A PrT net is a special HPrT where (1) graphically, there is only a single connected net without super (dotted) nodes and non-terminating arcs; and (2) algebraically, the hierarchy function becomes a map from each node to an empty set and thus can be dropped, and all labels are simple labels. Based on the above analysis, we can derive a PrT net for any given HPrT net.

For a given HPrT net $N = (P, T, F, \rho) + SPEC + (\varphi, L, R, M_0)$, in which $+$ denotes the disjoint union as in [2] and is overloaded with our label constructor (but does not create any ambiguity); we define a PrT net $N' = (P', T', F') + SPEC' + (\varphi', L', R', M_0')$ as follows:

(1) $\quad P' = \{p \mid p \in P \wedge \rho(p) = \varnothing\}$,

(2) $\quad T' = \{t \mid t \in T \wedge \rho(t) = \varnothing\}$,

(3) $\quad F' = (F \cap (P' \times T' \cup T' \times P')) \cup F_1 \cup F_2$, $\quad\quad\quad$ with

$F_1 = \{(p, t) \mid p \in P' \wedge t \in T' \wedge p' \in \bullet t \wedge p \in out^+(p') \wedge$

$$(slab(L(p,p\bullet)) \cap slab(L(p',t))) \neq \varnothing\},$$

$$F_2 = \{(t,p) \mid p \in P' \wedge t \in T' \wedge p' \in t\bullet \wedge p \in in^+(p') \wedge$$
$$(slab(L(\bullet p,p)) \cap slab(L(t,p'))) \neq \varnothing\}.$$

(4) $SPEC' = SPEC$,

(5) $\varphi' = \{p \mapsto \varphi(p) \mid p \in P'\}$,

(6) $L' = \{f \mapsto L(f) \mid f \in F \cap (P' \times T' \cup T' \times P')\} \cup$

$\{(p,t) \mapsto (slab(L(p,p\bullet)) \cap slab(L(p',t)))[2] \mid (p,t) \in F_1 \wedge p' \in \bullet t \wedge p \in out^+(p')\} \cup$

$\{(t,p) \mapsto (slab(L(\bullet p,p)) \cap slab(L(t,p')))[2] \mid (t,p) \in F_2 \wedge p' \in t\bullet \wedge p \in in^+(p')\}$,

(7) $R' = \{t \mapsto R(t) \mid t \in T'\}$,

(8) $M_0' = \{p \mapsto M_0(p) \mid p \in P'\}$.

In the above definition of the PrT net N', components (1), (2), (4), (5), (7), and (8) are basically restrictions of the corresponding components of the given HPrT net, and thus are always feasible. The feasibility of the definitions of components (3) and (6) is ensured by Theorem 1. Notice that L' is a mapping: $F' \to Exp_S(X_S)$, i.e. we have dropped the identifications in labels since they are no longer needed in PrT nets. From the definition N', we obtain the following corollary:

Corollary 1. An arc (p,t) is in F' of N' if and only if p is a part of the enabling condition of t in N , and the label of (p,t) is the same as that common to p and t in the enabling condition and the transition firing of t in N. An arc (t,p) is in F' of N' if and only if p is affected by the firing of t in N , and the label of (t,p) is the same as that common to p and t in the transition firing of t in N.

Proof. a simple comparision of relevant items in the definition of N' , and definitions of Rule 19 and 21 is adequate.   []

Let t be a transition in T', α be an occurrence mode instantiating all the variables related to t , and M be a marking, t enabled with α under M, written as $enabled(M[t /\alpha\!>)$, is defined by:

• **Rule 22:** $\forall p \in \bullet t.(\alpha(L'(p,t)) \subseteq M(p)) \wedge \alpha(R'(t))$

Similar to the extension of L, and extend the definition of L' as follows:

$$\overline{L}'(f) = \begin{cases} L'(f) & \text{if } f \in F \\ \varnothing & \text{if } f \notin F \end{cases}$$

Let t be a transition in T' enabled with occurrence mode α under marking M; the firing of t results in a following marking M', written as $M[t /\alpha\!>M'$, defined by:

• **Rule 23:** $\forall p \in \bullet t \cup t\bullet.(M'(p) = M(p) \cup \alpha(\overline{L}'(t,p)) - \alpha(\overline{L}'(p,t))) \wedge$
$\forall p \notin \bullet t \cup t\bullet.(M'(p) = M(p))$

Based on the Rule 22 and 23, we obtain the following theorem:

Theorem 2. The PrT net N′ is behaviorally equivalent to the given HPrT net N.

Proof. We only need to show the following fact:
(*) the set of execution sequences of N′ is identical to the set of essential execution sequences of N. Since the set of execution sequences are generated by firings of enabled transitions, we only need to prove that the enabling condition and firing effect of any elementary transition in N are the same as those of the corresponding transition in N′ with any occurrence mode under any reachable marking, which however is ensured by Corollary 1, transition enabling and firing Rules 19 and 21 in N, and transition enabling and firing Rules 22 and 23 in N′ (the detailed comparison is omitted here). □

The equivalent PrT net for the HPrT net in Fig. 1 is defined as follows:

$P' = \{p_2, p_3, p_4\}$,

$T' = \{t_1, t_3, t_4, t_5\}$,

$F' = \{(t_1, p_3), (p_3, t_4), (t_1, p_4), (p_4, t_5), (t_4, p_2), (p_2, t_3), (t_5, p_2)\}$,

$\varphi' = \{ p_2 \mapsto alpha,\ p_3 \mapsto alpha,\ p_4 \mapsto alpha \}$;

$L' = \{(t_1, p_3) \mapsto x,\ (p_3, t_4) \mapsto x,\ (t_1, p_4) \mapsto y,\ (p_4, t_5) \mapsto y,\ (t_2, p_2) \mapsto u,$

$\qquad (t_5, p_2) \mapsto v,\ (p_2, t_3) \mapsto w\}$

$R' = \{ t_1 \mapsto true,\ t_3 \mapsto true,\ t_4 \mapsto x = u,\ t_5 \mapsto y = v \}$;

$M_0' = \{ p_2 \mapsto \{\ \},\ p_3 \mapsto \{a\},\ p_4 \mapsto \{b\} \}$.

Graphical transformation is straightforward, in which all super nodes and non-terminating arcs are deleted. New arcs and labels are created according to (3) and (6) in the definition of a derived PrT net. The equivalent PrT net for the HPrT net in Fig. 1 is shown in Fig. 3.

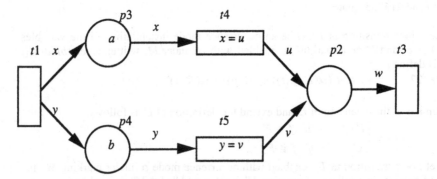

Fig. 3 - The equivalent PrT net of the HPrT net in Fig.1

6 Related Work

6.1 Hierarchical Colored Petri Nets (HCPN)

Hierarchical colored Petri nets were proposed in [12], which contain five hierarchical constructs: substituion of places, substitution of transitions, invocation of transitions, fusion of places, and fusion of transitions. A formal definition of hierarchical colored Petri nets with two of the above hierarchy constraints: substitution of transitions and fusion of places was given in [14], and further refined in [15]. In the following sections, we briefly compare major characteristics in the formal definition of hierarchical colored Petri nets with those in our formal definition of hierarchical predicate transition nets.

6.1.1 Graphical Notations

(1) HPrT nets:
Dotted transitions and predicates are used to represent super nodes, and a dotted boundary (circle or box) is used to enclose a subnet. Non-terminating arcs are used in subnets to indicate the proper data flows and labels associated with non-terminating arcs maintain the correct flow relationships between subnets with their external environments. Therefore, it is straightforward to identify and substitute a super node and its corresponding subnet.

(2) HCPN nets:
Different types of boxes and arcs are used to distinguish super nodes and ordinary nodes, and an enclosing box is used for a subnet called a page. Additional hierarchy inscriptions are used to define the types of hierarchy structures and types of flow directions. The environment nodes surrounding a substitution transition are called socket nodes, which are repeated and marked in the subnet (called port nodes). Declarations are introduced to define a family of structurally identical subnets, which is more descriptive than our approach but is more difficult to understand.

6.1.2 Algebraic Definition

(1) Hierarchy:
• HPrT nets:
A hierarchical function ρ defines the hierarchical relationships among all nodes, which ensures (1) non-recursiveness, (2) tree structure, (3) consistent interfaces, and (4) structuredness. It is easy to find the descedants and predecessors of a node by using the transitive closure and the inversed closure of ρ.
• HCPN nets:
A set SN of substitution transitions, a set S of corresponding pair-wise disjoint non-hierarchical colored Petri nets (pages), and a non-recursive page assignment function SA from SN to S are defined. It is clear SA is less general than ρ and does not have the properties of the inverse and transitive closures since SA maps a node to a net instead of a set of nodes in the subnet. A set PN of port nodes in S and an associated port type function indicating the flow directions of arcs are defined. The relationships between the set of port nodes and their correspondences (called socket nodes) are defined through a port assignment function, which also ensures that a pair of matching socket node and port node have the same type, color sets, and an equivalent initial marking.

(2) Place Sharing and Distinctive Instances:
- HPrT nets:

    No place sharing is allowed in HPrT nets since the hierarchy is a tree structure instead of a network structure. This restriction simplifies the net definition, maintains the distributed nature of Petri nets, and avoids the potential specification problems due to non-explicit sharing states.

- HCPN nets:

    A finite set FS of fusion sets are defined such that members of a fusion set have identical color sets and an equivalent initial marking. A fusion type function FT specifies three types of sharing including global, page, and instance fusions. A multiset PP of prime pages defines the distinctive instances of pages.

(3) Arc Labels:
- HPrT nets:

    Arc labels play an essential role in HPrT nets, denote the unique channels relating external environments with subnets through unique identifications, and indicate data flow patterns through label constructors $\times$ and $+$ (concurrency or non-determinancy). The transformation rules on labels in [10] facilitate proper stepwise refinement and abstraction of data flows.

- HCPN nets:

    Arc labels are expressions, which are not formally defined. It also seems that there are missing constraints in HCPN net definition to ensure that socket nodes and the corresponding port nodes have the same labels.

(4) Transition Constraints:
- HPrT nets:

    Constraints are logical expressions defined using the underlying algebra in HPrT nets.

- HCPN nets:

    Constraints are logical expressions, which are not formally defined.

(5) Markings
- HPrT nets:

    Tokens are defined by constant symbols of the underlying algebraic specification. The initial marking ensures the correct types and number of tokens among related interface predicates at different abstraction levels (the tokens in a super predicate are the sum of tokens in the predicates at the next level). The transition firing rule ensures the above property. Therefore successive refinement and abstraction of tokens are possible, and different abstraction levels of system states can be represented and viewed.

- HCPN nets:

    Tokens are defined by color sets. The initial marking ensures the correct types and number of tokens (1) in related port and socket nodes, and (2) shared places. However it is not clear how the marking concept can be extended to substitution places.

6.1.3 Dynamic Semantics

- HPrT nets:

    The enabling condition for an elementary transition is defined, which is quite different from the traditional definition for non-hierarchical nets as discussed in a previous section. The enabling condition for a super transition is defined in terms of enabling condition of its lower level elementary transition(s). The firing of an enabled transition results in a new marking defined by a new firing rule. The new firing rule defines token changes in all related elementary predicates, and thus all relevant predicates at all abstraction levels. Concurrent firings of two non-conflicting enabled transitions are possible.

- HCPN nets:
  The enabling condition for a step (essentially a multi-set of non-conflicting transitions) is defined using place instance groups characterizing the set of all equivalent place instances, which is basically a general description of the enabling condition for transition instances. Similarly, a general firing rule using place instance groups is given.

6.2 Hierarchical Petri Nets with Building Blocks (HPNBB)

In [4], a definition of hierarchical Petri nets was presented, which dealt with the net structure only without considering any semantic related concepts and issues. A hierarchical Petri net consisted of (1) a set of (elementary and super) places, (2) a set of (elemntary and super) transitions, (3) a set of arcs connecting only elementary nodes (the finest net), (4) a predecessor mapping relating descendent nodes to their ancesters, and (5) a top element (the root of the hierarchy tree). The connections among nodes at higher levels can be generated by cuts. The relationship between a hierarchical Petri net with sequences of refinements (net morphisms) was established. Based on the definition of hierarchical Petri nets, a new definition for hierarchical Petri nets with building blocks was given, which supported the reuse of existing subnets.

Some of the basic ideas in the above paper are similar to those of ours. For example, the predecessor mapping is actually the inverse of our hierarchy function ρ. However the above work only addressed the relationships of net structures during refinements without defining any semantic related concepts, thus it does not offer a technique for modeling and analyzing system behavior.

7 Conclusion

HPrT nets have been introduced to model large and complex concurrent and distributed systems ([10], [11]), but lacked a formal definition of syntax and semantics and thus may have the problems such as impreciseness and ambiguity of informal methods. This paper provides a formal syntax and semantics definition for HPrT nets. Our major contributions include (1) establishing the theoretical foundation of a very promising formal method - HPrT nets, and (2) providing the basis for the rigorous comparison of various hierarchical formal methods. The direct formal semantics definition of HPrT nets facilitates compositional net specification and verification [8].

In our formal definition of HPrT nets, we limited the hierarchical structure to be a tree, which has simplifed the formal semantics definition considerably but does not support additional structural sharing. It is easy to accommodate structural sharing by introducing a substitution relation in the syntax definition, but it is not simple to define a formal semantics for structural sharing since the meaning of the shared structure can be interpreted in two different ways: it is a single shared component (similar to the module concept in a typical high-level programming language) or it represents non-shared identical copies (similar to the macro mechanism in a typical assembly language). Both of them are useful under different circumstances, but cannot be defined concisely in a single semantic setting. Fig. 4 shows an HPrT net demonstrating the above ideas. Predicate p_4 is a refinement of both predicates p_1 and p_2 in Fig. 4(1). Under the first interpretation, an equivalent PrT net shown in Fig. 4(2) is obtained; and under the second interpretation, an equivalent PrT net shown in Fig. 4(3) is obtained. The above problem was informally addressed in [12], in which the first interpretation is achieved by combining fusion and substitution constructs and the second interpretation is obtained by using substitution constructs alone. In statecharts [9], only the first interpretation is used for the similar hierarchical structures. We will study different approaches and their impact of incorporating structural sharing into HPrT nets in our future work.

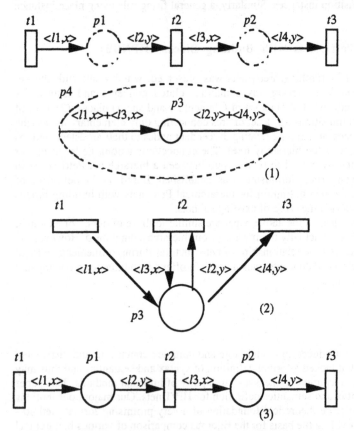

Fig.4 - An HPrT net and its two equivalent versions under different semantics

Acknowledgements

We thank three anonymous referees for their careful reading of an earlier version of this paper and for pointing out a few mistakes. This work was partially supported by the NSF under grant CCR-9308003 and under grant EHR-9108770 to the state of North Dakota.

229

References

1. J. Billington, G.R. Wheeler, and M.C. Wilbur-ham, PROTEAN: a high-level Petri net tool for the specification and verification of communication protocols. *IEEE Transactions on Software Engineering*, vol.14, no.3, 1988, 301-316.
2. H. Ehrig and B. Mahr, *Fundamentals of Algebraic Specification 1*, Springer-Verlag, 1985.
3. H. Ehrig and B. Mahr, *Fundamentals of Algebraic Specification 2*, Springer-Verlag, 1990.
4. R. Fehling, A concept of hierarchical Petri nets with building blocks, *Lecture Notes in Computer Science*, vol. 674, 1993, 148-168.
5. H.J. Genrich, and K. Lautenbach, System modeling with high-level Petri nets. *Theoretical Computer Science*, vol.13, 1981, 109-136.
6. D. Harel, On visual formalisms. *Communications of the ACM*, vol. 31, 1988, 514-530.
7. X. He, Integrating formal specification and verfication methods in software development, *Ph.D. dissertation*, Virginia Polytechnic Institute & State University, June, 1989.
8. X. He, A method for analysing properties of hierarchical predicate transition nets, *Proceedings of the 19th Annual International Computer Software & Applications Conference (COMPSAC'95)*, Dallas, TX, August, 1995, 50-55.
9. D. Harel, and C.-A. Kahana, On statecharts with overlapping. *ACM Transactions on Software Engineering and Methodology*, vol.1, no.4, 1992, 399-421.
10. X. He, and J.A.N. Lee, A methodology for constructing predicate transition net specifications. *Software - Practice and Experience*, vol.21, no.8, 1991, 845-875.
11. X. He, and C.H. Yang, Structured analysis using hierarchical predicate transition nets. *Proc. of the 16th Int'l Computer Software and Applications Conference (COMPSAC'92)*, Chicago, 1992, 212-217.
12. P. Huber, K. Jensen, and R.M. Shapiro, Hierarchies in colored Petri nets. *Lecture Notes in Computer Science*, vol.483, Spriner-Verlag, 1990, 313-341.
13. K. Jensen, Colored Petri nets and the invariant method. *Theoretical Computer Science*, vol.14, 1981, 317-336.
14. K. Jensen, Colored Petri nets: a high level language for system design and analysis, *Lecture Notes in Computer Science*, vol. 483, 1990, 342-416.
15. K. Jensen, *Colored Petri Nets*, vol.1, Springer-Verlag, 1992.
16. C. Kan and X. He, High-level algebraic Petri nets, *Information and Software Technology*, vol.37, no.1, 1995, 23-30.
17. Z. Manna and A. Pnueli, Completing the temporal picture, *Theoretical Computer Science*, vol. 83, pp.97-130, 1991.
18. Z. Manna and A. Pnueli, Models for reactivity, *Acta Informatica*, vol.30, pp.609-678, 1993.
19. W. Reisig, *Petri Nets - An Introduction*. EATCS Monographs on Theoretical Computer Science, vol.4, Springer-Verlag, 1985.
20. W. Reisig, Petri nets in software engineering. *Lecture Notes in Computer Science*, vol.255, Springer-Verlag, 1987, 63-96.
21. W. Reisig, Petri nets and algebraic specifications. *Theoretical Computer Science*, vol.80, 1991, 1-34.
22. E. Yourdon, *Modern Structured Analysis*, Yourdon Press, 1989.

Reduced State Space Representation for Unbounded Vector State Spaces

Kunihiko Hiraishi

School of Information Science
Japan Advanced Institute of Science and Technology
1-1 Asahi-dai, Tatsunokuchi, Ishikawa 923-12 Japan (email: hira@jaist.ac.jp)

Abstract. This paper presents a new method for computing reduced representation of vector state spaces consisting of infinitely many states. Petri nets are used as a model for generating vector state spaces, and the state space is represented in the form of semilinear subsets of vectors. By combining the partial order methods with the proposed algorithm, we can compute reduced state spaces which preserve some important properties, such as liveness of each transition and the existence of deadlocks. The state space of a finite capacity system can be viewed as that of an infinite capacity system projected to the states satisfying the capacity condition. We also show that the proposed algorithm is applicable to vector state spaces with finite capacities.

1 Introduction

Largeness of state space is a serious problem in analyzing systems that allow concurrent occurring of events. Even if the size of each subsystem is small, the state space generated by their composition becomes very large since independent events can be interleaved in many possible orders. This problem is called *state space explosion.*

To avoid this problem, several methods have been proposed, such as utilization of binary decision diagrams(BDDs)[11], symbolic model checking[3, 13], on-the-fly model checking[15], and the partial order methods[4, 5, 16]. Combination of these methods is also studied.

In the partial order methods, the partial order defined on the set of actions is used to compute reduced state spaces which preserves some important properties, such as liveness of each transition and the existence of deadlocks. In this paper, we aim to extend the partial order methods in the following two points.

1. Considering dynamically generated processes.
   We will consider the situation that new processes are dynamically generated during the execution. This means that the state space may contain infinitely many states. Such an infinite state space will be described in the form of semilinear subsets of vectors, in which every two states are equivalent to

each other. The equivalence is defined to at least satisfy that all states in each equivalence class have the same set of enabled transitions.

2. Treating the state space of a finite capacity system as an projection of the state space of an infinite capacity system.

   Since practical systems do not allow infinitely many resources, we usually give a capacity to each component of the system. The state space of a finite capacity system can be viewed as that of an infinite capacity system projected to the states satisfying the capacity condition. We will apply the state space generation algorithm for infinite capacity systems to finite capacity systems.

We will use Petri nets as a model for describing systems. The state space of a Petri net is represented by a labeled transition system in which every state is a k-dimensional nonnegative integer vector. We will refer such transition systems as *vector transition systems*. The vector transition systems generated by Petri nets have a characteristic property, which will be called *order-persistence*. The proposed algorithm uses this property to obtain a reduced representation of the state space. In addition, the stubborn set method[14], which is a kind of the partial order methods, will be combined with the proposed algorithm.

In Section 2, we first show the outline of the proposed algorithm. After that, we will show basic definitions and notations in Section 3. In Petri nets, the coverability tree was proposed to represent the infinite state space by a finite search tree. In Section 4, we will describe the proposed algorithm, comparing with that of the coverability tree. In Section 5, the algorithm will be modified to apply to the finite capacity systems. Discussions on the implementation and experimental results will be shown in Section 6. Some remarks and future work will be described in Section 7.

2 Outline of the Algorithm

Fig. 1 shows an unbounded Petri net, i.e., a Petri net that generates a state space containing infinitely many states. While processes are not explicitly described in Petri nets, process algebras are a model based on the behavior of processes. A CSP program[8] equivalent to this net is shown as follows:

$$
\begin{aligned}
P &= a \to (b \to P \mid c \to STOP), \\
Q &= (a \to Q_1 \mid d \to Q_1), \\
Q_1 &= (b \to Q_2 \mid e \to Q_2), \\
Q_2 &= f \to (Q \| Q), \\
R &= P \| Q.
\end{aligned}
\tag{1}
$$

Process Q is duplicated after the execution of action f. The state space of this Petri net contains infinitely many states. The coverability tree was proposed for representing such an infinite state space in the form of a finite search tree[9, 12]. In order to treating the infinity, the special symbol ω, representing "infinity", is introduced . By the depth-first search, the following sequence of states are generated in the coverability tree.

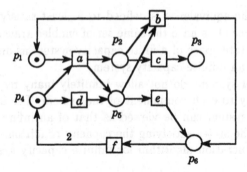

Fig. 1. An unbounded Petri net.

$$0.\ [100100] \xrightarrow{a}$$
$$1.\ [010010] \xrightarrow{b}$$
$$2.\ [100001] \xrightarrow{f}$$
$$3.\ [100\omega00] \xrightarrow{a}$$
$$4.\ [010\omega10] \xrightarrow{b}$$
$$5.\ [100\omega0\omega] \xrightarrow{a}$$
$$6.\ [010\omega1\omega] \xrightarrow{b}$$
$$7.\ [100\omega0\omega]$$

The coverability tree does not contain enough information to always solve the reachability problem, because the symbol ω is a loss of information, i.e., the individual numbers are discarded in this representation.

In the proposed algorithm, the following sequence of states is generated for the same Petri net.

$$0.\ [100100] \xrightarrow{a}$$
$$1.\ [010010] \xrightarrow{b}$$
$$2.\ [100001] \xrightarrow{f}$$
$$3.\ [100200] + (0..\infty) \cdot [000100] \xrightarrow{a}$$
$$4.\ [010110] + (0..\infty) \cdot [000100] \xrightarrow{b}$$
$$5.\ [100101] + (0..\infty) \cdot [000100] + (0..\infty) \cdot [000001] \xrightarrow{a}$$
$$6.\ [010011] + (0..\infty) \cdot [000100] + (0..\infty) \cdot [000001]$$

$$\begin{cases} 6a.\ [010011] + (0..\infty) \cdot [000001] \xrightarrow{b} \\ 6b.\ [010011] + (1..\infty) \cdot [000100] + (0..\infty) \cdot [000001] \end{cases}$$

$$7.\ [100002] + (0..\infty) \cdot [000001] \xrightarrow{f}$$
$$8.\ [100201] + (0..\infty) \cdot [000100] + (0..\infty) \cdot [000001]$$

The notation $(l..u)$ denotes any integer x such that $l \leq x \leq u$. Each node in the tree is associated with a set of markings described as above (we call this a *region*). At step 3, since $[100200] - [100100] = [000100] \geq 0$ holds, $(0..\infty) \cdot [000100]$ is added to the region. Each region is decomposed into a finite number of regions so that every two states in a region are equivalent to each other. The equivalence is defined to at least satisfy that every state in each region has the same set of enabled transitions. The region at step 6 is decomposed into two regions $6a$ and $6b$ since the set of enabled transitions in $6a$ is $\{b, c, e, f\}$, and the set in $6b$ is $\{b, c, d, e, f\}$. No information is lost in this representation.

3 Preliminaries

3.1 Vector State Spaces

We first give a representation of a state space consisting of nonnegative integer vectors. Let $\mathbb{N}$ denote the set of nonnegative integers. For a finite set A, we identify a function $f : A \to \mathbb{N}$ with a $|A|$ dimensional vector of nonnegative integers and write $f \in \mathbb{N}^A$.

Definition 1. *A k-dimensional vector transition system is a 4-tuple $VTS = (S, T, \tau, L)$, where $S \subset \mathbb{N}^k$ is a nonempty set of k-dimensional nonnegative integer vectors, T is a finite set of transitions, $\tau \subset S \times T \times S$ is a transition relation, and $L : T \to \Sigma$ is a labeling function.*

We remark that the labeling function L is not very important in the proposed method. However, it is necessary when we discuss the relationship to other models such as process algebra.

Let $\geq$ be a partial order defined on $\mathbb{N}^k$ such that $x = [x_1, \cdots, x_n] \geq y = [y_1, \cdots, y_n]$ if and only if $x_i \geq y_i$ for every $i \in \{1, \cdots, k\}$. VTS is called *bounded* if it consists of a finite number of vectors, otherwise it is called *unbounded*. Let T^* denote the set of all sequences (including the empty sequence λ) over T.

For a transition relation τ, let $\tau^* \subset S \times T^* \times S$ denote the smallest set containing τ such that (i) $\forall s \in S : (s, \lambda, s) \in \tau^*$ and (ii) $(s, t, s') \in \tau \wedge (s', \sigma, s'') \in \tau^* \Rightarrow (s, t\sigma, s'') \in \tau^*$. We write $(s, \sigma > \in \tau^*$ to indicate $\exists s' \in S : (s, \sigma, s') \in \tau^*$.

We say that state s' is *reachable* from state s if $\exists \sigma \in T^* : (s, \sigma, s') \in \tau^*$, and that a sequence $\sigma \in T^*$ is *enabled* in state s if $(s, \sigma > \in \tau^*$.

Let $EN(s)$ denote the set of all transition t enabled in s. For a state $s \in S$, let $VTS(s)$ denote the subsystem consists of all the states reachable from s.

The vector transition system studied in this paper is assumed to have the following property. This seems to be a reasonable assumption if we consider each value of vectors as resources.

Definition 2. *A vector transition system $VTS = (S, T, \tau, L)$ is called order-persistent if the following holds:*
(P1) $(s_1, \sigma, s_2) \in \tau^* \wedge s_1' \geq s_1 \Rightarrow (s_1', \sigma, s_1' + s_2 - s_1) \in \tau^*$.

From this definition, we immediately obtain the following.

Lemma 3. *Let $VTS = (S, T, \tau, L)$ be a vector transition system which is order-persistent. Then $(s, \sigma, s_1) \in \tau^* \wedge (s, \nu, s_2) \in \tau^* \wedge s_1 \geq s \wedge s_2 \geq s \Rightarrow (s, \sigma\nu, s') \in \tau^* \wedge (s, \nu\sigma, s') \in \tau^*$, where $s' = s_1 + s_2 - s$.*

3.2 Petri Nets

In this paper, Petri nets is used as a model for describing order-persistent systems.

Definition 4. *A* net *is a 4-tuple $N = (P, T, A, L)$, where P and T are disjoint finite sets, $A : (P \times T) \cup (T \times P) \to \mathbb{N}$ is a function, and $L : T \to \Sigma$ is a labeling function. Each element in P is called* a place, *and each element in T is called a* transition.

Let $N = (P, T, A, L)$ be a net. A state of net N is a function $s : P \to \mathbb{N}$, which is usually called *a marking*. In the following, we will use the word *state* instead of marking. A transition t is *enabled* in state s if and only if $\forall p \in P : s(p) \geq A(p, t)$. The state space of a net N is represented by a vector transition system $VTS_N = (\mathbb{N}^P, T, \tau_N, L)$, where $\tau_N \subset \mathbb{N}^P \times T \times \mathbb{N}^P$ is defined by

$$
(s, t, s') \in \tau_N \Leftrightarrow \\
\forall p \in P : [s(p) \geq A(p, t) \wedge s'(p) = s(p) + A(t, p) - A(p, t)]. \tag{2}
$$

For each transition t, let Δ_t denote the vector in $\mathbb{N}^P$ such that $\Delta_t(p) = A(t, p) - A(p, t)$ for each $p \in P$. The vector Δ_t indicates how the state changes by an occurrence of transition t. We can similarly define the vector Δ_σ for a sequence σ of transitions. A net N with an initial state s is called *bounded* if the set of states in $VTS_N(s)$ is bounded. Otherwise it is called *unbounded*. Petri nets are a resource conscious model, and have the following property.

Lemma 5. *VTS_N is order-persistent.*

3.3 Coverability Tree for Unbounded Nets

The coverability tree represents the state space of an unbounded Petri net in the form of a finite search tree. Let $VTS = (S, T, \tau, L)$ be order-persistent. If S is infinite, then there exist states s, s' and a sequence σ such that $(s, \sigma, s') \in \tau \wedge s \leq s'$. From Property (P1), the sequence σ can be repeated infinity many times from state s. Thus, we have the following.

Lemma 6. *Suppose that $(s, \sigma, s') \in \tau \wedge \Delta = s' - s \geq 0$. Then all states in the form $s + n \cdot \Delta \ (n \in \mathbb{N})$ are reachable from s.*

Finiteness of the coverability tree is accomplished by introducing the special symbol ω. The symbol ω represents infinity, and satisfies that for any $n \in \mathbb{N} : n < \omega \wedge n + \omega = \omega$. Each node in the tree is associated with a vector in $(\mathbb{N} \cup \{\omega\})^k$.

Let g and h be two distinct nodes in the tree. The we write $g \to h$ to indicate that there exists a path from g to h. In addition, we write $g \xrightarrow{\sigma} h$ to denotes the sequence of labels on the path is σ.

Now we show the algorithm to construct a coverability tree.

Algorithm 1: Coverability Tree

Input: a net $N = (P, T, A, L)$ with an initial state s_0.

Output: a coverability tree.

1. For the root r, let $l(r) = s_0$.

2. Let h be a node. If there exists a node g such that $l(h) = l(g)$, then h is a terminal node.

3. Otherwise, each son of node h corresponds to some enabled transition. Let h_t be the node for transition t. The edge from node h to node h_t is labeled with t, and node h_t is associated with a vector $l(h_t) \in (\mathbb{N} \cup \{\omega\})^P$ defined as follows:

(i) If there exists a node g such that

$$g \to h_t \wedge l(g) \le l(h) + \Delta_t \wedge l(g)(p) < (l(h) + \Delta_t)(p),$$

then let $l(h_t)(p) = \omega$.

(ii) Otherwise, let $l(h_t)(p) = (l(h) + \Delta_t)(p)$. $\qquad\qquad\square$

4 Algorithm for Unbounded Systems

In this section, we will show an algorithm to give a reduced representation for the state space of an unbounded Petri net. This algorithm is similar to that of the coverability tree, excepting the following points.

1. Each node of the tree is associated with a semilinear set instead of a vector in $(\mathbb{N} \cup \{\omega\})^k$. Therefore, no information is lost.

2. Reduced state space generation based on the partial order methods can be combined with the algorithm.

4.1 Semilinear Sets

It is known that some classes of Petri nets have semilinear reachability sets[6, 17]. Every semilinear set can be described by a finite expression. For finite subsets C and $D = \{d_1, \cdots, d_n\}$ of $\mathbb{N}^k$, let $L(C, D)$ denote the infinite set of vectors which have the form $c + k_1 \cdot d_1 + \cdots + k_n \cdot d_n$ ($c \in C$, $k_i \in \mathbb{N}$, $i = 1, \cdots, n$). For a nonnegative integer vector $c \in \mathbb{N}^k$ and a finite subset D of $\mathbb{N}^k$, a set which can be described in the form $L(\{c\}, D)$ is called a linear set. A finite union of linear set is called a semilinear set.

We use the following notation to represent a linear set $L(\{c\}, D)$.

$$c + (0..\infty) \cdot d_1 + \cdots + (0..\infty) \cdot d_n.$$

As described in Section 2, $(l..u)$ denotes any integer x such that $l \leq x \leq u$. When $u = \infty$, it is any integer x such that $l \leq x$. Therefore, an expression $c + (l_1..u_1) \cdot d_1 + \cdots + (l_n, u_n) \cdot d_n$ denote the following set.

$$\{c + k_1 \cdot d_1 + \cdots + k_n \cdot d_n \mid l_i \leq k_i \leq u_i (i = 1, \cdots, n)\}.$$

For example, an expression

$$c + (2..4) \cdot d_1 + (3..\infty) \cdot d_2$$

represents the semilinear set $L(\{c + 2d_1 + 3d_2, c + 3d_1 + 3d_2, c + 4d_1 + 3d_2\}, \{d_2\})$. Suppose that $(s, \sigma, s') \in \tau$ and $\Delta = s' - s \geq 0$. By Lemma 6, every state in the from $s + n \cdot \Delta$ ($n \in \mathbb{N}$) are reachable from s. The set of such reachable states is written as $s + (0..\infty) \cdot \Delta$.

We note that an expression $c + (l_1..u_1) \cdot d_1 + \cdots + (l_n..u_n) \cdot d_n$ is one of possible representations for the corresponding semilinear set. There may be many representations for the same set. For example, $[1, 0] + (3..5) \cdot [0, 1]$ represents the same set as that for $[1, 3] + (0..2) \cdot [0, 1]$. We call an expression $c + (l_1..u_1) \cdot d_1 + \cdots + (l_n..u_n) \cdot d_n$ *canonical* if $l_i = 0$ for every $i \in \{1, \cdots, n\}$.

As described in Section 2, we will call each expression $c + (l_1..u_1) \cdot d_1 + \cdots + (l_n..u_n) \cdot d_n$ a *region*. If we write $s \in r$ for a region r, then it means that s is in the semilinear set corresponding to the region r.

Let $r = c + (l_1..u_1) \cdot d_1 + \cdots + (l_n..u_n) \cdot d_n$ be a region. We call c *the constant* of r, and is denoted by $Const_r$. We also call each d_i a *periodic vector* of r, and let $Period_r$ denote the matrix $[d_1, \cdots, d_n]$. Each vector $s = c + k_1 \cdot d_1 + \cdots + k_n \cdot d_n$ of region r is uniquely specified by an integer vector $K = [k_1, \cdots, k_n]$, i.e., $s = Const_r + K \cdot Period_r$. Let $Coef_r = \{[k_1, \cdots, k_n] \in N^n \mid l_i \leq k_i \leq u_i (i = 1, \cdots, n)\}$. We note that $Const_r \in r$ is not true unless r is canonical.

4.2 Equivalence Relation on State Space

In the proposed algorithm, a region is associated with each node of the search tree. We will introduce some equivalence relation on the set of states. When a new region is obtained, it is decomposed into a finite number of regions in which every two states are equivalent to each other, and they are associated with new nodes respectively.

To handle each region in the same manner as an individual state, all states in a region should have the same set of enabled transitions. Then the equivalence is assumed to at least satisfy the following property.

(P2) If two states s_1 and s_2 are equivalent, then $EN(s_1) = EN(s_2)$.

4.3 The Algorithm

Now we show the algorithm for unbounded nets. We will call the tree generated by this algorithm a *semilinear reachability tree*.

Algorithm 2: Semilinear Reachability Tree
Input: a net $N = (P, T, A, L)$ with an initial state s_0.
Output: a semilinear reachability tree.
Each node h in the tree is associated with a region $l(h) = c + (l_1, u_1) \cdot d_1 + \cdots + (l_n, u_n) \cdot d_n$, and with a node label $a(h)$ which indicates the nearest ancestor of h at which a periodic vector is added.
1. For the root node r, let $l(r) = s_0$ and $a(r) = r$.
2. Let h be a node with a region

$$l(h) = c + (l_1..u_1) \cdot d_1 + \cdots + (l_n..u_n) \cdot d_n.$$

If for any vector $v \in l(h)$ there exists a node g such that $v \in l(g)$, then h is a terminal node.
3. Otherwise, each son of node h corresponds to some enabled transition. Let h_t be the node for transition t. The edge from node h to node h_t is labeled with t, and node h_t is associated with a region $l(h_t)$ defined as follows:
3-1. Let g be a node on the path from the root to node h. If (i) $\Delta = c + \Delta_t - Const_{l(g)} \geq 0, \Delta \neq 0$ and (ii) g is a descendant of $a(h)$, or $l(h)$ is canonical(i.e., $l_i = 0 (i = 1, \cdots, n)$), then let

$$l(h_t) = c' + (l_1..u_1) \cdot d_1 + \cdots + (l_n..u_n) \cdot d_n + (0..\infty) \cdot \Delta,$$

where $c' = c + \Delta_t$. Let $a(h_t) = h_t$.
3-2. If such node g does not exist, then let

$$l(h_t) = c' + (l_1..u_1) \cdot d_1 + \cdots + (l_n..u_n) \cdot d_n,$$

where $c' = c + \Delta_t$. Let $a(h_t) = a(h)$.
Using the equivalence relation, decompose the region $l(h_t)$ into a finite number of regions in which every two states are equivalent to each other. Add new nodes corresponding to these regions, and replace node h_t with these new nodes.   □

Remark. Either the depth-first or the breadth-first search is applicable when a new node h is selected. There is no guarantee that the algorithm terminates since the obtained set of reachable states has to be semilinear. It is known that there exists a Petri net whose reachability set is not semilinear.

The algorithm refers only the constant of each regions when it finds a new periodic vector. This is justified by the following lemma.

Lemma 7. *Construct a tree by Algorithm 2. Assume that for every node in the tree, all states in its associated region are reachable from the initial state. Let g and h be nodes of the tree such that*
(i) g is a node on the path from the root to node h such that $\Delta = Const_{l(h)} - Const_{l(g)} \geq 0, \Delta \neq 0$, and
(ii) g is a descendant of $a(h)$, or $l(h)$ is canonical.
Then every state in $l(h) + (0..\infty) \cdot \Delta$ is also reachable from the initial state.

Proof. Let σ be the sequence of transition labels on the path from node g to node h.

Suppose that g is a descendant of $a(h)$. This means that no new periodic vectors are added to any node on this path, i.e., the algorithm has applied the case 3-2. Let $s = c + k_1 \cdot d_1 + \cdots + k_n \cdot d_n$ be any state in $l(h)$. Then $s' = c' + k_1 \cdot d_1 + \cdots + k_n \cdot d_n$ is in $l(g)$ and $(s', \sigma, s) \in \tau_N$ holds by Property (P2). Since $c - c' = \Delta \geq 0$, all states in $s + (0..\infty) \cdot \Delta$ are reachable.

When $l(h)$ is canonical, every node on the path from node g to node h is also canonical. From Property (P2), $(Const_{l(g)}, \sigma, Const_{l(h)}) \in \tau_N$ holds. Since $Const_{l(h)}$ is the minimum vector in $l(h)$, the lemma holds.

We will prove that the algorithm correctly computes the state space if the following equivalence is used. Let EQ_{EN} be a binary relation on S defined by

$$(s, s') \in EQ_{EN} \Leftrightarrow EN(s) = EN(s'). \tag{3}$$

We first show the finiteness of the decomposition.

Theorem 8. *Let* $r = c + (l_1..u_1) \cdot d_1 + \cdots + (l_n..u_n) \cdot d_n$ *be a region. Then* r *is represented by the union of a finite number of regions* $r_1, \cdots, r_m$ *such that* $(s, s') \in EQ_{EN}$ *for every two states* s, s' *in each* r_i $(i = 1, \cdots, m)$.

Proof. We use the following two facts to prove the proposition.
(i) If $s \leq s'$, then $EN(s) \subset EN(s')$(Property (P1)).
(ii) $EN(s) \subset T$ and T is finite.

For any $s = c + k_1 \cdot d_1 + \cdots + k_n \cdot d_n \in r$, there exists an integer vector $v = [v_1, \cdots, v_n]$ such that every state in the region $c + (v_1..u_1) \cdot d_1 + \cdots + (v_n..u_n) \cdot d_n$ has the same set of enabled transitions. Let V_s be the set of such vector v, and let $V_r = Min \bigcup_{s \in r} V_s$. V_r is a finite set. Constructing a region $c + (v_1..u_1) \cdot d_1 + \cdots + (v_n..u_n) \cdot d_n$ for each $[v_1, \cdots, v_n] \in V_r$, the rest of the region r contains a finite number of vectors.

Now we show several properties of the semilinear reachability tree.

Theorem 9. *Using* EQ_{EN} *as the equivalence, construct a tree by Algorithm 2 for a net* $N = (P, T, A, L)$ *with an initial state* s_0. *Let* $VTS_N(s_0) = (S, T, \tau_N, L)$.

1. *For each state* $s \in S$, *there exists a node* h *in the tree such that* $s \in l(h)$.
2. *For each node* g *in the tree,* $l(g) \subset S$.
3. *If* $(s, t, s') \in \tau_N$, *then there exist nodes* g *and* h *in the tree such that* $s \in l(g)$, $s' \in l(h)$, *and* $g \xrightarrow{t} h$.
4. *Suppose that* $g \xrightarrow{t} <h_t^1, \cdots, h_t^m>$ *holds in the tree, where* $h_t^1, \cdots, h_t^m$ *are the nodes generated by the decomposition. Then for each* $K \in Coef_{l(g)}$, *there exist a region* h_t^i *and* $K' \in Coef_{l(h_t^i)}$ *such that*
    (i) $(Const_{l(g)} + K \cdot Period_{l(g)}, t, Const_{l(h_t^i)} + K' \cdot Period_{l(h_t^i)}) \in \tau_N$, *and*
    (ii) $K(i) = K'(i)(i = 1, \cdots, n)$ *and* $K'(j) = 0(j = n+1, \cdots, n')$, *where* n *is the number of periodic vectors in* $l(g)$ *and* n' *is that in* $l(h_t^i)$, *i.e.,* $Period_{l(g)}$ *has the form* $[d_1, \cdots, d_n]$ *and* $Period_{l(h_t^i)}$ *has the form* $[d_1, \cdots, d_n, \cdots, d_{n'}]$.

Proof. 1. Suppose that $(s, t, s') \in \tau_N$ holds for a state s on the tree and for a state s' not on the tree. Then the algorithm add a new node whose associated region contains s'. Hence, every state in S has to be contained in some region on the tree.

2. By Lemma 7, every state on the tree is reachable from s_0.

3. Suppose that $(s, t, s') \in \tau_N$. Then from 1, there exists a node h such that $s \in l(h)$. At lease one of such h is not terminal. Then a new node h_t is generated and $s' = s + \Delta_t \in l(h_t)$. Moreover, s' is also contained in the region associated with some node generated by decomposing h_t.

4. Let $s = Const_{l(g)} + k_1 \cdot d_1 + \cdots + k_n \cdot d_n$ be any state in $l(g)$. Then the state $s' = (Const_{l(g)} + \Delta_t) + k_1 \cdot d_1 + \cdots + k_n \cdot d_n$ is in $l(h_t^i)$ for some region h_t^i. □

4.4 Decomposing Regions

We now consider how to decompose each region for the equivalence relation EQ_{EN}. Let $r = c + (l_1, u_1) \cdot d_1 + \cdots + (l_n, u_n) \cdot d_n$ be a region. Let $K = [k_1, \cdots, k_i, \cdots, k_n]$ be a vector in $Coef_r$ such that $l_i < k_i$, and let K_{-i} denote the vector $[k_1, \cdots, k_i - 1, \cdots, k_n]$ which is also in $Coef_r$. Then the nonnegative integer k_i is called *a splitting point* for d_i if

$$EN(Const_r + K_{-i} \cdot Period_r) \neq EN(Const_r + K \cdot Period_r). \qquad (4)$$

The number of splitting points is finite. Then we have the following. The proof is omitted since it can be easily shown.

Proposition 10. *Let $r = c + (l_1, u_1) \cdot d_1 + \cdots + (l_n, u_n) \cdot d_n$ be a region. Let $k_i^1 < k_i^2 \cdots < k_i^{m_i}$ be splitting points for d_i, $k_i^0 = l_i$ and $k_i^{m_i+1} = u_i$. For $0 \leq j_i \leq m_i$ $(i = 1, \cdots, n)$, let*

$$r[j_1, \cdots, j_n] = c + (k_i^{j_1}, k_i^{j_1+1}) \cdot d_1 + \cdots + (k_n^{j_n}, k_n^{j_n+1}) \cdot d_n.$$

Then every two states in each region $r[j_1, \cdots, j_n]$ are equivalent to each other with respect to EQ_{EN}.

4.5 Stubborn Equivalence

The stubborn set method works effectively in generating reduced state space that preserves deadlocks, livelocks, and liveness of transitions[14]. The main idea of stubborn sets is the selective search, i.e., only some of enabled transitions are executed in each state. The set of transitions to be executed is called a stubborn set, which is defined as follows. Subset T_s of transitions is *stubborn* (in the strong sense) in state s if and only if:

S1. $\forall t \in T_s : [(s, t > \in \tau_N \Rightarrow \forall p \in P \; \forall t' \in T - T_s : min(A(t, p), A(t', p)) \geq min(A(p, t), A(p, t'))]$.

S2. $\forall t \in T_s : [(s, t > \notin \tau_N \Rightarrow \exists p \in P : s(p) < A(p, t) \wedge up_t(p) \subset T_s]$.

S3. $\exists t \in T_s : [(s, t > \in \tau_N \wedge \forall p \in P : down_t(p) \subset T_s]$.

Where

$$up_t(p) = \{t' \in T \,|\, A(t',p) > A(p,t') < A(p,t)\},$$
$$down_t(p) = \{t' \in T \,|\, A(p,t') > A(t',p) < A(p,t)\}.$$

Let $STUB(s)$ denote the family of all possible stubborn set in state s, and let

$$ESTUB(s) = \{T' \cap EN(s) \,|\, T' \in STUB(s)\}. \tag{5}$$

We can consider the following two stubborn-equivalence. Two states s and s' are called *stubborn-equivalent in strong sense* if $ESTUB(s) = ESTUB(s')$, and are called *stubborn-equivalent in weak sense* if $ESTUB(s) \cap ESTUB(s') \neq \emptyset$. We will use here the stubborn-equivalence in strong sense.

To use the stubborn-equivalence, Algorithm 2 is modified as follows: When expanding a new node at step 3, select one of stubborn sets in $ESUB(s)$, and only the transitions in the set is executed. Then the assumption (P2) on the equivalence has to be modified as follows.

(P2') If two states s_1 and s_2 are equivalent, then $ESTUB(s_1) = ESTUB(s_2)$.

The conditions $S1$ and $S3$ depend only on the structure of the net. Therefore, if $EN(s) = EN(s')$, then s and s' are equivalent with respect to Conditions $S1$ and $S3$. The condition $S2$ depends on the state.

We will show some results to prove the finiteness of the decomposition when the stubborn-equivalence is used.

Lemma 11. *Suppose that $s \leq s'$ and $EN(s) = EN(s')$. Then $STUB(s) \supset STUB(s')$.*

Proof. Let $T_{s'}$ be any stubborn set in s'. $T_{s'}$ satisfy the conditions $S1$ and $S3$ in s since $EN(s) = EN(s')$. Suppose that $t \in T_{s'}$ is not enabled. Then by $S2$, there exists a place p such that $s'(p) < A(p,t) \wedge up_t(p) \subset T_{s'}$. Since $s \leq s'$, t is not enabled in s and $s(p) < A(p,t)$ holds. Therefore, $T_{s'}$ satisfy the condition $S2$ also in s. Hence, $T_{s'}$ is stubborn in s. $\square$

Lemma 12. *Let r be a region. Then there exists a state $s \in r$ such that for any $s' \in r$: if $s \leq s'$, then $ESTUB(s) = ESTUB(s')$.*

Proof. Since the set of transitions is finite, there exists a state $s_1 \in r$ such that $EN(s_1) = EN(s')$ for any $s' \in r$ satisfying $s_1 \leq s'$. By applying Lemma 11, we will find a state $s_2 \in r$ such that $s_1 \leq s_2$ and $STUB(s_2) = STUB(s'')$ holds for any $s'' \in r$ satisfying $s_2 \leq s''$. $\square$

Using Lemma 12, we obtain the following theorem. The proof is similar to that of Theorem 8.

Theorem 13. *Any region can be decomposed into a finite number of regions in which every two states are stubborn-equivalent to each other.*

For the stubborn-equivalence, splitting points are defined as follows. Let $r = c + (l_1, u_1) \cdot d_1 + \cdots + (l_n, u_n) \cdot d_n$ be a region. Let $K = [k_1, \cdots, k_i, \cdots, k_n]$ be a vector in $Coef_r$ such that $l_i < k_i$. Then the nonnegative integer k_i is called a *splitting point* for d_i if

$$(Const_r + K_{-i} \cdot Period_r)(p) < A(p, t) \land$$
$$(Const_r + K \cdot Period_r)(p) \geq A(p, t). \tag{6}$$

We note that this is not an optimal decomposition. We can easily obtain a trivial upper bound of the number of generated regions. The number of splitting points for d_i is at most $max_{p,t} \lceil \frac{A(p,t)}{d_i(p)} \rceil$. Then the total number of generated regions is at most $(max_{p,t} \lceil \frac{A(p,t)}{d_i(p)} \rceil + 1)^n$, where n is the number of periodic vectors.

4.6 Termination of Algorithm

Algorithm 2 does not terminate for some classes of Petri nets. We will describe a sufficient condition for the algorithm to terminate.

Let $VTS(s_0) = (S, T, \tau, L)$ be order-persistent. Let $\Delta^+(s) = Min\{\Delta \mid \exists \sigma : (s, \sigma, s') \in \tau^* \land \Delta = s' - s \geq 0 \land \Delta \neq 0\}$. The set S of reachable states is semilinear if the following (P3) holds.

(P3) For any states s, s' such that $s' \geq s$, $s' \in L(\{s\}, \Delta^+(s))$.

Let us consider the set $U = \bigcup_{s \in S} \Delta^+(s)$. From the order-persistence, U is a projection of the following set.

$$Min\{[\Delta, s] \mid \exists \sigma : (s, \sigma, s') \in \tau^* \land \Delta = s' - s \geq 0 \land \Delta \neq 0\}.$$

Therefore, the set U is finite, and so is the set $\Delta^+ = \{\Delta^+(s) \mid s \in S\}$. Let $\Delta^+ = \{\Delta_1^+, \cdots, \Delta_n^+\}$ and $S_i = \{s \mid \exists \sigma : (s_0, \sigma, s) \in \tau^* \land \Delta^+(s) = \Delta_i^+\}$ $(i = 1, \cdots, n)$. Then, we can conclude that S is semilinear by the following fact.

$$S = \bigcup_{i=1,n} S_i \subset \bigcup_{i=1,n} L(Min\, S_i, \Delta_i^+) \subset S. \tag{7}$$

Since the set of periodic vectors $\Delta^+(s)$ is found within a finite number of iterations, Algorithm 2 always terminates. It is known that some classes of Petri nets, such as marked graphs and conflict-free nets[10], have this property[1].

5 Algorithm for Finite Capacity Systems

In this section, we will show that the idea used in Algorithm 2 is also applicable to finite capacity nets.

---

[1] In such nets, this property holds in the state space of extended markings - a combination of the marking and the firing counts.

5.1 Finite Capacity Nets

Practical systems does not allow infinity many resources. In order to represent finiteness, finite capacities are usually given to each component of the system.

Definition 14. *A finite capacity net* is a 5-tuple $CN = (P, T, A, L, Cap)$, where $N = (P, T, A, L)$ is a net and $Cap : P \rightarrow \mathbb{N}$ is *a capacity function*.

The transition relation for a finite capacity net is $\tau_{CN} \subset \mathbb{N}^P \times T \times \mathbb{N}^P$ such that

$$(s, t, s') \in \tau_N \Leftrightarrow$$
$$\forall p \in P : [s(p) \geq A(p, t) \wedge s(p) + A(t, p) \leq Cap(p) \wedge \tag{8}$$
$$s'(p) = s(p) + A(t, p) - A(p, t)].$$

5.2 Stubborn Sets for Finite Capacity Nets

For any finite capacity net CN, there exists an equivalent net N^c without capacities, where the equivalence is given by the isomorphism on the state graphs. N^c can be constructed by adding complementary places. For each place p, add a place p^c which connects to the same transitions as p, but each arc is drawn in the reverse direction, and the initial state is given by $s_0(p^c) = Cap(p) - s_0(p)$. Using this transformation, stubborn sets for finite capacity nets are defined as follows:

S1. $\forall t \in T_s : [(s, t >\in \tau_N \Rightarrow \forall p \in P \; \forall t' \in T - T_s : min(A(t, p), A(t', p)) = min(A(p, t), A(p, t')))].$
S2. $\forall t \in T_s : [(s, t >\notin \tau_N \Rightarrow \exists p \in P : (s(p) < A(p, t) \wedge up_t(p) \subset T_s)$
$\vee (s(p) + A(t, p) > Cap(p) \wedge up_i^c(p) \subset T_s)].$
S3. $\exists t \in T_s : [(s, t >\in \tau_N \wedge \forall p \in P : down_t(p) \cup down_p^c \subset T_s].$

Where,

$$up_t(p) = \{t' \in T | A(t', p) > A(p, t') < A(p, t)\},$$
$$up_i^c(p) = \{t' \in T | A(p, t') > A(t', p) < A(t, p)\},$$
$$down_t(p) = \{t' \in T | A(p, t') > A(t', p) < A(p, t)\},$$
$$down_i^c(p) = \{t' \in T | A(t', p) > A(p, t') < A(t, p)\}.$$

We note that this definition of stubborn set does not always give a minimum set.

5.3 Decomposing Regions

Algorithm 2 is applicable to finite capacity nets by modifying the definition of splitting points. Let $r = c + (l_1, u_1) \cdot d_1 + \cdots + (l_n, u_n) \cdot d_n$ be a region, and let $K = [k_1, \cdots, k_i, \cdots, k_n] \in Coef_r$ such that $l_i < k_i$.

We call k_i a splitting point for d_i if
(i) For the equivalence EQ_{EN}:

$$EN(Const_r + K_{-i} \cdot Period_r) \neq EN(Const_r + K \cdot Period_r)). \tag{9}$$

(ii) For the stubborn equivalence:

$$
\begin{aligned}
&((Const_r + K_{-i} \cdot Period_r)(p) < A(p,t) \wedge \\
&(Const_r + K \cdot Period_r)(p) \geq A(p,t)) \vee \\
&((Const_r + K_{-i} \cdot Period_r)(p) + A(t,p) < Cap(p) \wedge \\
&(Const_r + K \cdot Period_r)(p) + A(t,p) \geq Cap(p)).
\end{aligned}
\tag{10}
$$

We should also consider the fact that not all states in a region generated by the algorithm is feasible. Let $r = [1,0] + (0..4) \cdot [1,0] + (0..5) \cdot [1,1]$ be a region such that the capacity is 3 for the first place and 2 for the second. Then the feasible states are $\{[1,0],[2,0],[3,0],[2,1],[3,1],[3,2]\}$. That is, the set of feasible states are the projection of the region onto the states satisfying the capacity condition.

The feasible states of the region r can represented by the union of reasions r_1, r_2, r_3, where

$$
\begin{aligned}
r_1 &= [1,0] + (0..3) \cdot [1,0], \\
r_2 &= [1,0] + (0..2) \cdot [1,0] + (1..2) \cdot [1,1], \\
r_3 &= [1,0] + (0..1) \cdot [1,0] + (2..2) \cdot [1,1].
\end{aligned}
$$

We aim to find such a decomposition that every state in each resulted region is feasible. To obtain such a decomposition, it suffices to add the following k_i as a splitting point, and to remove regions which do not satisfy the capacity condition.

$$
\begin{aligned}
&(Const_r + K_{-i} \cdot Period_r)(p) < Cap(p) \wedge \\
&(Const_r + K \cdot Period_r)(p) \geq Cap(p).
\end{aligned}
\tag{11}
$$

5.4 The Algorithm

The modified algorithm for finite capacity nets is described as follows. It is expected that Algorithm 2 works effectively when the capacities become larger. In addition, the algorithm always terminates.

Algorithm 3: Semilinear Reachability Tree for Finite Capacity Nets
Input: a finite capacity net $CN = (P, T, A, L, Cap)$ with an initial state s_0.
Output: a semilinear reachability tree.
Each node h in the tree is associated with a region $l(h) = c + (l_1, u_1) \cdot d_1 + \cdots + (l_n, u_n) \cdot d_n$, and with a node label $a(h)$ which indicates the nearest ancestor of h at which a periodic vector is added.
1. For the root node r, let $l(r) = s_0$ and $a(r) = r$.
2. Let h be a node with a region

$$
l(h) = c + (l_1..u_1) \cdot d_1 + \cdots + (l_n..u_n) \cdot d_n.
$$

If for any vector $s \in l(h)$ there exists a node g such that $s \in l(g)$, then h is a terminal node.
3. Otherwise, each son of node h corresponds to some enabled transition. Let h_t be the node for transition t. The edge from node h to node h_t is labeled with t, and node h_t is associated with a region $l(h_t)$ defined as follows:

3-1. Let g be a node on the path from the root to node h. If (i) $\Delta = c + \Delta_t - Const_{l(g)} \geq 0, \Delta \neq 0$ and (ii) g is a descendant of $a(h)$, or $l(h)$ is canonical, i.e., $l_i = 0 (i = 1, \cdots, n)$, then let

$$l(h_t) = c' + (l_1..u_1) \cdot d_1 + \cdots + (l_n..u_n) \cdot d_n + (0..u_{n+1}) \cdot \Delta,$$

where $c' = c + \Delta_t$ and $u_{n+1} = min_p \lfloor \frac{Cap(p) - c'(p)}{\Delta(p)} \rfloor$. Let $a(h_t) = h_t$.
3-2. If such node g does not exist, then let

$$l(h_t) = c' + (l_1..u_1) \cdot d_1 + \cdots + (l_n..u_n) \cdot d_n,$$

where $c' = c + \Delta_t$. Let $a(h_t) = a(h)$.
Using the equivalence relation, decompose the region $l(h_t)$ into a finite number of regions in which every two states are equivalent to each other. Add new nodes corresponding to these regions, and replace node h_t with these new nodes. $\square$

6 Implementation and Experiments

We have implemented Algorithm 3 as a C program. In the following, we will first describe the current implementation, and then show the experimental results for some examples.

6.1 Checking Terminal Nodes

If every state in a region is already contained in other regions, then the node is set to be terminal. We should consider how to check the inclusion between two semilinear sets efficiently.

1. *Unbounded Nets (Algorithm 2):* Let $L = L(c, \{d_1, \cdots, d_n\})$ and $L' = L(c', \{d'_1, \cdots, d'_m\})$ be linear sets. Then $L' \subset L$ holds if and only if
(i) $c' \in L$, i.e., $c' - c \in L(\{0\}, \{d_1, \cdots, d_n\})$, and
(ii) $d'_i \in L(\{0\}, \{d_1, \cdots, d_n\}) \ (i = 1, \cdots, m)$.
Therefore, we have to solve the following system of equations:

$$\sum_{i=1}^n k_i d_i = d_0,$$

$$d_0 \in \mathbb{N}^P,$$
$$d_i \in \mathbb{N}^P, k_i \in \mathbb{N} \ (i = 1, \cdots, n). \tag{12}$$

Especially if $|P| = 1$, this problem is *a knapsack problem*. We can also prove that the problem for $|P| > 1$ is reducible in polynomial time to the problem for $|P| = 1$. This can be done by constructing a scalar value d_i^* for each d_i.

$$d_i^* = d_i(p_1) + 2^l d_i(p_2) + \cdots 2^{(m-1)l} d_i(p_m). \tag{13}$$

where $P = \{p_1, \cdots, p_m\}$ and $l = \lfloor log_2(max_{i,j} d_i(p_j)) \rfloor + 1$. Thus, this problem is hard to solve in general.

2. Finite Capacity Nets (Algorithm 3): Let r_i $(i = 1, \cdots, n)$ be regions which have been already found, and let r_{n+1} be a newly generated region. For each $s \in r_{n+1}$, we have to check $s \in r_i$ for each $i = 1, \cdots, n$. We can use Binary Decision Diagrams(BDDs)[1, 2] to represent each region. BDDs are widely used for manipulating Boolean function. It is known that most commonly encountered functions can be represented in a reasonable size. For each state s, let $BDD(s)$ denote the BDD such that it is true if and only if we assign it the values corresponding to state s. The BDD representing a region computed as the disjunction of all the $BDD(s)$ for each state $s \in r$. Let $R = \bigcup_{i=1,n} r_i$ be the set of regions which are already generated, and let $BDD(R)$ be the corresponding BDD. Then a new node associated with region r_{n+1} is terminal if $\neg BDD(R) \wedge BDD(r_{n+1}) = 0$. If it is not terminal, then we replace $BDD(R)$ with $BDD(R) \vee BDD(r_{n+1})$.

6.2 Restriction on Periodic Vectors

Let $r = 0 + (0..5) \cdot d_1 + (0..5) \cdot d_2$ be a region. If $d_1(p) = d_2(p) = 1$ and $A(p, t) = 5$ for some place p and transition t, then every $k = 1, \cdots, 5$ is a splitting point when the stubborn-equivalence is used. We should avoid such a case that $d_i(p) > 0 \wedge d_j(p) > 0$ for some $i, j (i \neq j)$. When d_1 and d_2 are the periodic vectors of some region, the vector $d_1 + d_2$ may be added as a new periodic vector. In many cases, the number of generate states can be reduced if we do not add such periodic vectors.

6.3 Avioding Unnecessary Splitting

The splitting points defined by (10) are too fine. That is, not every splitting points are necessary. In the implementation, such unnecessary splitting is avoidable by merging regions adjacent to each other.

In Fig.2, suppose that the following region is generated.

$$r = [4, 1, 1] + (0..5) \cdot [1, 1, 0] + (0..3) \cdot [2, 0, 1].$$

Given a capacity 10 for p_1, the splitting points for the stubborn-equivalence are obtained as $1, 2, 3, 4, 5$ for the first periodic vector, and $1, 2, 3$ for the second. In the implementation, only the following 9 regions will be generated.

$$r_1 = [4, 1, 1] + (0..3) \cdot [1, 1, 0],$$
$$r_2 = [4, 1, 1] + (0..1) \cdot [1, 1, 0] + (1..1) \cdot [2, 0, 1],$$
$$r_3 = [4, 1, 1] + (2..2) \cdot [2, 0, 1],$$
$$r_4 = [4, 1, 1] + (3..3) \cdot [2, 0, 1],$$
$$r_5 = [4, 1, 1] + (1..2) \cdot [1, 1, 0] + (2..2) \cdot [2, 0, 1],$$
$$r_6 = [4, 1, 1] + (2..2) \cdot [1, 1, 0] + (1..1) \cdot [2, 0, 1],$$
$$r_7 = [4, 1, 1] + (3..4) \cdot [1, 1, 0] + (1..1) \cdot [2, 0, 1],$$
$$r_8 = [4, 1, 1] + (4..4) \cdot [1, 1, 0],$$
$$r_9 = [4, 1, 1] + (5..5) \cdot [1, 1, 0].$$

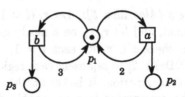

Fig. 2. A Petri net with two periodic vectors.

6.4 Depth-first . Breadth-first

The proposed algorithm works efficiently when periodic vectors are found in earlier stages of the algorithm. Therefore, the breadth-first may be better. We will compare these two strategies by experiments.

6.5 Experimental Results

We have executed the algorithm for several finite capacity nets, and have observed how the number of nodes in the tree can be reduced.

EX. 1. We first consider the net in Fig.1. The capacities are given by $Cap(p_3) = 100, Cap(p_4) = Cap(p_5) = Cap(p_6) = k$ and $Cap(p_1) = Cap(p_2) = 1$. We have executed the algorithm for several values of k. Table 1 shows the number of generated nodes. Where,

A: Algorithm 3 using stubborn-equivalence (breadth-first),
B: Algorithm 3 using stubborn-equivalence (depth-first),
C: The normal reachability tree using stubborn-equivalence (depth-first),
D: Algorithm 3 using EQ_{EN} (breadth-first),
E: Algorithm 3 using EQ_{EN} (depth-first),
F: The normal reachability tree using EQ_{EN} (depth-first).

The number of nodes generated by D is constant while the entire state space becomes larger. The number of nodes generated by A increases very slowly. We can observe that the breadth-first search gives better results. The reason why EQ_{EN} gives better results than the stubborn-equivalence is considered as follows. In the current implementation, the stubborn equivalence requires more splitting points that EQ_{EN}. Moreover, this net does not contain so many transitions occurring concurrently.

EX. 2. The finite capacity net in fig.3 causes a deadlock for some arc weights and capacities. Table 3 shows the results for such cases. The number of nodes generated by Algorithm 3 is constant for the (k_1, k_2) greater than $(27, 25)$.

Table 1. Experimental results 1

k	number of nodes					
	A	B	C	D	E	F
5	776	1149	1172	132	222	1772
6	695	1900	1903	132	222	2985
7	774	2645	2876	132	222	4652
8	776	3824	4121	132	222	6845
9	1051	5044	5677	132	222	9636
10	1071	6551	7573	132	222	13097

Table 2. Experimental results 2

(k_1, k_2)	number of nodes					
	A	B	C	D	E	F
(27,25)	146	223	244	154	267	841
(45,25)	153	251	464	157	390	1447
(75,55)	153	647	1487	157	442	5367
(105,85)	153	501	2227	157	442	7577

7 Conclusion and Future Work

As mentioned in Section 5, every finite capacity net CN has an equivalent net N^c without capacities. We should treat CN and N^c in a different manner, because we cannot find any periodic vectors from N^c. That is, CN is structurally unbounded, but N^c is structurally bounded. Algorithm 3 is based on such structural unboundedness of finite capacity nets.

As the future work, we will consider the following.

- We should find an optimal splitting of regions for the sutubborn-equivalence.
- The current implementation is not very efficient. Even if the size of the state space can be reduced, the computation time is not always reduced. Most of the time is spent on the stage of checking the inclusion of regions. We should improve each part of the algorithm.
- There are nets for which the proposed algorithm cannot reduce the size of the state space. We should find the condition in which the proposed algorithm works effectively.

References

[1] S.B. Akers: Binary Decision Diagrams, IEEE Trans. Computers, C-27-6, 509-516 (1978).

[2] R.E. Bryant: Graph-Based Algorithms for Boolean Function Manipulation, IEEE Trans. Computers, C35-8, 677-691 (1986).

[3] J.R. Burch, E.M. Clarke, K.L. McMillan: Sequential Circuit Verification Using Symbolic Model Checking, 27th ACM/IEEE Design Automation Conference, 46-51 (1990).

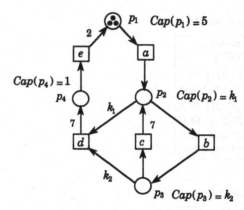

Fig. 3. A Petri net for EX2.

[4] P. Godfroid: Using Partial Orders to Improve Automatic Verification Methods, Lecture Notes in Computer Sience, No. 531, 176-185 (1990).

[5] P. Godefroid, P. Wolper: Using Partial Orderes for the Efficient Verification of Deadlock Treedom and Safety Properties, Lecture Notes in Computer Sience, No. 575, 332-342 (1991).

[6] K. Hiraishi, A. Ichikawa: On Structural Conditions for Weak Persistency and Semilinearity of Petri Nets, Theoretical Computer Science, Vol. 93, 185-199 (1992).

[7] K. Hiraishi, M. Makano: On Symbolic Model Checking in Petri Nets, IEICE Trans. Vol. E77-A, No. 10, 1602-1606 (1995).

[8] C. A. R. Hoare: Communicating Sequential Processes, Prentice Hall International Series in Computer Science, Prentice Hall, 1985.

[9] R. M. Karp and R. E. Miller: Parallel Program Schemata, J. Computer and System Science, Vol. 3, No. 2, 147-195, 1969.

[10] T. Murata: Petri Nets: Properties, Analysis and Applications, Proc. of the IEEE, Vol. 77, No. 4 (1989).

[11] E. Pastor, Oriol Roig, J. Cortadella, and R. Badia: Petri Net Analysis Using Boolean Manipulation, Lecture Notes in Computer Sience, No. 815, Springer-Verlag, 416-435 (1994).

[12] J. L. Peterson: Petri Net Theory and the Modeling of Systems, Prentice-Hall, 1981.

[13] M. Tiusanen: Symbolic, Symmetry, and Stubborn Set Searches, Lecture Notes in Computer Sience, No. 815, Springer-Verlag, 511-530 (1990).

[14] A. Valmari: Stubborn Sets for Reduced State Space Generation, Lecture Notes in Computer Sience, No. 483, Springer-Verlag, 491-515 (1990).

[15] A. Valmari: On-the-fly Verification with Stubborn Sets, Lecture Notes in Computer Sience, No. 697, Springer-Verlag, 397-408 (1993).

[16] P. Wolper, P. Godefroid, D. Pirottin: A Tutorial on Partial-Order Methods for the Verification of Concurrent Systems, Computer Aided Verification '93 Tutorial (1993).

[17] H. Yamasaki: On Weak Persistency of Petri Nets, Information Processing letters, Vol.13, No.3, 94-97, (1981).

Modelling and Analysis of Distributed Program Execution in BETA Using Coloured Petri Nets

Jens Bæk Jørgensen and Kjeld Høyer Mortensen

University of Aarhus, Computer Science Department
Ny Munkegade, DK–8000 Aarhus C, Denmark
{jbj,khm}@daimi.aau.dk

Abstract. Recently, abstractions supporting distributed program execution in the object-oriented language BETA have been designed. A BETA object on one computer may invoke a remote object, i.e., an object hosted by another computer. In this project, the formalism of Coloured Petri Nets (CP-nets or CPN) is used to describe and analyse the protocol for remote object invocation. In the first place, we build a model in order to describe, understand, and improve the protocol. Remote object invocation in BETA is modelled on the level of threads (lightweight processes) with emphasis on the competition for access to critical regions and shared resources. Secondly, the model is analysed. It is formally proved that it has a set of desirable properties, e.g., absence of dead markings.

Topics: System design and verification using nets; higher-level net models; computer tools for nets; experience with using nets, case studies; application of nets to protocols.

1 Introduction

In this project, the formalism of Coloured Petri Nets (CP-nets or CPN) [Jen92] is used to describe and analyse the protocol for remote object invocation in the object-oriented language BETA [MMPN93].

The project is divided into a construction stage and an analysis stage. In the construction stage a model is built in order to describe and understand the considered protocol. Several meetings are held between the modellers and the designer of the protocol. In the process, the designer increases his own understanding. As a consequence, a number of changes are made.

In the analysis stage, the protocol is verified. It is formally proved that it has a set of desirable properties. E.g., we prove that the protocol has no deadlocks, that certain BETA objects always have the chance to do remote object invocations (liveness), and that a monitor construction correctly ensures exclusive access to a critical region. We apply recently developed computer tools for formal analysis of CP-nets, an occurrence graph tool and a place invariant tool.

The rest of this paper is structured as follows: In Sect. 2, the system supporting distributed program execution in BETA and the protocol for remote object invocation are introduced. Sect. 3 describes the constructed CPN model and Sect. 4 its analysis. In Sect. 5 related work is discussed. Finally, in Sect. 6, we draw some conclusions.

2   Description of the DistBETA System

The system considered in this project will be called the *DistBETA system* [Bra94, BM93]. The DistBETA system is a framework for distributed program execution in BETA. It includes the protocol for remote object invocation. In this section we first introduce a set of concepts from the DistBETA system that are used to describe the protocol. Then we explain the protocol itself.

The following three concepts are relevant for the remote object invocation protocol:

- *Ensemble*: Is a representation of the operating system on a computer connected to a network.
- *Shell*: Is similar to a process. Shells exist inside ensembles. A shell can communicate with another shell in a remote ensemble or in its own ensemble. Moreover, a shell can communicate directly with its ensemble.
- *Thread*: Each shell contains at least one *user thread* executing the main program and exactly one *listener thread* taking care of incoming requests from the network.

The framework that the application programmer uses to support distributed program execution contains a class called the *RPC handler*. The RPC handler includes the necessary primitives for serialization (marshalling) and communication. The framework is more general than RPC[1]. An object can act both as a client and as a server, allowing arbitrarily long invocation-chains wandering through many different computers. The parameters passed in an invocation are not just values as for RPC. They may be objects or references to objects, to which messages can be sent resulting in object invocations.

In the following we describe the protocol for remote object invocation (see Fig. 1). Suppose that an active object o1 wants to invoke another object o2. The two objects are physically separated on two computers, Host1 and Host2 respectively. Each object has a unique object identifier (OID)[2]. The sequence of events is as follows:

1. o1 looks up the OID of the object to be invoked (o2) in a table containing OIDs of remote objects. The table is local to the shell of o1. o1 allocates

---

[1] RPC is an acronym for remote procedure call. For an introduction, see [Tan92]. In this paper the term "RPC" means remote object invocation.

[2] The purpose of an OID is to have a database key. Some objects may be persistent, i.e., they may survive between program executions, and are typically stored in a database on a permanent storage (as a hard-disk). The OID is then used to retrieve the object again or can even be used to get type information about the object.

Network

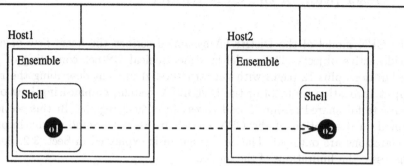

Fig. 1. A remote object invocation.

a resource containing the necessary primitives for serialization. All parameter objects are serialized. o1 invokes a method for sending remote object invocation requests in the RPC handler.

2. The remote object invocation together with its associated serialized parameter objects are sent to Host2. The invoking object o1 is blocked.

3. The RPC handler in the shell enclosing o2 receives the incoming request. A worker thread containing the necessary primitives for invoking objects is allocated. The worker thread unserializes the received parameter objects. While doing so it also looks up the OID of the receiver object in a table containing OIDs of local objects. The table is local to the shell of o2.

4. The object o2 is invoked with the unserialized parameter objects.

5. The worker thread gets the result which is serialized. Control is handed over to the RPC handler again. The worker thread is released and the result is sent back to Host1.

6. The result is unserialized by o1, the resource allocated is released, and finally the result is given to o1.

When objects are passing the boundary of shells their OIDs are looked up in tables of remote or local OIDs. Upon serialization it is checked if new objects cross the boundary. If a new OID is needed for an object not yet in any of the tables, it is necessary to communicate with the relevant ensemble, since it is the ensembles that generate unique OIDs.

Many shared resources and critical regions are involved in the sequence of events described above: Allocating and releasing resources, allocating and releasing worker threads, looking up OIDs in tables of remote and local objects, and communication with ensembles upon requests for new OIDs. Monitors and semaphores are used to grant exclusive access for competing threads. Threads compete with other threads belonging to the same shell. The DistBETA system ensures that no thread is starved, by associating queues with monitors and semaphores.

3 Description of the CPN Model

The CPN model of the DistBETA system describes the basic flow of control inside active objects. The model is a hierarchical CP-net consisting of global declarations plus 12 pages with net structure. It aims at describing the remote object invocation protocol of the DistBETA system, emphasising how objects compete for shared resources and access to critical regions. In this section we provide a description of the CPN model. In Sect. 3.1 some of the important declarations are outlined. The net structure is explained in Sect. 3.2. Sect. 3.3 addresses the limitations of the model.

The model is built with Design/CPN [JCHH96], a general editing, simulation, and analysis tool for CP-nets. Design/CPN uses the language CPN ML for declarations and inscriptions. CPN ML is an extension of the functional programming language Standard ML [Ull93].

3.1 Global Declarations

The basic components of the DistBETA system are ensembles, shells, and threads, which are all active objects, plus messages and packets, which are both passive objects.

Colour sets for these components are declared in a straightforward fashion in CPN ML. Most places in the model have colour set Thread. The movement of Thread-tokens describe the main flow of the model. A Thread-token being in a certain place is similar to a program counter having a certain value.

A token from colour set Thread is a pair. The first component identifies the thread. The colour set identifying a thread is called ThreadInfo. It is a record colour set with fields for identities of an ensemble, a shell, and a thread. The second component of a Thread-token is an environment holding information that the thread needs at certain points in its lifetime, e.g., values of local variables. It is modelled by the record colour set Environment.

There are different kinds of Thread-tokens used for different purposes: User threads are making requests for remote object invocations, listener threads are receiving requests, and worker threads are handling the requests and subsequently returning results.

Threads communicate by sending packets over a network. Packets are modelled in CPN ML by the colour set Packet, which is a record type holding a sender, a receiver, and a message. The set of messages is very coarse. Basically, a message is either a request or a result because we are only concerned with its direction. As an example a real BETA object may want to send a message like (2,5)->add to a remote object and expect the result 7 to be returned, but this level of description is not necessary for our purpose. We are concerned with communication patterns only, not with the data involved. In addition to the basic types of requests and results, a message may be a network error message.

3.2 Net Structure

The net structure can be seen from the hierarchy page of the CPN model, shown in Fig. 2.

The hierarchy page has a node for each page of the model. An arc between two nodes indicates that the source node contains a transition whose behaviour is described on the page of the destination node. Such a transition is called a substitution transition. The page of the destination node is called a subpage.

The model consists of four parts, each one with its own well-defined meaning: A top-level part, a network part, a sender part, and a receiver part. Some pages are shared between the sender and receiver part. This means that these pages have more than one page instance. Places, transitions, and arcs on a page with more than one instance, accordingly appear in more than one instance[3].

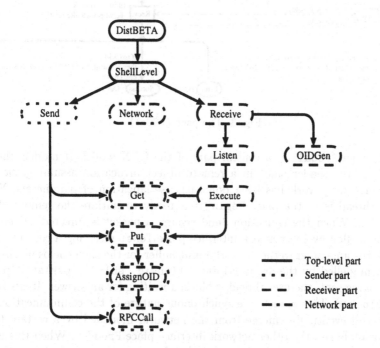

Fig. 2. The hierarchy page.

A detailed description of the model can be found in [JM95]. Below we describe two typical pages. The other pages are comparable with respect to the number of places, transitions, and arcs; and the complexities of the arc inscriptions.

Page RPCCall The page RPCCall shown in Fig. 3 is first described. The four places with thick border have colour set **Thread**. The two places with a dot-

---

[3] In this paper, when no confusion is possible, we say "page" instead of the proper term "page instance". Similar remarks apply to places, transitions, and arcs.

dashed border model the interface to the network. Conceptually they contain packets. Each one of them contains a list of packets for each shell. These lists act as input and output buffers to the network. To enhance readability, all colour sets are hidden. The variable thrinf is over colour set ThreadInfo and the variable envr is over colour set Environment.

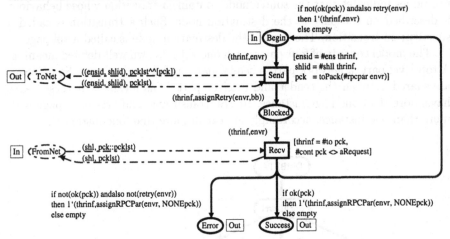

Fig. 3. The page RPCCall.

The page RPCCall is a central part of the CPN model. It models the behaviour of the sender side[4] in a remote object invocation assuming the necessary preceding work has been done, e.g., serialization of parameters. When a user thread is in the place Begin, it is ready to initiate the remote object invocation. When the transition Send occurs, a packet is directed to the network, modelled by the network interface place ToNet receiving a certain token. A packet is appended to the network input buffer for the shell enclosing the user thread, modelled by the arc inscription[5] ((ensid, shlid), pcklst^^[pck]). After having sent, the user thread is blocked waiting for an answer. It sits in the place Blocked. This models the synchronous nature of the communication. An answer will eventually emerge from the network, modelled by a certain token being available on the other network interface place FromNet. When this is the case, the transition Recv becomes enabled for the waiting user thread. In the normal case (the expression ok(pck) is true), an ordinary result comes back. In this case, the remote object invocation went well and the user thread ends up in the place Success. If an error appeared (ok(pck) is false), the user thread may end up in the place Error, or it may return to the place Begin and try to do the failed invocation once again depending on the value of the boolean variable retry in its environment.

---

[4] RPCCall is also included in the receiver part because a receiving object sometimes does an RPC with its ensemble in order to obtain an OID.

[5] The operator ^^ concatenates two lists.

Page `AssignOID` Now the page `AssignOID` shown in Fig. 4 is described. It models a monitor construction ensuring unique OIDs upon request from user threads.

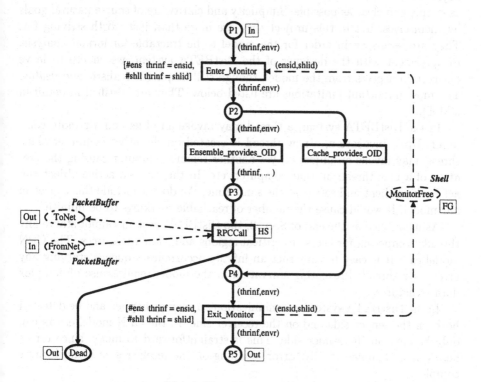

Fig. 4. The page `AssignOID`.

The monitor is controlled by the place `MonitorFree` which initially contains a token for each shell. A user thread is inside the monitor exactly when the corresponding token is in a place that is inside the monitor. The inside of the monitor is made up of places both on the page `AssignOID` itself and its subpage `RPCCall`. On `AssignOID`, the places P2, P3, P4, and Dead[6] are inside the monitor. In addition, the monitor includes the places on the subpage `RPCCall`. The entry to the monitor is modelled by the transition `Enter_Monitor`. When it occurs, a user thread enters the monitor, and the appropriate token is removed from `MonitorFree`. This disallows any other user thread from the same shell from entry before the appropriate token is redeposited in `MonitorFree`. The transition `Exit_Monitor` models exit from the monitor. When it occurs, the token corresponding to the shell of the exiting user thread is added to `MonitorFree`.

---

[6] A user thread is terminated, i.e., ends at the place Dead, if the ensemble fails to provide an OID upon request.

3.3 Limitations

The model explained in this section is an abstraction of the protocol for remote object invocation of the DistBETA system. We have aimed at keeping the model as simple and clear as possible. Simplicity and clarity are of course natural goals by themselves, but in this project they are more than just worth striving for. They are necessary in order for the model to be tractable for formal analysis. In agreement with the designer of the DistBETA system, we choose to leave out certain aspects from the model, though still keeping it realistic and usable. The most important limitations are listed below. They are justified in detail in [JM95].

In the DistBETA system, a thread may invoke an object on a remote computer. Upon invocation, a new thread is created on the other computer. This thread may invoke an object on yet another remote computer causing the creation of a new thread on that computer, etc. In this way an active object can act both as client and server at the same time. We do not include this aspect in our model. It would cause the number of reachable markings to be infinite.

As mentioned at the end of Sect. 2, the DistBETA system cannot starve any threads competing for access to critical regions and shared resources. The CPN model can. It is easy to construct an infinite occurrence sequence starving any given user thread. Preventing starvation in the model requires use of complex data structures.

In the DistBETA system, communication errors may happen and be detected both on the sender side and on the receiver side. In the CPN model errors can only happen on the sender side. This is straightforward to model since errors here can be handled locally. Error handling on the receiver side is much more complex.

4 Analysis of the CPN Model

Both informal and formal analysis methods are applied to the CPN model in this project. The informal analysis consists in simulating the model. Simulation is an important activity in any CPN modelling project. In this paper however, our focus is on describing the formal analysis of the CPN model.

This section consists of two subsections, one for each of the two formal analysis methods used. Occurrence graph analysis is described in Sect. 4.1 (occurrence graphs are also known as state spaces and reachability graphs). Place invariant analysis is described in Sect. 4.2. Using the formal methods, we are able to prove that the CPN model has certain properties. Both techniques are supported by computer tools. For a thorough introduction to formal analysis of CP-nets, see [Jen94].

4.1 Occurrence Graphs

In this section, first the main characteristics of the occurrence graph method are recalled. The goals of our analysis and the results obtained are described next.

Finally, an attempt to alleviate the state explosion problem using net reductions is discussed.

The Method An *occurrence graph* for a CP-net is a directed graph with a node for each reachable marking and an arc for each occurring binding element[7]. An arc is going from the node of the marking in which the associated binding element occurs to the node of the marking resulting from the occurrence.

All standard dynamic properties for a CP-net[8] can be derived from its occurrence graph, e.g., boundedness, home, liveness, and fairness properties. It is worthwhile also to construct the *SCC-graph* for the occurrence graph. The SCC-graph has a node for each of the strongly connected components of the occurrence graph. If there is an arc in the occurrence graph between two nodes from different strongly connected components, then the corresponding arc is in the SCC-graph. Investigating the SCC-graph instead of the full occurrence graph may significantly speed up the check of a dynamic property. Using Tarjan's algorithm (see, e.g., [Gib85]), the construction of the SCC-graph is an inexpensive operation. It is linear in the size of the occurrence graph.

The most serious drawback of the occurrence graph method is the *state explosion problem*: Very often, even for relatively small CP-nets the occurrence graphs get so big that they cannot be generated, even with the most powerful computers available. Another limitation inherent to the occurrence graph method is *dependency of the initial marking*: Each possible initial marking of a considered CP-net may yield a new occurrence graph. Thus verifying properties for all values (perhaps infinitely many) of some system parameter requires generation of an occurrence graph for each value.

The tool used for the occurrence graph analysis is called the Design/CPN Occurrence Graph Tool (Occ Tool) [CJK96]. It is an application that is integrated with the basic simulation tool offering functionalities to generate occurrence graphs, to draw them, and to do queries. All our analysis including generation of the occurrence graphs is conducted on a Sun Sparc 20 with 256 MB physical RAM.

Analysis Goals The state explosion problem and the dependency of the initial marking are general problems of the occurrence graph method. However, in a concrete application, we can try to find workarounds.

This specific CPN model is created with formal analysis in mind. Therefore, we have carefully tried to choose colour sets and to build the net structure so that the state explosion is controlled, at least for small initial markings, i.e., initial markings containing only a few user threads (and consequently only a few ensembles and shells).

Our model does have an infinite number of legal initial markings. What we will call a *configuration* is determined by the number of ensembles; for each

---

[7] A *binding element* is a pair containing a transition and a binding. The binding assigns values to all variables in the surroundings of the transition.

[8] Here we consider only CP-nets with a finite number of reachable markings.

ensemble, the number of shells; and for each shell, the number of user threads. Each configuration uniquely induces an initial marking. The model has an infinite number of legal configurations. Because of the dependency of the initial marking problem, we have no chance of verifying the model in general. However, if we show that the considered properties are satisfied for a number of configurations, our confidence in the CPN model is increased.

The occurrence graph analysis focusses on proving the two vital dynamic properties stated below.

1. The CPN model has no dead markings.
2. Each user thread can forever participate in the basic send/receive communication. Formally this can be stated as two specific sets of binding elements being live — one set of binding elements for each of the transitions Send and Recv on the page RPCCall (see Fig. 3) of the sender part.

If the CPN model has these two properties, there is strong evidence that the protocol is well-functioning. Property 2 means that each user thread remains active, i.e., it will always have the chance to request a remote object invocation, and a request will always be followed by a result (which might be an error notification). It is obvious that property 1 is a consequence of property 2. Thus it suffices to prove 2. As an aid we show:

3. The CPN model is always able to return to its initial marking, i.e., the initial marking is a home marking.
4. For any given user thread there is an occurrence sequence starting in the initial marking and containing an occurrence of each of the transitions Send and Recv on the page RPCCall of the sender part. Both occurrences with the variable thrinf bound to the given user thread.

Property 3 states that from any given reachable marking, it is possible to find an occurrence sequence leading back to the initial marking. Property 4 says that from the initial marking we can find an occurrence sequence containing a binding element from the set we are analysing for liveness. Hence, together 3 and 4 imply 2. It is well-known that 5 below implies 3. In summary, it is sufficient to prove 4 and 5.

5. The SCC-graph consists of exactly one component.

The original CPN model does not have the listed properties 1 and 2. The reason is rather technical and is explained in [JM95]. The violation is not a modelling error, but is inherent to the DistBETA system. Fortunately, we can easily construct a modified version of the model which we will prove to satisfy the properties. All we have to do is to make the communication between user threads asking for OIDs and their ensembles error-free. For this reason the CPN model analysed with the occurrence graph method is a slight variation of the original one. The modification is accomplished just by giving the transition Ensemble_provides_OID on the page AssignOID (see Fig. 4) the guard [false].

Analysis Results It is our aim to prove the properties 1 and 2 for as many configurations as possible. However, due to the state explosion problem, we cannot expect to be able to handle configurations with many user threads. The experiments will take us as far as we can get before the computer runs out of memory. Our results are summed up in Fig. 5 and are commented below. In the figure a configuration is visualised as in Fig. 1: Boxes represent ensembles, rounded boxes represent shells, and black filled circles represent user threads. E.g., configuration 1 has one ensemble with one shell containing two user threads. The time indicated in the figure is wall-clock time for the occurrence graph generation, not CPU time. The computer was only spending little time running other processes while the occurrence graph generations were going on.

Configuration	Time seconds	Nodes	Arcs
1	 96	5,501	13,725
2	 760	21,554	54,793
3	 653	21,554	54,793
4	 ≥ 5,247	≥ 75,018	≥ 183,827

Fig. 5. Statistics for generation of occurrence graphs.

Intuitively configuration 1 is more likely to lead to a dead marking than configurations 2 and 3. All monitors, resources, and critical regions are local to shells. Configuration 1 has two user threads running within the same shell, so conflicts will arise. In this case, it is possible to verify directly that property 5 holds. Property 4 is established in the simulator. As an alternative, property 4 may be proved by finding an appropriate path in the occurrence graph. Thus for configuration 1, the desired properties 1 and 2 hold.

For configurations 2 and 3, properties 1 and 2 are shown similarly. As an aside, we note that the two graphs have exactly the same size. This fact is no surprise, because in both cases, the two user threads are independent, i.e., they never have to wait for each other because they run in different shells.

For configuration 4, the full occurrence graph is too big to be generated. Thus this configuration cannot be analysed with the available tool and computers. This result is of course negative for the verification of our CPN model, but it does exhibit the state explosion problem very clearly. Increasing the number of user threads from two to three causes the occurrence graph to grow to a size that we presently cannot handle. We will return with an attempt to tackle configuration 4 at the end of this section. For now, we can only generate a partial occurrence graph, i.e., a graph where only a subset of the reachable markings and the occurring binding elements is included. Therefore, we only have lower bounds for the number of nodes and arcs, and for the generation time.

There are a number of configurations with a total of three user threads, e.g., three ensembles, one shell in each, and one user thread in each shell. Of these possibilities configuration 4 is the one with the lowest number of reachable markings: In a system with three user threads running in the same shell, the user threads are forced to wait for each other once in a while. Thus the behaviour is more restricted than, e.g., the system mentioned above. Obviously, generation of a full occurrence graph is not possible for a configuration with more than three user threads. Therefore, we have not tried to generate occurrence graphs for more configurations. In summary, with the software and hardware used for these experiments, configurations with less than three user threads can be analysed with the occurrence graph method. Larger configurations cannot.

Above the Occ Tool was used to verify properties for the final version of the CPN model. It is not the only way to use it. It is a convenient tool for debugging in the model creation stage as well. It provides a systematic way to investigate all occurrence sequences, in contrast to simulation where only one occurrence sequence at a time is considered. The process of finding and correcting modelling errors in this project was eased by using occurrence graphs. For more details see [JM95].

Alleviating State Explosion Using Reductions As mentioned at the beginning of this section, we make an attempt to alleviate the state explosion problem. We do so using *reductions*. Reductions of Petri nets are among the classical approaches to analysis. The basic idea is to derive a net from the original one in a systematic way using a set of rules that are known to preserve a selected set of dynamic properties. We certainly expect the derived net to have a smaller occurrence graph than the original.

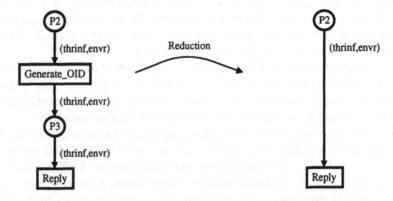

Fig. 6. Reduction of the CPN model using pre-agglomeration.

To illustrate how reductions are applied to our model, consider the extract of the model shown in Fig. 6. Removal of the transition Generate_OID and the place P3 results in a simpler model. This reduction is a special case of *pre-agglomeration of transitions* defined in [Had91]. In this reference it is proved that the reduced model has the same properties as the original with respect to the properties we are considering: Home markings, liveness, and dead markings.

We make a number of pre-agglomerations of transitions in the model. In total, five places and five transitions are discarded. The relevant statistics for the occurrence graph generation is shown in Fig. 7.

	Configuration	Time seconds	Nodes	Arcs
1	 ●●	47	2,541	6,877
2	 ●○ ○●	338	9,810	27,497
3	 ●○ ●○	345	9,810	27,497
4	 ●●●	≥ 10,340	≥ 75,001	≥ 245,794

Fig. 7. Statistics for generation of occurrence graphs for the reduced model.

The occurrence graphs for the reduced model are significantly smaller than for the original. For each of configurations 1, 2, and 3, the graph for the original model has approximately twice as many nodes as the graph for the reduced version. The same remark applies to the number of arcs. Smaller occurrence graphs for the reduced model is of course the expected result. Anyway, it is remarkable that removal of only five places and five transitions yields occurrence graphs of sizes half of the original. The original model has 70 places. Thus removing about 7% of the places gives decent reductions in the sizes of the occurrence graphs. The lesson learned is that it is sensible to apply as much as possible reduction strategies preserving the properties we are interested in before an occurrence graph is generated. Small reductions in the net structure may yield significant reductions. But alas, even with the reduction technique, we cannot handle configurations with more than two user threads. The occurrence graph for configuration 4 is partial, also for the reduced model.

4.2 Place Invariants

In this section, first the main characteristics of the place invariant method are recalled. The goals of our analysis and the results obtained are described next.

The Method The basic idea of the place invariant method is to find equations that are satisfied for all reachable markings of a considered CP-net. To explain how in more details, we recall the concepts of weights and sets of weights. In this context, a *weight* is a linear function associated with a certain place in the considered CP-net. Its domain is the place colour set. It can also be applied to multi-sets over the place colour set. A *set of weights* contains exactly one weight for each place, and all its weights have a common range. Given a set of weights, the weight of a place is computed by providing its marking as argument to the weight. The weight of a marking of the net is the multi-set sum of the weights of all the places. A set of weights can be viewed as a means to extract some specific information that we are interested in, from the complex information of the full marking. Typically, a set of weights will contain zero weights[9] for a large number of places and non-zero weights for a relatively small number of places. The set of places for which the weight is non-zero is called the *support* of the set of weights. A *place invariant* is a set of weights for which the weights of all reachable markings are identical.

Compared to the occurrence graph method, an advantage of the place invariant method is that it does not suffer from the state explosion problem. To check if a proposed set of weights is a place invariant does not require generation of all reachable markings. Instead, the check may be done statically and locally: For each binding element it is checked that the weight of the multi-set of tokens removed is identical to the weight of the multi-set of tokens produced upon occurrence. In this case, we will say that the binding element preserves the weight. Typically, all binding elements of a given transition may be checked simultaneously. A transition preserves the weight when all its binding elements preserve the weight. The set of all binding elements is known in advance, thus investigating all reachable markings is not needed. We just have to check that all transitions preserve the weight.

Another advantage of the place invariant method is that it is not dependent of the initial marking. When proving that all transitions preserve the weight, only the static inscriptions appearing in arcs and guards in the net are considered.

The place invariant analysis is conducted using the Design/CPN Place Invariant Tool (Inv Tool) [Tok93]. It is a research prototype. For a given set of weights, it checks if all transitions preserve the weight: The CPN ML expression for the weight of the net effect of a transition is translated into a lambda expression. The Inv Tool uses lambda reduction rules to rewrite this expression. When no more reductions are possible, it checks if the expression equals the zero function.

Analysis Goals The aim of the place invariant analysis is to increase our confidence in the CPN model. Throughout the model construction stage, we have in mind a set of important properties that a sensible model must satisfy. Using place invariants, we are able to prove that the final model actually has these properties. Most of them are very similar. They state that certain sets of tokens remain constant. Another property says that a monitor construction is correct. Specifically we prove the following:

---

[9] A zero weight maps each multi-set into the empty multi-set.

1. The set of user threads is constant, i.e., no user thread ever disappears and no new user thread is ever created.
2. The set of listener threads is constant.
3. The set of network input buffers is constant.
4. The set of network output buffers is constant.
5. The monitor ensuring unique OIDs on the page `AssignOID` (see Fig. 4) is correct, i.e., there are never two user threads from the same shell inside the monitor at the same time.

Analysis Results First we consider the verification of property 1: We are aiming at defining a suitable set of weights. We take the weight for any place that has colour set different from `Thread` to be zero. All places with colour set `Thread` get the same weight: If the argument is a user thread, it returns the identity of the thread, i.e., it ignores the environment. If the argument is a thread which is not a user thread, it returns the empty multi-set. The Inv Tool is able to verify that the weight set thus defined is a place invariant. Thus property 1 is proved. A set of weights corresponding to property 2 is constructed similarly.

Properties 3 and 4 are shown by constructing appropriate sets of weights whose support are the network buffer places `ToNet` and `FromNet` (see Fig. 3) respectively.

To establish property 5, we need a more sophisticated place invariant. Consider the page `AssignOID` shown in Fig. 4 in Sect. 3. The set of weights proving 5 contains the same weight for all places inside the monitor — a weight *shell* that maps a `Thread`-token to its enclosing shell. The set of weights contains the identity-function for `MonitorFree` and zero for all other places in the model. If this set of weights is a place invariant, it states that the places with non-zero weights in any reachable marking M together contain exactly one token from each shell (given the initial marking). In terms of an equation:

$$M(\texttt{MonitorFree}) + \sum_{p \in InMonitor} shell(M(p)) = AllShells$$

where *AllShells* is a multi-set containing exactly one appearance of each shell, and *InMonitor* is the set of all places inside the monitor. Thus assuming two or more user threads from the same shell inside the monitor at the same time is a contradiction: The multi-set on the left-hand side of the equation contains an element with coefficient at least two and the right-hand side does not.

The Inv Tool tool is not able to establish that the set of weights defined for property 5 is a place invariant. E.g., it cannot verify that the transition `Recv` on the page `RPCCall` (see Fig. 3) preserves the weight. This is a shortcoming in the tool. The involved arc expressions, e.g., on the `RPCCall` page are too complex for the present prototype. If we analyse the model that is modified by making the communication between user threads asking for OIDs and their ensembles error-free, property 5 can now be established by the Inv Tool. The modification must be done by explicitly deleting parts of the model. The reason is that it is thus not necessary to check the complex transitions on the page `RPCCall`, which were causing the problems in the original model. The model modified as described above has the same behaviour as the original one simplified by giving the transition Ensemble_provides_OID on the page AssignOID the guard [false], as we did for the occurrence graph analysis.

Thus all five properties are established (property 5 only for a modified version of the model).

An attempt was made to prove a place invariant capturing the following property: When a user thread is blocked after a send, either there is a packet from that user thread on its way on the network to the receiver side, or the receiver side is working on providing a result, or there is a packet addressed to the blocked user thread on the network. The set of weights defined in order to conduct the proof has a large support. Due to a shortcoming in the Inv Tool, it is unfortunately not possible to verify that this set of weights is a place invariant. However the tool is able to provide a partial check of the proposed place invariant. Partial in the sense that it can verify that some but not all transitions preserve the weight. In this way, the tool may be an aid in a semi-automatic checking of a proposed place invariant, by simply reducing the number of transitions the user has to consider manually.

It is possible to prove all of the listed properties in this section using occurrence graphs — for a fixed configuration. Each property can be verified by a traversal of the full occurrence graph where it is checked that every marking satisfies the considered property. In fact, we did this for property 5 above for the configurations for which the full occurrence graphs were generated.

5 Related Work

A large number of papers describing system modelling and simulation using Petri nets exist. Significantly fewer reports on formal analysis. The explanation is natural: Formal analysis of interesting models require tool support. So far, while having excellent tools for editing and simulation, the Petri nets community has been lacking high-quality tools for formal analysis of anything but very small models. Nevertheless, some papers documenting modelling and formal analysis projects do exist. In this section, we relate our work to a number of these.

In [CJ91] the modelling and simulation of a network management system using CP-nets is described. A large model was built and to some extent analysed using place invariants. However, the authors note that they only made very limited use of formal analysis due to the lack of tool support. They propose to use formal analysis during the model construction stage. Viewing our project in the light of these remarks, we note that formal analysis actually was used during our model construction: The Occ Tool assisted us finding and correcting errors in the process as described in Sect. 4.1. Moreover, we had a set of place invariants in mind throughout the model construction. These invariants were formally proved with the Inv Tool, thus increasing our confidence in the model.

The occurrence graph analysis of our model could only verify properties for a few initial markings. It is of course highly desirable if the model can be verified independently of some initial marking. Alas, this is prohibited in general because of the nature of the occurrence graph method (refer to the discussion of the dependency of the initial marking problem in Sect. 4.1). Sometimes, the problem may be overcome though. In [GS92], the modelling and analysis of a hardware chip (an arbiter cascade) is reported. The model is characterised with one single integer system parameter d, the depth of the cascade. The authors verify the model for all possible values of the system parameter in the following way: For $d = 0$ and $d = 1$, the number of reachable markings is small, and occurrence graphs are easily constructed. With them, the desired properties of the model are verified directly. Mathematical induction establishes the proof for all values $d > 1$. Using occurrence graphs in conjunction with induction is very appealing whenever applicable. This strategy solves the two most serious problems with occurrence graph analysis: The state explosion problem is simply eliminated because it is only necessary to generate occurrence graphs when the number of reachable markings is small. The dependency of the initial marking is elegantly overcome with the inductive step. However, this approach relies upon the considered system being regular in the sense that it is characterised by a natural number parameter, and there is a well-defined relation between the behaviour of the system with parameter d and the system with parameter $d + 1$. Unfortunately, our communication protocol does not adhere to these requirements. In fact, we think it will be hard in general to apply this technique to complex communication protocols. It seems that hardware designs are more regular, and thus more adequate for inductive proofs.

In our place invariant analysis, we concentrated on proving place invariants whose existence we presumed. Thus our place invariant analysis was a checking of properties that we expected the CPN model to have. An alternative is to try calculating all place invariants for a given model automatically. With a model of this size, we believe that the computational complexity of this approach is prohibitive. Moreover from some representation of all place invariants it is not necessarily easy to pick out the ones of interest, i.e., the ones that express relevant properties of the model. Our attitude towards place invariant analysis is shared by the authors of [BWW91].

In [JM95] our project is compared with three others that exploited formal analysis methods. They are described in [GCH91], [MV92], and [GHW94].

Finally in this section, we make clear the contribution of our project in comparison with the other projects mentioned here. We are convinced that the availability of suitable formal analysis tools is the reason why we obtained quite powerful analysis results compared to the earlier projects. Moreover, we exploit both occurrence graphs and place invariants. We are well aware that we do not use some combination of or interaction between the two methods. However, we believe that the two methods inspire the user to investigate different aspects of the model. Therefore it is sensible to apply both, if possible. It will typically produce a more comprehensive set of analysis results. It did in this specific example, where the two methods supplemented each other nicely.

6 Conclusion

In this project we have modelled and analysed the protocol for remote object invocation in BETA.

The first stage was to build the CPN model. It had some impact on the protocol. We had a number of meetings with the designer. The discussions gave both him and us a better insight into the behavioural aspects. As a consequence, a number of changes were made to the protocol. More specifically, some superfluous critical regions were removed.

The second stage was to analyse the model using formal methods. For this purpose we used two standard methods, occurrence graphs and place invariants. The results obtained were non-trivial. Careful choice of colour sets and net structure for the model implied a relatively successful use of the occurrence graph method: It was possible to prove important dynamic properties such as absence of dead markings and liveness of specific sets of binding elements for small initial markings. Place invariants were used to prove quite different dynamic properties of the model, e.g., that certain sets of threads remain constant, and that a monitor construction correctly ensures exclusive access to a critical region. We took advantage of two recently developed tools, the Occ Tool and the Inv Tool.

The analysis stage did not influence the protocol, because the design was already sensible before the analysis began. If the formal analysis had revealed, e.g., an unexpected deadlock, then of course this problem would have to be fixed in the design and implementation of the protocol. Thus the formal analysis would have had an impact. The designer of the protocol has the viewpoint that the formal analysis was mostly usable to increase his confidence in the CPN model. Therefore, the key question is if the model is a proper reflection of the design and implementation of the protocol. In this specific project, both the designer and ourselves are confident that the model sensibly captures relevant and important aspects of the protocol. The verification of the model thus does increase our confidence in the protocol. It is theoretically possible though that something left out from the model (e.g., error handling on the receiver side) is exactly what is causing a serious problem.

Advantages and drawbacks of both formal methods were discussed. The main drawbacks inherent to the occurrence graph method, state explosion and dependency of the initial marking, were exhibited. Although the problems are generally recognised in the theory, it is valuable to see their impact on a specific real-world example. Here we saw that only configurations with less than three user threads could be handled. When the number of user threads was increased, occurrence graphs could not be generated in full. We proposed net reductions as a means to alleviate the state explosion problem and demonstrated decent savings in the sizes of the occurrence graphs.

The model built in this project is complex. It is thus very hard to get certain information about its dynamic behaviour using informal methods such as simulation only. E.g., ruling out the possibility of a dead marking requires formal verification. Unfortunately we have to accept the fact that currently full occurrence graphs can only be generated for small initial markings. Thus, e.g., absence of dead markings is only established for a few markings. Hence on one hand, we saw an example of a model where formal analysis proved really useful when applicable. On the other, formal analysis was somewhat obstructed. There are two categories of sources limiting the applicability of formal analysis. One concerns the analysis methods, the other the tools. The most severe limitations are inherent to the methods, e.g., the state explosion problem. However, the present theoretical work is much more advanced than the present tool support. An important field for future work is development of better tools for formal analysis of CP-nets. With respect to the Occ Tool, generation of larger occurrence graphs will be possible when a version storing markings in a more economic fashion is implemented. The current version does not use memory in an optimal way. Moreover, only ordinary occurrence graphs are presently supported. It will definitely be valuable reconsidering the analysis done in this project when the Occ Tool is matured to support occurrence graphs with equivalences [Jen94]. With respect to the Inv Tool, it needs to be matured to handle more complex expressions.

Acknowledgements We thank Søren Brandt, Søren Christensen, Alexandre Valente Sousa, and Jan Toksvig for help in this project. Thanks to Kurt Jensen, Rikke Drewsen Andersen, Vincent Becuwe, Torben Bisgaard Haagh, Ludovic Joly, Lars Kristensen, and René Wenzel Schmidt for proof-reading various versions of this paper.

This work has been supported by grants from the Danish Research Councils SNF and STVF, and from the Faculty of Science at University of Aarhus.

References

[BM93]   S. Brandt and O. L. Madsen. Object-Oriented Distributed Programming in BETA. In R. Guerraoui, O.M. Nierstrasz, and M. Riveill, editors, *Object-Based Distributed Programming*, Lecture Notes in Computer Science, Kaiserslautern, Germany, 1993. Springer-Verlag.

[Bra94]   S. Brandt. Implementing Shared and Persistent Objects in BETA. Technical report, Computer Science Department, University of Aarhus, 1994.

[BWW91] J. Billington, G. R. Wheeler, and M. C. Wilbur-Ham. PROTEAN - A High-level Petri Net Tool for the Specification and Verification of Communication Protocols. In K. Jensen and G. Rozenberg, editors, *High-level Petri Nets, Theory and Application.* Springer-Verlag, 1991.

[CJ91]   S. Christensen and L. O. Jepsen. Modelling and Simulation of a Network Management System using Hierarchical Coloured Petri Nets. In E. Mosekilde, editor, *Proceedings of the 1991 European Simulation Multiconference*, Copenhagen, Denmark, 1991. Springer-Verlag.

[CJK96]   S. Christensen, K. Jensen, and L. Kristensen. *The Design/CPN Occurrence Graph Tool. User's manual version 3.0.* Computer Science Depart-

ment, University of Aarhus, 1996.
Online: http://www.daimi.aau.dk/designCPN/.

[GCH91]   C. Girault, C. Chatelain, and S. Haddad. Specification and Properties of a Cache Coherence Protocol Model. In K. Jensen and G. Rozenberg, editors, *High-level Petri Nets, Theory and Application*. Springer-Verlag, 1991.

[GHW94]   H. J. Genrich, H.-M. Hanisch, and K. Wöllhaf. Verification of Recipe-based Control Procedures by Means of Predicate/Transition Nets. In R. Valette, editor, *Proceedings of the 15th International Conference on Application and Theory of Petri Nets*, Lecture Notes in Computer Science, Zaragoza, Spain, 1994. Springer Verlag.

[Gib85]   A. Gibbons. *Algorithmic Graph Theory*. Cambridge University Press, 1985.

[GS92]   H. J. Genrich and R. M. Shapiro. Formal Verification of an Arbiter Cascade. In K. Jensen, editor, *Proceedings of the 13th International Conference on Application and Theory of Petri Nets*, Lecture Notes in Computer Science, Sheffield, UK, 1992. Springer Verlag.

[Had91]   S. Haddad. A Reduction Theory for Coloured Nets. In K. Jensen and G. Rozenberg, editors, *High-level Petri Nets, Theory and Application*. Springer-Verlag, 1991.

[JCHH96]   K. Jensen, S. Christensen, P. Huber, and M. Holla. *Design/CPN. A reference manual*. Computer Science Department, University of Aarhus, 1996. Online: http://www.daimi.aau.dk/designCPN/.

[Jen92]   K. Jensen. *Coloured Petri Nets — Basic Concepts, Analysis Methods and Practical Use. Volume 1, Basic Concepts*. EATCS Monographs on Theoretical Computer Science. Springer-Verlag, 1992.

[Jen94]   K. Jensen. *Coloured Petri Nets — Basic Concepts, Analysis Methods and Practical Use. Volume 2, Analysis Methods*. Monographs in Theoretical Computer Science. Springer-Verlag, 1994.

[JM95]   J.B. Jørgensen and K.H. Mortensen. Modelling and Analysis of Distributed Program Execution in BETA Using Coloured Petri Nets. Technical report, Computer Science Department, University of Aarhus, 1995.

[MMPN93]   O. L. Madsen, B. Møller-Pedersen, and K. Nygaard. *Object-Oriented Programming in the BETA Programming Language*. Addison Wesley, 1993.

[MV92]   W. M. McLendon, Jr. and R. F. Vidale. Analysis of an Ada System Using Coloured Petri nets and Occurrence Graphs. In K. Jensen, editor, *Proceedings of the 13th International Conference on Application and Theory of Petri Nets*, Lecture Notes in Computer Science, Sheffield, UK, 1992. Springer Verlag.

[Tan92]   A. S. Tanenbaum. *Modern Operating Systems*. Prentice-Hall International, 1992.

[Tok93]   J. Toksvig. Tool Support for Place Flow Analysis of Hierarchical CP-nets Version 2.0. Technical report, Computer Science Department, University of Aarhus, 1993.

[Ull93]   J. D. Ullman. *Elements of ML Programming*. Prentice-Hall, 1993.

Reachability Analysis Based on Structured Representations

Peter Kemper

Informatik IV, Universität Dortmund
44221 Dortmund, Germany
e-mail:kemper@ls4.informatik.uni-dortmund.de

Abstract. Exploration of the reachability set ($\mathcal{RS}$) is one of the crucial building blocks for various analysis methods ranging from model checking to Markov chain (MC) based performance analysis. In the context of MCs, structured representations of state transition matrices using tensor (Kronecker) algebra have been successfully employed to handle the impact of the state space explosion problem. In this paper such structured representations give rise to a new $\mathcal{RS}$ exploration algorithm for superposed generalized stochastic Petri nets and stochastic automata networks. The algorithm employs bitstate hashing with a perfect hash function, i.e. no collisions can occur. Two variations of this algorithm are discussed. Two examples are exercised to demonstrate the benefits of the new algorithm.

1 Introduction

Reachability analysis is one of the analysis methods applicable to Petri nets of various kinds: ordinary and colored, timed and untimed, hierarchical and non-hierarchical, etc. Furthermore it serves as an underlying method for several analysis techniques, e.g. model checking and Markov chain based analysis of stochastic Petri nets. Besides its usefulness the state space explosion problem has been recognized as its main drawback for long. Many approaches exist to handle the impact of large state spaces, among others there are:

Conventional reachability set ($\mathcal{RS}$) exploration is based on a straightforward search algorithm; G. Chiola [8] describes a sophisticated coding of markings and enabling tests exploiting causal dependencies to optimize this approach. The parallelization of conventional RS exploration for massively parallel machines by considering parallel transition firings and parallel member and insert operations for $\mathcal{RS}$ is discussed in [6, 7].

Reducing $\mathcal{RS}$ by avoiding unnecessary interleaving is the key idea for Valmari's stubborn set method [23, 24], Godefroid's sleep set method [14, 15], and combinations of these [15, 25], where many interesting qualitative properties can be analyzed from the reduced $\mathcal{RS}$.

Approximate methods, where certain parts of $\mathcal{RS}$ possibly remain unconsidered, have been developed. Among others, one very successful approach is the bitstate method by Holzmann of which different variations exist [16, 26, 17] and which allows to analyze extremely large state spaces for the price of completeness.

A modular approach is given by Christensen et al. [9] who consider a decomposition of a (colored) net into a set of components which interact via synchronized transitions (transition fusion). The main idea there is to compute reachability graphs for each component in parallel as far as independent, component specific transitions allow and then to check the enabling of synchronized transitions within all components. This yields one reachability graph per component plus an additional synchronization graph. The underlying assumption is that for a single component with an enabled synchronized transition its environment, given by the rest of components, is likely to disable this transition; hence the generation of the component reachability graphs is tightly coupled for synchronized transitions. The resulting modular representation of $\mathcal{RS}$ is shown to be useful for proving qualitative properties.

Structured representations of the generator matrix Q have been successfully developed in the context of MC based performance analysis. They are employed for the analysis of various hierarchical net formalisms with asynchronous communication by P. Buchholz [3]. For modeling formalism with synchronous interaction fundamental work refers to B. Plateau and coworkers for stochastic automata networks (SANs) [20, 21]. Her results where transferred to the Petri net world by S. Donatelli, who considers superposed stochastic automata in [12] and superposed generalized stochastic Petri nets (SGSPNs) in [13]. Structured representations are also known for stochastic process algebras [4, 5]. Within these structured representations often not all states are in fact reachable as stated in [10, 13, 18] such that the reachability problem arises again. The problem has been addressed in [19], where a solution for the numerical analysis of SGSPNs is demonstrated but the employed $\mathcal{RS}$ exploration algorithm is only briefly sketched. There the focus point is in the performance analysis based on the associated MC and computation of $\mathcal{RS}$ is used as a prerequisite.

In the following we describe the underlying algorithm of [19] for $\mathcal{RS}$ exploration in detail and for the extended context of SGSPNs and SANs. The algorithm follows a modular approach and exploits that (by definition) these modeling formalisms support a decomposition into components which interact via synchronized transitions. From this point of view the new algorithm is related to the modular approach of Christensen et al. [9]. The main difference is that reachability sets of components are explored in complete isolation assuming that synchronized transitions are not disabled by other components. This is advantageous because these state spaces can be explored in parallel and subsequent exploration of the overall $\mathcal{RS}$ can use bitstate hashing. Compared to the work of Holzmann [16, 26, 17] we also use a bit vector as a hash table but with a perfect(!) hash function instead, such that the resulting $\mathcal{RS}$ is exact. A by-product of this hash function is that a marking/state can be uniquely encoded into a single integer value, which reduces the memory requirements of other involved data structures dramatically. An additional improvement of the new algorithm is suggested, which reduces the impact of interleaving. The key idea is to consider the firing of transitions in certain orders without changing reachability. Similar ideas are discussed in [1] for the efficient elimination of vanishing markings and in [23] to reduce $\mathcal{RS}$. The main difference is that $\mathcal{RS}$ is not(!) reduced here and that sets of states are considered instead of single states. Note that for subsequent MC based performance analysis the complete $\mathcal{RS}$ is required; a reduced $\mathcal{RS}$ as produced by stubborn or sleep set methods is not sufficient for MC based performance analysis.

The paper is organized as follows: Sec. 2 gives the notational background for structured representations of SGSPNs and SANs; in Sec. 3 the new algorithm is described as a variation of the conventional search method, where data structures profit from a given structured representation. Improvements of this algorithm are presented in Sec. 4. Two examples for SGSPNs are taken from literature [7, 10] and exercised in Sec. 5 to demonstrate applicability and efficiency of the new algorithms. Conclusions and prospects to future work finish the paper.

2 Definitions

In this section we will briefly recall definitions and results for tensor algebra, SANs and SGSPNs, in order to fix the notation for subsequent sections and to clarify the similarities between these modeling formalisms at the level of state transition matrices. The notation for SGSPNs follows mainly [13], we assume that the reader is familiar with GSPNs and their dynamic behavior.

Definition 1. A GSPN is an eight-tuple $(P, T, \pi, I, O, H, W, M_0)$ where P is the set of places, T is the set of transitions such that $T \cap P = \emptyset$, $\pi : T \to \{0, 1\}$ is the priority function, $I, O, H : T \to Bag(P)$, are the input, output, and inhibition functions, respectively, where Bag(P) is the multiset on P, $W : T \to \mathbb{R}^+$ is a function that assigns a weight to each transition, $M_0 : P \to \mathbb{N}_0$ is the initial marking: a function that assigns a nonnegative integer value to each place.

Let $T_i := \{t \in T | \pi(t) = 1\}$ $(T_e := T \backslash T_i)$ denote the set of immediate (timed) transitions. Immediate transitions fire with priority over timed transitions. It is possible to define different levels of priorities for T_i but for simplicity we consider just $\pi : T \to \{0, 1\}$. $M[t> M'$ indicates that the system state/marking changes from M to M' due to firing of $t \in T$, $M[t>$ denotes that $t \in T$ is enabled in M and $M[>$ gives the set of enabled transitions. If marking dependent weights shall be considered W modifies to $W : M \times T \to \mathbb{R}^+$. Based on Def. 1 and the firing rule the reachability set $(\mathcal{RS})$, the reachability graph (RG), the tangible reachability set (TRS) and the tangible reachability graph (TRG) can be defined in the usual manner.

For GSPNs well-known techniques apply to derive a state transition matrix $\bar{Q}$ from the TRG, such that the generator matrix Q of the associated continuous time Markov chain (CTMC) is given by $Q = \bar{Q} - D$ with a diagonal matrix D, $D(i, j) := \sum_k \bar{Q}(i, k)$ if $i = j$ and 0 otherwise. Matrix $\bar{Q}$ represents the reachability relation and additional timing information. Since in this context we are only interested to compute $\mathcal{RS}$, resp. $\mathcal{TRS}$, we can exploit

$$\bar{Q}(M, M') \neq 0 \iff \exists t \in T_e, \sigma \in T_i^* : M[t\sigma> M',$$

272

where σ is a maximal firing sequence of immediate transitions, and do not further distinguish between numerical values of nonzero entries in $\bar{Q}$.

Superposed GSPNs are GSPNs, where additionally a partition of the set of places is defined, such that SGSPNs can be seen as a set of GSPNs which are synchronized by certain transitions.

Definition 2. A SGSPN is a ten-tuple $(P,T,\pi,I,O,H,W,M_0,\Pi,TS)$ where (P,T,π,I,O,H,W,M_0) is a GSPN, $\Pi = \{P^0,\ldots,P^{N-1}\}$ is a partition of P, $TS \subseteq \{t \in T|\pi(t) = 0\}$ is the set of synchronized transitions, that are timed by definition. Moreover Π induces on $T\backslash TS$ a partition of transitions. Such a SGSPN contains N components $(P^i,T^i,\pi^i,I^i,O^i,H^i,W^i,M_0^i)$ for $i \in IS :=$ $\{0,1,\ldots,N-1\}$, where $T^i := \bullet P^i \cup P^i \bullet$ and $\pi^i,I^i,O^i,H^i,W^i,M_0^i$ are the functions π,I,O,H,W,M_0 restricted to P^i, resp. T^i.

SGSPNs are naturally amenable for a modular $\mathcal{RS}$ analysis, since partition Π induces components, which are GSPNs themselves. In consequence state spaces of components can be explored independently and in parallel by a conventional $\mathcal{RS}$ exploration. The underlying assumption is that enabling of a synchronized transition depends only on the state of component i during generation of $\mathcal{RS}^i$, resp. $T\mathcal{RS}^i$. In [9] Christensen et al. argue that component state spaces $\mathcal{RS}^i$ can be infinite, while the $\mathcal{RS}$ of the complete model is finite. In general we cannot cure this problem; nevertheless if such a critical component i is not covered by P-invariants in isolation but covered by P-invariants within the SGSPN, these global P-invariants are suitable to achieve finiteness of $\mathcal{RS}^i$ by providing place capacities. P-invariants are similarly exploited in [19] to improve efficiency of a numerical method for SGSPNs by enforcing state spaces of components to obey place capacities deduced from global P-invariants. So in the following we assume that a SGSPN is given for which finite component state spaces $\mathcal{RS}^i$ are calculated via conventional $\mathcal{RS}$ exploration.

For GSPNs typically only so called tangible states are relevant and it is common and efficient practice to eliminate vanishing states during generation of TRG. In case of SGSPNs observe that synchronized transitions are timed by definition, which facilitates to obtain the TRG of a SGSPN from component state spaces TRG^i by eliminating vanishing states locally during generation of TRG^i [19]. Generation of TRG^i for an isolated component i yields a state transition matrix Q^i, which can be transformed into a term $Q^i = Q_l^i + \sum_{t\in TS} w(t)Q_t^i$ where Q_l^i contains all entries due to the firing of (timed) local transitions (possibly multiplied by the probability of subsequently firing a sequence of immediate transitions). In matrices Q_t^i for a synchronized transition $t \in TS$ the row sum of any row j is either 1 if t is enabled in M_j^i or 0 otherwise. Q_t^i contains the conditional probabilities of firing a (possibly empty) sequence of immediate transitions in component i under the condition that t is enabled and fires, i.e. $Q_t^i(j,k)$ gives the probability to reach M_k^i from $M_j^i[t >$ via a firing sequence of immediate transitions in component i.

These component matrices are used in a so-called structured representation to describe Q for the overall SGSPN by the help of tensor algebra [11], which itself is based on a mapping function using mixed radix number representation.

Definition 3. Mapping function mix

Let $\mathcal{RS}^i := \{0, 1, \ldots, k^i - 1\}$ be some finite sets with arbitrary but fixed constants k^i for all $i \in \{0, 1, \ldots, N - 1\}$ and $k = \prod_{i=0}^{N-1} k^i$. Let $\mathcal{PS} := \times_{i=0}^{N-1} \mathcal{RS}^i$ denote the product space. A mapping $mix : \mathcal{PS} \longrightarrow \{0, 1, \ldots, k - 1\}$ is defined by

$$mix(x^{N-1}, \ldots, x^1, x^0) := \sum_{i=0}^{N-1} x^i g_i$$

with weights $g_0 := 1, g_i := k^{i-1} * g_{i-1}$.

A vector $(x^{N-1}, \ldots, x^0) \in \mathcal{PS}$ is the mixed radix number representation of $x = mix(x^{N-1}, \ldots, x^0)$ with respect to basis $(k^{N-1}, \ldots, k^0)$. Like any number representation mix is bijective and its inverse mix^{-1} can be calculated from $x^i \equiv (mix(x^{N-1}, \ldots, x^1, x^0)/g_i) \mod k^i$ for each i. In the following, set $\mathcal{RS}^i$ of Def. 3 coincides with the set of reachable states in component i, such that mix induces a numbering on $\mathcal{PS}$ as well. Since such a numbering allows to identify states, we will not distinguish between a state $M_x = (M_x^{N-1}, \ldots, M_x^0)$ and its number, resp. component numbers $x = mix(x^{N-1}, \ldots, x^0)$ in order to preserve readability.

For the definition of tensor product we follow the notation in [11] but regard only the restricted case of square matrices to keep a concise notation, because only square matrices occur in the context of our modeling formalism.

Definition 4. Tensor product and sum for square matrices

Let $A^0, \ldots, A^{N-1}$ be square matrices of dimension $(k^i \times k^i)$ then their tensor product $A = \bigotimes_{i=0}^{N-1} A^i$ is defined by $a(x, y) := \prod_{i=0}^{N-1} a^i(x^i, y^i)$ where $x = mix(x^{N-1}, \ldots, x^0)$ and $y = mix(y^{N-1}, \ldots, y^0)$.

The tensor sum $B = \bigoplus_{i=0}^{N-1} A^i$ is then given by $\bigoplus_{i=0}^{N-1} A^i := \sum_{i=0}^{N-1} I_{l^i} \otimes A^i \otimes I_{r^i}$ where I_{l^i}, I_{r^i} are matrices of dimension $l^i \times l^i$, resp. $r^i \times r^i$ where $r^i = \prod_{j=0}^{i-1} k^j$, $l^i = \prod_{j=i+1}^{N-1} k^j$ and I(a,b) = 1 iff a = b and 0 otherwise.

As shown in [13, 19] the generator matrix of SGSPNs is given by:

$$Q = \bigoplus_{i=0}^{N-1} Q_l^i + \sum_{t \in TS} w(t) \bigotimes_{i=0}^{N-1} Q_t^i - D \qquad (1)$$

where D is a diagonal matrix providing row sums of $\bigoplus_{i=0}^{N-1} Q_l^i + \sum_{t \in TS} w(t) \bigotimes Q_t^i$. For $\mathcal{RS}$ exploration the structured representation of Q gives a valid state transition matrix for all states reachable from M_0. For other application areas additional restrictions might apply, e.g. for performance analysis TRG^i need to be strongly connected for ergodicity of the associated CTMC [13]. Based on Q we can reformulate the task of $\mathcal{RS}$ exploration: calculate the minimal, reflexive, and transitive closure of relation $reach(M_x, M_y)$ where $reach(M_0, M_0)$ and

$$reach(M_x, M_y) \iff Q(x, y) \neq 0.$$

Note that due to the elimination of vanishing states during generation of matrices Q^i, $reach(M_x, M_y)$ gives $\mathcal{TRS}$. Obviously $\mathcal{TRS} = \mathcal{RS}$ if $T_i = \emptyset$, hence computation of $\mathcal{RS}$ for untimed Place/Transition nets is simply possible by interpreting such nets as GSPNs where all transitions are timed. In this sense a strictly speaking $\mathcal{TRS}$ exploration algorithm serves also as a $\mathcal{RS}$ algorithm, such that we subsume such algorithms by "$\mathcal{RS}$ exploration algorithms" in the following. Before we describe such an algorithm, we clarify similarities to stochastic automata networks (SANs) [20].

A Side-glance on SANs On the level of component state spaces, a matrix Q^i can be seen as a matrix representation of a finite automaton i with additional edge labels specifying a Markovian timing. The set of component matrices can also be interpreted as a set of automata with synchronous communication. SANs follow this point of view and their structured representation coincides with Eq. 1, such that we can conclude that the RS algorithm given in Sec. 3 also applies to SANs. For more information on SANs the interested reader is referred to [20, 21, 22]. SANs have been analyzed in continuous and discrete time, here we consider continuous time.

Definition 5. A SAN consists of N stochastic automata (SAs) with index set $IS = \{0, \ldots, N-1\}$ such that SAs are numbered consecutively from 0 to $N-1$. Every automaton i is characterized by its finite state space $\mathcal{RS}^i$ containing k^i states and its transition function. States in $\mathcal{RS}^i$ are numbered consecutively form 0 to $k^i - 1$ starting from the initial state. All timing is Markovian. The following types of transitions are possible: local transitions which occur locally in $\mathcal{RS}^i$ without affecting other automata, synchronized transitions which have to occur synchronously in a set of automata, and functional transitions, where the transition rate is a nonnegative, real-valued function of the state of other automata. Local and synchronized transitions can be functional. Transitions which have a fixed rate are denoted as constant. A SAN includes TS different synchronized transitions (events); a subset of at least two automata in IS participates on a synchronization event $t \in TS$.

It is straightforward to show that the complete SAN specifies a CTMC, if the single automata observe Markovian timing. The complete CTMC can be described as a N-dimensional CTMC with state space $\mathcal{RS} \subseteq \mathcal{PS} := \mathcal{RS}^{N-1} \times \ldots \times \mathcal{RS}^1 \times \mathcal{RS}^0$. A single SA of a SAN is also referred to as a component. We assume for functional transitions, that their functions do not interfere with the logical behavior, i.e. the function does not evaluate to zero if the transition can occur. This means that the function determines the delay, but not the fact that the transition can occur. Hence the set of reachable states for a given SAN is independent of the selection of functions, such that we can safely replace any functional transition by a constant transition during state space exploration. Hence for $\mathcal{RS}$ analysis tensor operations of Def. 4 are sufficient and it is not necessary to use generalized tensor operations [22].

For the definition of a structured representation of generator matrix Q for SANs we start with the definition of some matrices considering a single automaton. Let A^i denote that a matrix A belongs to automaton i. Any such A^i is a $k^i \times k^i$ matrix. Let Q_l^i be a matrix containing local transition rates. For every synchronizing event t define Q_t^i as the transition matrix of automaton i containing transition rates for t. Every matrix Q_t^i contains only nonnegative elements. For automata that do not participate in event t we define $Q_t^i = I_{k^i}$. Usually one of the matrices belonging to event t contains the corresponding transition rates, all others have row sums equal to 0 or 1, depending whether event t is possible or not in the corresponding state. We additionally define $k^i \times k^i$ diagonal matrices $D_l^i = diag(Q_l^i e^T)$ and $D_t^i = diag(Q_t^i e^T)$ containing the row sums of the corresponding rows of Q_l^i and Q_t^i in the main diagonal, e is a row vector of appropriate size with all elements equal to 1. With these matrices the generator matrix Q of the SAN is described as follows:

$$Q = \bigoplus_{i=0}^{N-1} Q_l^i + \sum_{t \in TS} \bigotimes_{i=0}^{N-1} Q_t^i - D \qquad (2)$$

where $D := \bigoplus_{i=0}^{N-1} D_l^i + \sum_{t \in TS} \bigotimes_{i=0}^{N-1} D_t^i$. Eq. (2) is a slightly less elegant, but equivalent formulation of the structured representation given in [20, 22]. Obviously the structured representations of SGSPNs and SANs formally coincide if $w(t)$ is multiplied into the first term in the tensor product $\bigotimes Q_t^i$ for SGSPNs, such that we can in the following consider the $\mathcal{RS}$ problem on matrix level without distinguishing between SGSPNs and SANs.

3 $\mathcal{RS}$ Exploration Based on Structured Representations

Assume a structured representation is given for a certain model with initial state M_0. In this section we describe how successor states are calculated from the structured representation and formulate a search algorithm to compute $\mathcal{RS}$ (or $T\mathcal{RS}$ in case of SGSPNs with $T_i \neq \emptyset$). Since diagonal values D in Q are irrelevant for $\mathcal{RS}$ exploration we focus on state transition matrix $\bar{Q} := Q + D$ and let $\bar{Q}_l = \bigoplus_{i=0}^{N-1} Q_l^i$.

Successor states can be reached due to local or synchronized transitions. Considering local transitions, Def. 4 ensures that $\bar{Q}_l = \sum_{i=0}^{N-1} I_{l^i} \otimes Q_l^i \otimes I_{r^i}$ where $l^i = \prod_{j=i+1}^{N-1} k^j$ and $r^j = \prod_{j=0}^{i} k^j$. In consequence state transitions from a state M_x to a state M_y due to a local transition $t \in T^i \backslash TS$ in a component i are all specified in a single term $\bar{Q}_l^i := I_{l^i} \otimes Q_l^i \otimes I_{r^i}$ such that $\bar{Q}_l = \sum_{i=0}^{N-1} \bar{Q}_l^i$ and $M_x[t\sigma > M_y$ with a (possibly empty) firing sequence of immediate transitions σ if and only if $\bar{Q}_l^i(x, y) \neq 0$. Each term $\bar{Q}_l^i$ specifies a set of state transitions and the sum $\bar{Q}_l = \sum \bar{Q}_l^i$ behaves like a logical OR since all entries are nonnegative[1]. According to Def. 4 nonzero entries in $\bar{Q}_l$ are characterized by

$$\bar{Q}_l(x, y) \neq 0 \iff \exists i \in IS : Q_l^i(x^i, y^i) \neq 0 \land \forall j \in IS, j \neq i : x^j = y^j \qquad (3)$$

---

[1] In fact matrices Q_l^i, Q_t^i can also be mapped to boolean matrices and $+, *$ to $\land, \lor$ for $\mathcal{RS}$ analysis.

For a synchronized transition t let $IC(t) := \{i | i \in IS \land t \in T^i\}$ denote the set of involved components, then $\bar{Q}_t := \bigotimes_{i=0}^{N-1} Q_t^i$ follows with a similar argumentation as for (3):

$$\bar{Q}_t(x, y) \neq 0 \iff \forall i \in IC(t) : Q_t^i(x^i, y^i) \neq 0 \land \forall j \notin IC(t) : x^j = y^j \quad (4)$$

This follows from Def. 4, plus the fact that in general $\prod a_i \neq 0 \iff \forall i : a_i \neq 0$, and $Q_t^j = I \, \forall j \notin IC(t)$. Since the tensor product is based on a mixed radix number representation, value y of a successor state M_y for a given state M_x can be obtained from x as follows: in case of local transitions the value y for a $\bar{Q}_l(x, y) \neq 0$ is given by $y = mix(x^{N-1}, \ldots, x^{i+1}, y^i, x^{i-1}, \ldots, x^0)$. In case of a synchronized transition t, $y = mix(z^{N-1}, \ldots, z^0)$ where $z^i = y^i$ if $i \in IC(t)$ and $z^i = x^i$ otherwise. Since $x = mix(x^{N-1}, \ldots, x^0) = \sum_{i=0}^{N-1} x^i \cdot g_i$ and $y = \sum_{i \notin IC(t)} x^i \cdot g_i + \sum_{i \in IC(t)} y^i \cdot g_i$, y can be obtained from x by $y = x + \sum_{i \in IC(t)} (y^i - x^i) \cdot g_i$. Note that $(y^i - x^i) \cdot g_i$ is a local transformation such that in case of sparse matrix representations of Q_l^i, resp. Q_t^i, we can store Q_l^i with $Q_l^i(x^i, (y^i - x^i) \cdot g_i)$ resp. $Q_t^i(x^i, (y^i - x^i) \cdot g_i)$ on the same place instead, such that calculation of a successor state requires $|IC(t)|$ additions and no multiplications.

The basic algorithm follows the standard search algorithm for state space exploration by traversing the reachability graph, e.g. [8], but the calculation of successor markings is adapted to the context of structured representations:

Input: Matrices of a structured representation
Program:
Init: $\mathcal{RS} = \{M_0\}$, S=$\{M_0\}$
begin
    while not empty S
        take M_x out of S
        decode M_x into $(M_x^{N-1}, \ldots, M_x^0)$ by mix^{-1}
        foreach component i in IS
            foreach $Q_l^i(M_x^i, M_y^i) \neq 0$
                $M_y = M_x + (M_y^i - M_x^i) \cdot g_i$
                if $M_y \notin \mathcal{RS}$
                then insert M_y in $\mathcal{RS}$ and S
        foreach $t \in TS$
            if $\forall i \in IC(t) : \exists Q_t^i(M_x^i, M_y^i) \neq 0$
(*)   then foreach combination of elements $Q_t^i(M_x^i, M_y^i) \neq 0$ over IC(t)
                $M_y = M_x + \sum_{i \in IC}(M_y^i - M_x^i) \cdot g_i$
                if $M_y \notin \mathcal{RS}$
                then insert M_y in $\mathcal{RS}$ and S
end

In line (*) only one combination occurs if each of the corresponding rows $Q_t^i(M_x^i, .)$ of IC(t) contains exactly one nonzero entry. Due to subsequently firing immediate transitions, several nonzero entries per row are possible, such that for the general case one has to consider all combinations of such nonzero entries to derive all successor states.

Correctness follows from the fact, that the conventional method [8] is employed and only calculation of successor states (according to Eqs (3) and (4))and data structures are especially adapted. Successor states are obtained from the state transition matrix part of Q. Coding and decoding of states into vectors of component states follows mix, resp mix^{-1}, which coincides with the tensor products in the structured representation of Q.

It is well known, that the crux of the conventional method is the choice of appropriate data structures for $\mathcal{RS}$ and S. Conflicting interests are that due to the number of tests, member and insert operations for $\mathcal{RS}$ have to be as fast as possible but on the other hand due to the size of $\mathcal{RS}$ the memory spent per element of $\mathcal{RS}$ must be minimized to keep the method applicable for large state spaces. Additionally the size of $\mathcal{RS}$ is unknown in the general case. In the context of structured representations $\mathcal{RS}$ is a subset of $\mathcal{PS} = \times_{i \in IS} \mathcal{RS}^i$. We suggest to use hashing and a hash table v of boolean entries with $v(x) = true$ if $M_x \in \mathcal{RS}$ and $v(x) = false$ otherwise. Since $mix : \mathcal{PS} \to \{0, 1, \ldots, k-1\}$ is bijective, it gives a perfect hash function, no collisions can occur by $x = mix(M_x)$. This results in a bit-vector of length $\mathcal{PS}$ with 1 bit per state for v. Member and insert operations for $\mathcal{RS}$ are in $O(1)$, since the costs for evaluating mix are already included in the computation of successor states. Due to function mix the set S can be represented by a stack which contains just integer values. This is memory efficient compared to storing vectors of component states or markings. Push and pop operations on stacks are in $O(1)$. Decoding of a state into component states takes N division and modulo operations. Coding of component states in the calculation of a successor markings due to firing of a transition requires as many additions as components are involved (at most $IC(t)$) as discussed above.

The length of the bitvector to represent $\mathcal{RS}$ can be extreme if $|\mathcal{PS}| >> |\mathcal{RS}|$. Compared to other representations of $\mathcal{RS}$ note, that typically the memory for 1 pointer or 1 integer is about 32 bits for state representations based on marking vectors. If coding techniques and tree-type data structures according to [8] are used the amount of memory used for $\mathcal{RS}$ depends on the number of places, the maximum number of tokens estimated for each place and the reachability relation itself, such that a comparison is difficult to draw in general. In the context of bitstate hashing, Holzmann [17] assumes 64 bits per state in case of the hashcompact method [26] or allows for about 100 bits per state for an optimal working area of the double bit hashing method; for values less than 100 bits per state double bit hashing gradually looses coverage, i.e. the quality of results is reduced. This is different in our approach: mix gives a perfect hash function, such that the new algorithm is exact. In summary we can conclude that compared to any alternative state representation which uses at least 64 bits per state, the bitvector representation used here requires less(!) memory for $\mathcal{RS}$, if $|\mathcal{RS}|/|\mathcal{PS}| \geq 1/64 \approx 1.5\%$. Surely for a large set of models holds that 1.5% of $\mathcal{PS}$ is reachable. As a rule of thumb, a bitvector representation is applicable if it fits into the available main memory of a given hardware configuration, e.g. on a 32 MB machine $\mathcal{PS}$ can be up to 256 million states. For larger values of $\mathcal{PS}$ it is model dependent, because for a lot of models *mix* implies a certain locality which goes well with the locality assumption of virtual memory concepts.

4 Improvements for state space exploration

Interleaving of independent, concurrent transitions has been recognized as one source of the state space explosion problem and as a source for computational overhead during state space exploration, e.g. during elimination of vanishing markings the exploitation of extended conflict sets (ECS) can yield significant savings in computation time by avoiding unnecessary interleavings [1, 2]. In this context priorities are assigned to independent sets of transitions in order to reduce the degree of concurrency, i.e. to reduce the number of firing sequences taken into consideration. A more general approach in state space exploration is given by Valmari [23, 24] with the stubborn set method; the point of interest is, whether certain states are reachable or not. From this point of view interleavings caused by independent, concurrent transitions introduce a lot of "redundant" states which can safely be omitted, i.e. $\mathcal{RS}$ is reduced.

In our context the goal is to consider all nodes of RG but not all arcs. A spanning tree on RG would be ideal, however here we establish only a way to omit certain arcs of RG. In SGSPNs and SANs[2] the partition into components implies that transitions which are local and belong to different components are independent of each other. Interleaving of independent, concurrent transitions has the effect that a lot of states are reached over and over again if a straightforward search algorithm is employed. However it is not necessary to consider all permutations of these independent transitions. A selection of just one ordered sequence is sufficient if it is applied on sets of states. The example in Fig. 1 illustrates the idea by a net with components A, B, and C which have only local transitions. Not all arcs of the RG in Fig. 1 b) have to be considered for $\mathcal{RS}$ exploration. One method is to order the local transitions by components and apply them on sets of states. We start from state abc, apply the transition of component A and derive a'bc. As a second step the transition of component B is applied on set { abc, a'bc }, since all states in this set have the same component state for component B. This yields the set { abc, a'bc, ab'c, a'b'c}. Finally on all of these states we apply the local transition of component C with local state c. This procedure considers for the reachability graph of Fig. 1 b) the graph of Fig. 1 c) instead, where the number of arcs is significantly reduced. The method gains in efficiency if not only one local transition per component is fired as in Fig. 1 but sequences of local transitions per single component. This requires an additional termination criterion for such a local search. A natural criterion inherited from the conventional search method is to stop a local search when the global state is already element of $\mathcal{RS}$, e.g. if a'bc is already in $\mathcal{RS}$, no successors { a'b'c, a'b'c', a'bc'} would be considered in this search. This results in a massive pruning of such a search. Validity of this approach is proved later in the paper. Extending the search of local successors to sets of states allows that only some

---

[2] In fact this is typical for all modular models which communicate via fused transitions.

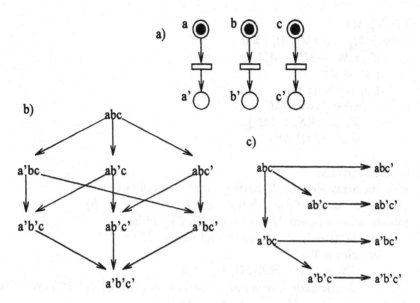

Fig. 1. An example net with 3 components, its reachability graph RG and the considered subgraph of RG

states require a complete decoding with N division and modulo operations, some states require only a partial decoding and some states require only a decoding for one component in the following algorithm. The basic idea is to keep track of the differences, resp. similarities between states and their successor states.

Input: Matrices of a structured representation
Program:
Init: x = 0 ; $\mathcal{RS}_x = \{M_0\}$; $S_x = \{M_0\}$
begin
    while not empty S_x
        take M_x out of S_x (step x)
        start = stop = last = 1
        copy state M_x to array(1)
        decode M_x into $(M_x^{N-1}, \ldots, M_x^0)$
        check TS on M_x like check-TS(0,last) but without extra decoding
        foreach component $i \in IS$ in increasing order
            foreach array element M between start and stop
                start-TS = last + 1
                search(i,M_x^i,M)
                check-TS(i,start-TS)
            stop = last   (last element in array)
        $\mathcal{RS}_{x+1} = \mathcal{RS}_x$; $S_{x+1} = S_x$; $x = x + 1$;
end

search(i,M_x^i,M)
    foreach M_y^i : $Q_l^i(M_x^i, M_y^i) \neq 0$
        $M' = M + (M_y^i - M_x^i) \cdot g_i$
        if $M' \notin \mathcal{RS}_x$
        then last = last + 1
            array(last) = M'
            $\mathcal{RS}_x = \mathcal{RS}_x \cup \{M'\}$
            search(i,M_y^i,M')

check-TS(h,start-TS)
    pick one array element M between start-TS and last
    decode M into M^j for $j < h$ (exploit $M_x^j = M^j$ for $j > h$)
    foreach array element $\bar{M}$ between start-TS and last
        decode $\bar{M}$ only at position h and update $M^h = \bar{M}^h$
        foreach $t \in TS$
            if $\forall i \in IC(t) : \exists Q_t^i(M^i, M_y^i) \neq 0$
            then foreach combination of nonzero elements $Q_t^i(M^i, M_y^i)$ over IC(t)
                $M' = M + \sum_{i \in IC}(M_y^i - M^i) \cdot g_i$
                if $M' \notin \mathcal{RS}_x$
                then insert M' in $\mathcal{RS}_x$ and S_x

Compared to the algorithm of Sec. 3, local transitions are only considered by a component specific call for function search, which stores new states in a special array, while synchronized transitions are only considered in function check-TS, which stores new states in stack S_x as before. Apart from the sophisticated decoding, function check-TS corresponds to the 2nd foreach-loop in the previous algorithm. A call for function search causes a component specific depth-first-search (DFS) via local transitions whose termination depends on $\mathcal{RS}_x$. Since array entries are caused by local transitions and new states are inserted behind position stop, all entries between positions start and stop have equivalent values M_x^i when component i is considered. A byproduct of this calculation is that a set of successor states is successively stored in the array (between positions check-TS and last), which differs only in component i, hence this structural information can be exploited to speed up decoding of states and checking synchronized transitions. In summary this technique avoids unnecessary interleavings, saves division and modulo operations for decoding, and simplifies enabling tests for synchronized transitions; surely these effects occur only if the model under consideration allows for a sufficient amount of independent, local transition firings.

Termination follows from the observation that any state is inserted at most once into the array or S_x. To ensure that the algorithm actually considers all reachable states, we have to show that any successor of a state in RS_x is either already element of RS_x or it still has a predecessor on stack S_x:

Lemma 6. *After every step x holds:* $\forall M \in \mathcal{RS}_x \backslash S_x$: *if* $\exists t \in T : M[t >$ $M' \land M' \notin \mathcal{RS}_x$ *then* $\exists M'' \in S_x, \sigma \in T^* : M''[\sigma > M'$.

Proof. Proof by induction, initially $\mathcal{RS}_0 = S_0 = \{M_0\}$ so lemma is trivially fulfilled by $\mathcal{RS}_x \backslash S_x = \emptyset$.

Induction step: indirect proof by contradiction, assume lemma does hold for step x but does not hold in step x+1. In this case for any counterexample $M \in \mathcal{RS}_{x+1} \backslash S_{x+1}$ holds $M_x[\sigma_x > M$ so we choose a suitable M with shortest σ_x.

   case $M_x = M$ Since all transitions enabled by M_x are considered by the algorithm $\not\exists t \in T : M[t > M'$ such that $M' \notin \mathcal{RS}_{x+1}$.

   case $M \in \mathcal{RS}_{x+1} \backslash (S_{x+1} \cup \{M_x\})$ i.e. M was reached in step x+1 for the first time and inserted into $\mathcal{RS}_{x+1}$ but not in S_{x+1}. Consider the kind of transition t_M by which M was reached: if $t_M \in TS$ then according to the algorithm any marking reached for the first time is inserted into $\mathcal{RS}_{x+1}$ and S_{x+1} which contradicts the above assumption, hence $t_M \notin TS$. So $t_M \in T^i \backslash TS$ is local to a certain component $i \in IS$. Since M is reached for the first time searching from M_x, it is reached by the search algorithm following a sequence $\sigma_M \in (T \backslash TS)^* :$ $M_x[\sigma_M > M$ of local transitions in the order of components, so $\sigma_M = \sigma_0 \cdot \sigma_1 \ldots \sigma_{N-1}$ with $\sigma_i \in (T^i \backslash TS)^*$ forall $i \in IS$.

Consider now the critical transition t for $M[t > M' : t \in TS$ is not possible, because M was reached for the first time, so according to the algorithm all synchronized transitions are considered for M. So $t \in T^j \backslash TS$ must be local to some component $j \in I$.

If $i < j$ then $t \in T^j \backslash TS$ is considered within a DFS-search for M, since M is stored in the array as any new element reached by local transitions and a search is called as for all elements between start and stop. In consequence M' would be element of $\mathcal{RS}_{x+1}$.

If $i = j$ then M is reached within a DFS-search of the same component i=j. According to the algorithm every $t \in T^i \backslash TS = T^j \backslash TS$ is considered such that M' would be reached and inserted into $\mathcal{RS}_{x+1}$.

If $i > j$ then $M_x[\sigma_M t > M'$ and the firing sequence does not follow the order of I due to transition t. Due to the independence of the firing of local transitions which belong to different components, firing sequence $\sigma_M t$ has permutation $\sigma_{perm} = \sigma_1 \cdots \sigma_j t \sigma_{j+1} \ldots \sigma_N$ which can be fired and yields M' as well. According to our assumption $M_x \in \mathcal{RS}_{x+1}$ and $M' \notin \mathcal{RS}_{x+1}$, hence this firing sequence has a first break where $M_1[t_1 > M_2$ and $M_1 \in \mathcal{RS}_{x+1}$ and $M_2 \notin \mathcal{RS}_{x+1}$. Let σ_1 denote the firing sequence reaching M_1 being a prefix of σ_{perm}. Since $t_1 \in T \backslash TS$ must be a local transition and M_2 is not reached by a DFS-search, DFS must terminate at M_1 because $M_1 \in \mathcal{RS}_{x+1}$ holds already. There are 2 possibilities: Firstly $M_1 \in \mathcal{RS}_x$ was reached in one of the previous steps. Then due to $M_2 \notin \mathcal{RS}_x \subseteq \mathcal{RS}_{x+1}$ either $M_1 \in S_x$ or $\exists M'' \in S_x, \sigma_2 \in T^* : M''[\sigma_2 > M_2$.

If $M_1 \in S_x$, then due to $M_2 \notin \mathcal{RS}_{x+1}$ must be $M_x \neq M_1$. Hence $M_1 \in S_{x+1}$, which in turn implies M_1 is a suitable element in $S_{x+1}, \exists \sigma_1 \in T^* : M_1[\sigma_1 > M'$. Otherwise if $\exists M'' \in S_x, \sigma_2 \in T^* : M''[\sigma_2 > M_2$ and if $M'' = M_x$ is the only suitable element of S_x from which M_2 can be reached then M_1 is already a suitable element for a counterexample like M_x but with a shorter sequence than σ contradicting our assumption. Hence $\exists M'' \neq M_x \in S_x, \sigma_2 \in T^* : M''[\sigma_2 > M_2$. Since M_x is the only element removed from S_x in step x+1, M'' is element of

S_{x+1} as well. Finally the first break cannot be at M, since this is ensured already above by $M_x \neq M$.

The second possibility is that $M_1 \in \mathcal{RS}_{x+1} \backslash \mathcal{RS}_x$, i.e. M_1 was reached before but within step x+1. Then M_1 is itself a starting point for DFS and t_1 follows in the order of components ($\sigma_1 t$ is an ordered sequence), the transition t_1 is checked at $M_1[t> M_2$. $\qquad\qquad\qquad\qquad\qquad\qquad\qquad\qquad\qquad\qquad\qquad\qquad\qquad$ □

Correctness of the algorithm simply follows from the lemma above in combination with the termination criterion $S_x = \emptyset$.

Breadth-First-Search (BFS) can be used instead of DFS for the price of an additional decoding at one component. The main advantage from a practical point of view is that the array entries can be reused as a BFS-queue such that no extra memory is required and recursion as in DFS can be avoided.

5 Examples

As examples for SGSPNs we consider the "benchnet model" of Caselli et al. [6, 7] and the flexible manufacturing system given by Ciardo et al. [10].

The Benchnet model This model has a simple net structure shown in Fig. 2, whose obvious symmetries are ignored by the implemented approach. The size of the state space is increased by modifying arc weights, i.e. by increasing values for parameter k. The net is partitioned into 4 components to obtain a SGSPN, namely A,B,C, and D as denoted in Fig. 2. Finiteness for the state space of component A is achieved by exploiting P-invariants [19], i.e. a P-invariant of the SGSPN implies that p_0 contains at most 3 tokens. Caselli et al. state that main memory is the main bottleneck for state space exploration. Since the SGSPN approach does not solve the state space explosion problem, but reliefs its impacts, it is interesting to see how far state space exploration can be pushed on a given hardware configuration. Figure 3 shows the CPU-time and the elapsed time in seconds for the basic and the improved method as a function of $|\mathcal{RS}|$; the CPU-time does not differ significantly from the elapsed time for $|\mathcal{RS}| \leq 43$ million states. These results are obtained on a Sparc 5 with 70 MHz CPU and 32 MB main memory.

Fraction $\mathcal{RS}/\mathcal{PS} = 25\%$ is constant for all values of k in this model; nevertheless it shows clearly that $|\mathcal{PS}| >> |\mathcal{RS}|$. The computation times for the component state spaces are negligible, i.e. they are less than 1 sec for the state spaces considered here. Component state spaces have very moderate cardinalities, i.e. $|\mathcal{RS}^A| = 4$ and $|\mathcal{RS}^B| = |\mathcal{RS}^C| = |\mathcal{RS}^D| = k + 2$. The number of nonzero entries which have to be stored for the structured representation remains less than 1600 for $k \leq 498$. The benchnet model is a kind of best case example for the improved method because large values of k imply a high degree of enabling for local transitions, such that a lot of interleaving occurs, which in turn is successfully treated by the improved method. The results clearly indicate that the improved method is much more efficient than the basic method for this example. Compared to computation times given in [7] note that the conventional approach in a sequential implementation fails for $|\mathcal{RS}| > 10^6$ on a Sparc 2 with 64 MB main memory; computation times for a parallel implementation on a

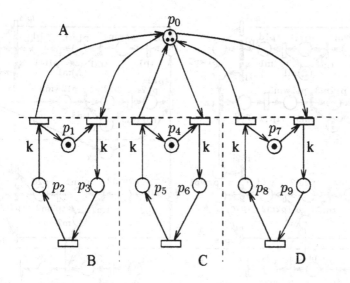

Fig. 2. The benchnet model with a partition into components A,B,C, and D

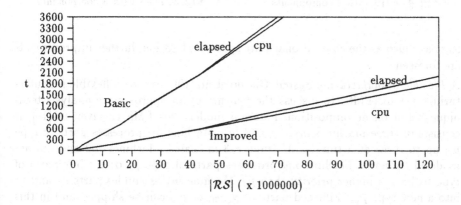

Fig. 3. Computation time t in seconds for benchnet model up to 125 million states

CM-5 are similar to the improved method up to the state space cardinalities considered in [7], i.e. up to 1.7 million states.

Comparisons with results in [6, 7] should be drawn carefully, since the new algorithm relies on the partition into components inherent to SGSPNs and SANs, but the conventional method (sequential or parallel) for GSPNs can be applied on arbitrary GSPNs without such an information. Hence we only conclude that for this example the SGSPN method allows to outperform the conventional sequential approach by far and furthermore to outweigh advantages obtained from parallel computation. Since the new algorithm is amenable to parallelization at

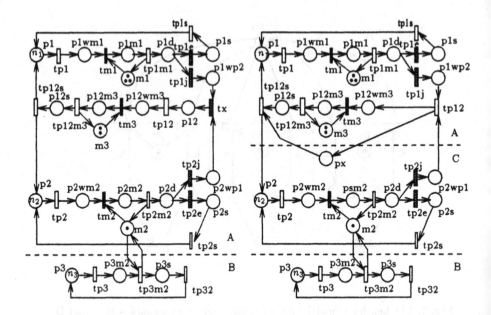

Fig. 4. FMS with 2 components   **Fig. 5.** FMS with 3 components

least as much as the conventional state space exploration, further improvements are foreseen.

A Flexible Manufacturing System Ciardo et al. [10] describe a flexible manufacturing system (FMS) to discuss the benefits of an approximate analysis technique based on decomposition. Figure 4 models this FMS as given in [10], it consists of three machines m_1, m_2, and m_3. Machine m_1 processes parts of type p_1, up to three at a time as $M_0(m_1) = 3$ indicates. Machines $m2$ and $m3$ are modeled similarly. Machine m_2 processes parts of type p_2 or p_3, but parts of type p_2 have a higher priority than p_3. Machine m_3 assembles parts p_1 and p_2 into a new type p_{12}. Finished parts p_1, p_2, p_3, or p_{12} can be shipped and in this case the same number of rough parts enters the system, to maintain a constant inventory. Since we are only interested in $\mathcal{RS}$ exploration here, we omit the additional assumptions considering aspects of time in [10] apart from the fact that $T_i := \{tm1, tm2, tm3, tp1, tp2, tp1j, tp2e\}$ are immediate transitions and all other transitions are timed. In [10] "flushing" arcs connect corresponding transitions tp1s,tp2s,tp3s, and tp12s; they have a marking dependent cardinality equal to the number of tokens in the input place for the transition. This exceeds the GSPN definition underlying the SGSPNs introduced here. For simplicity we assume constant arc weights for transitions tp1s,tp2s,tp3s, and tp12s. The FMS model is exercised for 2 different partitions:

FMS with partition into 2 components A decomposition into 2 components by separating processing of parts of type p_1, p_2 and p_{12} from parts p_3. In this model transition tp3m2 is the only synchronized transition. Table 1 shows the corre-

N	TRS^A	TRS^B	$PS = TRS$	$\sum NZ(Q^i)$	Comp.-expl. in sec	TRS-expl. in sec
3	652	10	6520	3391	2	2
4	2394	15	35910	14088	5	6
5	7272	21	152712	46719	19	22
6	19206	28	537768	131769	66	77
7	45540	36	1639440	328830	259	296

Table 1. Results of FMS model with partition into 2 components

sponding results for increasing values of $N = n_1 = n_2 = n_3$: the columns TRS^A and TRS^B give the sizes of component state spaces. TRS^A contains processing of parts p_1,p_2, and p_{12}; TRS^B processing of parts p_3. Column four shows the size of PS which equals TRS for this partition. Column $\sum NZ(Q^i)$ gives the number of matrix elements which are explicitly stored in the structured representation. Column six gives the elapsed time used for the state space exploration of components; the total elapsed time in seconds for state space exploration using the improved method is given in column seven, note that the values include values of column six.

Obviously the synchronization of components A and B via transition tp3m2 does not restrict reachability, all elements of the cross-product of component state spaces are reachable in this model. As a consequence the TRS-exploration of the SGSPN can safely be omitted for analysis purposes, nevertheless it is an interesting extreme case, in which the hash table is completely filled. The results demonstrate that most of the time in state space exploration is used to explore the state space of component A which uses the conventional approach. Once generation of component state spaces is finished, exploration of the overall TRS is extremely fast, e.g. 296-259=37 seconds for exploration of a TRS with more than 1.6 million states. Trivially, the implementation with hashing into a bit vector using a perfect hash function is insensitive to the filling of the hash table.

Since the dimension of component state spaces are of different orders of magnitude, such that their generation times differ significantly, a further decomposition of component A seems to be worthwhile.

FMS with partition into 3 components Component A of this model can be split into 2 components if transition tx is merged with transition tp12. The new model is partitioned into three components A,B,C as shown in Fig. 5, where transitions tp12s, tp12 and tp3m2 are synchronized transitions. Finiteness of TRS^C is ensured by taking place capacities/limits into consideration which follow from P-invariants of the complete model [19]. Nevertheless introducing an additional place px derived from a P-invariant is helpful to avoid useless states in TRS^C. Informally px aggregates the processing of p12. Result values are given in Table 2 and Fig. 6. Note that for this partition $|PS| >> |TRS|$ and the fraction TRS/PS is rapidly decreasing for increasing values of N.

The results demonstrate that the time for state space exploration is signif-

N	TRS^A	TRS^B	TRS^C	PS (x 1000)	TRS (x 1000)	TRS/PS	$\sum NZ(Q')$
3	56	10	35	19	6	0.3327	289
4	126	15	70	132	35	0.2714	697
5	252	21	126	666	152	0.2290	1458
6	462	28	210	2716	537	0.1980	2758
7	792	36	330	9408	1639	0.1742	4838
8	1287	45	495	28667	4459	0.1556	8001
9	2002	55	715	78728	11058	0.1405	12619
10	3003	66	1001	198396	25397	0.1280	19140
11	4368	78	1365	465060	54682	0.1176	28095

Table 2. Results of FMS model with partition into 3 components

icantly reduced if component state spaces are chosen to be small, e.g. for N=7 the elapsed time for the generation of component state spaces is 1 second instead of 259 for the model with 2 components and the complete TRS exploration requires 68 (52) seconds for the basic (improved) algorithm instead of 296 before. On the other hand it is quite clear that components should not be decomposed arbitrarily fine since their "local" behavior is one source of an efficient TRS-exploration, cf. Sec. 4 and $|PS| \gg |TRS|$ is the price paid for superposition. For N=11 the bitvector requires about 58 MB and exceeds the available 32 MB by far, such that a significant amount of paging activities increase the elapsed time, i.e. it takes 9109 seconds for the improved method to explore 54 million states while the CPU-time is only 2442 seconds. This effect is well known from the conventional approach and is experienced there for much smaller state spaces.

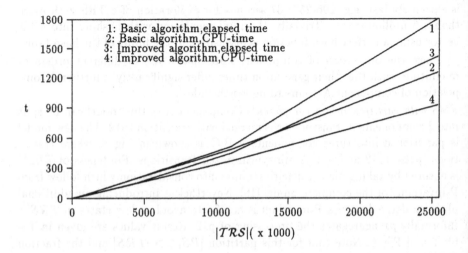

Fig. 6. Computation time t in seconds for FMS model up to 25 million states

6 Conclusions

In this paper a state space exploration algorithm is described, which computes the set of tangible reachable states ($T\mathcal{RS}$) for superposed generalized stochastic Petri nets (SGSPNs) and stochastic automata networks (SANs). Computation of the reachability set ($\mathcal{RS}$) for untimed Place/Transition nets is included as a special case. The algorithm exploits that SANs and SGSPNs by definition provide a decomposition of a model into components and profits from the structured representation known for the generator matrix Q of the associated CTMC. Q is trivially suitable as a state transition matrix for reachability analysis. Additionally the algorithm is based on hashing with a bitvector for the cross-product of component state spaces $\mathcal{PS}$ and a perfect(!) hash function derived from the structured representation. An improvement of this basic algorithm which reduces the impact of interleaving is also presented. The applicability of the new method is demonstrated by two examples for SGSPNs taken from literature [7, 10]. For the given decomposition the new approach outperforms the conventional algorithm in a sequential implementation by far in terms of the size of state spaces and computation times. Both examples show that state space cardinalities of several million states can be handled on a workstation with 32MB main memory.

Although originally designed to improve iterative numerical analysis of SANs and SGSPNs, efficiency of the algorithm motivates its application for any purely functional analysis which is based on state space exploration, e.g. to decide reachability, liveness etc. Since the algorithm does not(!) rely on stochastic timing its application to colored Petri nets with transition fusion [9] is straightforward. In [4, 5] structured representations similar to SANs are given for Markovian process algebras (MPAs), a modeling formalism where a set of processes interact via synchronized transitions as well. The approach given here can be adapted to MPAs as well.

So far the method requires that a decomposition of a model into components with finite component state spaces is given as a prerequisite. Ongoing work is dedicated to clarify the role of P-invariants for the derivation of a suitable decomposition. Another promising perspective is parallelization, since the generation of component state spaces is trivially possible in parallel and the computation of successor states based on component matrices shows at least as much potential for parallelization as the parallel computation of transition firings in GSPNs.

References

1. G. Balbo, G. Chiola, G. Franceschinis, and G. Molinar-Roet. On the efficient construction of the tangible reachability graph of generalized stochastic Petri nets. In *Int. Work. Petri Nets and Performance Models*. IEEE Computer Society, 1987.
2. A. Blakemore. The cost of eliminating vanishing markings from generalized stochastic Petri nets. In *3rd Int. Work. Petri Nets and Performance Models*. IEEE Computer Society, 1989.
3. P. Buchholz. Numerical solution methods based on structured descriptions of Markovian models. In *5th Int. Conf. Modeling Techniques and Tools*, 1991.

4. P. Buchholz. Markovian process algebra: Composition and equivalence. In *2nd Work. Process Algebras and Performance Modelling*, 1994.

5. P. Buchholz. On a Markovian process algebra. Technical Report 500, Universität Dortmund, 1994.

6. S. Caselli, G. Conte, F. Bonardi, and M. Fontanesi. Experiences on SIMD massively parallel GSPN analysis. In *Computer Performance Eval.*, Springer, 1994.

7. S. Caselli, G. Conte, and P. Marenzoni. Parallel state space exploration for GSPN models. In *16th int. Conf. Application and Theory of Petri Nets*, Springer, 1995.

8. G. Chiola. Compiling techniques for the analysis of stochastic Petri nets. In *4th Int. Conf. on Modeling Techniques and Tools*, Mallorca, 1989.

9. S. Christensen and L. Petrucci. Modular state space analysis of coloured Petri nets. In *16th int. Conf. Application and Theory of Petri Nets*, Springer, 1995.

10. G. Ciardo and K.S. Trivedi. A decomposition approach for stochastic Petri net models. In *4th Int. Work. Petri Nets and Performance Models*. IEEE Computer Society, 1991.

11. M. Davio. Kronecker products and shuffle algebra. *IEEE Transactions on Computers*, C-30(2):116–125, February 1981.

12. S. Donatelli. Superposed stochastic automata: a class of stochastic Petri nets with parallel solution and distributed state space. *Performance Eval.*, 18:21–26, 1993.

13. S. Donatelli. Superposed generalized stochastic Petri nets: definition and efficient solution. In *15th int. Conf. Application and Theory of Petri nets*, Springer, 1994.

14. P. Godefroid. Using partial orders to improve automatic verification methods. In *2nd int. Work. Computer Aided Verification*, LNCS 531. Springer, 1990.

15. P. Godefroid and D. Pirottin. Refining dependencies improves partial-order verification methods. In *5th int. Conf. Computer Aided Verification*, LNCS 697. Springer, 1993.

16. G.J. Holzmann. On limits and possibilities of automated protocol analysis. In *7th int. Work. Protocol Specification, Testing, and Verification*. North-Holland, 1987.

17. G.J. Holzmann. An analysis of bitstate hashing. In *15th int. Symp Protocol Specification, Testing and Verification*, IFIP. Chapman & Hall, 1995.

18. P. Kemper. Closing the gap between classical and tensor based iteration techniques (extended abstract). In *Computations with Markov Chains*. Kluwer, 1995.

19. P. Kemper. Numerical analysis of superposed GSPNs. In *6th Int. Work. Petri Nets and Performance Models*. IEEE Computer Society Press, 1995.

20. B. Plateau and K. Atif. Stochastic automata network for modelling parallel systems. *IEEE Trans. on Software Engineering*, 17(10):1093–1108, 1991.

21. B. Plateau and J.M. Fourneau. A methodology for solving Markov models of parallel systems. *Journal of Parallel and Distributed Computing*, 12, 1991.

22. W.J. Stewart. *Introduction to the numerical solution of Markov chains*. Princeton University Press, 1994.

23. A. Valmari. Error detection by reduced reachability graph generation. In *9th Europ. Work. Application and Theory of Petri Nets*, Venice, Italy, 1988.

24. A. Valmari. Stubborn sets for reduced state space generation. In *Advances in Petri Nets*, LNCS 483. Springer, 1990.

25. P. Wolper and P. Godefroid. Partial order methods for temporal verification. In *4th int. Conf. Concurrency Theory*, LNCS 715. Springer, 1993.

26. P. Wolper and D. Leroy. Reliable hashing without collision detection. In *5th int. Conf. Computer Aided Verification*, Elounda, Greece, 1993.

Arc-Typed Petri Nets*

Ekkart Kindler   and   Rolf Walter
Humboldt-Universität zu Berlin
Institut für Informatik
D-10099 Berlin

Abstract

We formally introduce *arc-typed nets* as a model for *causality based specifications*. The new feature of arc-typed nets is the distinction of different arc-types; the different arc-types carry over to a process of the net. Therefore, there are different types of causalities in a *run* of an arc-typed net.

Arc-typed nets have informally been used for modelling and verifying consistency protocols. It turned out that arc-typed nets provide an adequate level of abstraction for these kind of applications. We demonstrate the application of arc-typed nets by specifying, modelling, and verifying a simple consistency protocol.

Keywords: Arc-typed net; causality; partial order semantics; process; specification; verification; consistency protocol.

1 Introduction

Abstraction is the only way to model and verify industrial size applications. Usually, it is a difficult task to find an adequate level of abstraction. The reason is that the abstraction must be sufficiently detailed to reflect the essential behaviour of the application and sufficiently abstract to allow a formal verification. Formal verification requires that there is a formal model of a system which operationally describes a set of runs. Moreover, there must be a specification for a set of legal runs. Then, verification means to provide a proof that all runs of the system are legal with respect to the specification.

It turned out that for some specific applications an extension of Petri nets called *arc-typed nets* provide an adequate level of abstraction. The new feature of arc-typed nets is the distinction of different arc-types. The different arc-types carry over to the processes [2] of the underlying net. Thus, there are different types of causalities in a *run* of an arc-typed net. The different arc-types can be used for specifying legal runs; we call this style *causality based specification*.

The use of causalities for the description of distributed systems is not new and is usually attributed to Lamport [14]. Today, there are many approaches which exploit

---

*Supported by the Deutsche Forschungs Gemeinschaft, Projects: Verteilte Algorithmen and SFB 342 TP YE1.

causalities (e.g. [8, 11]). The distinction of causalities[1] by different arc-types, however, is new. We feel that arc-typed nets provide an adequate level of abstraction for applications which are concerned in some kind of data consistency. For example, arc-typed nets and causality-based specifications were used [5] to specify, model and verify a protocol of a weakly coherent memory management system [1]. The same idea can be used in the field of transaction systems [12, 13]. Here, we illustrate the benefit of arc-typed nets by specifying, modelling, and verifying a simple consistency protocol.

Up to now, arc-typed nets have been used in a more or less informal way. A precise mathematical model has not yet been introduced. In this paper we present the formal foundation of arc-typed nets and formalize some verification rules. Arc-typed nets are an extension of high-level nets. Therefore, we formalize high-level nets and their processes, first. Then, we will equip a high-level net with different arc-types and a labelling of transitions. We show how the arc-types and labels carry over to a process. For clearness we call a process equipped with different arc-types and transition labels *run*.

The paper is organized as follows: Section 2 motivates the use of arc-typed nets and causality based specifications by an example. The example is a simplified version of the model presented in [5]. In Sect. 3 we formalize arc-typed nets and their runs. Section 4 introduces some typical verification techniques, which will be applied to verify the example from Sect. 2.

2 An example: Distributed memory management

First of all, we present an example of a causality based specification. The example is taken from the world of distributed memory management: different sequential agents of a distributed system may hold copies of the same object (caching) for efficiency reasons. Then, a consistency protocol must guarantee that a read event reads valid data, only. Of course, the notion of valid data is subject to a formal specification. To this end, we specify the concept of *weakly coherent memory* [1] in Sect. 2.1. In Sect. 2.2 we present a simple consistency protocol — modelled as an arc-typed net — which guarantees weak coherence for two agents.

2.1 A causality based specification: Weak coherence

In our approach a specification characterizes a set of legal runs. Therefore, we first introduce the concept of runs. Then, we specify those runs which are weakly coherent.

A run consists of a set of *events* and a partial order on these events, which is called *causality*. Each event is labelled by an *action* of some system. Then, the causality represents the order in which the corresponding actions occur. Figure 1 shows an example of a run. The events e_1–e_8 are represented as boxes and correspond to occurrences of write and read actions on some objects o1 or o2 and send and receive actions. If an event e is labelled by some action A, we write e is A; e.g. we write $e1$ is W.o1.

---

[1]Note that our distinction of causalities has nothing to do with the distinction of AND- and OR-causalities [7].

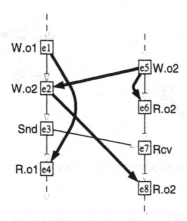

Figure 1: A run

The causality between the events is represented by different arc-types. In the example the bold-faced arcs represent the flow of data between write and read events. This causality is called *data causality*. The arcs with white arrowheads represent the control flow of some distributed program and is called *program causality*. If there is a data causality from event e to event e', we write $e \leq_D e'$; if there is a program causality from event e to event e' we write $e \leq_P e'$; e.g. in the run of Fig. 1 we have $e_5 \leq_D e_8$, $e_2 \not\leq_D e_7$, and $e_2 \leq_P e_7$.

Since there may be different copies of the same object, we must specify some consistency notion characterizing obsolete values, which must not be read any more. We choose the concept of *weak coherence* [1], which consists of two requirements. Informally, the first requirement guarantees that write events on the same object happen on the same instance of the object[2]. The second requirement guarantees that a read event returns a value which is as least as new as all write events which precede the read event with respect to program causality. We use data causality and program causality to make these requirements precise: a run is weakly coherent (with respect to program causality $\leq_P$) iff

1. for all events e and e' with e is W.o and e' is W.o holds: $e \leq_D e'$ or $e' \leq_D e$, and

2. for all events e and e' with e is W.o, e' is R.o and $e \leq_P e'$ holds: $e \leq_D e'$.

Figures 2 and 3 show two runs which are not weakly coherent. The run from Fig. 1, however, is weakly coherent. A more detailed motivation of this specification can be found in [5]. For our purpose it is sufficient to get a feeling for the style of a causality based specification: a causality based specification establishes requirements on the different causalities of a run and their interrelation.

2.2 An arc-typed net: A simple coherence protocol

Now, we present a system which meets the above specification. Basically, the system is modelled as a high-level net. We assume that D is some finite set $D = \{o_1, \ldots, o_n\}$

---

[2]Usually, this particular instance is called the original of the object.

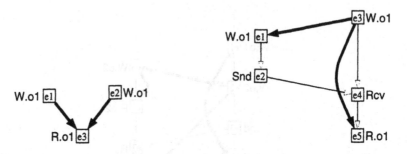

Figure 2: A run violating requirement 1   Figure 3: A run violating requirement 2

of objects and o is a variable with domain D. The symbol D is an abbreviation for the multiset $o_1 + \ldots + o_n$. Figure 4 shows the arc-typed net modelling the system. The

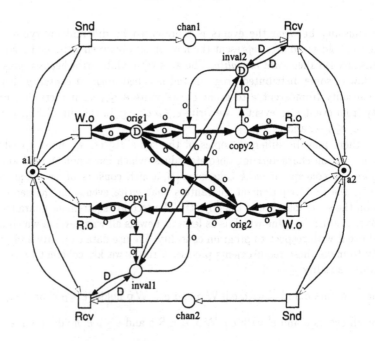

Figure 4: A protocol for weak coherence

model shows a protocol for two sequential agents a_1 and a_2, which can execute write, read, send and receive events. The send and receive events establish synchronizations between the two agents. The write and read events access objects residing as an original or as a copy at the corresponding agents. At each time there exists exactly one original of each object; the other agent may hold a copy of this object. Initially all originals D reside at agent a_1 and agent a_2 holds no copies. Only the original of an object is modified by a write event. In order to exclude read events on obsolete copies a receive event may only happen when all objects D are invalid. Of course, this is

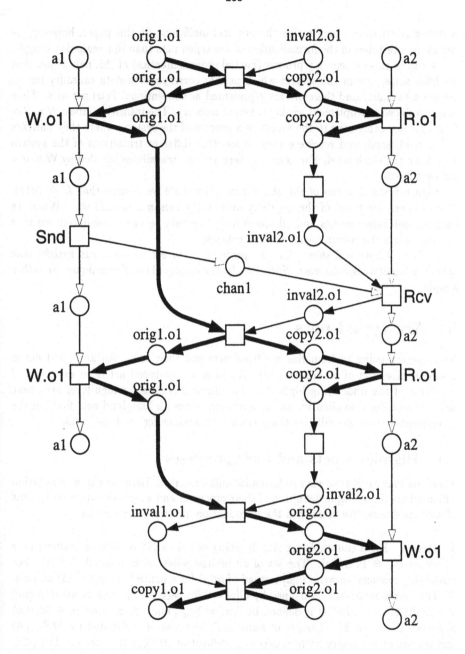

Figure 5: A run of the protocol

a strong restriction, which makes the protocol inefficient. In this paper, however, we are more interested in the formalization of arc-typed nets than in a realistic example.

A run of this system is a process (partial order semantics) of this net. Note, that we have some arcs in the system which neither correspond to data causality nor to program causality and therefore are represented as conventional Petri net arcs. They are introduced to implement weakly coherent memory. The transition labels W.o, R.o, Snd, and Rcv carry over to the events of a process of the system, where they indicate write, read, send, and receive events. Note, that different transitions of the system may have the same label. For example there are two transitions labelled by W.o, one for each agent.

Figure 5 shows a run of the above protocol (where we assume that $D \triangleq \{o1\}$). The different arc-types of the net carry over to the run in a natural way. When we omit all unlabelled events and all places from this run, we can immediately see that the run meets the requirements of weak coherence.

In Sect. 4.2 we will show, that the protocol is correct; i.e. each run satisfies the specification of weak coherence. Surprisingly, the employed proof techniques are rather simple.

3 Arc-typed nets

Now, we are going to formalize *arc-typed nets* and their runs. An arc-typed net is a simple extension of a high-level net. A *run* of an arc-typed net is an extension of a *process* of the underlying high-level net. Since processes for high-level nets have never been defined explicitly[3], we introduce processes for high-level nets first. In the subsequent section we will give the necessary extensions for arc-typed nets.

3.1 High-level nets and their processes

There are many different ways to formalize high-level nets. Here, we choose a notation which allows for a simple definition of their processes and a smooth extension to runs of arc-typed nets. We start with the introduction of some basic notions.

Multisets For a countable set A a mapping $m : A \to \mathbb{N}^\omega$ is called a *multiset over* A, where $\mathbb{N}^\omega \triangleq \mathbb{N} \cup \{\omega\}$. The set of all multisets over A is denoted $MS(A)$. For readability reasons we write $m[a]$ instead of $m(a)$ for a multiset $m \in MS(A)$ and $a \in A$. The *empty multiset* $\mathcal{O}$ is defined by $\mathcal{O}[a] = 0$ for each $a \in A$. The *cardinality* $\|m\|$ of a multiset $m \in MS(A)$ is defined by $\|m\| \triangleq \sum_{a \in A} m[a]$. A multiset $m \in MS(A)$ is *finite*[4], if $\|m\| \in \mathbb{N}$. The set of finite multisets over A is denoted by $MS_{fin}(A)$ and the set of *nonempty finite multisets* is defined by $MS_{fin}^+(A) \triangleq MS_{fin}(A) \setminus \{\mathcal{O}\}$. For two multisets m and m' we define the relation $m \geq m'$ and the addition $m + m'$ elementwise: $m \geq m'$ iff for each $a \in A$ holds $m[a] \geq m'[a]$; $(m+m)[a] \triangleq m[a] + m'[a]$. For $m \geq m'$ and $m' \in MS_{fin}^+(A)$ we define the difference $m - m'$ elementwise by $(m - m')[a] \triangleq m[a] - m'[a]$.

---

[3]For example, [16] defines processes of a high-level net as the processes of the unfolded high-level net.

[4]Note that no multiset m has a greater cardinality than countable infinity ω, because A is countable.

Algebras A pair $\mathcal{A} = (A, O)$ is an *algebra*, if A is some set and O is a finite set of mappings $f : A^n \to A$ for some $n \in \mathbb{N}$. The set A is called *base set* of the algebra; a mapping $f \in O$ is called *operation* of the algebra $\mathcal{A}$. In particular, an operation $f : A^0 \to A$ is called a *constant*. An algebra $\mathcal{A} = (A, O)$ is *countable*, if its base set A is countable.

A countable set X (disjoint from O), where each $x \in X$ is associated with some set $A_x \subseteq A$, is called *set of variables* for $\mathcal{A}$. For $x \in X$ the set A_x is called the *domain* of x. For a set of variables X a mapping $\beta : X \to A$ is an *assignment for X in $\mathcal{A}$*, if $\beta(x) \in A_x$ for each $x \in X$. The set of all assignments for X in $\mathcal{A}$ is denoted by $ASS(X, \mathcal{A})$.

By $T_{\mathcal{A}}(X)$ we denote the set of *terms* over algebra $\mathcal{A}$ and variables X, which is inductively defined by:

1. $X \subseteq T_{\mathcal{A}}(X)$

2. If $f : A^n \to A$ is an operation of $\mathcal{A}$ and $u_1, \ldots, u_n \in T_{\mathcal{A}}(X)$, then $f(u_1, \ldots, u_n) \in T_{\mathcal{A}}(X)$

Note, that we distinguish between an operation f and the corresponding *operation symbol* f by using a different type face in the definition of terms. But, we relax this distinction, where clear from the context.

An assignment $\beta : X \to A$ can be extended to the set of all terms $\overline{\beta} : T_{\mathcal{A}}(X) \to A$ inductively over the structure of terms:

1. $\overline{\beta}(x) \triangleq \beta(x)$ for $x \in X$

2. $\overline{\beta}(f(u_1, \ldots, u_n)) \triangleq f(\overline{\beta}(u_1), \ldots, \overline{\beta}(u_n))$

Moreover, we extend an assignment β to multisets of terms $\tilde{\beta} : MS(T_{\mathcal{A}}(X)) \to MS(A)$ in a straight-forward way:

$$\tilde{\beta}(m)[a] = \sum_{u \in T_{\mathcal{A}}(X) : \overline{\beta}(u) = a} m[u]$$

for $m \in MS(T_{\mathcal{A}}(X))$ and $a \in A$.

Petri nets As usual, a net $N \triangleq (S, T; F)$ consists of two disjoint sets S and T, where the elements of S are called *places* and the elements of T are called *transitions*. The places and transition are related by the *flow relation* $F \subseteq (S \times T) \cup (T \times S)$. For an *element* $x \in S \cup T$ of net N we define the *preset* ${}^\bullet x$ and the *postset* $x^\bullet$ of x by

$${}^\bullet x \triangleq \{y \in S \cup T \mid (y, x) \in F\} \qquad x^\bullet \triangleq \{y \in S \cup T \mid (x, y) \in F\}$$

In order to exclude some pathological cases, we consider only nets for which the preset ${}^\bullet t$ and postset $t^\bullet$ of each transition $t \in T$ are not empty. These are often called *T-restricted* nets.

High-level nets A high-level net is a finite net $N = (S, T; F)$ equipped with some countable algebra $\mathcal{A} = (A, O)$ and a set of variables X. Each arc is inscribed by a finite nonempty multiset of terms over $\mathcal{A}$ and X. A *marking* is a mapping $M : S \to MS(A)$. The set of markings is denoted by $\mathcal{M}(S, A)$. Again, the relation $\geq$ and the operations $+$ and $-$ can be extended to markings place-wise. For $M, M' \in \mathcal{M}(S, A)$ we define: $M \geq M'$ iff for each $s \in S$ holds $M(s) \geq M'(s)$; $(M + M')(s) \triangleq M(s) + M'(s)$. For $M \geq M'$ we define $M - M'$ by $(M - M')(s) \triangleq M(s) - M'(s)$.

Definition 1 (High-level net)

Let $N = (S, T; F)$ be a T-restricted net, $\mathcal{A} = (A, O)$ a countable algebra and X a variable set for $\mathcal{A}$. If W is a mapping $W : F \to MS^+_{fin}(T_{\mathcal{A}}(X))$ and $M_0 \in \mathcal{M}(S, A)$ is a marking, then $\Sigma = (N, \mathcal{A}, X, W, M_0)$ is a *high-level net*.

The mapping W is called *arc inscription* and M_0 is called the *initial marking* of Σ. In order to get a simple definition of processes of high-level nets we introduce some conventions: for $f \notin F$ we define $W(f) \triangleq O$. Then, we define for each transition $t \in T$ and each assignment $\beta : X \to A$ the markings t^-_β and t^+_β by $t^-_\beta(s) \triangleq \tilde{\beta}(W(s, t))$ and $t^+_\beta(s) \triangleq \tilde{\beta}(W(t, s))$ for each $s \in S$.

Now, we define the firing rule for high-level nets. A transition t is *enabled in mode* β *at marking* M, if $M \geq t^-_\beta$. Then, t can *fire in mode* β resulting in *follower marking* $M' \triangleq (M - t^-_\beta) + t^+_\beta$ and we write $M \longrightarrow M'$. The transitive closure $\overset{*}{\longrightarrow}$ of the relation $\longrightarrow$ is called the *reachability relation* of Σ and we say M' is *reachable* from M, if $M \overset{*}{\longrightarrow} M'$.

Processes of high-level nets A process of a high-level net will be defined as an inscribed occurrence net. A net $K = (B, E; <)$ is an *occurrence net* iff

1. for each $b \in B$ holds $|{}^\bullet b| \leq 1$ and $|b^\bullet| \leq 1$.

2. $<$ contains no cycles; thus, the transitive and reflexive closure of $<$ is a (partial) order on $B \cup E$, which we denote by $\leq$.

3. for each $b \in B$ the set $\{b' \in B \mid b' \leq b\}$ is finite.

For readability reasons, we call the places of an occurrence net $K = (B, E; <)$ *conditions* and the transitions *events*. The *initial state of K* is defined by ${}^\circ K \triangleq \{b \in B \mid {}^\bullet b = \emptyset\}$.

For an occurrence net $K = (B, E; <)$ and two sets $Q, Q' \subseteq B$ we define the relation $\longrightarrow$ by: $Q \longrightarrow Q'$ iff there exists an event $e \in E$ such that $Q \supseteq {}^\bullet e$ and $Q \cap e^\bullet = \emptyset$ and $Q' = (Q \setminus {}^\bullet e) \cup e^\bullet$. Again, we call the transitive closure $\overset{*}{\longrightarrow}$ of $\longrightarrow$ the *reachability relation* of K and for $Q \overset{*}{\longrightarrow} Q'$ we say Q' is *reachable* from Q. A set $Q \subseteq B$ is called a *state* of K, if Q is reachable from the initial state ${}^\circ K$.

An event of an occurrence net is supposed to represent some occurrence of a transition of a system in a particular mode. To formalize this, we use an inscription $\rho_E : E \to T \times ASS(X, \mathcal{A})$. A condition of an occurrence net corresponds to an element of the algebra at a particular place. Again, this can be represented by an inscription $\rho_B : B \to S \times A$. Then, a subset $Q \subseteq B$ corresponds to a marking of Σ; in particular, each state of K corresponds to a marking of Σ. This correspondence is formalized by a mapping $\hat{\rho} : 2^B \to \mathcal{M}(S, A)$ which is defined by

$$\hat{\rho}(Q)(s)[a] \triangleq |\{b \in Q \mid \rho_B(b) = (s, a)\}|$$

The union of the two mappings ρ_E and ρ_B is denoted by $\rho : B \cup E \to (S \times A) \cup (T \times ASS(X, \mathcal{A}))$ and defined by $\rho(b) \triangleq \rho_B(b)$ for $b \in B$ and $\rho(e) \triangleq \rho_E(e)$ for $e \in E$.

Definition 2 (Process of a high-level net)

Let Σ be a high-level net as in Def. 1, let $K = (B, E; \prec)$ be an occurrence net, and let $\rho_E : E \rightarrow T \times ASS(X, \mathcal{A})$ and $\rho_B : B \rightarrow S \times A$ be mappings such that

1. $\widehat{\rho}(^\circ K) = M_0$ and

2. for each $e \in E$ with $\rho_E(e) = (t, \beta)$ holds $\widehat{\rho}(^\bullet e) = t_\beta^-$ and $\widehat{\rho}(e^\bullet) = t_\beta^+$.

Then, (K, ρ) is a *process of* Σ, where ρ is the union of ρ_B and ρ_E.

In a graphical representation of a process we omit the inscription of events. An inscription (s, a) of a condition will be written $s.a$.

Up to now, we have defined high-level nets and processes as their partial order semantics. Though we use a slightly different notation, the definition of high-level nets is not new (cf. [15, 9]). Since high-level nets can be easily unfolded to a P/T-net, the notion of processes for P/T-Systems [6, 2] carries over to high-level nets [16]. Here, we have introduced an explicit representation of processes for high-level nets.

3.2 Arc-typed nets and their runs

In this section we will equip high-level nets with different arc-types and a labelling for transitions. The different arc-types are formalized as a family of arcs $(F_j)_{j \in J}$ for some finite index set J. The transition labels extract the relevant information about actions. For example, for a variable o the label W.o of a transition denotes a write action; the object affected by the write action is depended on the mode β in which transition t occurs. Formally, the *labelling* is a mapping $l : T \rightarrow L \times T_\mathcal{A}(X)$, where L is some set of labels.

Definition 3 (Arc-typed net)

Let $\Sigma = (N, \mathcal{A}, X, W, M_0)$ be a high-level net, l be a mapping $l : T \rightarrow L \times T_\mathcal{A}(X)$ for some set L, and $(F_j)_{j \in J}$ a finite family such that for each $j \in J$ holds $F_j \subseteq F$. Then $(\Sigma, (F_j)_{j \in J}, l)$ is an *arc-typed net*.

In Sect. 2 we have already seen a graphical representation of an arc-typed net (Fig. 4). The algebra of this net is $\mathcal{A} = (D \cup \{\bullet\}, O)$, where $D = \{o_1, \ldots, o_n\}$ is a finite set of objects and O contains a constant for each element of $D \cup \{\bullet\}$. Remember that D is an abbreviation for the multiset of terms $o_1 + \ldots + o_n$. In the graphical representation of the net some arcs are not inscribed; we assume that these arcs are inscribed by $\bullet$. The different arc-types are indicated by different graphical representations of arcs. A label $l(t) = (A, u)$ is represented as A.u. Note that we omitted labels for those transitions, which are not relevant because these labels do not occur in the specification. In the labels Snd and Rcv we omitted the second component because it is irrelevant.

A run of an arc-typed net $(\Sigma, (F_j)_{j \in J}, l)$ is a process of Σ, which is equipped with a family of arcs and a labelling of the transitions over $L \times A$ such that the arc-types and the labelling of the transitions correspond to the arc-types and the transitions of the system.

Definition 4 (Run of an arc-typed net)

Let $\Lambda = (\Sigma, (F_j)_{j \in J}, l)$ be an arc-typed net and (K, ρ) be a process of Σ. Moreover, let $(\prec_j)_{j \in J}$ be a family of relations such that $\prec_j \subseteq \prec$ for each $j \in J$ and l' be a mapping $l' : E \to L \times A$ such that

1. for each $e \in E$ with $\rho_E(e) = (t, \beta)$ and $l(t) = (c, u)$ holds $l'(e) = (c, \overline{\beta}(u))$ and

2. for each $b \in B$, $e \in E$ with $\rho_B(b) = (s, a)$ and $\rho_E(e) = (t, \beta)$ holds for each $j \in J$:

   (a) $b \prec_j e$ iff sF_jt, and

   (b) $e \prec_j b$ iff tF_js.

Then, we call $(K, \rho, (\prec_j)_{j \in J}, l')$ a *run of* Λ.

Figure 5 shows a run of the arc-typed net of Fig. 4.

4 Verification

In this section we will present verification techniques for proving that an arc-typed net meets a causality based specification. First, we formalize some rules which allow to derive properties of a run of a system from the structure of the system. Then, we apply these rules to prove that the model of Fig. 4 meets the specification of weak coherence.

4.1 Verification Rules

Causality based specifications often require the existence of a particular causality between two events of a given run. The verification rules use simple properties of an arc-typed net in order to deduce such causalities. The rules are based on two observations; the first relates reachable markings of a high-level net with states[5] of its processes.

Proposition 5

Let Σ be a high-level net. A marking M of Σ is reachable from M_0, iff there exists a process (K, ρ) of Σ with a state Q of K such that $\hat{\rho}(Q) = M$.

The second observation characterizes pairs of *unordered conditions* of an occurrence net (i.e. two conditions b and b' such that neither $b \leq b'$ nor $b' \leq b$ holds) in terms of the states of the occurrence net.

Proposition 6

Let $K = (B, E; \prec)$ be an occurrence net and $b, b' \in B$ two different conditions of K. Conditions b and b' are unordered in K, iff there exists a state Q of K such that $b, b' \in Q$.

---

[5]Remember, that a state of an occurrence net K is a set $Q \subseteq B$ which is reachable from the initial state $°K$.

Construction of causalities Now, we will combine the above observations to a verification rule. For motivating the rule, let us consider the arc-typed net from Fig. 4 again. For each reachable marking M of this model holds $M(inval1)[o1] + M(copy1)[o1] + M(orig1)[o1] \leq 1$, which can be easily proven by an S-invariant[6]. Now, consider the piece of a run which is shown in Fig 6. Note, that we represent those conditions in the pre- and postsets of the events, only, which are relevant for a particular purpose. By Proposition 5 and 6 we conclude $b_4 \leq b_5$ or $b_6 \leq b_1$ as follows:

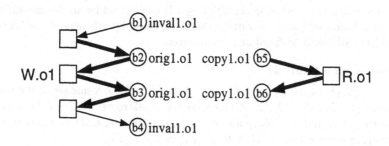

Figure 6: Motivation of Rule 1

otherwise, b_1 and b_5 (and the other conditions) would be unordered in this run; by Prop. 6 this implies that there exists a state of the occurrence net which contains both, b_1 and b_5; then, Prop. 5 implies that there is a reachable marking M of the system with $M(inval1)[o1] + M(orig1)[o1] \geq 2$ — a contradiction to the above property of reachable markings of Λ.

To formalize this argumentation, we introduce the notions of a *safe set* and a *chain*.

Definition 7 (Safe sets and I-chains)

Let Σ be a high-level net and $I \subseteq S \times A$.

1. The set I is a *safe set of* Σ, if for each reachable marking M of Σ holds:

$$\sum_{(s,a)\in I} M(s)[a] \leq 1$$

2. A nonempty sequence $b_1 b_2 \ldots b_{n+1}$ is an I-*chain* of a process (K, ρ) of Σ with $K = (B, E, <)$, if $b_1, \ldots, b_{n+1} \in B$ and for each $i \in \{1, \ldots, n\}$ holds $b_i^\bullet = {}^\bullet b_{i+1}$ and for each $i \in \{1, \ldots, n+1\}$ holds $\rho(b_i) \in I$.

   Two I-chains $b_1 \ldots b_{n+1}$ and $b'_1 \ldots b'_{m+1}$ are *separate*, if $b_{n+1} = b'_1$, or $b'_{m+1} = b_1$, or $\{b_1, \ldots, b_{n+1}\} \cap \{b'_1, \ldots, b'_{m+1}\} = \emptyset$.

With these concepts we can easily formalize the above idea.

---

[6]For verification techniques for S-invariants (place invariants) of high-level nets we refer to [15, 10].

Proposition 8 (Rule 1)

Let $\Lambda = (\Sigma, (F_j)_{j \in J}, l)$ be an arc-typed net and I be a safe set of Σ. Moreover, let $(K, \rho, (<_j)_{j \in J}, l)$ be a run of Λ and $b_1 \ldots b_{n+1}$ and $b'_1 \ldots b'_{m+1}$ be two separate I-chains of the underlying process (K, ρ). Then we have $b_{n+1} \leq b'_1$ or $b'_{m+1} \leq b_1$.

Sometimes, we do have to construct a specific causality in a given piece of a run. In our example, we want to show that between two conditions b and b' of a run which are inscribed by $orig1.o1$ and $orig2.o1$ there is a data causality. This can be easily proven since $M(orig1)[o1] + M(orig2)[o1] \leq 1$ is guaranteed by an S-invariant which only covers bold-faced arcs. Formally, this is captured by safe sets of the high-level net in which all non bold-faced arcs are omitted.

Proposition 9 (Rule 2)

Let $\Lambda = (\Sigma, (F_j)_{j \in J}, l)$ be an arc-typed net and $I \subseteq S \times A$ a safe set of the high-level net $((S, T; F_k), \mathcal{A}, X, W, M_0)$ for some $k \in J$. Moreover, let $(K, \rho, (<_j)_{j \in J}, l)$ be a run of Λ and $b_1 \ldots b_{n+1}$ and $b'_1 \ldots b'_{m+1}$ be two separate I-chains of the underlying process (K, ρ). Then $b_{n+1} \leq_k b'_1$ or $b'_{m+1} \leq_k b_1$.

An application of this rule will be shown in Sect. 4.2.

Construction of events Now, we show how we can construct some events for a given piece of a run. First, we consider an example, again. Figure 8 shows a piece of a run of the arc-typed net in Fig. 4. Condition b_1 is inscribed by $copy1.o1$; since $M_0(copy1)[o1] = 0$, we know that b_1 is not in the initial state of the occurrence net. Therefore, there must be an event producing b_1; this must be one of the two events shown in Fig. 9.

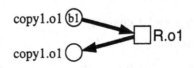

$copy1.o1$ (b1)
$copy1.o1$ ◯ R.o1

Figure 7: A piece of a run

Proposition 10 (Rule 3)

Let $\Lambda = (\Sigma, (F_j)_{j \in J}, l)$ be an arc-typed net and $(K, \rho, (<_j)_{j \in J}, l')$ a run of Λ. If $b \in B$ is a condition of K with $\rho_B(b) = (s, a)$ and $M_0(s)[a] = 0$, then there exists an event $e \in {}^{\bullet}b$ such that $\rho_E(e) = (t, \beta)$ and $t_\beta^+(s)[a] \geq 1$.

Though these rules are very simple, they are the gist of proofs for arc-typed nets, as we will see in the following example.

4.2 Verification of the coherence protocol

Now, we verify that the runs of the arc-typed net from Fig. 4 are weakly coherent by applying the above rules. We have to verify the two requirements given in Sect. 2.1.

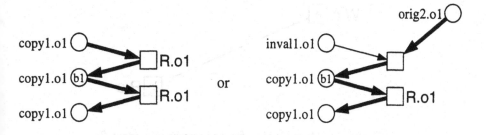

Figure 8: Two extensions of the run of Fig. 8

1. Requirement: First, we show that each two write events on the same object o are totally ordered by data causality. To this end, we consider the system with bold-faced arcs, only, and call it Σ_D. For each $o \in D$ the set $I = \{(orig1, o), (orig2, o)\}$ is a safe set of Σ_D (which can be easily proven by an S-Invariant of Σ_D). Now, consider two write events as shown in Fig. 10: $b_1 b_2$ and $b_3 b_4$ are two separate I-chains of the underlying process. By Rule 2, we know $b_2 \leq_D b_3$ or $b_4 \leq_D b_1$. Thus, we know that there is a data causality between these two write events. The above situation concerns write events of two different agents, only; the proof for two write events of the same agent is very similar.

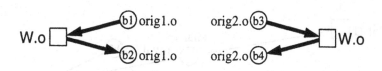

Figure 9: Two write events of a run

2. Requirement: The proof of the second requirement is more involved. We consider a write event e and a read event e' on the same object with $e \leq_P e'$ as indicated in Fig. 11. We have to show, that $e \leq_D e'$. Let us again assume that the write event is on agent a_1 and the read event is on agent a_2 (the other cases are similar, again). Then, by the structure of the system there must be a send and a receive event between e and e', since this is the only way to establish program causality between events of different agents[7]. Moreover, e is on the original and e' is on a copy. This situation is shown in Fig. 12. From the structure of the system we know, that the receive event may only happen, when object o is invalid at agent a_2.

Since b_3 is inscribed by $copy2.o$, we know by Rule 3 that there must be an event in the preset of b_3; this can be either a read event or the event which makes a copy from the original; since the read event again has a condition in its preset which is inscribed by $copy2.o$ there is a chain of events which eventually will be ended by a copy event e_1 as shown in Fig. 13.

---

[7]This step is not explicitly formalized by a verification rule.

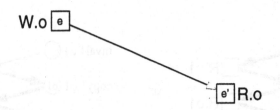

Figure 10: Assumption for proving requirement 2

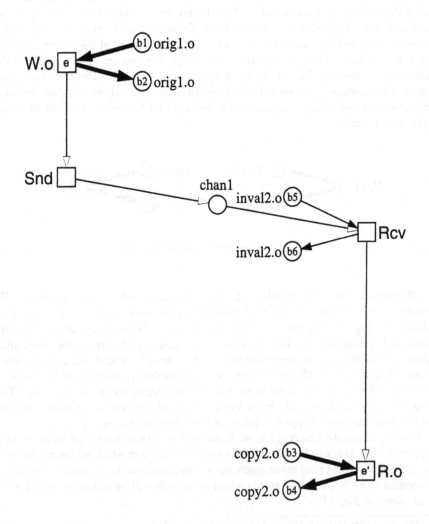

Figure 11: Program causality between events of different agents

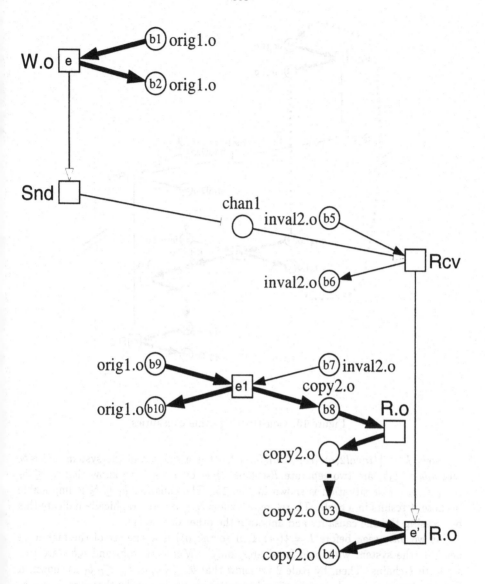

Figure 12: Constructing a sequence of events

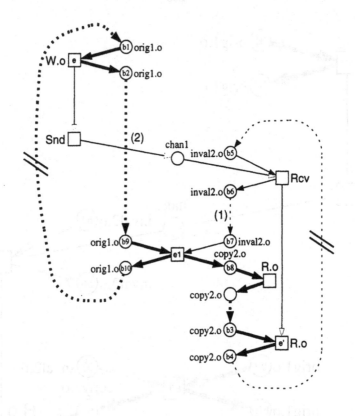

Figure 13: Constructing some causalities

Now, $I' = \{(inval2, o), (copy2, o), (orig2, o)\}$ is a safe set of the system and $b_5 b_6$ and $b_7 b_8 \ldots b_3 b_4$ are two separate I'-chains; thus by Rule 1 we know that $b_6 \leq b_7$ or $b_4 \leq b_5$. This situation is shown in Fig. 14. The causality $b_4 \leq b_5$ is impossible because it results in a cycle. Therefore, we know $b_6 \leq b_7$: we graphically indicate this by crossing the one causality and annotate the other one by (1).

As we have seen before $I = \{(orig1, o), (orig2, o)\}$ is a safe set of the high-level net Σ_D (the system with bold-faced arcs, only). Moreover, $b_1 b_2$ and $b_9 b_{10}$ are two separate I-chains. Then, by Rule 2 we know that $b_2 \leq_D b_9$ or $b_{10} \leq_D b_1$ as shown in Fig. 14. Again, $b_{10} \leq_D b_1$ is impossible because it results in a cycle; therefore, $b_2 \leq_D b_9$ (indicated by (2)) holds true. Now, we have constructed a path of bold-faced arcs from e to e'. Thus, we have $e \leq_D e'$, which finishes the proof.

5 Conclusion

We have formalized arc-typed nets and their runs. Moreover, we have introduced rules for proving that an arc-typed net meets a causality based specification. Thus, we have provided a formal foundation for an idea which was applied in an informal way before [4, 3, 5, 12, 13]. For these applications arc-typed nets have provided an adequate level of abstraction, such that formal verification is still possible. The simplified example used in this paper catches only a glimpse of these applications.

Still, the idea of causality based specifications and arc-typed nets is not yet a method. The development of a method is the goal of our ongoing research. The presented formalization of arc-typed nets is a sound foundation for this research.

References

[1] J. K. Bennet, J.B. Carter, and W. Zwaenepoel. Munin: Distributed shared memory based on type-specific memory coherence. In *2nd ACM SIGPLAN Symposium on Principles and Practise of Parallel Programming*. ACM, March 1990.

[2] Eike Best and César Fernández. *Nonsequential Processes, EATCS Monographs on Theoretical Computer Science* 13. Springer-Verlag, 1988.

[3] Dominik Gomm and Ekkart Kindler. Causality based specification and correctness proof of a virtually shared memory scheme. SFB-Bericht 342/6/91 B, Technische Universität München, August 1991.

[4] Dominik Gomm and Ekkart Kindler. A weakly coherent virtually shared memory scheme: Formal specification and analysis. SFB-Bericht 342/5/91 B, Technische Universität München, August 1991.

[5] Dominik Gomm and Ekkart Kindler. Causality based proof of a distributed shared memory system. In A. Bode and M. Dal Cin, editors, *Parallel Computer Architectures: Theory, Hardware, Software, Applications, LNCS* 732, pages 131–149. Springer-Verlag, 1993.

[6] Ursula Goltz and Wolfgang Reisig. The non-sequential behaviour of Petri nets. *Information and Control*, 57:125–147, 1983.

[7] Jeremy Gunawardena. Geometric logic, causality and event structures. In J. Baeten and J. Groote, editors, *CONCUR '91, LNCS* 527, pages 266–280. Springer-Verlag, 1991.

[8] Jeremy Gunawardena. Causal automata. *Theoretical Computer Science*, 101:265–288, 1992.

[9] Kurt Jensen. *Coloured Petri Nets, Volume 1: Basic Concepts*. EATCS Monographs on Theoretical Computer Science. Springer-Verlag, 1992.

[10] Kurt Jensen. *Coloured Petri Nets, Volume 2: Analysis Methods*. EATCS Monographs on Theoretical Computer Science. Springer-Verlag, 1995.

[11] Joost-Pieter Katoen. Causal behaviours and nets. In Giorgio De Michelis and Michel Diaz, editors, *Application and Theory of Petri Nets 1995, LNCS* 935, pages 258–277. Springer-Verlag, June 1995.

[12] Ekkart Kindler, Andreas Listl, and Rolf Walter. Kausalitätsbasierte Beweismethoden für parallele Datenbanksysteme: Eine Fallstudie. Informatik-Berichte 30, Humboldt-Universität zu Berlin, Institut für Informatik, June 1994.

[13] Ekkart Kindler, Andreas Listl, and Rolf Walter. A specification method for transaction models with data replication. Informatik-Berichte 56, Humboldt-Universität zu Berlin, March 1996.

[14] Leslie Lamport. Time, clocks, and the ordering of events in a distributed system. *Communications of the ACM*, 21(7):558–565, July 1978.

[15] Wolfgang Reisig. Petri nets and algebraic specifications. *Theoretical Computer Science*, 80:1–34, May 1991.

[16] Einar Smith and Wolfgang Reisig. The semantics of a net is a net, an exercise in general net theory. In K. Voss, H.J. Genrich, and G. Rozenberg, editors, *Concurrency and Nets*. Springer-Verlag, 1987.

The SEA Language for System Engineering and Animation

B. Kleinjohann, E. Kleinjohann and J. Tacken

C-LAB,
Joint R&D Institute of Universität-GH-Paderborn
and Siemens Nixdorf Informationssysteme AG,
D-33094 Paderborn
e-mail: {bernd, lisa, theo}@cadlab.de

Abstract. This paper describes the hierarchical, graphical SEA Language. The SEA Language relies on extended Predicate Transition Nets that unambiguously define the semantics of graphical system specifications. These nets are also used as a basis for simulating/animating system specifications. The SEA Language allows to easily incorporate several user defined or standardized graphical symbols and their behaviour. Via this approach a unified semantic basis for different graphical specification formalisms/techniques can be provided. These techniques may be application independent like data flow graphs, statecharts or block diagrams for differential equations or application dependent like logical gates for hardware systems. Hence, engineers can specify system (parts) using the graphical symbols they are used to. Due to the underlying common semantics defined by extended Predicate Transition Nets, a heterogeneous system can be specified and simulated as a whole at very early stages of design before some system components are already constructed. This approach and its benefits will be explained using the well known elevator system as an example of a heterogeneous system.

1 Introduction

During the last years interest in graphical languages increased steadily. However, the claim "A picture is worth a thousand words" is no longer accepted without contradiction. This is mainly due to the fact that pictures can easily be misinterpreted without any comment on their semantics. Nevertheless, simple pictures like pictograms are an appropriate means e.g. to create an intuitive user interface of a complex system or to show its status in an intuitive, abstract fashion, that can also be understood by non experts. An example is the instrument panel of a car that builds an intuitive user interface for this complex mechatronic system and also shows its actual state in an abstract, yet very intuitive way e.g. by the speedometer or the temperature display. Such pictograms are an important feature for system specification and well suited to illustrate the system structure and an actual state. But if the information to be transported by a picture is more complex, an unambiguous semantics is very evident. Usually this is the case, if the picture should give information about the behaviour of the depicted elements and not only about their user interface or actual status.

Graphical languages that describe the behaviour of systems in specific application domains or in an application independent way are already existing for several years. In hardware engineering e.g. schematic diagrams or logical gates (like nands, nors etc.) are used to describe the structure respectively the logical behaviour of a circuit. Software is often visualized by SA [Ross77] or at lower level of detail by Nassi-Shneiderman Charts [YSch78]. Differential equations that are used to specify analogue behavior are visualized by block diagrams. As examples for application independent graphical languages Pictorial Janus [KaSa90], State Charts [Hare87], Data Flow Graphs [Denn85], or Petri Nets [Pete81] may serve. In contrast to pictograms that allow freely defined graphics, these graphical languages rely on a set of predefined symbols from which complex structures can be constructed with a more or less formally defined behavioural semantics. Currently, none of these graphical languages is accepted as *the* language for specification of complex heterogeneous systems. Reasons for this may be the restricted applicability of application specific languages and the lack of intuitive appeal in the more general languages.

Hence we have a situation where many isolated graphical languages are used according to personal preferences or applicability for specific domains. And none of these languages can cope with the intuitiveness of pictograms. The question that now arises is, how this situation can be improved, in such a way that intuitiveness, unambiguous semantics and the user preferences are regarded. One way would be the definition of a "universal intuitive graphical language". But this goal certainly cannot be reached. Another more promising way is to overcome the "isolation" of the different specification "formalisms". This can either be done by providing "adapters" between the different "formalisms", which will be a never ending task. Alternatively, a unified semantical basis for the graphical "formalisms" and a method for integrating their predefined resp. freely defined graphical symbols including their behavioural semantics into one specification can be provided.

The latter approach is followed by the SEA Language presented in this paper. The SEA Language allows to hierarchically specify the structure and the behaviour of a heterogeneous system in an application oriented way. Freely defined graphical elements as well as predefined elements e.g. lend from existing graphical languages can be used. As a unified formal semantical basis extended Predicate Transition Nets (Pr/T Nets) are used. They build the bottom level of hierarchy in each specification. Via this unified formal semantics the integration of specification elements from different application domains, as it is needed for the engineering of heterogeneous systems, can be reached. Furthermore, the underlying Pr/T Net allows a simulation / animation of the graphical system specification at arbitrary levels of hierarchy. Even the freely defined graphical elements like pictograms can be animated.

Only the STATEMATE tool [State92] is known to us that has done a first step in combing a formal model (in this case statecharts) with arbitrary, freely defined graphics. STATEMATE provides a mock-up panel representing the user interface and the actual status of the final system but at the top most hierarchical level.

Yet, the SEA Language provides construction principles that allow to construct more complex specifications from a set of these freely defined graphics including the underlying behaviour at arbitrary level of hierarchy.

The remainder of the paper is structured as follows. In Section 2 the SEA Language is described in more detail. First some of our extensions to Pr/T Nets and afterwards the graphical aspects of the SEA Language are explained. The modeling and simulation of heterogeneous system using the SEA Language is described in Section 3 by the well known example of an elevator system. After a short overview over the realization of the SEA Environment in Section 4 some conclusions are presented in Section 5.

2 Overview of the SEA Language

The SEA Language relies on extended Pr/T Nets to describe the semantics of a specification. In this Section we describe our extensions to the Pr/T Nets introduced by Genrich and Lautenbach [GeLa81]. Firstly we have extended the formal model to support an easier specification. Secondly we defined several extensions to achieve a comfortable embedding of extended Pr/T Nets in the "external world" and allow a user dependent graphical representation.

2.1 Extensions to the formal model

Time dependent firing of transitions In extended Pr/T Nets an *enabling delay* and a *firing delay* [Star90] can be defined for a transition by a triple with time values (*min, typ, max*). For the enabling delay this triple determines the minimum (min) time delay before a transition may become active after it has been enabled for any substitution, the maximum (max) time delay after which a transition has to become active, and a typical (typ) time delay which is useful for analysis. The time is not bound to a special substitution, so the substitution for which the transition is enabled may change during the delay period. In a similar way the firing delay specifies how long a transition in an extended Pr/T Net is active. If a transition is active, the tokens from the input places are removed but the tokens for the output places are not yet produced. For the firing delay time values with $0 \leq min \leq typ \leq max < \infty$ are valid and for the enabling delay also $max \leq \infty$ is allowed. Both time concepts (enabling delay and firing delay) are necessary for the specification of system parts which work continuously, like parts that are specified by differential equations (cf. Section 3.2).

Hierarchy Extended Pr/T Nets allow a hierarchical specification. Hierarchical specifications are useful to handle complexity in large designs and allow the reuse of predefined nets in several models. This is a necessity for the definition of libraries with subnets for special purposes. Furthermore, a hierarchical specification supports both a top down design as well as a bottom up design during the system specification.

In extended Pr/T Nets transitions and places can be refined by subnets. Such nodes are called structured nodes. The subnet of a structured node is itself an extended Pr/T Net which may also contain structured nodes. The same subnet can be used to refine several structured nodes. In this case the structure of the subnet is not copied, instead structured nodes only contain a reference to their subnet. This mechanism is called instantiation. Only the dynamic information, e.g. markings generated during simulation of the net, is copied. A subnet of a structured node may also have connections to nodes in the instantiating net. The places and transitions which are connected to nodes in the instantiating net are called port-places or port-transitions. The structured nodes are not replaced by their subnets. They have a special semantics which is defined via the activity of their subnets. Only if a subnet is active its transitions may fire:

Structured places: The subnet of a structured place is active as long as the structured place contains at least one token. If the last token is removed, the subnet becomes inactive, which means that no transition in the subnet can start a firing cycle. All transitions which are active in this state can finish their firing cycle.

Structured transitions: The subnet of a structured transition is active as long as the structured transition is in its active phase. If the structured transition enters its marking phase the subnet becomes inactive and only the active transitions of the subnet can finish there firing cycle.

For the semantic behaviour of structured transitions it must be specified when an active structured transition becomes inactive again. For structured transitions with no firing delay we use the philosophy of structured nets as described in [ChKo81]. Otherwise the firing delay determines the active phase.

The concept for structured places is similar to the hierarchical concept in statecharts ([Hare87]). A structured hierarchy without semantics can be specified using initially marked structured places with no connections to other nodes.

Recursive Nets can be specified due to the instantiation mechanism. A net can be used as subnet of a structured node which is defined in the net itself. This is possible because the structured node only contains a reference to the subnet and the dynamic information is copied only if necessary (a structured place is marked or a structured transition becomes active).

2.2 Extensions for the integration of textual and graphical representations

Transitions in an extended Pr/T Net are annotated with a condition, a preaction and a postaction. The condition is a Boolean expression which is evaluated to determine if a transition is enabled. The preactions are executed after the demarking of the input places and the postactions are executed right before the marking of the output places. The pre- and postactions of transitions in Pr/T Nets can be used to compute output values (values in output tokens) from input values (values in input tokens). These computations may be simple mathematical operations but in the case of modeling complex heterogeneous systems they may be the result of complex functions or differential equations. Therefore it is useful that existing modeling techniques for different domains can be easily integrated in extended Pr/T Nets. Graphical modeling techniques can be incorporated by defining corresponding libraries. Examples for block diagrams and data flow graphs can be found in Sections 3.2 and 3.3. In order to integrate textual languages the annotations in extended Pr/T Nets should support complex computations.

To support complex computations during the firing of transitions it is possible to allow the annotation of programming languages (C, C++, ML, Pascal, ...) or modeling languages (DSL [Schr91], ...). In extended Pr/T Nets in addition to the standard mathematical operations the definition of C or C++ – code and the invocation of shell commands and tools is allowed. For the analysis of results from shell commands and tools the standard C or C++ file operations can be used.

In extended Pr/T Nets it is possible to change the time values for the enable and firing delay of a transition (cf. Section 2.1) with the annotation of its pre- or postactions. This supports for example the specification of variable time steps for the discretization of continuous system parts. If someone changes the firing delay within the preactions the new delay is valid for the actual firing cycle. If it is changed within the postactions the change is valid for the next firing cycle.

Abstract graphical representation Petri Nets and also Pr/T Nets have a standard graphical representation with places as circles, transitions as bars, and edges as arrows. For an engineer who has to decide whether a model works correctly this graphical representation is not easy to read. Therefore, an abstract, intuitive graphical representation which reflects the structure and state of a defined net would gain more acceptance. For example, if someone wants to specify the functionality of a seven bit integer display with the seven segments a - g as depicted in figure 1 a) he could define a Pr/T Net as shown in figure 1 b). The places A - G are port-places which can be connected to nodes in the surrounding net if this net is used to refine a structured node. If, for example, the place A is marked the transition a fires inverting the state of the segment a (a becomes visible if it was invisible and vice versa).

Figure 1 c) shows a possible abstract graphical representation of the net structure as it can be specified in the SEA Language. This representation is called an interface of an extended Pr/T Net whereas the net structure and the initial marking is called the content description. For every net one or more interfaces may be defined. Such an interface description may contain, in addition to the abstract graphical representation, the information about the allowed connections from instantiating nets to local nodes (ports), which are also represented in the interface (thick lines labelled a - g). Although the net is very simple the graphical abstraction is more understandable for non expert users.

An abstract graphical representation would be pretty useless if it only reflects

312

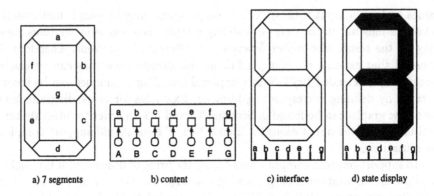

a) 7 segments   b) content   c) interface   d) state display

Fig. 1. An extended Pr/T Net for a seven segment display

the structure of a defined net but not its actual state during the simulation. Therefore, the SEA Language allows to annotate transitions with commands that manipulate the graphical representation. If, for example, the transition *a* in figure 1 b) fires, an annotation like "show segment a in the graphical representation" or "hide segment a in the graphical representation" is needed. In the SEA Language we have implemented a simple mechanism to manipulate the graphical representation. If a net is defined as subnet of a structured node the abstract graphical representation is displayed within the structured node. Such a structured node has a vector of so called "visibility bits" which determines which graphical objects of the interface description are visible. When defining an interface each graphical object also has a vector of visibility bits. Only if the vector of the structured node and the vector of the graphical object have common bits the graphical object is displayed. Within the annotation of a transition the vector of the instantiating structured node can be manipulated. For example the graphical objects which represent the structure of the seven segment display net in figure 1 b) have a visibility bit vector with the default bit set which should be the initial vector for a structured node refined by this net. In addition the interface contains graphical objects for the filled segments *a - g*. So the transition *a* in figure 1 b) has an annotation "visibility ~a" which inverts the bit a in the visibility bit vector of the instantiating structured node. The bit a can be explicitly set with "visibility +a" and unset with "visibility -a". If, for example, the actual state of the net in figure 1 b) reflects the integer value 3 the resulting abstract graphical representation would look as shown in figure 1 d).

If a net *n1* is used as subnet in a net *n2* which is used as subnet in a net *n3* the SEA Language allows to display the interface of *n1* in *n3*. This is called transparent instantiation mechanism. An example for the transparent instantiation mechanism is shown in Figure 6.

Input/Output fields Very often it is also useful to display the content of variables during the firing of transitions in the graphical representation. For this purpose the SEA Language allows the definition of *output fields* in the interface description. Figure 2 a) shows the interface and content of a net modeling a display element which displays the value of a variable x in an output field with name *value*. The annotation "write "value" x" of the transition will display the value of x in the output field *value* within the firing cycle of the transition.

For the communication with the user it should also be possible to accept values entered by the user during simulation. For this purpose *input fields* can be defined in the graphical representation. In the SEA Language an *input field* may have one of two different types, *button* or *text*. Button input fields can be used to accept mouse clicks in special areas within the simulation and text input fields to accept textual inputs. If, for instance, the value of an input field with name *init* should be read during the firing of a transition the annotation could be: "x = read("init")". By the condition of such a transition it can be checked whether an input has been made. "armed("init")" is true in case of a user input in the field "init". In the SEA Language also the combination between input and output fields is allowed. Figure 2 b) shows the interface and content of a subnet which models a delay component which is defined in a library for block diagrams for differential equations (see Section 3.2). For the delay component an initial value can be set with the input field *init* as described above.

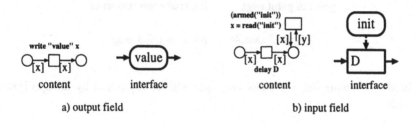

a) output field   b) input field

Fig. 2. Different input/output fields

For button input fields the function "armed" has an optional second parameter which specifies from which mouse button a click is expected ($1 \equiv$ left, $2 \equiv$ middle, $3 \equiv$ right). An example for the use of button input fields is the modeling of a cabin panel for an elevator as shown in Figure 6.

Complex edges During the design of a heterogeneous system with the SEA Language subnets with interfaces for different domains are defined which have to be connected to model the system. To support these connections in the SEA Language complex edges can be used. Complex edges are edges which are refined by special subnets. These edges can be used to model broadcast (*point to point, point to many*) or synchronization (*many to point, many to many*) concepts. Figure 3 shows the four types with their basic subnets.

The basic subnets are automatically extended when using complex edges. For example, someone may use a point to point complex edge to connect two places. The basic subnet for a point to point edge consists of one place. To connect

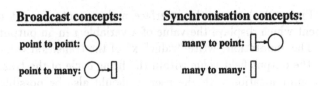

Fig. 3. Complex edges

two places two transitions have to be added. Figure 4 shows at the left side a point to point edge connecting a place P1 with another place P2 and a transition T1. On the right side the internal realization is depicted. The place PE is the basic subnet of the point to point edge and the transitions Tx and Ty are added because of the places P1 and P2.

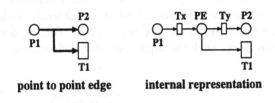

point to point edge   internal representation

Fig. 4. Example for point to point edge

In the following Section these concepts will be explained by an application example.

3 Application Example

As an example we take the well known *elevator system* consisting of a set of *elevators*, a *central control system*, and an *elevator requests* module that gathers requests sent by *floor panels* installed at each floor outside the elevator. In the remainder of this section we will develop a hierarchical SEA Language model of an elevator system in a stepwise manner. We will start with the top level view of the *elevator system*.

3.1 Top level specification of the elevator system

Top level function In Figure 5 the hierarchical decomposition of the *elevator system* is depicted. Intermediate levels of hierarchy, which are not represented by boxes, are indicated by dashed arrows.

The *central control system* consists of a *central panel* displaying the actual floors and moving directions of all elevators and a *central control* unit. The *central control* unit gathers all *elevator requests* that are received via the *floor*

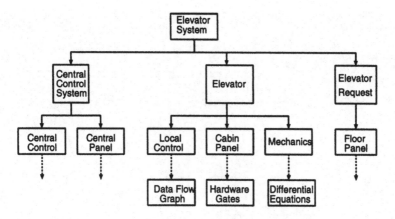

Fig. 5. Elevator System Hierarchy

panels installed at each floor outside the elevators and distributes them to an appropriate *elevator*.

For each *elevator* a *cabin panel* exists where passengers enter their requests and where the actual floor level is displayed. The *cabin panel* is realized by discrete hardware that is specified at the gate level via elements of a logical gate library represented by the *hardware gates* box.

In order to accelerate and brake an elevator in a comfortable manner for the passengers the *mechanics* of the elevator have to be considered. The *mechanics* of the *elevator* show analogue behaviour and are modelled via differential equations represented by the box *differential equations* (see Section 3.2).

In addition to the *central control system* for each *elevator* also a *local control* exists which is responsible for the correct execution of requests received from the *central control*. The *local control* acts mainly as an interface between the analogue part (*mechanics*), and the discrete software and hardware parts (*central control*, *cabin panel*). It is specified via elements of a data flow library represented by the box *data flow graph* (see Subsection 3.3).

Top level SEA model The hierarchy depicted in Figure 5 can be directly mapped into a Pr/T Net hierarchy. Figure 6 shows the user defined graphical representation of the elevator system.

Each of the boxes from Figure 5 (except *data flow graph*, *hardware*, *differential equations*) is modelled via a structured place. The graphical interfaces and interconnections of the subnets instantiated by these structured places are depicted in Figure 6. In this figure the transparent instantiation mechanism is used extensively, i. e. the graphical interfaces of lower levels of hierarchy are visible at higher levels of hierarchy. The ports of the subnets are represented by solid rectangles. They are also "inherited" from lower hierarchical levels to higher ones.

The *floor panels* and *cabin panels* may send *requests* (integers specifying the desired floor level) to the *central control*. The *central control* decides in which

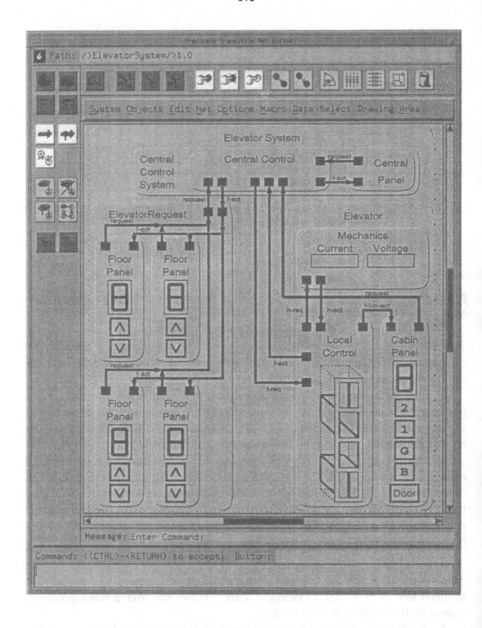

Fig. 6. Top level graphics of elevator system net

order the requests are processed and sends the floor level to be served next *f-req* to the *local control* of the *elevator*. The *requests* can be interactively entered by the user of the SEA environment during simulation/animation of the entire system. This can be done simply by clicking on the *input buttons* in the *floor panels* (input fields ∧, ∨) or the *cabin panel* (input fields *G, B, 1, 2*). In our example for each click a sound is generated, but it would also be possible to highlight the corresponding input field. Several *requests* can be entered before execution of the first one, like in real elevator systems. Via this feature, for instance, the correctness of the *central control* algorithm can be interactively checked, before it is installed.

When the *local control* receives a desired floor level *f-req*, it transforms this integer value into a continuous height *h-req* for the *mechanics*. The *local control* contains an output panel that shows (from top to bottom) a closed door without elevator, an open door with elevator, an open door without elevator, and a closed door with elevator. These output fields are highlighted alternately during simulation depending on the actual height of the elevator *mechanics h-act* and the status of the door controlled by the *local control*. However, the third output field (open door without elevator) should never be highlighted during normal operation of the *elevator*, otherwise the algorithm of the *local control* contains an error.

In addition, the *local control* calculates the integer and binary representation of the actual floor level (*f-act*, *f-bin-act*) from the actual height *h-act* continuously generated by the elevator *mechanics*. The binary representation is needed for display of the floor level by a hardware seven segment display in the *cabin panel* (see Figure 1). The current status of this hardware element is displayed during simulation like depicted in Figure 1 d).

The elevator *mechanics* interface contains two output fields that show the actual status of the motor driving the elevator cabin, i. e. *current* and *voltage*. These values allow the engineer to check during the simulation, whether the *mechanics* work correctly or if e. g. the motor is overcharged.

This top level SEA representation shows the structure of the *elevator system* in an application oriented way. The behaviour is specified at lower hierarchical levels using extended Pr/T Nets. Via the animation of the top level graphical representation during simulation of the underlying extended Pr/T Net an application oriented illustration of the system behaviour is possible before the system is constructed. Also non experts (regarding the internal elevator details and the net formalism), for instance customers who ordered the elevator system are able to understand this graphical animation. Hence, they can check whether the specified system meets their requirements or whether changes are necessary. All critical operations can be checked and visualized through appropriate graphical means. This is especially valuable for the interfaces between different realization technologies (software, mechanics, or hardware), because these interfaces cannot be checked by pure software tests or pure hardware respectively mechanical simulators. And although the top level view is quite abstract, the system and its behaviour are unambiguously specified due to the underlying Pr/T Net semantics. Hence, a sound basis for the real system construction is available, avoiding expensive changes after (partial) construction of the system.

In order to show how extended Pr/T Nets can be used for the specification of heterogeneous, hybrid (mixed analogue, digital) systems we will concentrate on the specification of the mechanics and software parts of the elevator in the next subsections. The modeling of discrete hardware by high level nets is already de-

318

scribed in [GeSh93] and in [Klei94, KlMi93] the Petri Net approach for hardware modeling presented by [Kinn 81] is extended.

3.2 Elevator Mechanics

Function of elevator mechanics The function of mechanical system parts is usually specified via a mathematical model using differential equations. In order to develop a mathematical model for the mechanics of one elevator we simplify it to the relevant parts. We model the elevator mechanics as a mass system that is moved via an electric motor. The electric motor is controlled via a controller that calculates the input voltage for the motor from the difference between the required height h_{req} and the actual height h_{act}. This system is depicted in Figure 7.

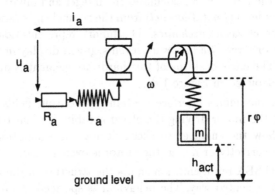

Fig. 7. Elevator mechanics

The electric motor drives via an axle a cylinder that winds or unwinds a cable to which a mass representing the elevator cabin is connected. The motor itself is represented by the resistance R_a of the anchor cycle and the self inductivity of the anchor cycle L_a. Its movement is determined via the input voltage u_a. ω denotes the angular velocity of the motor, r the radius of the cylinder, Θ the moment of inertia for the cylinder (including the axle), m the mass of the elevator cabin, h_{act} the height of the mass in relation to a specific zero level, c_s the spring constant of the unwinded part of the cable. Hence, the elevator (including motor) may be viewed as a transfer system with the voltage u_a as input and the height h_{act} relative to the ground as output.

Following the conservation principle, for this system the balance equations for the voltage is given by $u_a = R_a i_a + L_a \dot{i}_a + u_i$ where $u_i = k_1 \omega$ denotes the voltage part corresponding to the mass, k_1 is a constant. From these equation follows:

$$\dot{i} = -\frac{R_a}{L_a} i_a - \frac{k_1}{L_a} \omega + \frac{u_a}{L_a} \tag{1}$$

The balance equation for the moment of inertia is $M_{mot} = M_{mass} + k_3\omega + \Theta\dot{\omega}$ with $M_{mot} = k_2 i_a$ and $M_{mass} = Fr$ where F is the force to the mass m that calculates to $F = mg + m\ddot{h}_{act} = c_s\Delta h_{act} = c_s(r\varphi - h_{act})$. $\Delta h_{act} = r\varphi - h_{act}$ denotes the resilient part of the cable to which the spring constant c_s applies. From these equations follows:

$$\dot{\omega} = \frac{k_2 i_a}{\Theta} - \frac{c_s r^2 \varphi}{\Theta} + \frac{rc_s h_{act}}{\Theta} - \frac{k_3\omega}{\Theta} \tag{2}$$

With v denoting the velocity of the mass m and g denoting the acceleration of the mass due to gravity the following equations hold:

$$\dot{h}_{act} = v \tag{3}$$
$$\dot{v} = \ddot{h}_{act} = \frac{c_s r\varphi}{m} - \frac{c_s}{m}h_{act} - g \tag{4}$$
$$\dot{\varphi} = \omega \tag{5}$$

From the differential equations 1 to 5 the block diagram of Figure 8 can be constructed setting the constants c_1 to c_9 and I_1 to $I - 5$ as follows:

$c_1 = 1/L_a$ $c_2 = R_a/L_a$ $c_3 = k_2/\Theta$ $c_4 = k_1/L_a$ $c_5 = k_3/\Theta$ $c_6 = r^2 c_s/\Theta$ $c_7 = rc_s/\Theta$
$c_8 = c_s r/m$ $c_9 = c/m$ $\quad I_1 = i_a(0)$ $I_2 = \omega(0)$ $I_3 = \varphi(0)$ $I_4 = v(0)$ $I_5 = h(0)$

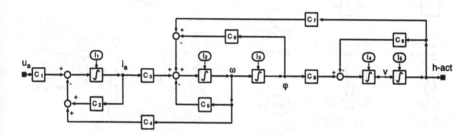

Fig. 8. Block diagram of motor driven elevator

In analogy the block diagram of the controller can be constructed.

SEA library for block diagrams In this subsection we develop a net model for the standard block diagram representation of differential equations. In [BrKl93, BrKl93b, Briel95] is already described how differential equations can be discretized and may be transformed into extended Pr/T Nets using the Z-transform or a state based approach. Since this paper focuses on the graphical aspects of the SEA Language, we restrict ourselves to a straight forward approach using the rectangle backward integration. We also neglect the problems that arise during discretization of differential equations, e.g. for determining appropriate sampling rates. Rather we show how in a systematic manner the application specific predefined graphics of block diagrams like the one in Figure 8 can be also provided using the SEA Language and how the underlying semantics are defined using extended Pr/T Nets.

For each symbol needed in the block diagram a library element is developed. This library element specifies the behaviour of the element in a corresponding subnet and the graphical appearance in the interface description of the subnet.

Library elements can be instantiated in several block diagrams. The mechanical engineer can construct block diagrams using only the graphical interfaces without knowledge on extended Pr/T Nets.

Figure 9 shows the subnets and the graphical interfaces for some elements of the block diagram library (BD-lib). The interfaces look exactly like the graphical symbols used in Figure 8. The interfaces focus on the structural aspect of the system (part) and give only an intuitive idea of the underlying behavioural semantics, that is clearly defined by the underlying extended Pr/T Net. This library is a good example to show how also application specific predefined graphical elements can be incorporated into the SEA Language. The elements of the BD-lib can be instantiated in several block diagrams. The mechanical engineer can construct block diagrams using only the graphical interfaces. Hence, he or she needs no know how on Pr/T Nets.

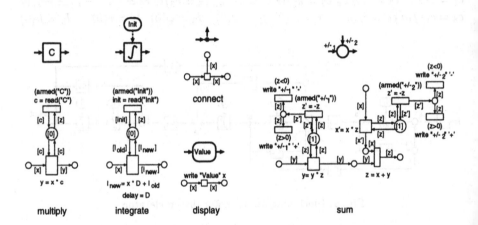

Fig. 9. Library for differential equations

In Figure 9 transition annotations are depicted above or below the transition. The firing delay is specified if needed as *delay* otherwise it is 0. The enable delay is 0.

multiply - calculates the output (y) by multiplying the input (x) with a constant (c). The value of c is entered on demand of the user via an input field that is connected to the upper transition. For this purpose the transition has the enable condition $armed(C)$, that causes the transition to fire each time a user input is entered, and the preaction $read(C)$, that reads the value entered by the user into the net.

integrate - integrates x over the time via rectangle backward integration. The new output value I_{new} is the sum of the old value from the previous time step I_{old} and the rectangle for the actual time period of length D ($x \times D$). The initial value is read from an input field connected to the upper transition.

sum - calculates the output (z) by summing up the inputs (x and y). This element provides the opportunity to correct the signs of the inputs manually if necessary also during simulation. Because sign errors may frequently happen during the elaboration of a differential equation, this feature is very useful. As default the sign + is chosen. Via clicking on the corresponding input button the alternate sign (compared to the actual one) is selected and will be displayed.

connect - distributes one input (x) to two outputs (x).

In all these library elements for each input and output a corresponding place is defined as port. Hence, the elements need to be connected via complex edges, because in Pr/T Nets it is not allowed to directly connect two places (or transitions) .

After corresponding calculations the initial values and the constants can be entered interactively via input fields connected to the transitions annotated with the *read* statement. This can also be done during simulation, if it is necessary.

SEA model for elevator mechanics Using the library elements described above the block diagrams of the net model for the motor driven mass system is specified in the same form as in Figure 8. The *controller* can be specified similarly.

The specification at the next higher level of hierarchy (the subnet of the structured place *mechanics* in Figure 6) is depicted in Figure 10. The structured place *engine* instantiates the subnet corresponding to Figure 8. The output field *current* is "inherited" from this subnet.

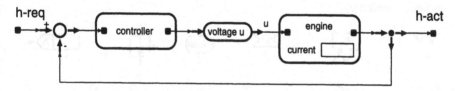

Fig. 10. SEA representation of elevator *mechanics*

This net contains the structured places *controller* and *engine* which are coupled via the *display* element *voltage u* from the block diagram library above. The structured place *engine* instantiates the subnet corresponding to Figure 8. The output field *current* is "inherited" from this subnet. The feedback loop is realized via a *connect* element and a *sum* element. All edges (thin lines) are assumed to be complex point to point edges with annotation [x].

3.3 Local control

Function of local control The task of the *local control* is threefold (see also Figure 6).

- Firstly, the *local control* transforms the movement requests received as the required floor level *f-req* by the *central control* into a continuous value denoting the required height of the mechanical system *h-req*. This calculation is done in a loop. The required floor level *f-req* is decremented by 1 until 0 is reached and for each decrement the *floor height* is added to the required height of the mechanical system *h-req* (see Figure 12).

- Furthermore, it reports the actual floor level calculated from the continuously specified actual height of the mechanical system to the *central control*. This calculation is done in a loop again. The *floor height* is subtracted from the actual height *h-act* until 0 is reached and for each subtraction cycle the integer value denoting the floor level *f-act* is incremented by 1.

- Lastly, the *local control* transforms the actual floor level *f-act* into a binary representation of the floor level *f-bin-act* to be displayed by the *cabin panel*.

SEA library for data flow graphs In order to model the functions described above, data flow graphs [Denn85] were chosen. To provide the engineer with the well known appearance of data flow graphs the following data flow library (DFG-lib) partly depicted in Figure 11 was developed.

The DFG-lib is an example how application independent predefined graphical specification techniques can be integrated into the SEA Language. This approach would also work for other application independent graphical specification 'formalisms' like state charts, that are less or equal powerful than extended Pr/T Nets [Suff90].

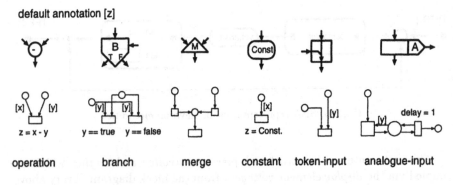

default annotation [z]

operation   branch   merge   constant   token-input   analogue-input

Fig. 11. Data flow library

The DFG-lib assumes a very simple behavioural model of data flow graphs. Only one token may be present on each link (corresponding to 1-safety in nets). Hence, no tag fields have to be considered. Of course, also the management of tag fields can be included into the SEA library if desired.

Figure 11 shows the SEA models for some of the usual node types of data flow graphs: operations including assignment (only subtraction is depicted), branch, merge, and constant. As default annotation for the edges [z] is assumed.

In addition to the usual node types the elements *token-input* and *token-output* (not depicted) are introduced to guarantee that the calculation specified by a data flow graph can be activated once only (1-safety). In these elements an initial token in the topmost place with value 0 is assumed.

Furthermore two elements acting as "interface" between the analogue and the digital system parts are necessary. The *analogue-input* acts as a hold element that provides an input to the analogue system part as long as it is needed. The *analogue-output* (not depicted) performs the task of a sampling element. It provides the actual output of the analogue part as input for the digital part and deletes all intermediate values generated by the analogue part. The central places in the elements *analogue-input* contain an initial token containing the default input or output value (0).

The "interface" between digital software and hardware is realized by the elements *binary-input* and *binary-output* (not depcited). These elements convert a decimal representation into a binary representation and vice versa.

The DFG-lib follows a slightly different philosophy than the library for block diagrams (BD-lib) regarding the ports of its elements. In the BD-lib for each input and output a place port was introduced. This had the consequence that complex edges were needed to directly connect the library elements, because in Pr/T Nets neither two places (nor two transitions) may be directly connected. In the DFG-lib only the inputs are represented by port places. For the outputs corresponding transitions are defined as ports. Hence, the library elements can be directly connected via simple edges annotated with [z]. The only exception are the elements acting as interface to analogue or hardware parts.

SEA model for local control The data flow graph for the first task of the *local control* is depicted in Figure 12.

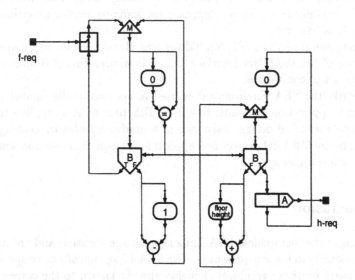

Fig. 12. Floor level to height conversion of *local control*

In the left part of the data flow graph the loop for decrementing the floor level *f-req* is depicted. The right part contains the loop for adding the *floor height* in order to calculate the required height *h-req*.

By specifying all subnets instantiated by the places in Figure 6 down to the level of predefined library elements or self defined extended Pr/T Nets the behaviour of the *elevator system* is clearly defined and can be simulated and animated at an arbitrary hierarchical level of the specification.

4 Realization of the SEA Environment

The SEA Environment is implemented in the C++ language under the UNIX operating system. The main parts of the SEA Environment are a Pr/T Net Interface Editor for the definition of abstract graphical representations and a Pr/T Net Editor for a comfortable editing of extended Predicate/Transition Nets, that build the underlying semantics of the abstract graphical interfaces of a SEA specification. In order to allow the change of a SEA specification also during its simulation the Pr/T Net Editor is responsible for the simulation of the Pr/T Nets, too. As already mentioned above, the animation of the graphical interfaces is specified via some annotations to transitions in the extended Pr/T Nets. Hence, via simulation of the Pr/T Net its abstract graphics is animated accordingly and represents the net state (marking of places and activation of transitions) in an abstract way. The simulation/animation can be observed at arbitrary levels of hierarchy by opening and showing the subnets of the corresponding structured nodes during simulation. For example a customer who ordered the elevator system may only see the top level view (Fig. 6), whereas a mechanical engineer may be interested in the block diagram level (Fig. 8), and the overall system engineer who has (among others) to check the interfaces between the modules from different application domains needs simulation results at all levels of hierarchy.

A screendump of the Pr/T Net Editor was already shown in Figure 6. The appearance of the Pr/T Net Interface Editor is similar but, of course, provides a different set of commands.

Currently the SEA Environment is used in two nationally funded projects in different application domains: In the SYDIS project it is applied for work flow management and design assistance in a hardware/software codesign environment. In the METRO project it is applied for the specification and simulation of mechatronic systems.

5 Conclusion

In this paper the hierarchical SEA Language for specification and animation of heterogeneous systems was presented. The SEA Language allows to specify each system (part) by those graphical elements already known to the corresponding system engineers. These graphical 'formalisms' may either be application domain specific 'formalisms' like block diagrams or hardware gates, or they may be general graphical 'formalisms' like data flow graphs or statecharts. Furthermore, graphical system specifications may also use freely defined elements (see e. g. Figure 6).

Via the underlying extended Pr/T Nets that build the bottom level of each SEA specification, a unified and unambiguous semantics for the behaviour of these different graphical specification "formalisms" can be provided.

Predefined graphical symbols and the Pr/T Nets representing their semantics can be stored in libraries, as for instance a data flow graph or block diagram library described above. These elements can then be instantiated in several specifications. Hence, system engineers may only work with the graphical interfaces of the library elements giving them an intuitive understanding of the specification, because they are used to these graphical symbols. System engineers need no knowledge in designing extended Pr/T Nets, nevertheless in cases of doubt the semantics of the graphical symbol is unambiguously defined.

Since libraries can be extended or newly generated arbitrarily, new application domains or specification formalisms can be easily made accessible in the SEA Environment. And even more important, the "interfaces" between different application domains or specification formalisms (like data flow graphs, block diagrams of differential equations or statecharts, SDL) need to be defined only once in a corresponding library. This will certainly increase the productivity of system engineers.

By the simulation/animation of SEA specifications of heterogeneous systems, errors, particularly those between system parts from different application domains (e. g. mechanics, software, hardware), can be detected at very early stages in the engineering process and not only after (partial) construction of the system.

Currently the SEA Environment realizes an interpretative approach for simulation of the extended Pr/T Nets. In future also code generation for simulation will be realized. Another focus of our future work will be the integration of other existing specification formalisms into the SEA Language.

Acknowledgements

First of all we would like to thank all students who did the hard work of implementing the SEA Environment, especially Thomas Pusch whose diploma thesis is the basis for the graphical net representation. Furthermore, we acknowledge the work of Maria Brielmann on modeling of analogue systems with extended Pr/T Nets.

References

[BrKl93] M. Brielmann, B. Kleinjohann: A formal model for coupling computer based systems and physical systems; in Proceedings of EURO-DAC, 1993.

[BrKl93b] M. Brielmann, B. Kleinjohann: Petri Nets as a Common Model for Combined Physical and Computer Based Systems; in Petri-Netze im Einsatz für Entwurf und Entwicklung von Informationssystemen, 1993.

[Briel95]   M. Brielmann: Modelling Differential Equations by Basic Information Technology Means; in Proceeding of EUROCAST'95, Innsbruck, 1995.

[CCIT88]   CCITT Recommendation Z.100: Specification and Description Language SDL. AP IX-35, 1988.

[ChKo81]   L.A. Cherkasova, V.E. Kotov: Structured Nets; Lecture Notes of Computer Science 118, Springer Verlag, 1981.

[Denn85]   J. B. Dennis: Models of Data Flow Computation; in M. Broy (ed.), Control Flow and Data Flow: Concepts of Distributed Programming, Springer Verlag, 1985.

[GeLa81]   H. J. Genrich, K. Lautenbach: System Modelling with High-Level Petri Nets; Theoretical Computer Science, 13, North Holland, 1981.

[GeSh93]   H.J. Genrich, R. M. Shapiro: A Design of a Cascadable Nacking Arbiter; MetaSoft TR 93-1, Meta Software Corporation, Cambridge, MA, 1993.

[Hare87]   D. Harel: Statecharts - A visual formalism for complex systems; Science of Computer Programming, 8, 1987.

[Kinn 81]   D. J. Kinniment: Regular Programmable Control Structures; Proc. VLSI Conference, 1981.

[Klei93]   E. Kleinjohann: Integrierte Entwurfsberatung auf der Basis erweiterter Prädikat/Transitionsnetze; Dissertation, Universität-Gesamthochschule Paderborn, Fachbereich 17, 1993.

[Klei94]   B. Kleinjohann: Synthese von zeitinvarianten Hardware Modulen; Dissertation, Universität-Gesamthochschule Paderborn, Fachbereich 17, 1994.

[KlMi93]   B. Kleinjohann, R. Milczewski: Ein einheitliches formales Modell zur Schnittstellenspezifikation und Hardwarebeschreibung; in GI/ITG–Workshop: Formale Methoden zum Entwurf korrekter Systeme, 1993.

[KaSa90]   K. Kahn, V.A. Saraswat: Complete visualization of concurrent programs and their execution; in IEEE Workshop on Visual Languages, Oct. 1990.

[Pete81]   J. L. Peterson: Petri Net Theory and the Modelling of Systems; Prentice Hall, 1981.

[Petri62]   C.A. Petri: Kommunikation mit Automaten; Schriften des IIM Nr. 2, Institut für Instrumentelle Mathematik, Bonn, 1962.

[Ross77]   D. T. Ross: Structured Analysis (SA) - A language for Communicating Ideas; in IEEE ToSE Se-3:1, 1977.

[Schr91]   J. Schröer: A short description of a model compiler/interpreter for supporting simulation and optimization of nonlinear and linearized dynamic systems; in 5th IFAC/IMACS Symposium on CADCS, 1991.

[Star90]   P. H. Starke: Analyse von Petri-Netz-Modellen; Teubner Verlag, 1990.

[State92]   STATEMATE, When you hit the switch ...., i-Logix product information, i-Logix Inc., Burlington, MA, USA, 1992.

[Suff90]   U. Suffrian: Vergleichende Untersuchungen von State-Charts und strukturierten Petri-Netzen; Diplomarbeit, Universität-Gesamthochschule Paderborn, Fachbereich 17, 1990.

[Tack92]   J. Tacken: Steuerung und Überwachung von Entwurfssystemen mit Hilfe von Prädikat/Transitions-Netzen; Diplomarbeit, Universität-Gesamthochschule Paderborn, Fachbereich 17, November 1992.

[YSch78]   C. M. Yoder, M. L. Schrag: Nassi-Shneiderman Charts: An Alternative to Flowcharts for Design; in Proceedings of ACM SIGSOFT/SIGMETRICS Software Quality and Assurance Workshop, Nov. 1978.

A New Iterative Numerical Solution Algorithm for a Class of Stochastic Petri Nets

Werner Knaup

Informatik IV, Universität Dortmund
44221 Dortmund, Germany
email: knaup@ls4.informatik.uni-dortmund.de

Abstract. In the area of the numerical analysis of stochastic Petri Nets we present an algorithm which enables the reduction of storage requirements for generator matrices of the underlying Markov chain. We show that neither the generator matrix nor "parts" of them need be generated and stored. The solvable model class contains the superposed stochastic automatas defined by Donatelli [2] as a special case. The state spaces of the underlying Markov chains in the examples range from about 10^7 up to 10^8 states with up to 10^9 matrix entries and we show that for such models a solution is possible. Further, this algorithm can be easily integrated in tools which contain iterative numerical solution techniques.

1 Introduction

The numerical analysis of Stochastic Petri Nets (SPN) is based on the computation of the equilibrium state probabilities of a continuous time Markov chain. These describe the behaviour of a SPN. The computation of the equilibrium state probabilities as the solution of $Q^t \pi = 0$, where Q is the generator matrix and π the vector of equilibrium state probabilities, involves typically the generation of the state space and the generator matrix. The concept of generating these stuctures is not a theoretical problem. Problems occur if the state space of the Markov chain has so many states that the Q matrix cannot be stored in the computer's available main memory.

There have been many attempts to handle this problem. At this time many exact and approximate solution techniques are being discussed for more or less large model classes. The algorithm presented in this paper is related to methods which are based on a tensor algebraic approach to reduce the storage requirement for the generator Q. A method of this sort was first presented in [7] and [8] for networks of stochastic automatas. The transformation of these models in SPNs using the same solution method is given in [2]. The model class is called superposed stochastic automata and an extension of this model class is given in [3]. A further application of the tensor approach can be found in [1] for generalized colored SPNs if a hierarchical model description is given.

In contrast to the tensor based methods, we present an algorithm in the area of iterative numerical solution techniques which is not based on the generation and storing of "parts" of the generator matrix Q. As in the tensor based methods the algorithm makes use of the idea of defining a net as connection of "subsystems" of the net. The difference to other methods is that relevant information about the behaviour of the "subsystems" are not defined by matrices.

The presented algorithm is a result of a theory about the algebraic structure of generator matrices. Since the algorithm can be explained in a (relatively) simple way, we do not refer to the underlying theory. This theory has only been formulated for closed multichain queueing networks and can be found in [4]. A numerical solution algorithm for single chain queueing networks is given in [5] and the algorithm presented in this paper is a generalization of this algorithm.

The great advantage of the presented algorithm is the fact that the generation and storing of Q is not necessary and so the solvability of the linear equation system depends only on the size of the iteration vector. A further advantage is the easy integration of this algorithm in tools which have a numerical solver.

We will discuss some details and technical aspects of the algorithm, especially aspects which are connected to counting and enumeration of finite sets. The consideration of such additional aspects forces us to introduce some restrictions. We restrict the algorithm to the model class which contains the model class given in [2] as special case. We have chosen this restricted model class to make the underlying concepts clear. The extension of the algorithm to a more general model class (see [3] for example) will be introduced in a following paper.

This paper is organized in the following way. In Section 2 we define the model class and we consider a simple example to explain the underlying ideas. The basic concept and the kernel of the algorithm are given in Section 3, and in Section 4 we extend this kernel to analyse the defined model class. In Section 5 we discuss the integration of iterative numerical solution techniques. We have chosen the Gauss-Seidel method, and the integration of this technique in the algorithm is given. In Section 6 we present experimental results relating to time requirements of the algorithm. Section 7 concludes the paper.

2 Model class and description of the method

As mentioned above, the model class considered here contains the superposed stochastic automata defined in [2] as special case. The definition in [2] can be used for small changes.

Definition 1. The considered model class N is defined by a tupel (P, T, A, Π, f) where

- P denotes the non empty set of m places, T is the non empty set of s transitions and $A \subseteq \{P \times T\} \cup \{T \times P\}$ is the set of arcs with $dom(A) \cup codom(A) = P \cup T$.
- The set of input and output places for transition t_j, $j = 1, \ldots, s$ are denoted by $\bullet t_j$ and $t_j \bullet$ respectively. Then we have the condition

$$| \bullet t_j | = |t_j \bullet| \geq 1, \quad \text{for } j = 1, \ldots, s. \tag{1}$$

- The net consists of c partitions $\Pi = \{\pi_q\}$ of P so that

$$|\pi_q \cap \bullet t_j| = |\pi_q \cap t_j \bullet| \leq 1 \quad \text{for} \quad q = 1,\dots,c, \quad j = 1,\dots,s. \tag{2}$$

- Each partition is strongly connected. A partition is strongly connected if there are two paths in opposite directions between any two places in the partition, where a path is an alternating non repeating sequence of places, arcs and transitions.
- The number of places in partition π_q is denoted by m_q.
- Let any initial marking of places in π_q with n_q tokens be given. The number of all reachable markings is given by

$$\prod_{q=1}^{c} \binom{m_q + n_q - 1}{m_q - 1} \tag{3}$$

- The function $f : T \to \Re^+$ with $f(t_j) = \mu_j$, $j = 1,\dots,s$ is the parameter of the exponential distribution associated with transition t_j.

The difference between the model class in Definition 1 and the definition given in [2] is that the number of tokens in each partition can be greater than one. In the example in Figure 1 we have two partitions, $\pi_1 = \{p_1, p_2, p_3, p_4\}$ and $\pi_2 = \{p_5, p_6\}$. The partitions are interpreted as subnets where the interaction

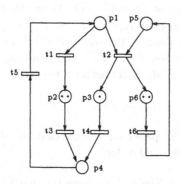

Fig. 1. A net with two partitions

of these two subnets is given by transition t_2. In the algorithm presented here it is important to see on which principle the analysis of such synchronized nets is based when applying the tensor calculus. The main idea is to consider each subnet in isolation and to interpret all transitions as one-to-one transitions, i.e. each transition has exactly one input place and exactly one output place. In the example above, two matrices are constructed where each matrix is the generator matrix of the corresponding subnet. The generator matrix of the whole net is the Kronecker sum of these two matrices. The format of the resulting matrix is correct if we interpret each row and column index as a state of the Markov chain. The error in this matrix results from transition t_2. Obviously transition t_2 is a one-to-one transition in each subnet and a two-to-two transition in reality. Hence t_2 can only fire if at the most one token is at place p_1 as well as at place p_5. To eliminate this error, so called correcting matrices are defined to set such entries in Q to zero.

The algorithm introduced in this paper also makes use of the following method: the state space is constructed for each partition of a net. The state space of the whole system is then given by the cross product over the state spaces of the subnets.

To explain the method, we omit at first the problem of which entries in the matrix are zero or not. In the first step we will construct and define the state space of the corresponding Markov chain.

First we introduce some notation that will be used subsequently. Let the positive integers k and n be given. Then we define the index set

$$\hat{\Gamma}_n^k = \{\hat{\alpha} \mid \hat{\alpha} = \hat{\alpha}_1 \ldots \hat{\alpha}_k, 0 \leq \hat{\alpha}_i \leq n, i = 1, \ldots, k, \sum_{i=1}^{k} \hat{\alpha}_i = n\}. \tag{4}$$

Each $\hat{\alpha} \in \hat{\Gamma}_n^k$ denotes a positive integer sequence of length k for which the sum over all components $\hat{\alpha}_i$ is equal to n.

As stated in Definition 1, each partition π_q consists of m_q places and in each partition there are n_q tokens. From (3) and (4) we can immediately see that the set of all reachable markings is given by the set $\hat{\Gamma}_{n_q}^{m_q}$, if we interpret an $\hat{\alpha} \in \hat{\Gamma}_{n_q}^{m_q}$ as a marking, where $\hat{\alpha}_i$ denotes the number of tokens at place i of partition π_q. Using the short notation $\hat{\Gamma}^q := \hat{\Gamma}_{n_q}^{m_q}$ and denoting each element of $\hat{\Gamma}^q$ by $\hat{\alpha}^q = \hat{\alpha}_1^q \ldots \hat{\alpha}_{m_q}^q$, the set of all markings for a net with c partitions is given by

$$\hat{\Gamma} := \hat{\Gamma}^1 \times \ldots \times \hat{\Gamma}^c \tag{5}$$

Remark. In the rest of the paper we assume that the elements of a set of type (5) are listed in lexicographical order.

Example 1. For the net in Figure 1 we have two partitions and, hence, the set $\hat{\Gamma}_3^4$ for partition one and the set $\hat{\Gamma}_2^2$ for partition two:

$$\hat{\Gamma}_3^4 = \{0003, 0012, 0021, 0030, \ldots, 2001, 2010, 2100, 3000\}, \tag{6}$$
$$\hat{\Gamma}_2^2 = \{02, 11, 20\}.$$

A sequence $\hat{\alpha}_1^2 \hat{\alpha}_2^2 \in \hat{\Gamma}_2^2$ means that there are $\hat{\alpha}_1^2$ tokens at place five and $\hat{\alpha}_2^2$ tokens at place six. Note that place five is the first place and place six is the second place in partition two.

For each partition we need some information about the connection between places and transitions and about the type of transition:

Definition 2. For each partition π_q with m_q places and n_q tokens, the $m_q \times m_q$ matrices V^q and R^q are defined as follow:

$$r_{ij}^q = \begin{cases} k, & \text{if } i \in \bullet t_k, j \in t_k \bullet \\ 0, & \text{else} \end{cases}$$

$$v_{ij}^q = \begin{cases} n, & \text{if } r_{ij}^q \neq 0 \text{ and } r_{ij}^q \text{ is a n-to-n transition} \\ 0, & \text{if } r_{ij}^q = 0 \end{cases}$$

for all $i, j \in \{1, 2, \ldots, m_q\}$, $i \neq j$.

A non-zero entry r_{ij}^q refers to the upper index of a transition and, hence, to the "name" of the transition for which the place $i \in \pi_q$ is the input place and $j \in \pi_q$ the output place. The corresponding entry v_{ij}^q contains the information about the type of transition. This will be important information when we want to compute non-zero entries in the Q matrix.

Example 2. Referring to the net in Figure 1, for partition π_1 we have

$$R^1 = \begin{array}{|c|c|c|c|} \hline 0 & 1 & 2 & 0 \\ \hline 0 & 0 & 0 & 3 \\ \hline 0 & 0 & 0 & 4 \\ \hline 5 & 0 & 0 & 0 \\ \hline \end{array} \quad V^1 = \begin{array}{|c|c|c|c|} \hline 0 & 1 & 2 & 0 \\ \hline 0 & 0 & 0 & 1 \\ \hline 0 & 0 & 0 & 1 \\ \hline 1 & 0 & 0 & 0 \\ \hline \end{array} \tag{7}$$

Let a marking $\hat{\alpha}^q \in \hat{\Gamma}^q$ for partition π_q be given. If we interpret each non-zero entry in V^q as a one-to-one transition, then we must know the marking $\hat{\beta}^q \in \hat{\Gamma}^q$ which follows directly after marking $\hat{\alpha}^q$ if a transition $v_{ij}^q \neq 0$ fires.

Definition 3. For each partition π_q and each sequence $\hat{\alpha}^q \in \hat{\Gamma}^q$, a $m_q \times m_q$ matrix C^q is defined by $c_{ij}^q = \hat{\beta}^q$, $\hat{\beta}^q \in \hat{\Gamma}^q$ and each component $\hat{\beta}_k^q$, $k = 1, \ldots m_q$ of c_{ij}^q is defined by

$$\hat{\beta}_k^q := \begin{cases} \hat{\alpha}_k^q, & \text{if } k \neq i, k \neq j \\ \hat{\alpha}_i^q + 1, & \text{if } v_{ij}^q \neq 0, k = i \\ \hat{\alpha}_j^q - 1, & \text{if } v_{ij}^q \neq 0, k = j \end{cases}$$

for all $i, j \in \pi_q$.

Let a sequence $\hat{\alpha}^q$ be given. Then each non-zero entry $c_{ij}^q \in C^q$ can be interpreted as a marking. If transition r_{ij}^q fires, then we subtract one token at place i and add one token at place j. Obviously in this case we come directly from marking c_{ij}^q to marking $\hat{\alpha}^q$.

Example 3. For partition π_2 of the net in Figure 1 we have the following C^2 matrix for a given row index $\hat{\alpha}^2$:

$$C^2 = \begin{array}{|c|c|} \hline - & \hat{\alpha}_1^2 + 1, \hat{\alpha}_2^2 - 1 \\ \hline \hat{\alpha}_1^2 - 1, \hat{\alpha}_2^2 + 1 & - \\ \hline \end{array} \tag{8}$$

The entries $c_{11}^2 = -$ and $c_{22}^2 = -$ mean that the corresponding entries v_{11}^2 and v_{22}^2 are equal to zero and, hence, a sequence is not defined in this case.

On the assumption that neither matrix Q^t nor "parts" of it are stored, we will discuss the example in Figure 1 in an informative way to show the underlying ideas and concepts to perform a matrix vector multiplication of the form $Q^t v$.

Let the transposed generator matrix Q^t and an iteration vector v, stored as linear array, be given. We assume that the row and column numbering of the corresponding Q^t matrix is given by elements of the index set $\hat{\Gamma} = \hat{\Gamma}^1 \times \hat{\Gamma}^2$ in lex order (lexicographical order):

$$\hat{\Gamma}^1 \times \hat{\Gamma}^2 = \{000302, 000311, 000320, \ldots, 300002, 300011, 300020\}.$$

Let a row index word $\hat{\alpha} = \hat{\alpha}^1 \hat{\alpha}^2 \in \hat{\Gamma}$ of Q^t be given. We want to know which column index words $\hat{\beta} \in \hat{\Gamma}$ refer to possible non-zero entries $q_{\hat{\alpha}, \hat{\beta}} \in Q^t$, $\hat{\alpha} \neq \hat{\beta}$. We can construct all column indices in three steps. In the first step, we construct all column indices $\hat{\beta}$ for the given row index $\hat{\alpha}$ which must fulfill the following property:

$$\hat{\beta} = c_{ij}^1 \circ \hat{\alpha}^2 \quad \text{for all } i, j \in \pi_1 \text{ with } v_{ij}^1 = 1. \tag{9}$$

The operation "$\circ$" stands for the concatenation of sequences of the sets $\hat{\Gamma}^1$ and $\hat{\Gamma}^2$. All column indices $\hat{\beta}$ in (9) define a state change from $\hat{\beta}$ to $\hat{\alpha}$ provided that a one-to-one transition in partition π_1 fires and that no transition from partition π_2 fires. Obviously all these column indices are constructed from $\hat{\alpha}^2$ and from the matrix entries c_{ij}^1 of C^1 for which $v_{ij}^1 = 1$. Furthermore, the corresponding matrix entry $q_{\hat{\alpha}\hat{\beta}} \in Q^t$ for each $\hat{\beta}$ from (9) is given by $f(r_{ij}^1)$. In the same way we construct column indices $\hat{\beta}$ which refer to state changes based on the condition that only one-to-one transitions in partition π_2 fire. Thus we have the following column indices in the second step:

$$\hat{\beta} = \hat{\alpha}^1 \circ c_{ij}^2 \quad \text{for all } i, j \in \pi_2 \text{ with } v_{ij}^2 = 1. \tag{10}$$

For the column indices $\hat{\beta}$ in (10) we can directly compute the corresponding matrix entry $q_{\hat{\alpha}\hat{\beta}} \in Q^t$ as $f(r_{ij}^2)$. In the third step we consider the two-to-two transition t_2:

$$\hat{\beta} = c_{ij}^1 \circ c_{m,n}^2 \quad \text{for all } i, j \in \pi_1 \text{ and } m, n \in \pi_2 \text{ with } v_{ij}^1 = v_{mn}^2 = 2. \tag{11}$$

The column indices $\hat{\beta}$ in (11) are constructed from sequences from matrix C^1 and C^2 for all $i, j \in \pi_1$ and $m, n \in \pi_2$ for which $v_{ij}^1 = v_{mn}^2 = 2$. The corresponding matrix entry $q_{\hat{\alpha}\hat{\beta}} \in Q^t$ is, in this case, given by $f(r_{ij}^2)$ if $v_{ij}^1 = 2$.

Any iterative numerical algorithm based on the assumptions above must carry out the following steps: compute all index words $\hat{\alpha} \in \hat{\Gamma} = \hat{\Gamma}^1 \times \ldots \hat{\Gamma}^c$ in lexicographical order, compute column indices $\hat{\beta} \in \hat{\Gamma}$ referring to non-zero matrix entries for each row index $\hat{\alpha} \in \hat{\Gamma}$, compute the matrix entries $q_{\hat{\alpha}\hat{\beta}}$ and, finally, compute the position of component $v_{\hat{\beta}}$ in the linear array v for each $\hat{\beta} \in \hat{\Gamma}$ and perform the multiplication $q_{\hat{\alpha}\hat{\beta}} v_{\hat{\beta}}$.

The performance of any algorithm based on these steps depends on the fast computing of indices $\hat{\alpha} \in \tilde{\Gamma}$ and their corresponding position in lex order. Based on (5) must compute the function values of function

$$pos : \tilde{\Gamma} \to \{0, \ldots, |\tilde{\Gamma}| - 1\}, \qquad (12)$$

where $pos(\hat{\alpha})$ defines the position of the index $\hat{\alpha}$ in lex order, if the elements of the set $\tilde{\Gamma}$ are listed in lex order. Referring to the example above, in each iteration step must compute the positions $pos(\hat{\beta})$ for all column indices $\hat{\beta}$ (for each row index $\hat{\alpha}$) so that we have the address of the components $v[pos(\hat{\beta})]$ of array v. Function (12) is called "ranking" in combinatorics (see [9]). We do not use a standard solution and so we must first introduce and solve our "ranking problem". In the next two sections we introduce a computation method for function values of (12).

3 Nets with only one partition

In this section we consider nets which consist of only one partition and in which all transitions are of type one-to-one. From the previous section we see that each net is treated as a connection of such one-to-one subnets. The solution for such nets will be extended to nets with c partitions in the following section.

Let a net with m places and n tokens be given. Then we define the following index set:

$$\tilde{\Gamma}_n^m = \{\tilde{\alpha} \mid \tilde{\alpha} = \tilde{\alpha}_1 \ldots \tilde{\alpha}_m, 0 \leq \tilde{\alpha}_i \leq n \text{ for } i = 1, \ldots, m, \ \tilde{\alpha}_1 \leq \ldots \tilde{\alpha}_m\} \quad (13)$$

$\tilde{\Gamma}_n^m$ denotes the set of all sequences $\tilde{\alpha}_1 \ldots \tilde{\alpha}_m$ of integers of length m with $\tilde{\alpha}_1 \leq \ldots \leq \tilde{\alpha}_m$ and $0 \leq \hat{\alpha}_i \leq n$ for all $i \in \{1, \ldots, m\}$.

We start with the generation of index words $\hat{\alpha} \in \tilde{\Gamma}_n^m$ in lex order. One method of doing this is given by the following algorithm:

```
(01)        procedure row_index(k)
(02)           if k = 1 then
(03)              for i = 0 (1) n   do
(04)                 α̂₁ = i;   α̃₁ = i;
(05)                 row_index(k + 1);
(06)           else                        **k > 1**
(07)              if k < m − 1 then
(08)                 for i = 0 (1) n − α̃_{k−1} do            (14)
(09)                    α̂_k = i;   α̃_k = α̃_{k−1} + α̂_k;
(10)                    row_index(k + 1);
(11)              else                     **k = m − 1**
(12)                 for i = 0 (1) n − α̃_{m−2} do
(13)                    α̂_{m−1} = i;   α̂_m = n − α̃_{m−2} − α̂_{m−1};
```

If we start with the procedure call $row\_index(1)$, then we need $m-2$ procedure calls to compute the first sequence $\hat{\alpha} \in \hat{\Gamma}_n^m$ in lex order. From the for statements in lines (03),(08) and (12) it is also clear that all index words $\hat{\alpha} \in \hat{\Gamma}_n^m$ are computed in lex order. Furthermore, one can see that in line (04) and (09) index words $\tilde{\alpha} \in \tilde{\Gamma}_n^{m-2}$ are computed to control the for statement in line (08). From line (04) and (09) it follows that:

$$\tilde{\alpha}_k = \sum_{i=1}^{k} \hat{\alpha}_i \text{ for } k \in \{1, \ldots m-2\}. \tag{15}$$

and, therefore, the connection between the sets $\hat{\Gamma}_n^m$ and $\tilde{\Gamma}_n^{m-2}$ is given by (15). Note that the elements $\tilde{\alpha} \in \tilde{\Gamma}_n^{m-2}$ in algorithm (14) are constructed in lex order. In lines (04),(09) and (13) of the algorithm, the construction of an index word $\hat{\alpha}$ starts with $\hat{\alpha}_1$, in the next step with $\hat{\alpha}_2$ and so on. Hence at each level k, for $k < m$, we have constructed an index word $\hat{\alpha}_1 \ldots \hat{\alpha}_k$.

Consider the procedure call $row\_index(k)$ on the condition that $\tilde{\alpha}_k = i$. We now need information about how many sequences are computed if we start with the sequence $\tilde{\alpha}_k \ldots \tilde{\alpha}_{m-2}\hat{\alpha}_{m-1}$ provided that $\tilde{\alpha}_k = i$.

Definition 4. Let the positive integers m and n be given. For the algorithm row_index a function $d : \{1, \ldots m-2\} \times \{0, 1, \ldots, n\} \to \mathcal{N}$ is defined by

$$d(k, i) = |\{\tilde{\alpha}_k \ldots \tilde{\alpha}_{m-2}\hat{\alpha}_{m-1} \,|\, \tilde{\alpha}_k = i\}|$$

for all $k \in \{1, \ldots, m-2\}$ and $i \in \{0, \ldots, n\}$.

For a Petri net with m places and n tokens and one partition with only one-to-one transitions, the corresponding Q matrix is a $|\hat{\Gamma}_n^m| \times |\hat{\Gamma}_n^m|$ matrix. In the following we use the function values $d(k, i)$ to compute, for a given row index $\hat{\alpha} \in \hat{\Gamma}_n^m$, the positions of the corresponding column indices $\hat{\beta} \in \hat{\Gamma}_n^m$ which refer to non-zero entries $q_{\hat{\alpha}\hat{\beta}} \in Q$ for $\hat{\alpha} \neq \hat{\beta}$. From this fact we see that we need a fast method to compute the function values $d(k, i)$. The following theorem shows that a fast and simple recursive computation is possible:

Theorem 5. *Let the function* $d : \{1, \ldots m-2\} \times \{0, 1, \ldots, n\} \to \mathcal{N}$ *from Definition 4 be given. The function values* $d(k, i)$ *can be recursively computed as follows:*

$$d(k, i) = \begin{cases} 1, & \text{if } i = n \\ d(k+1, i) + d(k, i+1), & \text{if } k < m-2, i < n \\ |\{0, \ldots, n-i\}|, & \text{if } k = m-2, i < n \end{cases}$$

for all $k \in \{1, \ldots, m-2\}$ *and* $i \in \{0, 1, \ldots, n\}$.

The proof of this theorem is given in [5]. We consider an example to make Definition 4 and Theorem 5 clear:

Example 4. For $m = 5$ and $n = 3$, the function values $d(k, i)$, $k \in \{1, \ldots, 3\}$, $i \in \{0, \ldots, 3\}$ are given by the following matrix D, where a matrix entry d_{ij} denotes the function value $d(i, j)$:

$$D = \begin{array}{|c|c|c|c|} \hline 20 & 10 & 4 & 1 \\ \hline 10 & 6 & 3 & 1 \\ \hline 4 & 3 & 2 & 1 \\ \hline \end{array} \qquad (16)$$

For example, entry $d(1, 0)$ means that we have twenty sequences $\tilde{\alpha}_1 \tilde{\alpha}_2 \tilde{\alpha}_3 \hat{\alpha}_4$ with $\tilde{\alpha}_1 = 0$ and entry $d(3, 2)$ means that we have two sequences $\tilde{\alpha}_3 \hat{\alpha}_4$ with $\tilde{\alpha}_3 = 2$.

Following the argumentation in Section 2, we must now compute the function values of function (12) for the case $\hat{\Gamma} = \hat{\Gamma}_n^m$. The following theorem is based on function d from Definition 4:

Theorem 6. *The function values of pos* $: \hat{\Gamma}_n^m \to \{0, \ldots, \binom{m+n-1}{m-1}) - 1\}$ *can be computed as follows:*

$$pos(\hat{\alpha}) = \sum_{i=0}^{\hat{\alpha}_1 - 1} d(1, i) + \sum_{k=2}^{m-2} \sum_{i=1}^{\hat{\alpha}_k} d(k, \tilde{\alpha}_{k-1} + i - 1) + \hat{\alpha}_{m-1}$$

for all $\hat{\alpha} \in \hat{\Gamma}_n^m$. For $q < p$ we use the convention $\sum_{i=p}^{q} d(\ldots) = 0$.

The proof of this theorem is given in [5]. Returning to Section 2, we are now able to compute the positions of the corresponding column indices $\hat{\beta} \in \hat{\Gamma}_n^m$ for each row index word $\hat{\alpha}$ (of Q^t). We integrate these computations in algorithm (14). In the algorithm we compute the positions of the corresponding column indices $\hat{\beta}_1$ directly after line (04) for $\hat{\alpha}_1$ and $\hat{\beta}_1 \ldots \hat{\beta}_k$ directly after line (09) for $\hat{\alpha}_1 \ldots \hat{\alpha}_k$.

Definition 7. Let the $m \times m$ matrices R and V be given. The $m \times m$ matrices C_k and S_k for $k = 1, \ldots, m - 1$ are defined as follow:

$$c_k(i, j) = \hat{\beta}_k := \begin{cases} \hat{\alpha}_k, & \text{if } k \neq i, k \neq j \\ \hat{\alpha}_k + 1, & \text{if } v_{ij} \neq 0, k = i \\ \hat{\alpha}_k - 1, & \text{if } v_{ij} \neq 0, k = j \end{cases} \qquad (17)$$

$$s_k(i, j) = \begin{cases} c_1(i, j), & \text{if } k = 1 \\ s_{k-1}(i, j) + c_k(i, j), & \text{if } k > 1 \end{cases} \qquad (18)$$

for $i, j \in \{1, \ldots, m\}$, $i \neq j$. Based on (17), (18) and Theorem 6, the $m \times m$ matrices A_k, $k = 1, \ldots, m - 1$ are defined as follows:

$$a_k(i, j) = \begin{cases} \sum_{r=0}^{c_1(i,j)-1} d(1, r), & \text{if } k = 1 \\ a_{k-1}(i, j) + c_{m-1}, & \text{if } k = m - 1 \\ a_{k-1}(i, j) + \sum_{r=1}^{c_k(i,j)} d(k, s_{k-1}(i, j) + r - 1), & \text{if } 1 < k < m - 1 \end{cases} \qquad (19)$$

for $i, j \in \{1, \ldots, m\}$.

In (17) the matrix entries $c_k(i,j)$ of matrix C_k define the component $\hat{\beta}_k$ of a column index $\hat{\beta}_1 \ldots \hat{\beta}_k$ for a given row index $\hat{\alpha}_1 \ldots \hat{\alpha}_k$. The sum $\sum_{p=1}^{k} c_p(i,j)$ is then the component $\tilde{\beta}_k$. Applying (15) we see that $\tilde{\beta}_k = \sum_{p=1}^{k} \hat{\beta}_p$.

In (19) we split the computation of function values of function *pos* from Theorem 6 into $m-1$ steps. Each step k refers to a column index $\hat{\beta}_1 \ldots \hat{\beta}_k$ and to the corresponding component $\tilde{\beta}_k$. For each k we compute, for a given index $\hat{\alpha}_1 \ldots \hat{\alpha}_k$, the position for the column index $\hat{\beta}_1 \ldots \hat{\beta}_k$ based on Theorem 6. In the following algorithm we have integrated the computation of column indices in algorithm *row_index*. According to Theorem 6, at each recursion level we compute the corresponding "parts" of the function *pos* based on (19).

$$
\begin{aligned}
&(01)\quad procedure\ row\_col\_index(k)\\
&(02)\quad\ \ if\ k=1\ then\\
&(03)\quad\ \ \ \ for\ i=0\ (1)\ n\quad do\\
&(04)\quad\ \ \ \ \ \ \hat{\alpha}_1=i;\quad \tilde{\alpha}_1=i;\\
&(05)\quad\ \ \ \ \ \ <\text{compute } C_1,S_1,A_1\text{ for }k=1\text{ (Def. 7) }>\\
&(06)\quad\ \ \ \ \ \ row\_col\_index(k+1);\\
&(07)\quad\ \ else\qquad\qquad\qquad **k>1**\\
&(08)\quad\ \ \ \ if\ k<m-1\ then\\
&(09)\quad\ \ \ \ \ \ for\ i=0\ (1)\ n-\tilde{\alpha}_{k-1}\ do\\
&(10)\quad\ \ \ \ \ \ \ \ \hat{\alpha}_k=i;\quad \tilde{\alpha}_k=\tilde{\alpha}_{k-1}+\hat{\alpha}_k;\\
&(11)\quad\ \ \ \ \ \ \ \ <\text{compute } C_k,S_k,A_k\text{ for }1<k<m-1\text{ (Def. 7) }>\\
&(12)\quad\ \ \ \ \ \ \ \ row\_col\_index(k+1);\\
&(13)\quad\ \ \ \ else\qquad\qquad\qquad **k=m-1**\\
&(14)\quad\ \ \ \ \ \ for\ i=0\ (1)\ n-\tilde{\alpha}_{m-2}\ do\\
&(15)\quad\ \ \ \ \ \ \ \ \hat{\alpha}_{m-1}=i;\quad \hat{\alpha}_m=n-\tilde{\alpha}_{m-2}-\hat{\alpha}_{m-1};\\
&(16)\quad\ \ \ \ \ \ \ \ <\text{compute } C_{m-1},S_{m-1},A_{m-1}\text{ (Def. 7) }>
\end{aligned}
\tag{20}
$$

In line (11) of the algorithm we compute entries $a_k(i,j)$ of A_k for the case $1<k<m-1$. The computation of entries $a_k(i,j)$ in line (11) is controlled by the *for* statement in line (09) and, therefore, we must add only $d(k,\ldots)$ to compute position $a_k(i,j)$ and not the sum as described in equation (19) for the case $1<k<m-1$.

Example 5. Let a net with $m=5$ places and $n=3$ tokens be given. The corresponding d function is given by (16). For the 5×5 matrix V, we have $v_{ij}\neq 0$ for all $i,j\in\{1,\ldots,5\}$ with $i\neq j$. Further let the index sequence $\hat{\alpha}_1\hat{\alpha}_2=11$ be given. Applying (17) and (19), the C_1 and the A_1 matrix are given by the following matrices for $\hat{\alpha}_1=1$:

$$
C_1=
\begin{array}{|c|c|c|c|c|}
\hline
- & 2 & 2 & 2 & 2\\
\hline
0 & - & 1 & 1 & 1\\
\hline
0 & 1 & - & 1 & 1\\
\hline
0 & 1 & 1 & - & 1\\
\hline
0 & 1 & 1 & 1 & -\\
\hline
\end{array}
\quad
A_1=
\begin{array}{|c|c|c|c|c|}
\hline
- & 30 & 30 & 30 & 30\\
\hline
0 & - & 20 & 20 & 20\\
\hline
0 & 20 & - & 20 & 20\\
\hline
0 & 20 & 20 & - & 20\\
\hline
0 & 20 & 20 & 20 & -\\
\hline
\end{array}
\tag{21}
$$

In the first row of the C_1 matrix we have $c_1(1,j) = 2$ for $j = 2, \ldots m$ and, using (16), in the first row of A_1 we have the entries $a_1(1,j) = d(1,0) + d(1,1) = 30$ for $j = 2, \ldots m$. Now we consider the computation of C_2 and A_2 for $\hat{\alpha}_2 = 1$:

$$
C_2 =
\begin{array}{|c|c|c|c|c|}
\hline
- & 0 & 1 & 1 & 1 \\
\hline
2 & - & 2 & 2 & 2 \\
\hline
1 & 0 & - & 1 & 1 \\
\hline
1 & 0 & 1 & - & 1 \\
\hline
1 & 0 & 1 & 1 & - \\
\hline
\end{array}
\tag{22}
$$

The computation of entries of C_2 follows from (17) and the A_2 matrix is given by

$$
A_2 =
\begin{array}{|c|c|c|c|c|}
\hline
- & 30 & 33 & 33 & 33 \\
\hline
16 & - & 29 & 29 & 29 \\
\hline
10 & 20 & - & 26 & 26 \\
\hline
10 & 20 & 26 & - & 26 \\
\hline
10 & 20 & 26 & 26 & - \\
\hline
\end{array}
=
\begin{array}{|c|c|c|c|c|}
\hline
- & 30 & 30 & 30 & 30 \\
\hline
0 & - & 20 & 20 & 20 \\
\hline
0 & 20 & - & 20 & 20 \\
\hline
0 & 20 & 20 & - & 20 \\
\hline
0 & 20 & 20 & 20 & - \\
\hline
\end{array}
+
\underbrace{
\begin{array}{|c|c|c|c|c|}
\hline
- & 0 & 3 & 3 & 3 \\
\hline
16 & - & 9 & 9 & 9 \\
\hline
10 & 0 & - & 6 & 6 \\
\hline
10 & 0 & 6 & - & 6 \\
\hline
10 & 0 & 6 & 6 & - \\
\hline
\end{array}
}_{B}
\tag{23}
$$

The matrix A_2 is computed with equation (19) for the case $k = 2$. The first matrix after the equation sign is matrix A_1 and the matrix elements b_{ij} of matrix B are given by $\sum_{r=1}^{c_2(i,j)} d(2, c_1(i,j) + r - 1)$.

The goal is to extend procedure *row_col_index* to nets with more than one partition. This will be done in the following section.

4 Extension to multiple partitions

In this section we consider nets with $c > 1$ partitions. Referring to Section 2, we have m_q places and n_q tokens in each partition π_q for $q = 1, \ldots, c$. First we must construct all elements $\hat{\alpha} \in \hat{\Gamma}$ for $\hat{\Gamma} = \hat{\Gamma}^1 \times \ldots \times \hat{\Gamma}^c$, where each $\hat{\Gamma}^q$ is the short notation for the index set $\hat{\Gamma}_{n_q}^{m_q}$. The following algorithm is a small extension of algorithm *row_index* given in Section 3:

```
(01)        procedure extend_row_index(k, q)
(02)          if k = 1 then
(03)            for i = 0 (1) n_q  do
(04)              α̂₁^q = i;   α̃₁^q = i;
(05)            extend_row_index(k + 1, q);
(06)          else                    **k > 1**
(07)            if k < m_q − 1 then
(08)              for i = 0 (1) n_q − α̃_{k−1}^q  do          (24)
(09)                α̂_k^q = i;   α̃_k^q = α̃_{k−1}^q + α̂_k^q;
(10)              extend_row_index(k + 1, q);
(11)            else                  **k = m_q − 1**
```

(12) $\qquad for \ \ i = 0 \ (1) \ n_q - \tilde{\alpha}^q_{m_q-2} \ do$

(13) $\qquad \hat{\alpha}^q_{m_q-1} = i; \quad \hat{\alpha}^q_{m_q} = n_q - \tilde{\alpha}^q_{m_q-2} - \hat{\alpha}^q_{m_q-1};$

(14) $\qquad if \ \ q < c \ then \ extend\_row\_index(1, q+1);$

Parameter q of procedure *extend_row_index* refers to partition π_q, and parameter k has the same meaning as in algorithm (14). If we start with the procedure call *extend_row_index*$(1,1)$ then we compute the first index word $\hat{\alpha}^1 \in \hat{\Gamma}^1$ (in lex order) after $m_1 - 2$ procedure calls at line (13). In line (14) we continue the computation for the first sequence $\hat{\alpha}^2 \in \hat{\Gamma}^2$ in lex order. The procedure call *extend_row_index*$(1,1)$ needs $\sum_{i=1}^{c}(m_i - 2)$ procedure calls to compute the first index $\hat{\alpha} \in \hat{\Gamma}$ in lex order. It is easy to see that algorithm (24) computes all $\hat{\alpha} \in \hat{\Gamma}$ in lex order. The elements $\tilde{\alpha} \in \tilde{\Gamma}$ for $\tilde{\Gamma} = \tilde{\Gamma}^1 \times \ldots \times \tilde{\Gamma}^c$ with $\tilde{\Gamma}^q := \tilde{\Gamma}^{m_q-2}_{n_q}$ for $q = 1, \ldots, c$ are also constructed in lex order. The connection between sets $\hat{\Gamma}$ and $\tilde{\Gamma}$ is given by

$$\tilde{\alpha}^q_k = \sum_{i=1}^{k} \hat{\alpha}^q_i \quad k \in \{1, \ldots m_q\}, q \in \{1, \ldots, c\}. \qquad (25)$$

Given that set $\hat{\Gamma} = \hat{\Gamma}^1 \times \ldots \times \hat{\Gamma}^c$, we know from Section 2 that we must compute the function values of (12). Referring to Theorem 5 the functions $d^q : \{1, \ldots m_q\} \times \{0, \ldots n_q\} \to N$, $q = 1, \ldots, c$ are defined by

$$d^q(k,i) = \begin{cases} 1, & if \ i = n_q \\ d^q(k+1,i) + d^q(k,i+1), & if \ k < m_q - 2, i < n_q \\ |\{0, \ldots, n_q - i\}|, & if \ k = m_q - 2, i < n_q. \end{cases} \qquad (26)$$

Based on Theorem 6, the function $pos^q : \hat{\Gamma}^q \to \{0, \ldots, |\hat{\Gamma}^q| - 1\}$, $q = 1, \ldots, c$ is given by

$$pos^q(\hat{\alpha}^q) = \sum_{i=0}^{\hat{\alpha}^q_1-1} d^q(1,i) + \sum_{k=2}^{m_q-2} \sum_{i=1}^{\hat{\alpha}^q_k} d^q(k, \tilde{\alpha}^q_{k-1} + i - 1) + \hat{\alpha}_{m_q-1} \qquad (27)$$

for each partition π_q. Using (27), the function values of $pos : \hat{\Gamma} \to \{0, \ldots, |\hat{\Gamma}| - 1\}$ can be computed as follows:

Theorem 8. *Let the set* $\hat{\Gamma} = \hat{\Gamma}^1 \times \ldots \times \hat{\Gamma}^c$ *be given. The function values of* $pos : \hat{\Gamma} \to \{0, \ldots, \prod_{k=1}^{c} \binom{m_k+n_k-1}{m_k-1} - 1\}$ *can be computed as follows:*

$$pos(\hat{\alpha}) = \sum_{i=1}^{c}(pos^i(\hat{\alpha}^i) \prod_{k=i+1}^{c} |\hat{\Gamma}^k|) \qquad (28)$$

for all $\hat{\alpha} \in \hat{\Gamma}$. *For* $q < p$ *we use the convention* $\prod_{i=p}^{q}(\ldots) = 1$.

Proof. The proof is based on an induction over the number c of partitions. The case $c = 1$ refers to a net with one partition and, hence, we have Theorem 6.

Suppose the result holds for $c - 1$ partitions. Then we transform the right side of equation (28):

$$\underbrace{(\sum_{i=1}^{c-1}(pos^i(\hat{\alpha}^i)\prod_{k=i+1}^{c-1}|\hat{\Gamma}^k|))|\hat{\Gamma}^c|}_{t_1} + pos^c(\hat{\alpha}^c)$$

Term t_1 gives the position of a sequence $\hat{\alpha}^1\ldots\hat{\alpha}^{c-1}$ if we list the elements of $\hat{\Gamma}^1 \times \ldots \times \hat{\Gamma}^{c-1}$ in lex order. Let a sequence

$$\hat{\alpha}^1\ldots\hat{\alpha}^{c-1} \tag{29}$$

be given. Then we construct a sequence

$$\hat{\alpha}^1\ldots\hat{\alpha}^{c-1} \circ \hat{\alpha}^c \tag{30}$$

with $\hat{\alpha}^c \in \hat{\Gamma}^c$. We have $|\hat{\Gamma}^c|$ different sequences of this type. Obviously sequence (30) is an element of the set $\hat{\Gamma} = \hat{\Gamma}^1 \times \ldots \times \hat{\Gamma}^c$. The first sequence of type (30) starts with sequence (29) and $\hat{\alpha}^c = \hat{\alpha}_1^c\ldots\hat{\alpha}_{m_c}^c = 0\ldots0n_c$. This sequence is the first sequence which comes after listing all smaller sequences in lex order. The position (in lex order) of this sequence is then given by term t_1 multiplied by $|\hat{\Gamma}^c|$. In the case $\hat{\alpha}^c > 0\ldots0n_c$ we must add $pos^c(\hat{\alpha}^c)$ to this product. □

For a Petri net according to Definition 1, let the matrices R^q, V^q for $q = 1,\ldots c$ and the the set $\hat{\Gamma} = \hat{\Gamma}^1 \times \ldots \times \hat{\Gamma}^c$ be given. Then the corresponding generator matrix Q^t is a $|\hat{\Gamma}| \times |\hat{\Gamma}|$ matrix and we assume that the row and column numbering is given by the elements of $\hat{\Gamma}$ in lex order. Further, let a row index $\hat{\alpha} \in \hat{\Gamma}$ be given. As in Section 2, we want to know which column indices $\hat{\beta} \in \hat{\Gamma}$ refer to possible non-zero entries $q_{\hat{\alpha},\hat{\beta}}$, $\hat{\alpha} \neq \hat{\beta}$ in Q^t. We use (27) to compute positions for column indices $\hat{\beta}$ for a given row index $\hat{\alpha}$.

The integration of (27) in algorithm *extend_row_index* is based on the same idea as that already discussed in Section 3. Analogous to Definition 7 we define the following matrices for each procedure call *extend_row_index*(k,q):

Definition 9. Let the $m_q \times m_q$ matrices R^q and V^q for $q = 1,\ldots,c$ be given. Then the $m_q \times m_q$ matrices C_k^q and S_k^q for $k = 1,\ldots,m_q - 1$ are defined as follows:

$$c_k^q(i,j) = \hat{\beta}_k^q := \begin{cases} \hat{\alpha}_k^q, & \text{if } k \neq i, k \neq j \\ \hat{\alpha}_k^q + 1, & \text{if } v_{ij}^q \neq 0, k = i \\ \hat{\alpha}_k^q - 1, & \text{if } v_{ij}^q \neq 0, k = j \end{cases} \tag{31}$$

$$s_k^q(i,j) = \begin{cases} c_1^q(i,j) & \text{if } k = 1 \\ s_{k-1}^q(i,j) + c_k^q(i,j) & \text{if } k > 1 \end{cases} \tag{32}$$

for $i,j \in \{1,\ldots,m_q\}$, $i \neq j$ and $q \in \{1,\ldots,c\}$.

Based on (31),(32) and (27), the $m_q \times m_q$ matrices A_k^q, $k = 1, \ldots, m_q - 1$ are defined as follows:

$$a_k^q(i,j) = \begin{cases} \sum_{r=0}^{c_1^q(i,j)-1} d^q(1,r), & \text{if } k = 1 \\ a_{k-1}^q(i,j) + c_{m-1}^q, & \text{if } k = m_q - 1 \\ a_{k-1}^q(i,j) + \sum_{r=1}^{c_k^q(i,j)} d^q(k, s_{k-1}^q(i,j) + r - 1), & \text{if } 1 < k < m_q - 1 \end{cases} \quad (33)$$

for $i, j \in \{1, \ldots, m_q\}$.

The matrices from Definition 9 are integrated in the procedure *extend_row_index* in an analogous way to that discussed for the single partition case:

(01) *procedure single_it_step(k, q)*

(02)   *if $k = 1$ then*

(03)   *for $i = 0$ (1) n_q do*

(04) $\hat{\alpha}_1^q = i; \quad \tilde{\alpha}_1^q = i;$

(05) $< \text{compute } C_1^q, S_1^q, A_1^q \text{ for } k = 1 \text{ (Def. 9) } >$

(06)   *single_it_step$(k + 1, q)$;*

(07)   *else* $* * k > 1 * *$

(08)   *if $k < m_q - 1$ then*

(09)   *for $i = 0$ (1) $n_q - \tilde{\alpha}_{k-1}^q$ do*   (34)

(10) $\hat{\alpha}_k^q = i; \quad \tilde{\alpha}_k^q = \tilde{\alpha}_{k-1}^q + \hat{\alpha}_k^q;$

(11) $< \text{compute } C_k^q, S_k^q, A_k^q \text{ for } 1 < k < m_q - 1 \text{ (Def. 9) } >$

(12)   *single_it_step$(k + 1, q)$;*

(13)   *else* $* * k = m_q - 1 * *$

(14)   *for $i = 0$ (1) $n_q - \tilde{\alpha}_{m_q - 2}^q$ do*

(15) $\hat{\alpha}_{m_q - 1}^q = i; \quad \hat{\alpha}_{m_q}^q = n_q - \tilde{\alpha}_{m_q - 2}^q - \hat{\alpha}_{m_q - 1}^q;$

(16) $< \text{compute } C_{m_q - 1}^q, S_{m_q - 1}^q, A_{m_q - 1}^q \text{ (Def. 9) } >$

(17) $a^q(r, s) = a_{m_q - 1}^q(r, s) \prod_{p=q+1}^c |\hat{\Gamma}^p|;$

(18) $row\_pos[q] = row\_pos[q] + \prod_{p=q+1}^c |\hat{\Gamma}^p|;$

(19)   *if $q < c$ then single_it_step$(1, q + 1)$;*

(20) $< \text{procedure call } gauss\_seidel >$

The integration of computing matrices from Definition 9 is given in lines (05), (11) and (16) of algorithm *single_it_step*. Note that these matrices contain information about positions of column indices for each partition separately. That means each partition is analysed independently of each other partition. The connection between the column positions and, therefore, the connection between the partitions is given in line (17). The entries $a^q(r, s)$ of the $m_q \times m_q$ matrices A^q contain the position of column indices $\hat{\beta}^q \hat{\beta}^{q+1} \ldots \hat{\beta}^c = \hat{\beta}^q \hat{\alpha}^{q+1} \ldots \hat{\alpha}^c$ for the given tail $\hat{\alpha}^{q+1} \ldots \hat{\alpha}^c$ of a row index where the computation is based on (28). The array elements $row\_pos[q]$ refer to the position of the first row index starting with

$\hat{\alpha}^1 \ldots \hat{\alpha}^q$. These array elements are initialized with zero before the procedure call $single\_it\_step(1,1)$. In line (20) for a row index $\hat{\alpha} \in \hat{\Gamma} = \hat{\Gamma}^1 \times \ldots \times \hat{\Gamma}^c$ the positions of all possible column indices $\hat{\beta} \in \hat{\Gamma}$ are given by the matrices A^q. Hence, we can perform a row vector multiplication for row $\hat{\alpha}$ if an iterative numerical method is given. In the algorithm we have chosen the Gauss-Seidel method. The name $single\_it\_step$ reflects the fact that after each constructed row with the corresponding column indices and their positions, a row vector multiplication step is possible. In the following section we present the kernel of the procedure $gauss\_seidel$.

5 The Gauss-Seidel procedure

We have chosen the forward Gauss-Seidel iteration $(D - L)v^{i+1} = Uv^i \quad i = 0, 1, \ldots$ which is based on splitting $Q^t = D - L - U$ into a diagonal matrix D, a strictly lower triangular matrix L and a strictly upper triangular matrix U (for details see [6]). Special aspects relating to the chosen iteration method or the convergence criteria are not considered in detail.

The procedure $gauss\_seidel$ is based on some data structures and initial settings. Referring to Section 2, we have defined the R^q matrix for each partition π_q. Further, we have seen that for each partition the positions of column indices are computed separately and so for all n-to-n transitions with $n > 1$ we have the same transition in n different partitions. The computation of non zero entries in the matrix Q for such n-to-n transitions is carried out in the first partition π_q in which this transition is named in the R^q matrix. So we must eliminate this transition in all other R^p $(p = q + 1, \ldots, c)$ matrices.

For all s transitions let an array t with s components be given where each array element $t[j]$ refers to a record for transition j consisting of two arrays, in_p and out_p. The elements $in\_p[i]$ and $out\_p[i]$ can be referenced by $t[j].in\_p[i]$ or $t[j].out\_p[i]$ for $i \in \{1, \ldots, c\}$. The element $in\_p[i]$ denotes the input place for transition j in partition i and $out\_p[i]$ the output place for transition j in transition i. Using this notation we can eliminate transitions in R^p as follows:

```
(01) procedure  multiple_entries
(02)    for q = 1   (1)   c do
(03)      for i = 1   (1)   m_q do
(04)        for j = 1   (1)   m_q do
(05)          if v^q(i,j) > 1 then
(06)            for p = q + 1   (1)   c do
(07)              r^p(t[q].in_p[p], t[q].out_p[p]) = 0
```

Note that the procedure call $multiple\_entries$ cames after the initialization of the R^p and V^p matrices.

The array it_vec denotes the iteration vector, where the numbering of components is given by the index set $\{0, \ldots, |\hat{\Gamma}| - 1\}$. As mentioned above, the sum

$\sum_{i=1}^{c} row\_pos[i]$ refers to the row position of Q^t. This is also the index of the vector component of the "new" computed iteration vector.

The variable *sum* stands for the result of the row vector multiplication. For each row index $\hat{\alpha} \in \hat{\Gamma}$ of the Q matrix, the diagonal element $q_{\hat{\alpha}\hat{\alpha}}$ is stored in the variable *diag*. The kernel of the procedure *gauss_seidel* is then given by

```
(01) procedure  gauss_seidel
(02)     sum = 0.0; diag = 0.0
(03)     for  q = 1  (1)   c  do
(04)      for  i = 2  (1)   m_q  do          ** row of lower matrix L **
(05)       for  j = 1 (1)  i - 1   do
(06)        if  r^q(i,j) > 0  then
(07)         if  v^q(i,j) = 1  then
(08)          diag = diag + f(r^q(i,j));
(09)          sum := sum + f(r^q(i,j)) * it_vec[a^q(r,j) + ∑_{k=1}^{q-1} row_pos[k]]
(10)         else
(11)          for  p = q + 1  (1)  c    do
(12)           c_pos = a^q(i,j);
(13)           if  t[q].in_p[p] ≠ 0  then
(14)            c_pos = c_pos + a^p(t[q].in_p[p], t[q].out_p[p]);
(15)           diag = diag + f(r^q(i,j));
(16)           sum = sum + f(r^q(i,j)) * it_vec[c_pos];
(17)      for  i = 1  (1)   m_q - 1  do       ** row of upper matrix U **
(18)       for  j = i + 1  (1)  m_q   do
(19)        < repeat steps (06) - (16) for row of upper matrix >
(20)     vec[∑_{i=1}^{c} row_pos[i]] = sum/diag;
```

The statements in lines (04)-(19) are controlled by the *for* statement in line (03). Let a row index $\hat{\alpha}$ be given. For each partition q we then compute the row vector multiplication for matrix entries $q_{\hat{\alpha}\hat{\beta}}$ with $\hat{\alpha} > \hat{\beta}$ in lines (04)-(16). The *if* case in line (07) refers to the 1-to-1 transitions in partition q and the *else* case in line (10) to the n-to-n transitions with $n > 1$. In both cases the diagonal entry is computed and the multiplication of matrix entries with the corresponding vector component is carried out. Lines (17)-(19) refer to the case $\hat{\alpha} < \hat{\beta}$ which contains the same computation steps as in the case $\hat{\alpha} > \hat{\beta}$. Finally, in line (20) we have stored the result of the row vector multiplication in the corresponding vector component.

An algorithm that is based on the methods given here is called an **mfi** (matrix free iteration) algorithm. The last two sections show that the procedures *single_it_step* and *gauss_seidel* can be integrated in tools which contain numerical solution techniques without much effort.

6 Experimental Results

The forthgoing sections show that only the size of the iteration vector determines the main storage requirements. Therefore we only consider the time requirements of the algorithm. The question of interest now is how long does it take to solve a model. One aspect is the number of iteration steps. The number of iteration steps depends on the chosen model parameters and the convergence criteria. These aspects must be considered in each algorithm which contains a iterative solution method. Further, we know that the presented algorithm only differs from other methods in the handling of the Q matrix. So the best information about the "costs" of this new algorithm is given by the time needed for one iteration step.

Our basic model consists of two partitions similar to the example given in Figure 1 but with five places in the first partition and three places in the second partition. In this model we have one 2-to-2 transition and all other transitions are of type 1-to-1.

We consider two experiments. In the first experiment we have a maximum of transitions, i.e. all $v_{ij}^1 > 0$ for $i \neq j, v_{ij}^1 \in V^1$ and $v_{ij}^2 > 0$ for $i \neq j, v_{ij}^2 \in V^2$. This is the "worst case" for the mfi algorithm because, in this case, we have a maximum of non zero entries in Q and, therefore, the highest possible expense for computing positions in a linear array. In the second experiment we have chosen a minimum of transitions so that we have as few entries in Q as possible.

The description of the model for the first experiment is given by the R^1 and R^2 matrices:

$$R^1 = \begin{array}{|c|c|c|c|c|} \hline - & 1 & 2 & 3 & 4 \\ \hline 5 & - & 6 & 7 & 8 \\ \hline 9 & 10 & - & 11 & 12 \\ \hline 13 & 14 & 15 & - & 16 \\ \hline 17 & 18 & 19 & 20 & - \\ \hline \end{array} \qquad R^2 = \begin{array}{|c|c|c|} \hline - & 6 & 21 \\ \hline 22 & - & 23 \\ \hline 24 & 25 & - \\ \hline \end{array} \tag{35}$$

An entry $r_{ij}^q \in R^q$ ($q = 1, 2$) denotes the name of the transition between places i and j in partition q and the 2-to-2 transition is $r_{2,3}^1 = r_{1,2}^2 = 6$. In the second experiment we consider the following model:

$$R^1 = \begin{array}{|c|c|c|c|c|} \hline - & 1 & - & - & - \\ \hline - & - & 6 & - & - \\ \hline - & - & - & 11 & - \\ \hline - & - & - & - & 16 \\ \hline 17 & - & - & - & - \\ \hline \end{array} \qquad R^2 = \begin{array}{|c|c|c|} \hline - & 6 & - \\ \hline - & - & 23 \\ \hline 24 & - & - \\ \hline \end{array} \tag{36}$$

In each experiment the number of tokens in partition two is three and in partition one we change the number of tokens to generate larger Q matrices. The experiments were run on a Sun SS 10 workstation under unix. For the time measurement we used the unix *time* command. In the following table we have listed some results of the first experiment. Column two contains the number of states and column three the number of non-zero entries in Q^t. Note that the number of diagonal entries is not contained in column three. Column entry *it_time* means the *user time* of the *time* command for only one iteration step respectively. Column entry *init_time* also means the *user time* but in this case from the

tokens	n. o.	n. o. tran-	mfi $[sec]$	
n_1	states	sitions	it_time	init_time
55	4 551 260	96 810 700	371	5
70	11 506 260	247 792 260	906	11
85	24 416 260	530 299 220	1 938	26
100	45 981 260	1 004 513 680	3 659	44
110	66 728 760	1 462 179 320	5 330	70
121	96 913 750	2 129 466 446	7 872	116

Table 1. Results for the first experiment

beginning of the program up to the first iteration step. This time includes the generation of some data structures and the initialisation of the iteration vector. The following table shows results of the second experiment:

tokens	n. o.	n. o. tran-	mfi $[sec]$	
n_1	states	sitions	it_time	init_time
55	4 551 260	24 977 932	218	4
70	11 506 260	63 861 492	567	11
85	24 416 260	136 566 452	1 171	26
100	45 981 260	258 656 172	2 211	44
110	66 728 760	376 256 552	3 283	70
121	96 058 751	547 834 046	4 808	116

Table 2. Results for the second experiment

If we compare Table 1 and Table 2, then we can see that the mfi algorithm is about a factor of 1.6 faster in the second experiment than in the first experiment for a single iteration step. We find that the times in both experiments are tolerable because the mfi method has a great deal to do when performing an iteration step: for each row of Q we must compute positions in a linear array corresponding to non-zero entries and, further, the diagonal entries in Q and, finally, we must divide the resulting row vector multiplication by the diagonal entry. In a following paper we will report on a detailed and comprehensive analysis of the time requirements.

We would like to mention that we have had no chance to compare the results with any other tool as there is probably no tool available at the moment to solve models with about 10^7 up to 10^8 states and 10^9 matrix entries.

7 Conclusions and further research

In this paper a new iterative numerical solution algorithm was introduced for a class of SPNs. The algorithm presented computes, for each row of the transposed generator matrix Q^t, the corresponding column entries and their positions and then performs the row vector multiplication. Thus the algorithm is not based on the storing of the generator matrix Q or "parts" of the matrix.

The algorithm is an attractive tool for very large models because only the size of the iteration vector determines the main storage requirements and thus the time needed for an iteration step as well. Furthermore, integrating the algorithm in tools which contain numerical solution techniques can be done without much effort.

The examples discussed in Section 6 show that very large models with about 10^8 states and 10^9 entries in Q can be solved. The presented algorithm is based on ideas which are not directly comparable with other existing methods and makes it possible to handle such large models for the first time.

In the near future the model class will be extended to m-to-n transitions with $m \neq n$. A further goal is to speed up the time needed for a single iteration step. Some experiments in this direction have been carried out and we believe that this is an attainable goal.

References

[1] Buchholz,P. *Hierarchies in colored GSPNs.* Application and Theory of Petri Nets, Springer LNCS 691, 1993,pp. 106-125.

[2] Donatelli, S. *Superposed Stochastic Automata: A class of Stochastic Petri nets amenable to parallel solution.* Proc. of the Fourth Intern. Workshop on Petri Nets and Performance Models (PNPM91), Melbourne (Australia), 1991, pp. 54-63.

[3] Donatelli, S. *Superposed stochastic automata: a class of stochastic Petri nets with parallel solution and distributed state space.* Performance Evaluation, North-Holland,1993, Vol.18, pp. 21-36.

[4] Knaup, W. *Algebraische Strukturen in einfachen Warteschlangen-Netzen.* Verlag Hänsel-Hohenhausen, Egelsbach (Germany), 1994.

[5] Knaup, W. *A New Iterative Numerical Solution Algorithm for Markovian Queueing Networks.* In H. Beilner, F. Bause (eds.); Quantitative Evaluation of Computing and Communication Systems, Springer LNCS 977, 1995, pp. 194-208.

[6] Krieger, U.R., Müller-Clostermann, B. and Sczittnick, M. *Modelling and Analysis of Communication Systems Based on Computational Methods for Markov Chains.* IEEE J. Sel. Areas in Com., Vol.8, No.9, December 1990, pp. 1630-1648.

[7] Plateau, B. *On the stochastic structure of parallelism and synchronization models for distributed models.* ACM Sigmetrics Conference on Measurement and Modelling of Computer Systems, Austin, August 1985.

[8] Plateau, B., Fourneau, J.M. and Lee, K.H. *PEPS: A package for solving complex Markov models of parallel systems.* Fourth Intern. Conf. on Modelling Techniques and Tools for Computer Performance Evaluation, Palma (Mallorca), Sept., 1988.

[9] Williamson, S.G. *Combinatorics for Computer Science.* Comp. Sc. Press, 1985.

A Structural Approach for the Analysis of Petri Nets by Reduced Unfoldings

Alex Kondratyev, Michael Kishinevsky, Alexander Taubin, Sergei Ten

The University of Aizu, Aizu-Wakamatsu, 965-80, Japan,
e-mail: {kondraty,kishinev,taubin,ten}@u-aizu.ac.jp

Abstract. This paper suggests a way for Petri Net analysis by checking the ordering relations between places and transitions. The method is based on unfolding the original net into an equivalent acyclic description. In an unfolding the ordering relations can be determined directly by the structure of an underlying graph. We improved on the previously known cutoff criterion for truncating the unfolding [10]. No restrictions are imposed on the class of general PNs. The new criterion significantly reduces the size of unfolding obtained by PN. The PN properties for analysis can be various: boundedness, safety, persistency etc. The practical example of the suggested approach is given in application to the asynchronous design. The circuit behavior is specified by an interpreted Petri net, called Signal Transition Graph (STG) which is then analyzed for the implementability by asynchronous hazard-free circuit. The implementability conditions are formulated in such a way that they can be checked by analysis of ordering relations between signal transitions rather than by traversal of states. This allows avoiding the state explosion problem for highly parallel specifications. The experimental results show that for highly parallel STGs checking the implementability by unfolding is one to two orders of magnitude less time-consuming than checking it by symbolic BDD traversal of the corresponding State Graph.

1 Introduction

There are several well-known techniques to avoid the "state explosion problem" in behavioural analysis of Petri Nets (PN). Stubborn sets help to compress redundancy in the reachability graph which occurs due to concurrency between transitions[21]. Symbolic Binary Decision Diagram (BDD) traversal of a reachability graph allows its implicit representation that is more compact than by using explicit enumeration of states [16]. Methods based on the partial orders avoid generation of the corresponding reachability graph [10, 3, 5]. Instead of the reachability graph, a finite prefix (called unfolding) of the equivalent occurrence net (acyclic net where all places have not more than one input transition) is generated.

Different methods cover different areas of application and demonstrate results that are often incomparable for certain properties and subclasses of PNs. For example, the symbolic model checking technique based on BDDs is useful[16], but the size of BDDs might still be exponential to the size of the original PN. Also, the BDD traversal performs efficiently only if the property can be formulated by a characteristic predicate, which covers a subset of markings, not an individual marking. A check for boundedness is problematic since one needs to guess what the upper bound for the token count in a place is, in order to obtain an efficient encoding of markings in the BDD. The stubborn sets approach is intended for checking a limited set of properties (e.g., deadlocks) and can guarantee compression of the reachability marking space only in some cases (see an example of a concurrent queue in [3]). It cannot be used without supplementing methods for checking speed-independence in the applications of PNs to asynchronous circuit design. Unfoldings might be less efficient for checking deadlocks. Hence, these should be viewed as complementary techniques, rather than completely disjoint methods.

This paper concentrates on the methods of using the ordering relations in the unfoldings for checking properties of PNs. It is shown how to analyze safeness, boundedness and persistency on-the-fly while the unfolding is generated, and the ordering relations (precedence, concurrency and conflict) are iteratively calculated.

We further apply the verification by unfoldings to asynchronous designs specified with Signal Transition Graphs (STGs) [1, 9, 11]. STGs are PNs whose transitions are labeled with falling and rising signal transitions. It is shown how to check the implementability of STGs by speed-independent circuits based on the ordering relations.

To justify the use of unfoldings an efficient method for the unfolding generation is given. This method improves the technique developed in [10] and results in a compression of the unfolding's size (see Section 5). Another successful attempt to improve the technique of [10] was done be in [4]. The method in [4] allows to get the unfolding of minimal size for safe PNs. Our approach can be applied to a wider class of general PNs, however it cannot guarantee the minimal size of obtained unfolding (though it always less than in [10]). It means that in comparison to [4] some efficiency is sacrificed to get more generality. This is dictated by the applications to asynchronous design. The widest subclass of asynchronous circuits that are speed-independent are semimodular circuits [13]. In [22] it was proved that it is impossible to specify semimodular circuits by safe PNs. Thus the class of safe PNs is too narrow for the description of speed-independent circuits because it specifies only the subclass of the so called distributive circuits.

The experimental results demonstrate that for highly parallel PNs checking properties by unfolding is one or two orders of magnitude less time-consuming than checking it by symbolic BDD traversal of the corresponding State Graph.

The paper is further organized as follows. We present basic definitions and terminology in Section 2. In Section 3, a theory of unfoldings is given. We prove soundness of the enhanced cutoff criterion for the general PNs. In Section 4, it is shown how to check safeness, boundedness, and persistency based on the ordering relations between places and transitions of the unfolding. Then we develop algorithms (Section 5) and present experimental results on using the unfolding technique (Section 6).

2 Basic Notions

Let $N = \langle P, T, F, m_0 \rangle$ be a Petri net (PN) [14], where P is the set of places, T is the set of transitions, $F \subseteq (P \times T) \cup (T \times P)$ is the flow relation, and m_0 is the initial marking.

A transition $t \in T$ is enabled at marking m_1 if all its input places are marked. An enabled transition t may fire, producing a new marking m_2 with one less token in each input place and one more token in each output place (this is denoted by $m_1 \xrightarrow{t} m_2$ or $m_1 \to m_2$). The new marking m_2 can again make some transitions enabled. We can therefore talk about sequences of transitions that fire under the markings reachable from the initial marking m_0. Such sequences of transitions will be called *feasible traces* or shortly *traces*. The set of input places of transition t is denoted by $\bullet t$ and the set of output places by $t \bullet$. Similarly, $\bullet p$ and $p \bullet$ stand for the sets of input and output transitions of place p. A place p is called a *choice* place if it has more than one output transition. A PN is *free-choice* if any output transition of a choice place has only one input place (this place is called a free-choice place).

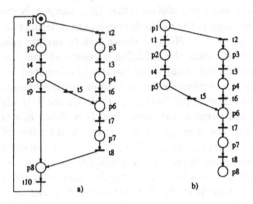

Fig. 1. Cyclic (a) and acyclic PNs (b).

The set of all markings reachable in N from the initial marking m_0 is called the Reachability Set of N. Its graphical representation is called the Reachability Graph (RG).

A PN is *acyclic* if there are no cycles in the graph of the PN. In an acyclic PN there are some places without input transitions. We assume that all these places are initially marked with one token and no other places are initially marked. Therefore, the initial marking for an acyclic PN is given by default and one does not need to define it explicitly. Under the default initial marking agreement, all properties of the acyclic net are completely determined by the net structure. Note that we do not allow transitions without input places in acyclic PNs. The examples of a cyclic and an acyclic PNs are shown in Figure 1,a,b. There are two choice places in the cyclic PN, $p1$ and $p5$, both of them are free-choice places and consequently the PN is free-choice.

A PN is called:

- *k-bounded*, if for every reachable marking the number of tokens in any place is not greater than k (the place is called k-bounded if for every reachable marking the number of tokens in it is not greater than k),
- *bounded*, if there is a finite k for which it is k-bounded,
- *safe*, if it is 1-bounded (a 1-bounded place is called a safe place),

Figure 2,a shows an example of an unbounded PN. The place $p3$ is unbounded because the trace $t1, t2, t1, \ldots$ can generate unbounded number of tokens in $p3$.

A transition t_i is called *non-persistent* if t_i is enabled in a reachable marking m together with another transition t_j and t_i becomes disabled after firing t_j. Non-persistency of t_i with respect to t_j is also called *a direct conflict* between t_i and t_j. A PN is *persistent* if it contains no non-persistent transitions.

In a PN from Figure 1,a transitions $t1$ and $t2$ are both enabled in the initial marking.Firing each of them disables the other, hence this PN is non-persistent.

Since we will use equivalent transformations for PNs an equivalence notion must be defined. The behaviors of PNs can be compared by the languages they realize, where a language of a PN is a set of its traces.

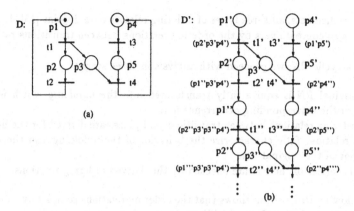

Fig. 2. Unbounded PN (a) and its occurrence net (b).

Definition 1 (Trace equivalence). [18] PNs $N1$ and $N2$ with sets of transitions $T1$ and $T2$ are trace equivalent with respect to partition $r = \{T1\} \times \{T2\}$ iff for any trace $s = t_1, \ldots, t_k, \ldots$ feasible in $N1$ there exists a trace $p = t'_1, \ldots, t'_k, \ldots$, feasible in $N2$ such that $t'_i \; r \; t_i$ for every $i = 1, \ldots, k, \ldots$

Although traces are convenient for defining properties of PNs they are impractical for checking these properties due to the explosion in the number of traces for parallel PNs. Our aim is to define the ordering relations between places and transitions in acyclic PNs and to use these ordering relations for checking properties of PNs, in particular for checking equivalence, without extracting traces.

Definition 2 (Ordering relations). Let $N = \langle P, T, F \rangle$ be an acyclic PN and $x_1, x_2 \in P \cup T$.

- x_1 *precedes* x_2 (denoted by $x_1 \Rightarrow x_2$) if (x_1, x_2) belongs to the reflexive transitive closure of F, i.e., there is a path in the graph of a PN between x_1 and x_2.
- x_1 and x_2 are in *conflict* (denoted by $x_1 \# x_2$), if there exist distinct transitions $t_1, t_2 \in T$ such that $\bullet t_1 \cap \bullet t_2 \neq \emptyset$, and $t_1 \Rightarrow x_1$, and $t_2 \Rightarrow x_2$. If $x \# x$ (where $x \in P \cup T$), then x is in *self-conflict*.
- x_1 and x_2 are *concurrent* (denoted by $x_1 \| x_2$), if they neither in precedence, nor in conflict.

Let us consider the introduced relations for the example of an acyclic PN in Figure 1,b. Directly by Definition 2 one can find that $t1 \Rightarrow t4 \Rightarrow t5$, $t2 \Rightarrow t3 \Rightarrow t6$, $t1 \# t2$, $t1 \# t3, \ldots$. There are no concurrent places and transitions in this example.

The concurrency relation is defined to be disjoint with the conflict and precedence relations. However, this is not the case for conflict and precedence relations. We will further restrict acyclic nets making conflict and precedence relations disjoint.

Definition 3 (Occurrence net). An occurrence net is an acyclic net $N = \langle P, T, F \rangle$ in which every place $p \in P$ has at most one input transition $| \bullet p | \leq 1$.

This definition corresponds to the definition of occurrence nets given in [15, 3]. Note that we consider the occurrence net as a particular case of acyclic PN. Therefore, the initial marking for occurrence nets is uniquely defined by the places having no input transitions.

In occurrence nets, all three types of ordering relations are disjoint [15]. It gives the opportunity to compare nets by the ordering relations between transitions rather than by traces.

Given a cyclic PN, we proceed with analysis in three stages.

- The original PN is equivalently transformed into the *unfolding* which is a finite prefix of the corresponding occurrence net.
- The ordering relations between transitions and places are derived for the unfolding. These relations fully characterize the behavior of the unfolding and therefore the behavior of the original net.
- The properties of the PN are checked by the derived ordering relations.

The following Proposition shows that the ordering relations completely characterize occurrence nets and, therefore, unfoldings. Its proof can be found in [7].

Proposition 4. *Two occurrence nets $N1 = \langle P1, T1, F1 \rangle$ and $N2 = \langle P2, T2, F2 \rangle$ with a bijection r between sets of transitions $T1$ and $T2$ are trace equivalent iff the ordering relations between the corresponding reachable transitions in $N1$ and $N2$ coincide.*

3 Unfolding

This section presents basic theory of unfoldings and gives an enhanced criterion to truncate an occurrence net into an unfolding. Figure 3 shows an occurrence net for the PN from Figure 1,a. Each transition ti of the initial PN has a set of the corresponding transitions in the occurrence net $ti', ti'', ti''', \ldots$, that are called *instantiations* of ti. Similarly, for each place pj the occurrence net contains the set of the corresponding instantiations $pj', pj'', pj''', \ldots$. It can be shown that under the partition r that associates each transition t of PN with its instantiations $t', t'', \ldots$, the original PN is trace equivalent to its occurrence net. Further we will refer to any object in the occurrence net equivalent to the object in the original cyclic PN by adding one or more apostrophes (or by adding a superscript) to its name. For example, t' and t are the corresponding transitions in an occurrence net and a PN, m' is a marking in the occurrence net and m is the corresponding marking in the PN, etc.

For analysis of a cyclic PN an equivalent occurrence net is used. Although, the occurrence net for a cyclic PN can be infinite it is always possible for a bounded PN to truncate the occurrence net up to a finite "complete" subgraph which possesses all information about original PN (e.g., contains all reachable markings of the original net). This complete subgraph is called an *unfolding*. Criteria to generate unfoldings for different classes of PNs were suggested in [10, 8, 19, 4]. We will present notions related to the unfolding and discuss two criteria to generate unfoldings for general PNs.

To truncate an occurrence net let us introduce several notions.

Definition 5 (Configurations). [10].

- A set of transitions $C' \subset T'$ is a *configuration* in an occurrence net if: (1) for each $t' \in C'$ the configuration C' contains t' together with all its predecessors, and (2) C' contains no mutually conflict transitions.
- The minimal configuration that contains t' and all the transitions preceding t' is called a *local configuration* of the transition t' (denoted $\{\Rightarrow t'\}$)

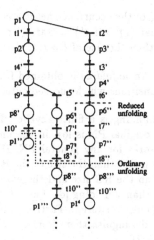

Fig. 3. Unfolding of a cyclic PN into the occurrence net

Each configuration C' corresponds to a marking that is reachable from m_0 after all the transitions from C' have been fired. This marking is called the *final marking* of C' and is denoted by $FM(C')'$ [10] (following our notation the corresponding PN marking will be depicted by omitting the apostrophe – $FM(C')$).

Definition 6 (Final and basic markings). Let C' be a configuration of an occurrence net. A *final marking* of C', denoted $FM(C')'$, is a marking reachable from the initial marking after all transitions from C' and only those transitions are fired. A final marking of a local configuration of t' is called a *basic marking* of t' and denoted $BM(t')'$.

In the occurrence net a configuration exists for any reachable marking. In other words, each marking serves as a final marking to some configurations. A basic marking $BM(t')'$ is a marking reachable after t' and all its predecessors have been fired.

In the occurrence net of the unbounded PN from Fig. 2,b, the local configuration $\{\Rightarrow t4'\}$ for transition $t4'$ is equal to $\{t1', t3', t4'\}$. The basic marking for $t4'$ is $BM(t4')' = \{p2', p4''\}$. It indicates that $p2'||p4''$, since all places of a marking are concurrent. Marking $\{p1'', p4''\}$ does not correspond to any local configuration, however it corresponds to the configuration $\{t1', t2', t3', t4'\}$.

The size of local configuration plays an important role in the construction of unfolding. It defines the order in which different transitions are generated. To fix the metrics in the order of transition generation let us introduce several notions.

Let C' be a configuration in an occurrence net, then $|C'|$ stands for the *size* (number of transitions) of the configuration C'. $|\Rightarrow t'|$ stands for the size of a local configuration $\{\Rightarrow t'\}$ and is also called a *depth* of transition t'. A depth for a set of transitions T' (denoted by $||T'||$) is defined as the maximal depth among transitions in T', i.e. $||T'|| = max_{\forall t' \in T'} |\Rightarrow t'|$. A set of transitions of an occurrence net with the same depth will be called a *tier* of the occurrence net. As will be shown in Section 5 the occurrence net is generated following an increasing depth of transitions, tier by tier.

A *cutoff criterion* is needed for truncating an occurrence net. Such a criterion was introduced in [10].

Definition 7. A transition t_i' of the occurrence net is a *cutoff* transition, if there exists another transition t_j' such that (1) $BM(t_i') = BM(t_j')$ and (2) $| \Rightarrow t_i'| > | \Rightarrow t_j'|$, i.e., the size of $\{\Rightarrow t_i'\}$ is greater than the size of $\{\Rightarrow t_j'\}$.

Definition 8 (Unfolding). An *unfolding* is obtained from the occurrence net by removing all the places and transitions which succeed cutoffs.

For a bounded PN an unfolding is a finite acyclic PN. It was shown in [10] that no reachable marking of the original PN is lost in the unfolding. The cutoff criterion introduced by Definition 7 works for any general PN. However, it cannot guarantee that the size of the unfolding is less than the size of the RG [8]. For some PNs the size of the unfolding is greater than the RG size and therefore symbolic BDD traversal of RGs might be more efficient than unfolding [6]. The size of unfolding can be reduced on the basis of the enhanced cutoff criterion. Further cutoffs obtained by Definition 7 will be called *GT-cutoffs* [1] to distinguish them from the *EQ-cutoffs* [2], obtained by our enhanced criterion (Definition 9).

Definition 9 (EQ-cutoff). A transition t_i' of the occurrence net is a EQ-cutoff transition if it is not a GT-cutoff transition and there exists another transition t_j' such that (1) $BM(t_i') = BM(t_j')$, (2) $| \Rightarrow t_i'| = | \Rightarrow t_j'|$, (3) t_i' is not parallel to t_j', (4) there are no EQ-cutoffs among transitions t_k' such that $t_k' || t_j'$ and $| \Rightarrow t_k'| \le | \Rightarrow t_j'|$.

If a transition is either a GT-cutoff or a EQ-cutoff, then it is called an *enhanced cutoff* or simply a *cutoff*. Transition t_j' from Definitions 7 and 9 is called an *image* of cutoff t_i'. The GT-cutoff definition allows to choose for a cutoff image only transitions with a smaller size of local configuration. The enhanced cutoff definition relaxes this condition and allows to choose for a cutoff image another transition, t_j', with an equal to t_i' size of local configuration. It can be done if every previously generated parallel to t_j' enhanced cutoff t_k' has an image with a smaller size of local configuration and therefore t_k' is a GT-cutoff. Therefore, a transition might or might not be a EQ-cutoff depending on the order in which transitions of the occurrence net are generated. We will further show that there is an order of generation of the unfolding which allows to use EQ-cutoffs without ambiguity. An unfolding generated on the basis of the enhanced cutoff criterion will be called *a reduced unfolding*, while McMillan's unfolding will be called *an ordinary*.

The major advantage of the enhanced cutoff criterion is the typically much smaller size of the reduced unfolding in comparison with the size of the RG for a PN. For example, for the 60-users distributed mutual exclusive (DME) arbiter (see Section 5) the RG contains approximately 7.0×10^{19} markings, while the reduced unfolding – only 240 transitions. The reduced and ordinary [10] unfoldings for PN from Figure 1 are shown in Figure 3 by dashed and dotted lines correspondingly.

The major reduction in the size of unfoldings is achieved for PNs with several sequential conflicts. For such nets the size of ordinary unfoldings truncated by the GT-cutoffs only can grow exponentially to the number of conflicts, while the reduced unfolding grows linearly. The latter is demonstrated by Figure 4,a,b, where all the places $p1, \ldots, p_{n-1}$ make free-choice between alternative trajectories of equal length and therefore the ordinary unfolding will contain 2^{n-1} instantiations of the place p_n,

---

[1] GT stands for "greater than" and refers to the size of local configuration for a cutoff and its image.

[2] EQ stands for "equal" and again refers to the size of local configuration.

while in the reduced unfolding each place is represented only once. Basic markings used for cutoffs are shown in brackets for some of the transitions in Figure 4,b.

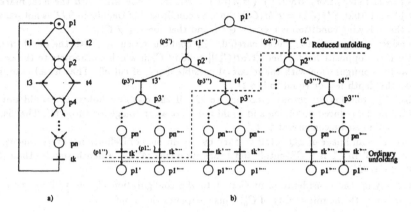

Fig. 4. PN with sequential choices (a) and its unfolding (b)

The reduced unfolding contains complete information about the original PN. To prove this, let us show that the reduced unfolding presents complete information on the reachability of markings for a general PNs. A similar proposition for the ordinary unfoldings was proved in [10].

Proposition 10. *Let C' be a configuration in an occurrence net and $FM(C')'$ be its final marking. Then there is a configuration $C1'$ in the reduced unfolding N' such that $FM(C1') = FM(C')$.* [3]

Proof. If configuration C' contains no cutoffs, then $C' \subset N'$ and Proposition is trivial.

Else, let $C'_1, ..., C'_m$ be all configurations in the occurrence net such that: 1. $FM(C') = FM(C'_1) = ... = FM(C'_m)$ and 2. $C'_1, ..., C'_m$ are of minimal size. This set is always finite. Assume that neither of them is included in the reduced unfolding N'. Let us seek for contradiction.

The only reason why $C'_i \not\subset N'$ is that it contains at least one cutoff. Let $Cut(C'_i) = \{t'_{i1}, ..., t'_{ik}\}$ be a set of cutoffs for each C'_i. Let C'_j be a configuration from $C'_1, ..., C'_m$ with a maximal depth for the set of cutoffs, i.e., for any other C'_i: $||Cut(C'_j)|| \geq ||Cut(C'_i)||$. Assume that $t1' \in C'_j$ is the cutoff with the maximal depth. Assume also that $t2' \in N'$ is an image for cutoff $t1'$, and therefore $BM(t1') = BM(t2')$.

Two cases are possible:

Case 1. $| \Rightarrow t1'| > | \Rightarrow t2'|$, i.e., $t1'$ is a GT-cutoff. Marking $FM(C'_j)'$ is reached from $BM(t1')'$ in $|C'_j| - | \Rightarrow t1'|$ steps. Therefore, in the original PN $FM(C'_j)$ is also reached from $BM(t1')$ in $|C'_j| - | \Rightarrow t1'|$ steps. Since $BM(t1') = BM(t2')$, in the occurrence net there is another marking corresponding to $FM(C'_j)$ which is reached from $BM(t2')'$ in the same $|C'_j| - | \Rightarrow t1'|$ steps. Therefore, in the occurrence net there exists another configuration $C*'$ with a final marking $FM(C*')'$, such that $t2' \in C*'$ and $FM(C*') = FM(C'_j) = FM(C')$. Configuration $C*'$ differs from C'_j [4]. Since $| \Rightarrow t2'| < | \Rightarrow t1'|$, the following condition for the sizes of configurations holds: $|C*'| < |C'_j|$. This contradicts the assumption that C'_j is of minimal size.

---

[3] $FM(C1')$ and $FM(C')$ are PN markings corresponding to $FM(C1')'$ and $FM(C')'$.
[4] cf. the proof of Lemma 5 in [10]

Case 2. $| \Rightarrow t2'| = | \Rightarrow t1'|$, i.e., $t1'$ is a EQ-cutoff. Let us choose among $C_1', \ldots, C_m'$ configuration C_k' which contains transition $t2'$. Such configuration C_k' exists in the occurrence net because if $FM(C_j')'$ is reached from $BM(t1')'$ in $|C_j'| - | \Rightarrow t1'|$ steps then by making $|C_j'| - | \Rightarrow t1'|$ steps from $BM(t2')'$ $(BM(t2') = BM(t1'))$ one will reach the final marking $FM(C_k')'$ such that $FM(C_k') = FM(C_j')$. Since by condition 3 of Definition 9 $t2'$ is not parallel to $t1'$, the following condition is met: $t1' \notin C_k'$ and therefore $C_k' \neq C_j'$.

Consider configuration C_k' more carefully. C_k' contains no cutoffs with the depth $n, n > | \Rightarrow t1'|$. If the opposite holds, then $||Cut(C_k')|| > ||Cut(C_j')||$ which contradicts to the choice of C_j' as the configuration with the maximal possible depth of cutoffs. That is, all cutoffs in C_k' have the depth less or equal to $| \Rightarrow t1'|$.

Also, C_k' contains no cutoffs preceding to $t2'$. If the opposite holds, $t2'$ would not be generated in the reduced unfolding and could not serve as an image for cutoff $t1'$. That is, all cutoffs in C_k' must be concurrent to $t2'$.

Since $t2'$ is an image of EQ-cutoff $t1'$ then by Condition 4 of Definition 9 every concurrent to $t2'$ cutoff $t3'$ in C_k' is a GT-cutoff and hence its image $t4'$ has a depth smaller than $t3'$, $| \Rightarrow t4'| < | \Rightarrow t3'|$.

By applying the consideration of Case 1 to the configuration C_k' we will come to the contradiction with the minimality of C_k'. This completes the proof.

Corollary 11. *For a PN reachability of a marking, a transition, or a place can be determined by the reduced unfolding.*

From Corollary 11 follows that all problems related to the marking reachability in a PN can be solved directly by the unfolding. However, the information on reachability of markings is implicitly stored in the unfolding since each place of the PN has many corresponding instantiations in the unfolding. Hence, it is not easy to find a configuration for a particular marking. A more promising approach is to analyze a PN based on the ordering relations in the unfolding.

4 Checking properties of PNs

In this Section we will show how to analyze PNs by checking the ordering relations in their unfoldings. The properties chosen for analysis are: safeness, boundedness, and persistency. These are the key properties for applying PNs for the specification of asynchronous circuits. Boundedness guarantees finiteness of the specification and therefore confirms that the specification might be implemented with a finite circuit. Safeness and persistency simplify the implementation of a PN with a circuit significantly. For example, a direct translation of safe and persistent PNs into self-timed circuits is proposed in [17]. This method gives correct-by-construction implementations, that need not be verified afterwards. Persistency is also related to hazard-freedom of asynchronous circuits.

Safeness

Proposition 12. *A PN is safe if and only if each place p has no concurrent instantiations in the reduced unfolding.*

For example, in the reduced unfolding in Figure 2,b instantiations $p3'$ and $p3''$ are concurrent. Therefore, the PN in Figure 2,a is unsafe.

Boundedness

The following proposition extends the results of [8] to the whole class of PNs. The proof can be found in [7].

Proposition 13. *Let N' be a reduced unfolding of PN N. N is unbounded if and only if there is a transition $t \in T$ that has two instantiations in N', t' and t'', such that $t' \Rightarrow t''$ and $BM(t') < BM(t'')$.*

Return to our example of Figure 2. As soon as the transition $t1''$ will be generated in the unfolding one can conclude that the original PN is unbounded. Indeed, both $t1'$ and $t1''$ are the instantiations of the same transition $t1$, and $t1' \Rightarrow t1''$. $BM(t1'') = p2p3p3p4 > BM(t1') = p2p3p4$, so the markings of place $p3$ can grow infinitely and this place is unbounded.

Persistency

The persistency property (see Section 2) is defined in terms of markings but it is closely related to the conflict relations between transitions.

The unfolding gives an explicit representation of conflicts. However, even in unfoldings, a finer consideration of conflict relations is needed for determining the non-persistency of transitions. Not every pair of conflict transitions are in direct conflict. For example, conflict transitions $t3', t4'$ from PN in Figure 5 do not share any input place and thus cannot disable each other. Conflict transitions $t1', t4'$ are also not in direct conflict (although they share the input place $p2'$), because both transitions are never enabled simultaneously.

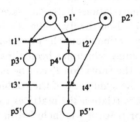

Fig. 5. PN with direct and non-direct conflicts

The structural properties of direct conflicts in a reduced unfolding are given by the following Proposition.

Proposition 14. *Let the set of transitions $I(t) = \bullet(\bullet t)$ be called the set of direct predecessors of t. ($I(t)$ contains all transitions whose output places are input for t). Transitions $t1'$ and $t2'$ of an unfolding are in direct conflict iff they share the same input place and there is no conflict pair $\{t'_i, t'_j\}$, ($\{t'_i, t'_j\} \neq \{t1', t2'\}$), such that $t'_i \in I(t1') \cup t1'$ and $t'_j \in I(t2') \cup t2'$.*

The proof of this proposition can be found in [7].

For example, in PN Figure 5 transitions $t1'$ and $t4'$ are not in the direct conflict because $t2'$ is the direct predecessor of $t4'$ and $t2'$ is in conflict with $t1'$. However, for the pair $\{t1', t2'\}$ all the conditions of Proposition 14 are satisfied and therefore $\{t1', t2'\}$ are in direct conflict.

Proposition 15. *Safeness, boundedness and persistency of a PN can be analyzed on the basis of the ordering relations in the PN unfolding.*

Application to asynchronous design

Signal Transition Graphs (STGs) model is popular in asynchronous design for specifying control circuits[1, 9, 11, 5]. STGs are PNs whose transitions are interpreted as changes of the circuit signals. Checking that given an STG it is possible to implement it with a hazard-free asynchronous circuit requires analyzing certain properties of implementability. In [7] it is shown how checking of these properties can be done on-the-fly while constructing the unfolding.

5 Generating unfoldings and ordering relations

The unfolding will be constructed by a breadth-first traversal, tier by tier. A tier contains transition instantiations of the same depth, i.e., tiers are ordered by the size of local configurations for included transitions. Only two tiers are needed to be stored: a *Current-tier* is used for generation of a *New-tier*. Each tier contains a set of transitions, *Current-T-tier*, and a set of outputa places from these transitions, *Current-P-tier*. All transitions which are ready to be fired are collected in a *T-Fired*. Those transitions which have the minimal size of local configuration are moved from *T-Fired* to a *New-tier*.

An initial tier (with number 0) has no transitions and contains all the places p' that represent the initial marking m_0 in PN N. Figures 6,a,b describe algorithms for generating an unfolding and for constructing a tier.

Two matrices of the ordering relations between places and transitions, *Relations-T* and *Relations-P* are constructed on-the-fly, while the unfolding is being generated. These matrices contain the information about the precedence, conflict and concurrency relations in the part of the unfolding which has been already generated. They play the major role in checking whether the transition t_i' can be included into the unfolding (a subroutine $Ready(t_i')$). Due to the symmetry of the conflict and concurrency relations and asymmetry of the precedence relation the matrices can be kept triangle.

After a new transition t_i' (or place p_j') is included in the unfolding the matrix of relations is updated by adding the ordering relations of t_i' (p_j') with all other transitions (places) generated before. This can be done by simply inheriting these relations from the transitions (places) that serve as direct predecessors of t_i' (p_j'). Consider this inheritance by an example of the ordering relations for transitions:

Precedence $- t_j' \Rightarrow t_i'$ if:

1. $t_j' \in \bullet(\bullet t_i')$ (t_j' is a direct predecessor of t_i')
2. $t_j' \Rightarrow t_k'$ and $t_k' \in \bullet(\bullet t_i')$ (inheriting of $\Rightarrow$ from the direct predecessors)

Conflict $- t_j' \# t_i'$ if:

1. $\bullet t_i' \cap \bullet t_j' \neq \emptyset$ (direct conflict: t_i' and t_j' share an input place)
2. $t_j' \# t_k'$ and $t_k' \in \bullet(\bullet t_i')$ (inheriting of conflicts from the direct predecessors)
3. $t_k' \Rightarrow t_j'$ and t_k' is in direct conflict with t_i' (conflicts spread over transitions succeeding a direct conflict).

Similar conditions can be given for places. Procedures *Update_Relations-T* and *Update_Relations-P* for generating the ordering relations are given in Figure 7.

```
generate_unfolding (D') {
    Reached = ∅; T-Fired = ∅;
    Current-T-tier = ∅;
    Current-P-tier = {M'_0};
    do{
        Reached = Reached ∪ Current-tier;
        generate_New-tier;
        Current-T-tier = New-T-tier − Cutoffs;
        Current-P-tier = New-P-tier;
        is_unfolding_correct(Current-tier)
    while (Current-P-tier ≠ ∅);}
    return Reached;  }
```

(a)

```
generate_New-tier {
    New-T-tier = ∅; New-P-tier = ∅;
    for each p' ∈ Current-P-tier do {
        for each t_i ∈ p• do
            if Ready(t'_i) then T-Fired = T-Fired ∪ t'_i;
        /* T-Fired is kept sorted or a hash-table by | ⇒ t'_i|*/
    }
    New-T-tier = {t'_j ∈ T-Fired: ∀t'_k ∈ T-Fired| ⇒ t'_j| ≤ | ⇒ t'_k|};
    T-Fired = T-Fired − New-T-tier;
    for each t'_i ∈ New-T-tier do Update_Relations-T(t'_i);
    Check_cutoff(New-T-tier);
    for each t'_i ∈ New-T-tier − Cutoffs do {
        New-P-tier = New-P-tier ∪ t'_i•;
        Update_Relations-P(t'_i) }
}
```

(b)

Fig. 6. Algorithms for generating unfolding (a) and unfolding tier (b)

A new transition t'_i can be included in the *T-Fired* after a place p' from *Current-P-tier* if, in the generated part of the unfolding, there exists a set of places that can be matched as input places to t'_i. This set cannot be used on the earlier stages for the generation of another instantiation of t_i. The concurrency relation allows one to reduce significantly the size of the set *Candidates* that is used as a search space for the input places for t'_i. Indeed, input places for t'_i must be pairwise concurrent and thus to obtain the *Candidates* it is sufficient to choose the places that are concurrent to p' and are matched with $•t_i$. In most cases, *Candidates* contains one copy for each place from $•t_i$. However, sometimes (e.g., for the unsafe PNs) there can be several places $p'_j, p''_j, \ldots$ that can be matched with the input of t'_i. In this case we need to check all the possible combinations *Inputs* for $•t'_i$. The number of these combinations grows exponentially from the size of the *Candidates* set. However, typically the number of instantiations of a place concurrent to an input place of a transition is small and this exponential dependency can be neglected. Function *Ready* tests conditions for t'_i to be fired.

After a new tier is generated in the unfolding, one has to check: (1) The cutoff conditions, (2) The correctness of unfolding (boundedness, safeness, non-autoconcurrency).

$Update\_Relations\-T(t_i')$ {
  for each $t_j' \in N'$ do
    Write "$t_j' \| t_i'$" in Relations-T;
  for each $t_j' \in \bullet(\bullet t_i')$ do
    Write "$t_j' \Rightarrow t_i'$" in Relations-T;
    for each $t_k' \Rightarrow t_j'$ do
      Write "$t_k' \Rightarrow t_i'$" in Relations-T;
    for each $t_k' \# t_j'$ do
      Write "$t_k' \# t_i'$" in Relations-T
  end for;
  for each t_j', $\bullet t_j' \cap \bullet t_i' \neq \emptyset$ & $t_j \neq' t_i'$ do
    Write "$t_j' \# t_i'$" in Relations-T;
    for each t_k', $t_j' \Rightarrow t_k'$ do
      Write "$t_k' \# t_i'$" in Relations-T
  end for }

(a)

$Update\_Relations\-P(t_i')$ {
  for each $p_j' \in N'$ do
    Write "$p_j' \| p_r'$" in Relations-P;
  for each $p_r' \in t_i'\bullet$ do
    for each $p_j' \in \bullet t_i'$ do
      Write "$p_j' \Rightarrow p_r'$" in Relations-P;
      for each $p_k' \Rightarrow p_j'$ do
        Write "$p_k' \Rightarrow p_r'$" in Relations-P;
      for each $p_k' \# p_j'$ do
        Write "$p_k' \# p_r'$" in Relations-P
      for each p_k', $p_j' \Rightarrow p_k'$ do
      if $p_k' \notin t_i'\bullet$
      Write "$p_k' \# p_r'$" in Relations-P;
    end for
  end for }

(b)

Fig. 7. Algorithms to update Relations-T (a) and Relations-P (b)

For this check, the basic markings are constructed for each of the transitions. They can be constructed by iterative formulas using basic markings for the direct predecessors. Assume that $Consumed(t_i')$ is the set of places in the unfolding that precede the transition t_i'. Then the following iterative formulas are valid.

(1) The set of consumed places is equal to that of the places consumed by the directly preceding transitions plus input places of t_i':

$Consumed(t_i') = \sum_{t_j' \in \bullet(\bullet t_i')} Consumed(t_j') + \bullet t_i'$

(2) The basic marking is given by the basic markings of the directly preceding transitions that are not consumed by t_i' plus output places of t_i':

$BM(t_i') = \sum_{t_j' \in \bullet(\bullet t_i')} BM(t_j') - Consumed(t_i') + t_i'\bullet$.

Procedures *Ready* and *Is_unfolding_correct* are given in Figure 8a,b. Procedure *Check_cutoffs* for detecting GT-cutoffs and EQ-cutoffs is given in Figure 9. This detection is made following the definitions of GT- and EQ-cutoffs (Definitions 7 9).

```
Ready(t'_i) {
    Candidates = p';
    for each p'_j || p' & p_j ∈ •t_i do
        Candidates = Candidates ∪ p'_j;
    for each p_j ∈ •t_i do
        if p'_j ∉ Candidates return false;
    repeat
        Create new Inputs ⊆ Candidates;
    /* Inputs has instantiations of all places from •t_i*/
    /* These instantiations are pairwise concurrent */
        Used = false;
        for each t''_i ∈ T-Fired /*t''_i instantiation of t_i*/ do
            if •t''_i = Inputs then Used = true;
        if not(Used) then {
            •t'_i = Inputs;
            Calculate_BM(t'_i);
            return true
        }
    until No new Inputs ⊆ Candidates;
    return false    }
```

(a)

```
Is_unfolding_correct(Current-tier) {
    for each t'_i ∈ Current-T-tier do {
    /* boundedness check */
        for each t''_i ∈ Reached
        /*t''_i instantiation of t_i*/ do {
            if BM(t''_i) < BM(t'_i) then unbounded;
    /* transition autoconcurrency check */
            if t'_i || t''_i then autoconcurrent
        }
    }
    for each p' ∈ Current-P-tier do {
    /* safeness of place p' ∈ Current-P-tier */
        for each p'' ∈ Reached   /*p'' instantiation of p*/
            if p' || p'' then unsafe
    }
}
```

(b)

Fig. 8. Analysis of transition readiness (a) and unfolding correctness (b).

Let us evaluate the complexity of the suggested algorithm for the unfolding generation and construction of the ordering relations. Let N_T and N_P be the number of transitions and places in the unfolding and O_p be the maximum fan-out for a place.

The procedure _generate_New-tier_ checks for each place p' in the unfolding the possibility to add a new transition t'_i for which $p' \in •t'_i$. For the same place p', there can exist several patterns of _Inputs_ that can be matched with t'_i. Assume that r is the upper bound for the number of instantiations of one place from $•t'_i$ concurrent to p'. Then the number of different _Inputs_ patterns cannot exceed $R \equiv r^{|•t_i|}$. Thus in

```
Check_cutoffs(New-T-tier) {
/* GT-cutoffs */
for each t'_i ∈ New-T-tier do {
    for each t'_j ∈ Reached do
        if BM(t''_j) = BM(t'_i) then
            GT-cutoffs = GT-cutoffs ∪ t'_i;
}
/*EQ-cutoffs*/
Cutoffs=Cutoffs ∪ GT-cutoffs;
for each t'_i ∈ New-T-tier − Cutoffs do
    repeat
        Find t'_j ∈ New-T-tier − Cutoffs such that
        t'_j#t'_i & BM(t'_j) = BM(t'_i);
        if there is such t'_j then  {
            Is_Image = true;
            if there is t'_k ∈ EQ-Cutoffs & t'_k||t'_j then Is_Image = false;
            if Is_Image then {
            Create Ready_cut⊆ New-T-Tier − Cutoffs;
            /* Ready_cut contains all t'_m: t'_m#t'_j & BM(t'_j) = BM(t'_m)*/
            /* In particular, t'_i ∈ Ready_cut*/
            for each t'_m ∈ Ready_cut do {
                if there are no t'_r ∈ EQ-cutoff-images ∩ New-T-tier & t'_r||t'_m then {
                EQ-cutoffs = EQ-cutoffs ∪ t'_m;
                Cutoffs = Cutoffs ∪ t'_m; }
            Mark t'_j as image for all EQ-cutoffs t'_m;
            }
        }
    until t'_i ∈ Cutoffs ∨ (No new t'_j found);
}
```

Fig. 9. Detecting cutoffs.

generate_New-tier for one place p' we perform $Ready(t'_i) - O(R * O_p)$ times, and for all places $- O(R * O_p * N_P)$ times.

All properties in the procedure *Is_unfolding_correct* are analyzed by a linear search over the transitions (or places) in the unfolding, the complexity of this procedure is $O(N_T)$. The complexity for detecting cutoffs does not exceed $O(N_T^2)$.

It gives the overall complexity of the unfolding generation and ordering relations construction as $O(R * O_p * N_P * N_T + N_T^2)$. Considering that $N_T \leq N_P$, R is small for typical PNs and $O_p << N_T$, the simplified estimation for the upper worst case bound is $O(N_P * N_T)$. Note that our algorithm is compared quite favorably with the results reported in [3], where unfoldings for a much more restricted class of PNs, safe Marked Graphs[5] are constructed in $O(N_P^2 * N_T^2)$. It is also worth noting that similar to [3], our approach allows checking reachability of a certain state in polynomial time for pipeline PNs, while the stubborn sets approach [21] does not solve this problem in the polynomial time.

In Sections 3 it was shown how properties of PNs can be analyzed by checking the

---

[5] A Marked Graph is a Petri Net whose places have at most one input and at most one output transition.

ordering relations in the unfolding. It follows that the complexity of analysis for PNs is the same as for generating an unfolding, i.e., $O(N_P * N_T)$.

6 Experimental results

We have implemented the method for PNs and STGs unfolding presented in this paper inside the SIS tool [20]. This allows us to combine the verification method with the synthesis methods for STGs implemented in the SIS. The efficiency of the presented approach was evaluated in comparison to the rather efficient technique of a PN traversal based on BDDs [6]. The results for three examples are presented Table 1: a PN for the dining philosophers [10], an n-stage Muller pipeline [12] with $n/3$ portions of information moving concurrently (see Figure 10,a,b, where the 4-stage pipeline and its STG specification are shown) and an n user distributed mutual exclusion (DME) arbiter [6] (see Figure 11, where the STG for a two-user arbiter is shown). All examples are scalable, in such a way that the number of states of the system can be exponentially increased by iteratively repeating a basic pattern. For the case of the dining philosophers, the check was done for the boundedness and safety only, while for Muller pipeline and DME arbiter the implementability of the corresponding STGs was analyzed as well.

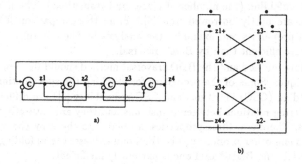

Fig. 10. 4-stage Muller pipeline: circuit (a), corresponding STG (b)

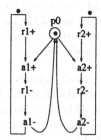

Fig. 11. A two-user mutual exclusion element

Table 1 shows the results of experiments. The run-time is given for a Sparc 10 workstation. The table gives the size of the initial PN (STG) specification, the number

Example	n	# of places	# of transitions	# of markings	BDD size peak	BDD size final	BDD CPU (sec.)	Unfolding # of places	# of trans.	CPU (sec.)
n dining	20	140	100	2.2×10^{13}	–	3091	10	140	100	1
philosophers	40	280	200	2.9×10^{19}	–	251839	455	280	200	1
	50	350	250	–	–	1870847	> 4 hours	350	250	1
	60	420	300	–	–	–	–	420	300	1
n-stage	30	120	60	6.0×10^{7}	7897	4784	132	490	240	1
Muller's	45	180	90	6.9×10^{11}	23590	10634	740	1035	510	2
pipeline	60	240	120	8.4×10^{15}	53446	18788	3210	1780	880	4
n-user	20	81	80	2.2×10^{7}	1688	1688	11	81	80	1
DME	40	161	160	4.5×10^{13}	6568	6568	101	161	160	1
arbiter	60	241	240	7.0×10^{19}	14648	14648	342	241	240	1

Table 1. Verification by unfolding and BDD-traversal

of markings in the Reachability Graph, the number of nodes for the peak and the final reduced and ordered BDD representation of the reachable marking space and the size of the reduced unfolding (the number of places and transitions). The numbers for the BDD traversal are partly borrowed from [6].[6] From the comparison follows that for the considered properties and examples the analysis by the ordering relations in the unfolding is more efficient than by BDD traversal.

The unfolding method and the BDD traversal method should be viewed as complementary techniques. Some of the properties for checking the specifications (like Complete State Coding (CSC) violations and CSC reducibility [6]) are difficult to express in terms of ordering relations between transitions and they are naturally formulated in terms of states. Therefore, these properties are better to check by the symbolic BDD traversal. For some other properties, like the boundedness, the unfolding technique is very efficient while the BDD-based one is extremely inefficient.

Table 2 shows the results on unfolding using different cutoff criteria for the STGs from the known set of benchmarks [20]. This table indicates only those examples where the size of the reduced unfolding is smaller than the size of McMillan's unfolding. This experiment allows to conclude that EQ-cutoffs reduce the size of unfolding in comparison with GT-cutoffs if the choice and concurrency structure is relatively complex. The third column of the table presents the size of the initial specification (the number of places, transitions and places plus transitions). The fourth column presents the size of the unfoldings using the following cutoff criteria.

1. A simple cutoff criterion, which totally ignores comparison of configuration sizes. The algorithm cut the constructing unfolding each time a transition with an already generated basic marking occurs. Although, this criterion is not correct in general, it gives a useful lower bound on the size of any correct unfolding which loose no reachable markings of the initial PN.
2. The enhanced cutoff criterion for generating a reduced unfolding. Both GT-cutoffs and EQ-cutoffs are used.
3. The GT-cutoff criterion for generating McMillan's unfolding.

---

[6] The data on the dining philosophers were kindly provided by Oriol Roig.

Example	Size	Unf. Size (Pl/Tr/Total (time-sec.))			# Cutoff (GT/EG/Tot.) or GT		
	Pl/Tr/Total	Simple	Enhanced	McM	Sim.	Enhan.	MM
t2.g	66/67/133	66/67/133	68/69/137	922/982/ 1904 (5 s.)	16	11/7/16	232
slave-j25.g	68/67/135	68/67/135	70/69/139	2221/2103/ 4314 (31 s.)	15	9/6/15	328
trimos-s.g	30/18/48	70/42/112	70/42/112	268/171/ 439	5	2/3/5	16
fairarb.g	13/20/33	13/20/33	13/20/33	19/26/45	8	4/4/8	8
fairar-j.g	20/32/52	20/32/52	20/32/52	49/62/111	13	4/9/13	14
unsafe2.g	9/6/15	50/56/106	60/64/124	81/99/180	25	21/7/28	48
josephs.s	20/16/36	44/36/80	44/36/80	46/36/82	6	4/2/6	4
low-new.g	23/18/41	56/44/100	56/44/100	70/50/120	7	3/4/7	4
low-good.g	21/16/37	60/46/106	60/47/107	75/58/133	9	5/5/10	10
low-true.g	25/20/45	72/58/130	74/61/135	97/84/181	9	4/6/10	14
low-uns.g	39/30/69	135/100/235	135/100/235	205/151/359		2/4/6	8
nak-good.g	36/33/69	182/142/ 324	175/152/ 327	451/366/ 817	33	21/24/45	80
n-reset.g	25/20/45	70/52/122	70/52/122	88/58/146	11	6/5/11	8
dags55-o.g		412/120/532	412/120/532	—	12	1/11/12	—
sl-j25.fc.g	74/57/133	87/61/148	87/61/148	722/465/ 1187	11	2/9/11	30
din20m.g	180/120/300	2816/2220 5036(24s)	3641/2910 6551(37s)	3979/3178 7157(44s)	586	640/51/ 691	723
din30m.g	270/180/450	6326/4985 11311(135s)	8576/6885 15461(232s)	9114/7313 16427(259s)	1331	1550/81/ 1631	1683
v-pet.fc.g	66/62/128	104/86/190	104/86/190	479/346/825	12	12	15
v-w-n.fc.g	49/36/85	49/36/85	49/36/85	50/38/88	2	2	3
cp-pipe.g	116/100/ 216	1065/763/ 1828(3s)	1234/879/ 2113(3s)	10910/7656/ 18566(547s)	46	47/34	316
cp-pipe.g.o	71/56/127	996/763/ 1759(2s)	1167/878/ 2045(3s)	10334/7656/ 17990(363s)	46	49/36	316
pe-send.fc.g	109/208/ 317	126/217/ 343	126/217 343	—	106	106/97	—

Table 2. Experimental results

This column also includes a run-time if it is more than one second. The next column presents cutoff counts. If the program run out of time for a selected time-out a cell of the table contains "–".

For checking how possible equivalent transformations of a PN can influence the size of the unfolding we have used tool *petrify* [2] for generating different PNs/STGs for the same initial PN/STG. For example, *slave-j25* is the initial specification from the SIS benchmark of asynchronous circuits, *sl-j25.fc* is an equivalent free-choice PN. One can observe that for *slave-j25* and *sl-j25.fc* the size of the reduced unfolding is significantly smaller than the size of the McMillan's unfolding. However for the minimized version of the same PN obtained with *petrify* both methods give equal size of unfoldings.

In [4] Esparza et. all suggested another refinement of McMillan's criteria for cutoffs. This refinement is in some sense exact because it allows to get the total order between the transitions with the same basic markings, i.e. each time two transitions with the same basic marking are met in unfolding one of them will be defined as a cutoff. The unfoldings obtained by method [4] have the minimal size (similar to those obtained by simple cutoffs). The shortcoming of the approach [4] is the restriction to the class of safe PNs. Enhanced cutoffs can work for general PNs though they cannot guarantee the generation of unfolding with minimal size. The slotted ring protocol example considered in [4] was checked for the enhanced cutoff criteria. As expected the size of obtained unfolding was in between the criteria by [4] and the McMillan's one. However the possibility to handle the unsafe specifications seems rather important for the asynchronous design because the OR-casuality as was shown in [22] can be modeled only by unsafe PNs. It is worth to mention that seven from the examples of Table 2 were detected to be unsafe.

7 Conclusion

We have presented an approach for checking properties of a PN on the basis of the ordering relations for transitions and places in the unfolding equivalent to the original PN. The unfolding is a finite prefix of the occurrence net. To truncate the occurrence net, we use the so-called, enhanced cutoff criterion. Contrary to the cutoff criterion presented in [10], our criterion allows to truncate occurrence net using cutoffs with images of equal size of local configuration, i.e., the occurrence net is truncated earlier in the generation process. Therefore, the size of the reduced unfolding is typically much smaller than the size of the reachability graph of the PN. We further applied this theory to asynchronous design. An implementability of Signal Transition Graphs, signal interpreted PNs, might be efficiently checked with the unfolding method. The experimental results show that for the safeness, boundedness, persistency and implementability conditions our method compares favorably with the symbolic BDD traversal of the PNs [16, 6]. We have implemented the method of PNs and STGs unfolding as a new command in the SIS tool [20].

Checking deadlocks by unfolding is exponential in general. However, McMillan suggested a practically efficient algorithm for this problem [10]. Using the enhanced cutoff criterion will further improve efficiency of checking deadlocks.

Acknowledgments.

We are greatly indebted to Javier Esparza for his successful efforts in finding flaws in the earlier versions of our cutoff criterion. We are also grateful to Alex Yakovlev and Alex Semenov for fruitful discussions on the early stage of this work. We highly appreciate the work on PN unfolding by Ken McMillan that was the impact for our research. We thank Luciano Lavagno for helping us to understand the SIS tool.

References

1. T.-A. Chu. *Synthesis of Self-timed VLSI Circuits from Graph-theoretic Specifications.* PhD thesis, MIT, June 1987.
2. J. Cortadella, M. Kishinevsky, L. Lavagno, and A. Yakovlev. Synthesizing Petri nets from state-based models. In *Proceedings of the International Conference on Computer-Aided Design*, November 1995. to appear.
3. J. Esparza. Model checking using net unfoldings. In M.-C. Gaudel and J.-P. Jouannaud, editors, *TAPSOFT'93: Theory and Practice of Software Development. 4th Int. Joint Conference CAAP/FASE*, volume 668 of *Lecture Notes in Computer Science*, pages 613–628. Springer-Verlag, 1993.

4. T. Esparza, S. Romer, and W. Vogler. An improvement of mcmillan's unfolding algorithm. Technical Report TUM: 19599, Technische Universitat Munchen, August 1995.

5. M. A. Kishinevsky, A. Y. Kondratyev, A. R. Taubin, and V. I. Varshavsky. *Concurrent Hardware. The Theory and Practice of Self-Timed Design.* John Wiley and Sons Ltd., 1994.

6. A. Kondratyev, J. Cortadella, M. Kishinevsky, E. Pastor, O. Roig, and A. Yakovlev. Checking Signal Transition Graph implementability by symbolic BDD traversal. In *Proceedings of the European Design and Test Conference (ED&TC)*, pages 325–332, Paris, France, March 1995.

7. A. Kondratyev, M. Kishinevsky, A. Taubin, and S. Ten. Analysis of Petri nets by ordering relations in reduced unfoldings. Technical Report TR:95-2-002, The University of Aizu, June 1995.

8. A. Kondratyev and A. Taubin. Verification of speed-independent circuits by STG unfoldings. In *Proceedings of the Symposium on Advanced Reserch in Asynchronous Cirsuits and Systems*, pages 64–75, Utah, USA, November 1994.

9. L. Lavagno and A. Sangiovanni-Vincentelli. *Algorithms for synthesis and testing of asynchronous circuits.* Kluwer Academic Publishers, 1993.

10. K. McMillan. A technique of state space search based on unfolding. *Formal Methods in System Design*, 6(1):45–65, 1995.

11. T. H.-Y. Meng. *Synchronization Design for Digital Systems.* Kluwer Academic Publishers, 1991. Contributions by David Mes David Dill.

12. D. E. Muller. Asynchronous logics and application to information processing. In *Proc. Symp. on Application of Switching Theory in Space Technology*, pages 289–297. Stanford University Press, 1963.

13. D. E. Muller and W. C. Bartky. A theory of asynchronous circuits. In *Annals of Computing Laboratory of Harvard University*, pages 204–243, 1959.

14. T. Murata. Petri nets: Properties, analysis and applications. *Proceedings of IEEE*, 77(4):541–580, April 1989.

15. M. Nielsen, G. Plotkin, and Winskel G. Events structures and domains. *Theoretical Computer Science*, 13(1):85–108, 1980.

16. E. Pastor, O. Roig, J. Cortadella, and R. Badia. Petri net analysis using boolean manipulation. In *15th International Conference on Application and Theory of Petri Nets*, pages 416–435, Zaragoza, Spain, June 1994.

17. S. Patil and J. Dennis. The description and realization of digital systems. In *Proceedings of the IEEE COMPCON*, pages 223–226, N.Y., 1972.

18. L. Pomello, G. Rozenberg, and C. Simone. A survey of equivalence notions for net based systems. *Lecture Notes in Computer Science*, 609:410–472, 1993.

19. A. Semenov and A. Yakovlev. Event-based framework for verifying high-level models of asynchronous circuits. Technical Report TR No.487, Computing Science, University of Newcastle upon Tyne, June 1994.

20. E. M. Sentovich, K. J. Singh, L. Lavagno, C. Moon, R. Murgai, A. Saldanha, H. Savoj, P. R. Stephan, R. K. Brayton, and A. Sangiovanni-Vincentelli. SIS: A system for sequential circuit synthesis. Technical Report UCB/ERL M92/41, U.C. Berkeley, May 1992.

21. A. Valmari. State of the art report: Stubborn sets. *Petri Nets Newsletter*, (46):6–14, 1994.

22. A. Yakovlev, M. Kishinevsky, A. Kondratyev, and L. Lavagno. OR causality: modelling and hardware implementation. In *Proceedings of the 15th International Conference on Application and Theory of Petri Nets*, volume 815 of *Lecture Notes in Computer Science*, pages 568–587, Zaragosa, Spain, June 1994. Springer-Verlag.

An $O(|S| \times |T|)$-Algorithm to Verify if a Net is Regular*

Andrei Kovalyov

Institut für Informatik, Technische Universität München ***

Abstract. In this paper we present an $O(|S| \times |T|)$-algorithm to decide if a given net is regular. Regular nets were introduced in [D1]. Their definition refers to the linear algebraic representation of nets. Regularity is a sufficient condition for an ordinary net to be live and bounded [D1]. Live and bounded extended free choice nets (EFC-nets for short) are a proper subset of regular nets [D1]. It is shown in [D1] that regular marked nets are not just a technical generalization of EFC-nets but have in fact more expressive power. We reduce the problem of regularity to the problem of liveness and boundedness of EFC-nets. To decide the well-formedness of a given EFC-net we give an inductive criterion of well-formedness. To prove the soundness of our algorithm we give a theorem on covering of strongly connected EFC-nets by minimal deadlock. An $O(|S| \times |T|)$-algorithm to decide the liveness of an initial marking for well-formed EFC-nets is given.

1 Introduction

In this paper we present an $O(|S| \times |T|)$-algorithm to decide if a given net is regular.

Regular nets were introduced in [D1]. They are those nets that satisfy the conditions of the Rank theorem [D1,D2], which refers to the linear algebraic representation of nets. Regularity is a sufficient condition for an ordinary net to be live and bounded [D1]. Liveness corresponds to the absence of global or local deadlock in a modeled system, boundedness to the absence of overflows in stores. Live and bounded EFC-nets are a proper subset of regular nets [D1]. It is shown in [D1] that regular marked nets are not just a technical generalization of EFC-nets but have in fact more expressive power.

In [D1] a simple transformation rule has been given, i.e. a mapping, which domain is the class of ordinary nets and range EFC-nets. We give a necessary and sufficient condition of regularity of a net. It is liveness and boundedness of the corresponding EFC-net. Hence we reduce the problem of regularity to the problem of liveness and boundedness of EFC-nets.

*** On leave from Institute of Engineering Cybernetics, Academy of Sciences, Minsk, Belarus

* This work was Partially supported by the Sonderforschungsbereich 342 of the Deutsche Forschungsgemeinschaft, Teilprojekt A3.

The EFC-nets are a small extension of free choice nets (FC-nets) [H]. They are attractive classes of Petri nets. They can model both conflict and concurrency and many problems of analysis can be decided for these classes easier than for wider classes of nets [DE1]. Even there exists a notion of free-choice hiatus [Be]. But the class of EFC-nets has the restrictive expressive power.

Well-formedness of a net is the existence of live and bounded marking for the net. A necessary and sufficient condition of liveness and boundedness of a Petri net is well-formedness of its structure and liveness and boundedness of its initial marking.

To decide the well-formedness of a given EFC-net we give an inductive criterion of well-formedness. This criterion is based on complete reducibility of the class of nets [DE2].

In [K1] we present an $O(|S| \times |T|)$-algorithm to decide if a given EFC-net is live and bounded. In this paper we refine this algorithm and apply to decide if a net is regular.

To prove the soundness of our algorithm we give a theorem on covering of strongly connected EFC-nets by minimal deadlocks. The theorem was not formulated and proved nowhere before, though for FC-nets the similar proposition follows from Theorem 4.2 [EBS]. The theorem from [EBS] uses an algorithm of constructing of minimal deadlock beginning from an arbitrary place. In our paper more formal proof is given.

It is well known from [H] that every initial marking of a well-formed EFC-net is bounded. Hence we need to check only liveness of a given marking.

To decide the liveness of an initial marking for well-formed EFC-nets, we define a simple transformation rule, i.e. a mapping, which domain is the class of EFC-nets and range 1-out nets (every place has at most one output transition). It proves to be useful twofold: at the checking whether an EFC-net is well-formed and at the checking whether an initial marking of a well-formed EFC-net is live. It is shown that a marked well-formed EFC-net is live iff (if and only if) the maximal unmarked deadlock of the corresponding 1-out net is the empty set. An $O(|S| \times |T|)$-algorithm to find the maximal unmarked deadlock of 1-out net is given.

2 Basic definitions and Preliminaries

A *net* is a triple $N = (S, T, F)$ with $S \cap T = \emptyset$ and $F \subseteq (S \times T) \cup (T \times S)$. S is the set of *places*, T is the set of *transitions*, $X = S \cup T$ is the set of vertices of N. For $x \in X$, the pre-set ${}^\bullet x$ is defined as ${}^\bullet x = \{y \in X | (y, x) \in F\}$ and the post-set $x^\bullet$ is defined as $x^\bullet = \{y \in X | (x, y) \in F\}$. For $Y \subseteq X$, ${}^\bullet Y = \cup_{x \in Y} {}^\bullet x$ and $Y^\bullet = \cup_{x \in Y} x^\bullet$.

On Fig.1 a net is shown. Places are graphically represented by circles, transitions by boxes and arcs of the relation F by arrows.

From now on we assume that all nets we deal with are *finite* (i.e. X is finite), *connected* (i.e. $X \times X = (F \cup F^{-1})^*$, where R^* denotes the reflexive and transitive

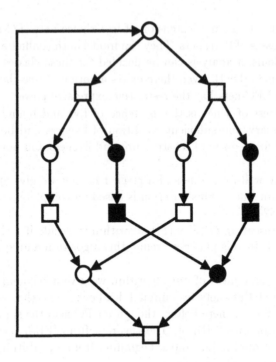

Fig. 1. A net

closure of a relation R), and have at least one place and one transition. N is *strongly connected* iff $X \times X = F^*$.

Let S and T be arbitrarily but fixed ordered. Let $Q \subseteq S (Q \subseteq T)$. Then we denote $\chi(Q)$ the characteristic vector of the set Q. The incidence matrix $C_N : S \times T \rightarrow \{-1, 0, 1\}$ of N is defined by $C_N(-, t) = \chi[t^\bullet] - \chi[^\bullet t]$. From now on we will write C instead of C_N for short. Let $\mathbf{0}$ denotes null vector. An *S-(T-)invariant* is a rational-valued solution of the equation $J \cdot C = \mathbf{0}$ $(C \cdot J = \mathbf{0})$.

For two vectors J and J', $J > J'$ iff for every i-th component $J(i) > J'(i)$. An invariant J is *semi-positive* iff $J \geq \mathbf{0}$ and $J \neq \mathbf{0}$. An invariant J is *positive* iff $J > \mathbf{0}$. The *support* of a semipositive invariant J is the set of elements x satisfying $J(x) > 0$; it is denoted by $\mathcal{S}(J)$.

For a set $Q \subseteq S$, J_Q denotes the space of such S-invariants J which satisfy $\forall s \notin Q$ $J(s) = 0$. $dim(Q)$ denotes the dimension of this space.

A net $N = (S, T, F)$ is called an *FC-net* iff $\forall s \in S$ $\forall t \in T$ $(s, t) \in F$ implies $s^\bullet = \{t\}$ or $^\bullet t = \{s\}$.

A net $N = (S, T, F)$ is called an *EFC-net* iff $\forall s \in S$ $\forall t \in T$ $(s, t) \in F$ implies $^\bullet t \times s^\bullet \subseteq F$.

A net $N = (S, T, F)$ is called an *1-out net* iff $\forall s \in S$ $|s^\bullet| = 1$.

A *marking* of a net $N = (S, T, F)$ is defined as a mapping $M : S \rightarrow \mathbb{N}$. A place is called *marked* by M iff $M(s) > 0$. Suppose $S' \subseteq S$; then $M(S') = \Sigma_{s \in S'} M(s)$. S' is *marked* iff $M(S') > 0$.

The expression $M_1[t > M_2$, where M_1, M_2 are markings of N, denotes that

M_1 enables transition t (e.g. t is fireable at M_1), and that the marking reached by the occurrence of t is M_2. The empty sequence ϵ is an occurrence sequence: we have $M[\epsilon > M$ for every marking M.

A transition t is *live* iff for every reachable marking M there exists a marking M' reachable from M which enables t. A marking M is called *live* iff every transition is live. A place $s \in S$ is *n-bounded* ($n \in \mathbb{N}$) iff for every reachable marking M, $M(s) \leq n$; s is *bounded* iff $\exists n \in \mathbb{N}$ such that s is n-bounded. A marking M is *bounded* iff $\forall s \in S$ s is bounded. A net N is *well-formed* iff there exists a live and bounded marking of N.

A *cluster* of a net N is a connected component of the relation $F_1 \cup F_1^{-1}$, where $F_1 = F \cap (S \times T)$. $[x]$ denotes the cluster containing $x \in S \cup T$). The set of all clusters is denoted by A.

In [D2] a necessary and sufficient condition of well-formedness of EFC-nets is given. It is so called the Rank Theorem.

Theorem 2.1 [D2]. *Let N be an EFC-net. Then*
(a) N is well-formed iff

$$there\ exists\ a\ positive\ S\text{-}invariant \tag{1}$$

$$there\ exists\ a\ positive\ T\text{-}invariant \tag{2}$$

$$rank(C) = |A| - 1 \tag{3}$$

(b) (N, M) is live and bounded iff N is well-formed and

$$for\ every\ semi\text{-}positive\ S\text{-}invariant\ I\ of\ N, I \cdot M > 0 \tag{4}$$

□

The Rank Theorem does not hold for arbitrary nets, but provides a sufficient condition for liveness and boundedness. In [D1] the corresponding class of regular nets is defined.

A marked net (N, M) (not necessarily EFC) is called *regular* [D1] iff it satisfies the conditions (1-4).

Let a be a cluster of a net. A *feedback* [DE1] of a is an arc $(t, s) \in F$ such that $s \in a, t \in a, (s, t) \notin F$. A net N is *feedback-free* [DE1] iff no cluster has feedbacks.

Let N be feedback-free. Its *EFC-representation* [D1] N' is defined by $N' = (S, T, F \cup \hat{F} \cup \hat{F}^{-1})$ where $\hat{F} = \{(s, t) \in S \times T \mid [s] = [t] \wedge (s, t) \notin F\}$ and $\hat{F}^{-1} = \{(t, s)|(s, t) \in \hat{F}\}$.

Lemma 2.2 [D1] *Every regular net is live, bounded and feedback-free.* □

Lemma 2.3 [D1] *Let N be feedback-free and N' its EFC-representation. Then*
(a) N' is an EFC-net;
(b) If (N, M) is regular then so is (N', M);
(c) $C = C'$, i.e. both nets have identical incidence matrices.
□

Now we give a necessary and sufficient condition of regularity.

Lemma 2.4 *Let N be a net. Then (N, M) is regular iff N is feedback-free and (N', M) is live and bounded, where N' is the EFC-representation of N.*

Proof. $\Rightarrow$. By Lemma 2.2 N is feedback-free. By Lemma 2.3a N' is an EFC-net. By Lemma 2.3b (N', M) is regular. Hence by Lemma 2.2 (N', M) is live and bounded.
$\Leftarrow$ By Lemma 2.1 (N', M) is regular. Since N is feedback-free, by Lemma 2.3c (N, M) is regular. □

Now we give a simple algorithm to construct the EFC-representation of N, which simultaneously checks whether a net is feedback-free.

Algorithm 2.5 *EFC-representation*

Input: a net N.
Output: the EFC-representation N' of N or the **message** "N is not feedback-free and hence not regular".
begin
    $N' := N$; $A_1 := A$ (the set of the clusters of N)
    while $A_1 \neq \emptyset$ **do**
        choose $a \in A_1$; $A_1 := A_1 \backslash a$; $S_a := S \cap a$; $T_1 := T \cap a$;
        while $S_a \neq \emptyset$ **do**
            choose $s \in S_a$; $S_a := S_a \backslash s$; $T_a := T_1$;
            while $T_a \neq \emptyset$ **do**
                choose $t \in T_a$; $T_a := T_a \backslash t$;
                if $(s, t) \notin F$ and $(t, s) \in F$ **then Stop, Output Message, endif**
                if $(s, t) \notin F$ and $(t, s) \notin F$ **then** $F' := F' \cup (s, t) \cup (t, s)$ **endif**
            endwhile
        endwhile
    endwhile
end

Proposition 2.6 *The problem to decide if net N is feedback-free and to construct EFC-representation of N has the complexity $O(|S| \times |T|)$.*

Proof. By definition a cluster is a connected (and also strongly connected) component of the relation $F_1 \cup F_1^{-1}$, where $F_1 = F \cap (S \times T)$. Hence the set of all clusters can be computed with the $O(|S| \times |T|)$-algorithm of [Ta]. Since the clusters are a partition on the set of vertices, the result follows. □
    Lemma 2.4 reduces the problem of regularity to the problem of liveness and boundedness of EFC-nets. Hence from now on we restrict ourselves to EFC-nets.

3 Complete reducibility of well-formed EFC-nets

In this section we give an inductive criterion of well-formedness.

It is well known that the theory of EFC-nets exhibits a nice duality [D2,H], strictly speaking, reverse-duality, i.e. when a proposition for $N = (S, T, F)$ remains true for *reverse-dual* net $N^{-d} = (T, S, F^{-1})$.

Lemma 3.1 [D2] *Let N be an EFC-net. Then N is well-formed iff the reverse-dual net N^{-d} is well-formed.* □

Let $N = (S, T, F)$ and $S' \subseteq S$, $T' \subseteq T$. Then (S', T', F') is a *subnet* of N iff $F' = F \cap ((S' \times T') \cup (T' \times S'))$. Since F' is completely defined by S' and T', we will write (S', T') instead of (S', T', F').

A net (S, T) is an *S-net* iff it satisfies two following conditions

$$\forall t \in T \quad |{}^{\bullet}t \cap S| = 1 \tag{5}$$

$$\forall t \in T \quad |t^{\bullet} \cap S| = 1 \tag{6}$$

A subnet (S_1, T_1) is an *S-component* iff it is a strongly connected S-net satisfying $S_1^{\bullet} = {}^{\bullet}S_1$

The following lemma gives a necessary condition of well-formedness of EFC-nets.

Lemma 3.2 [H] *If an EFC-net is well-formed then it is covered by S-components.* □

Let N' a subnet of N. We denote by $N \backslash N'$ the subnet generated by the nodes of $(S \cup T) \backslash (S' \cup T')$.

Let N be covered by S-components and $CV = \{N_1, \ldots, N_k\}, k > 1$ be a minimal cover of N. $\hat{N}$ is a *private subnet* of N_j iff it is a maximal subnet with

(i) $\hat{N}$ is nonempty and connected and

(ii) $\forall n \in \{1, \ldots, j - 1, j + 1, \ldots, k\}$ $N_n \cap \hat{N} = \emptyset$.

The following two lemmata are given in [DE2], where the notion of private subnets is defined using a cover of T-components. But due to the reverse-duality in well-formed EFC-nets (Lemma 3.1) we can formulate these results in reverse-dual way.

Lemma 3.3 [DE2] *Let $CV = \{N_1, \ldots, N_k\}, k > 1$ be a minimal cover of N by S-components. Then $\exists N_j \in CV$ such that for every private $\hat{N}$ of N_j, $\bar{N} = N \backslash \hat{N}$ is strongly connected.* □

From now on we will consider only such $N_j \in CV$ that for every private subnet $\hat{N}$ of N_j, $\bar{N} = N \backslash \hat{N}$ is strongly connected. By Lemma 3.3 such S-components exist.

We denote $S_o = {}^{\bullet}\bar{T} \cap \hat{S}$.

Lemma 3.4 *[DE2] Let N be a well-formed EFC-net and not an S-net, CV be a minimal cover of N by its S-components, $N_j \in CV$. Then for every private subnet $\hat{N}$ of N_j the following two conditions hold.*

$$\bar{N} = N \backslash \hat{N} \text{ is well-formed} \tag{7}$$

$$|\hat{S}_o| = 1 \tag{8}$$

□

The net on Fig.1 is a well-formed EFC-net, and one of its private subnets (black vertices) satisfies (8).

Lemma 3.4 (complete reducibility of well-formed EFC-nets) means that we can remove private subnets until we obtain a strongly connected S-net as the resulting net (an S-component of the initial net). The lemma gives us a necessary condition of well-formedness of N. We will prove that it is also a sufficient condition. For the proof we need now some results for EFC-nets.

Lemma 3.5 *(a) [D2] Let N be an EFC-net. Then $rank(C) = |S| - dim(S)$;*
(b) [K2] N is well-formed iff it is covered by S-components and (3) holds;
(c) [K2] If N is covered by S-components then $rank(C) \geq |A| - 1$. □

We are ready now to prove the main result of this section.

Theorem 3.6 *Let N be an EFC-net, covered by S-components and not an S-net, CV be a minimal cover of N by its S-components, $N_j \in CV$ and $\hat{N}$ be one of private subnets of N_j. Then N is a well-formed net iff for every private subnet $\hat{N}$ of N_j (7) and (8) hold.*

Proof. $\Rightarrow$. Lemma 3.4.

$\Leftarrow$. Since N has a cover of S-components, and taking into acount Lemma 3.5b, it is sufficient to prove that (3) holds for N. Taking into account Lemma 3.5c we need only to prove: $rank(C) \leq |A| - 1$. Since N has a cover of S-components and $\bar{S} \subset S$, we have $dim(S) \geq dim(\bar{S}) + 1$. We have also $|S| = |\bar{S}| + |\hat{S}|$. By (8) $|A| = |\bar{A}| + |\hat{S}| - 1$. By (7), $rank(\bar{C}) = |\bar{A}| - 1$. By Lemma 3.5a $rank(C) = |S| - dim(S) \leq |\bar{S}| + |\hat{S}| - dim(\bar{S}) - 1 = rank(\bar{C}) + |\hat{S}| - 1 = |\bar{A}| - 1 + |\hat{S}| - 1 = |A| - 1$.
□

To apply the criterion of Theorem 3.6 we need to look for a cover of S-components. After a next S-component is found, we will check if the covered subnet is well-formed or not. To satisfy the conditions of Theorem 3.6 we need the strongly connectedness of net N at every step of finding of a cover of S-components. We propose to choose place s from set $^\bullet \bar{T} \backslash \bar{S}$. This guarantees strong connectedness of the net obtained. To decide well-formedness of N, it is left only to check (8) for every private subnet.

4 Cover of Minimal Deadlocks

In this section we will show that every strongly connected EFC-net is covered by minimal deadlocks.

A set $D \subseteq S$ is a *deadlock* (a *trap*) iff ${}^\bullet D \subseteq D^\bullet (D^\bullet \subseteq {}^\bullet D)$.

Lemma 4.1 [H] *Let N be a well-formed EFC-net. Then every minimal deadlock is an S-component.* $\square$

Suppose an EFC-net has a cover of minimal deadlocks. If a deadlock of the cover is not an S-component then by Lemma 4.1 N is not well-formed. Hence for our aim, it is sufficient to look for a cover of minimal deadlocks and to check every deadlock of the cover to be an S-component. But before our algorithm stops, we can not know if a given EFC-net is well-formed or not. Hence we know neither an arbitrary EFC-net is covered by S-components nor by minimal deadlocks. Hence for the proof of soundness of our future algorithm, we need a proposition on covering of every strongly connected EFC-net by minimal deadlocks. The proposition is not true for all EFC-nets. Strongly connectedness is not very strong restriction, and by Lemma 3.2 it is a necessary condition of well-formedness.

To simplify the finding of deadlocks in EFC-nets, we define a simple transformation rule ϕ, i.e. a mapping, which domain is the class of EFC-nets and range 1-out nets.

Let N be an EFC-net. The net $\phi(N)$ is defined as the result of transformation of the transitions of every cluster a into a single transition t of $\phi(N)$:

${}^\bullet t = {}^\bullet t_1, t_1 \in a;$

$t^\bullet = \bigcup_{t_1 \in a} t_1^\bullet.$

On Fig.2 we see a cluster $\{s_1, s_2, t_1, t_2\}$ of an EFC-net N together with output places, and on Fig.3 the corresponding transition t of 1-out net $\phi(N)$ together with input and output places.

Lemma 4.2 *Let $N = (S, T, F)$ be an EFC-net. Then*

(a) D is a deadlock of N iff D is a deadlock of $\phi(N)$;

(b) D is a minimal deadlock of N iff D is a minimal deadlock of $\phi(N)$;

(c) Let $R \subseteq S$. D is a maximal deadlock of N in R iff D is a maximal deadlock of $\phi(N)$ in R.

Proof.

(a) $\Rightarrow$. Let $t \in {}^\bullet D$ in N. Since D is a deadlock, $t \in D^\bullet$. By the EFC-property, all transitions of the same cluster have the same set of input places. Hence in $\phi(N)$, the corresponding transition $t_c \in D^\bullet$ and ${}^\bullet D \subseteq D^\bullet$. Hence D is a deadlock of $\phi(N)$.

$\Leftarrow$. Analogously.

(b) and (c) easy follow from (a). $\square$

A *path* of N is a nonempty sequence $x_1, \ldots, x_k$ of nodes satisfying (x_1, x_2), $\ldots, (x_{k-1}, x_k) \in F$. It is a *circuit* iff $x_k = x_1$. An *elementary* path (circuit) is a path (circuit) which does not contain any vertex twice (exept maybe $x_k = x_1$).

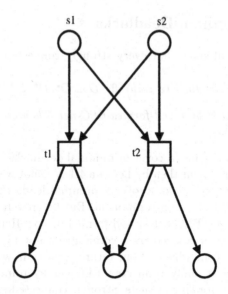

Fig. 2.

A subnet (S,T) is said to be *generated* by the set of places S iff $^\bullet S = T$.

The following lemma gives some structural properties of minimal deadlocks in EFC-nets.

Lemma 4.3 $[DE1]$ *Let D be a set of places of an EFC-net. Then D is minimal deadlock iff $(D, ^\bullet D)$ is strongly connected and (5) holds for $(D, ^\bullet D)$.* $\square$

We give now some structural property of minimal deadlocks in 1-out nets.

Lemma 4.4 *Let D be a set of places of an 1-out net. Then D is minimal deadlock iff $(D, ^\bullet D)$ is a maximal strongly connected subnet and (5) holds for $(D, ^\bullet D)$.*

Proof. $\Rightarrow$. By Lemma 4.3 $(D, ^\bullet D)$ is strongly connected and (5) holds for $(D, ^\bullet D)$. We continue indirect. Let $(D, ^\bullet D)$ be not maximal with those properties. Then there exists a maximal strongly connected subnet $(S_1, T_1) \supset (D, ^\bullet D)$ satisfying (5). We have two cases.

(a) $D = S_1$. Then $^\bullet D \subset T_1$. Hence $T_1 \backslash ^\bullet S_1 \neq \emptyset$ and (S_1, T_1) is not strongly connected. A contradiction.

(b) $D \subset S_1$. Since (S_1, T_1) is strongly connected, there exists a path $(S_1 \backslash D) \ni s_1 t s_2 \in D$. Since $t \in {}^\bullet D$, and $(D, ^\bullet D)$ is strongly connected, $|{}^\bullet t \cap S_1| \geq 2$. A contradiction to (5).

$\Leftarrow$. Lemma 4.3. $\square$

Lemma 4.5 *Let $N = (S, T, F)$ be a strongly connected 1-out net. Then it is covered by*

(a) elementary circuits;

(b) subnets generated by minimal deadlocks.

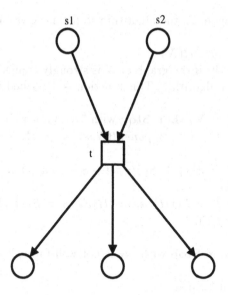

Fig. 3.

Proof. (a) trivial.

(b) By (a) N is covered by its elementary circuits. By definition every elementary circuit defines a strongly connected subnet of N with (5). By Lemma 4.4 N is covered by minimal deadlocks. □

Theorem 4.6 *Let $N = (S, T, F)$ be a strongly connected EFC-net. The arcs of $F \cap (T \times S)$ of N and hence the vertices $S \cup T$ are covered by subnets generated by minimal deadlocks.*

Proof follows from Lemmata 4.5 and 4.2b. □

5   The algorithm to decide well-formedness of EFC-nets and its complexity

Algorithm 5.1 *decide-well-formed (N)*

Input:
$N = (S, T, F)$ is a strongly connected EFC-net
Output is one of the following messages:
1: "N is well-formed "
2: "N is not well-formed "
Function: *get − S − component(N, s)*
This function finds an S-component of N containing place s. It is similar to the function mark-S-component [Ke2], but gives (as the output) the set $\bar{S}$ of places of an S-component containing s. If it finds a minimal deadlock not

generating an S-component, the algorithm stops and gives a message "N is not well-formed".

Function: $strongconnect(N)$

This function checks if the graph of N is strongly connected and is based on the depth-first search algorithm. This function is described in [Ta].

begin

if $strongconnect(N) = No$ **then Stop** with "N is not well-formed" **endif**

$choose\ s \in S;\ \bar{S} := get - S - component(N, s);\ \bar{T} := \bar{S}^\bullet;\ S^i := {}^\bullet\bar{T}\backslash\bar{S};$

while $(S^i \neq \emptyset)$ **do**

$\quad choose\ s \in S^i;\ S^i := S^i\backslash\{s\};\ S_1 := get - S - component(N, s);\ T' := {}^\bullet s\backslash\bar{T};$

$\quad$**while**$(T' \neq \emptyset)$ **do**

$\quad\quad choose\ t \in T';\ \bar{T} := \bar{T}\cup\{t\};\ s_i := {}^\bullet t \cap S_1;\ \hat{S} := \hat{S}\cup\{s_i\};\ S^i := S^i\cup({}^\bullet t\backslash s_i);$

$\quad\quad T' := T'\cup({}^\bullet s_i\backslash\bar{T});$

$\quad$**endwhile**

$\quad$**if** $(\hat{S}^\bullet\backslash\bar{T}) \neq \emptyset)$ **then Stop** with "N is not well-formed" **endif**

endwhile

Stop with "N is well-formed";

end

Proposition 5.2 *The problem to decide well-formedness of EFC-nets has the complexity $O(|S| \times |T|)$.*

Proof. The complexity of function "strongconnect" for a directed graph $G = (V, E)$ is $O(|V| + |E|)$ [Ta]. The complexity for function "mark-S-component" is given in [Ke2] ($O(|T|)$). Since the set of the private subnets (under consideration) defines a partition on the set of vertices of the net, the complexity of the finding and checking of the private subnets is $O(|S| \times |T|)$. Hence the upper bound of the time complexity of the algorithm is $O(|S| \times |T|)$. $\square$

6 Liveness of an initial marking

Since well-formedness is only a necessary condition for a net to be live and bounded, we need to decide if a given initial marking of a well-formed EFC-net is live and bounded or not.

Lemma 6.1 [H] *Let N be an EFC-net. Then*

(a) an initial marking M^0 of N is live iff every deadlock contains a marked trap;

(b) if N is well-formed then every initial marking of N is bounded. $\square$

Lemma 6.2 *Let N be a well-formed EFC-net and M be a marking of N. Then the following equivalent:*

(a) (N, M) is live and bounded;

(b) the maximal unmarked deadlock of (N, M) is the empty set;

(c) the maximal unmarked deadlock of $(\phi(N), M)$ is the empty set.

Proof. $(a) \Rightarrow (b)$ (Indirect.) Let the maximal unmarked deadlock D of (N, M) is not empty. From definition of a deadlock it follows easy that an unmarked deadlock stay unmarked at any reachable marking of N. Then all the transitions $D^{\bullet}$ can not be live. A contradiction.

$(a) \Leftarrow (b)$ Let D be a minimal deadlock of N. Since the maximal unmarked deadlock of N is the empty set, D is marked. By Lemma 4.1 D is an S-component, hence a trap. By Lemma 6.1a (N, M) is live. By Lemma 6.1b (N, M) is bounded.

$(b) \Leftrightarrow (c)$ follows from Lemma 4.2c. $\square$

Lemma 6.2 reduces the problem of liveness of a well-formed EFC-net N to the problem to decide if the maximal unmarked deadlock of $\phi(N)$ is the empty set.

The maximal unmarked deadlock of an 1-out net N can be obtained by the following algorithm, a modification of an algorithm in [S].

Algorithm 6.3 *get-max-unmarked-deadlock*

Input: a marked 1-out net (N, M).
Output: maximal unmarked deadlock D of (N, M).
begin
$\quad D := S \backslash \{s \in S | M(s) > 0\}; \ T_d := T \backslash D^{\bullet}; \ T_1 := T \backslash T_d$
$\quad$ **while** $T_d \neq \emptyset$ **do**
$\quad\quad D := D \backslash T_d^{\bullet}; \ T_d := T_1 \backslash D^{\bullet}; \ T_1 := T_1 \backslash T_d$
$\quad$ **endwhile**
end

Proposition 6.4 *The following problems has the complexity* $O(|S| \times |T|)$:
(a) to decide whether a strongly connected 1-out net (N, M) is live;
(b) to decide whether a well-formed EFC-net (N, M) is live;
(c) to decide whether a marked net (N, M) is regular.

Proof. (a) Let 1-out net N be presented as two functions $S_o : S \rightarrow T$, $T_o : T \rightarrow 2^S$, defined as $S_o(s) = s^{\bullet}$, $T_o(t) = t^{\bullet}$. Since the sets T_d do not intersect in any two different iterations of the algorithm, the complexity of computation:
$\quad D$ (in all iterations) $- O(|S| \times |T|)$;
$\quad T_d$ (in all iterations) $- O(|S| \times |T|)$;
$\quad T_1$ (in all iterations) $- O(|T| \times |T|)$.
Since $|T| \leq |S|$ in strongly connected 1-out nets, the upper bound of the time complexity is $O(|S| \times |T|)$.

(b) It is easy to see that the simple algorithm to construct $\phi(N)$ has the complexity $O(|S| \times |T|)$. Then the result follows from (a) and Lemma 6.2.

(c) follows from (b), Lemma 2.4, Propositions 2.6 and 5.2. $\square$

7 Conclusions

In [Ke1] $O(|S|^2 \times |T|)$-algorithm to decide if a given EFC-net is well-formed was given. This algorithm checks the conditions (1)-(3) of Theorem 2.1. To check the condition (1) (if a net has a positive S-invariant), this algorithm finds a cover of S-components. Checking of the condition (2) has been done analogously (finding a cover of S-component on reverse-dual net). Hence two main steps of the algorithm in [Ke1] are: to find a cover of S-components and to compute the rank of the incidence matrix. Finding a cover of S-components has been done in [Ke2] in $O(|S| \times |T|)$-time. Since the calculation of a matrix rank requires $O(|S|^2 \times |T|)$, the Rank Theorem can be checked in $O(|S|^2 \times |T|)$.

Our algorithm avoids the computation of rank. It results in a reduction by one order of magnitude, compared to the algorithm in [Ke1]. Hence the absolute minimum of the complexity has been reached, because the input information on the structure of the net takes the capacity $O(|S| \times |T|)$.

The mapping ϕ was used to decide the liveness of an initial marking. But it proves to be useful for another aim. In [Ke3] the net N itself is used to find a cover of S-components for the FC-net N. But the structure of EFC-net hampers to do that ([Ke2]). To overcome this difficulty, in [Ke2] the cluster graph is used as an auxiliery data to find a cover of S-components. We propose the mapping ϕ for this aim. Since 1-out nets are a subclass of FC-nets, Lemma 4.2b allows us to compute a cover of minimal deadlocks using only one graph $\phi(N)$ instead of two (N itself and cluster graph) in [Ke2]. But to check if every minimal deadlock is an S-component, we need again N itself.

References

[Be] E.Best: Structure theory of Petri nets: the free choice hiatus. LNCS 254. (1987) 168-205.

[D1] J.Desel: Regular marked Petri nets. Proc. of the 19th Int. Workshop on Graph-Theoretic Concepts in Computer Science/ Jan van Leeuween (ed.) LNCS 790, pp.264-275 (1993).

[D2] J.Desel: A proof of the Rank Theorem for extended free choice nets. Proc. of the 13th Int. Conf. on Appl. and Theory of Petri nets/Jensen,K.(ed.) LNCS 616, pp.134-153 (1992).

[DE1] J.Desel, J.Esparza: Free Choice Petri Nets. Cambridge Universty Press. 1995, 244p.

[DE2] J.Desel,J.Esparza: Reachability in cyclic extended free choice systems. Theoretical Computer Science 114(1993) 93-118.

[EBS] J.Esparza, E.Best, M.Silva: Minimal deadlocks in free choice nets. Hildesheimer Informatik-Berichte. N 1. July 89. Universitat Hildesheim.

[H] M.T.Hack: Analysis of production schemata by Petri nets, Cambridge, Mass.: MIT, Project MAC TR-94, (1972) pp.119. Corrections 1974.

[Ke1] P.Kemper: On Finding a cover of minimal siphons in extended free choice nets. Algorithmen und Werkzeuge für Petrinetze. Workshop der GI-Fachgruppe 0.0.1 "Petrinetze und verwandte Systemmodelle". Berlin, 10-11 Oktober 1994.

[Ke2] P.Kemper: $O(|P| \times |T|)$-algorithm to compute a cover of S-components in EFC-nets. Forschungsbericht 543, Universitaet Dortmund, 1994.

[Ke3] P.Kemper: Linear time algoritm to find a minimal deadlock in a strongly connected free-choice nets. Proc. of the 14th Int. Conf. on Appl. and Theory of Petri nets. Chicago, June 1993. LNCS 691.

[K1] A.V. Kovalyov: An $O(|S| \times |T|)$-Algorithm to Verify Liveness and Boundedness in Extended Free Choice Nets. Proceedings of the 10th IEEE International Symposium On Intelligent Control. August 27-29, 1995, Monterey, Califorornia, USA. pp.597-601.

[K2] A.V. Kovalyov: A new formulation of the Rank Theorem for live and bounded extended free choice nets. Submitted at the 1996 IEEE International Conference on Robotics and Automation. Minneapolis, Minnesota, USA.

[S] P.H.Starke: Analyse von Petri-Netz-Modellen. B.G.Teubner Stuttgart 1990 (in German).

[Ta] R. Tarjan. Depth-first search and linear graph algorithms. SIAM J. Comp., Vol.1, p.p.146-160.

The Consistent Use of Names and Polymorphism in the Definition of Object Petri Nets

Charles Lakos,
Computer Science Department,
University of Tasmania,
GPO Box 252C,
Hobart, TAS, 7001, Australia.
Email: Charles.Lakos@cs.utas.edu.au

Abstract: This paper seeks to present a more elegant and general definition of Object Petri Nets than previously. It is more general since it supports transition fusion as well as place fusion. It is more elegant because it captures all the notions of place substitution, transition substitution, place fusion, and transition fusion under the single notion of binding. This is achieved by explicitly supporting names in the formalism, in line with the π-calculus which recognises that names are pervasive and should be explicitly included in a formalism in order to model object mobility. The definition in this paper is also more consistent in its use of polymorphism and embodies a more obvious duality between states and changes of state. Object Petri Nets represent a complete integration of object-oriented concepts into Petri Nets. They have a single class hierarchy which includes both token types and subnet types, and which readily supports modelling systems with multiple levels of activity. Interaction between subnets can be synchronous or asynchronous depending on whether the subnet is defined as a super place or a super transition. While not presented in this paper, Object Petri Nets can be transformed into behaviourally equivalent Coloured Petri Nets, thus providing a basis for adapting existing analysis techniques.

Keywords: Theory of High-Level Petri Nets, Object-Orientation, Multiple Levels of Abstraction

1 Introduction

In recent years there has been considerable interest in applying object-oriented technology to Petri Nets. A measure of this interest is apparent from the decision to hold a special Workshop on Object Oriented Programming and Models of Concurrency as part of the 1995 International Conference on the Application and Theory of Petri Nets. A number of papers at that workshop included proposals for integrating object-oriented technology with Petri Nets, some of which are considered in §5.

One possible approach is to consider the enhancements required to progress from Coloured Petri Nets (CPNs) [13] to Object Petri Nets (OPNs) [19]. The enhancements include allowing token values to be object identifiers, the inclusion of inheritance (together with the support for polymorphism), and the inclusion of test and inhibitor arcs (together with functions for evaluating part of the state of a net without modifying that state). The definitions presented there were overly complex, partly because of a desire to demonstrate the continuity between CPNs and OPNs.

The current paper attempts to present a more elegant definition of OPNs by rationalising and generalising the concepts from the previous definition. These advances were motivated, in part, by the π-calculus [23, 24] which strives to present a semantics for concurrent systems as elegant as the λ-calculus is for sequential systems. It is not the intention of the current paper to rival the proposals of the π-calculus – the purpose of the two formalisms is distinct. In fact, Milner et. al. specifically acknowledge that there should be a range of formalisms for dealing with concurrent systems. Their primary concern is to arrive at a minimal semantics which supports object mobility, and consequently they avoid supporting different kinds of values apart from names. Further, while they intend to use the π-calculus to model concurrent object-oriented systems, they make no explicit provision for object-oriented features.

By contrast, this paper presents a formal definition for Object Petri Nets, which includes arbitrary values, which carefully integrates object-oriented features, and which is set in the context of Petri Net theory. However, it does adopt a key aspect of the π-calculus, namely the recognition of the pervasiveness of names, and their explicit inclusion in the formalism. In recognition of this, the current formalism for Object Petri Nets is abbreviated OΠNs (which also serves to distinguish it from the previous definition which was abbreviated OPNs). The explicit inclusion of names was also prompted by the recognition that the textual implementation of the OPN formalism (in the language LOOPN++ [16]) seemed to have a greater economy of ideas than the formal definition.

Another significant feature of this paper is the more consistent incorporation of polymorphism. Polymorphism goes hand in hand with inheritance, and the two constitute the characteristic features of object orientation [33]. The previous definition of OPNs explicitly distinguished between simple places (the traditional Petri Net places) and super places (subnets with place properties), and between simple transitions (the traditional Petri Net transitions) and super transitions (a variant of substitution transitions). The current formalism for OTNs achieves the same flexibility by only catering for simple places, simple transitions and the use of polymorphism. We recognise that super places can be defined as inheriting from the corresponding simple place type, and hence can be used polymorphically in simple place contexts. The same applies to transitions.

This paper motivates the formalism with a sequence of examples in §2 which shows that, as you seek to model increasingly complex systems, there is a natural progression from Place-Transition nets, to Coloured Petri Nets, to Hierarchical Coloured Petri Nets, to Object Petri Nets. The formal definition of OTNs is presented in §3, together with explanatory comments, while some aspects of the formal definition are highlighted in §4. In §5, we contrast the formalism of OTNs with others, while the conclusions are given in §6.

2 Examples

This section presents a sequence of examples which demonstrate the enhanced descriptive power of a number of Petri Net formalisms, leading to Object Petri Nets. The sequence culminates in an example motivated by Electronic Data Interchange (EDI). A simpler version of this example was presented in [19] and was adapted from the Z specification given by Swatman et. al. [30].

2.1 The four seasons or the four quarters

A simple example to demonstrate the concepts of Petri Nets is that of the four seasons or the four quarters (of the year). This can be modelled as the Place-Transition Net (PTN) of fig 2.1 [26].

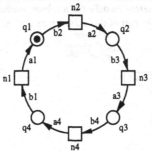

Fig 2.1: The four seasons or four quarters

This PTN has four places drawn as circles representing the four quarters of the year (and labelled q1, q2, q3, q4), and four transitions drawn as squares representing the changes between the quarters (and labelled n1, n2, n3, n4 to indicate the next quarter). The arcs are also labelled (b1, a1, etc.) to indicate the before and after state of each change. The initial marking indicates the presence of a token in place q1, implying that we begin in the first quarter. In this initial marking, only transition n2 is enabled, and when it is fired, the token is removed from place q1 and a token is added to q2.

If we wish to extend this PTN to indicate the year as well as the quarter, then catering for y years would require 4y places and 4y transitions in a spiral rather than circular pattern. The above simple structure will be lost, unless we move to a more expressive Petri Net formalism, such as that of Coloured Petri Nets (CPNs) [13]. A CPN can model the same extended system using the same structure of fig 2.1 simply by associating an integer colour with the token. Now the labelling of the arcs becomes significant, since this must indicate which coloured token is removed and added by each transition. (We have chosen to refer to the four quarters rather than the four seasons so that the change of year is easy to locate.) Two possible styles for indicating this evolution are given in parts (a) and (b) of fig 2.2 for each of transitions $n1$ and $n2$. For transition $n2$ (the next quarter is 2), the year does not change. For the transition $n1$ (the next quarter is 1), the year does change. In part (a), distinct identifiers are associated with distinct arcs and the relationship between the associated tokens is given by the guard associated with the transition (enclosed in brackets). In part (b), matching expressions are used to relate the associated tokens. The latter style is adopted in the tool Design/CPN [14]. Note, however, that from a formal perspective, the two approaches are equivalent.

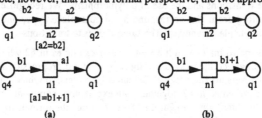

(a)   (b)

Fig 2.2: Annotations for coloured tokens

2.2 Token ring network

The example of §2.1 demonstrated the desirability of associating a data value with a token and thereby avoiding excessive duplication of net components. In modelling more complex systems, it is necessary to have more descriptive mechanisms. For example, we may wish to treat the net of fig 2.1 as a token ring network (as in [13]). In this case, the transitions $n1$, $n2$, $n3$, $n4$ would be the nodes on the network, and the places $q1$, $q2$, $q3$, $q4$ would model the cabling or quadrants joining the nodes.

Now, the tokens circulating round the ring are records which include the sending and receiving node numbers (together with the content of the message, which we ignore). Each node then involves a complex transformation, and is therefore modelled as a subnet, leading to the formalism of Hierarchical Coloured Petri Nets (HCPNs) [13]. Part of this subnet may appear as in fig 2.3.

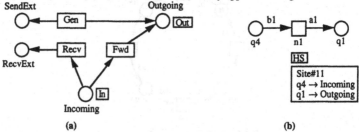

(a)   (b)

Fig 2.3: Subnet for a token ring node

Part (a) illustrates the subnet which is to replace each of the transition nodes from fig 2.1. Places *Incoming* and *Outgoing* are annotated to indicate that they are used as port nodes (for interfacing with the external environment) together with their mode of usage (for input or output). The transitions will have attached guards (which are not shown) to filter the messages: those which specify a receiver other than the current node will be forwarded (by transition *Fwd*); those destined for the current node will be received (by transition *Recv*); and those which are generated locally for other nodes are submitted to the net (by transition *Gen*). Part (b) illustrates how each transition node from fig 2.1, such as $n1$, is substituted by the subnet, as indicated by the *HS* flag (for *Hierarchy + Substitution*) and the associated inscription. In this case, the inscription indicates that the node is to be replaced by an instance of the subnet *Site#11*, with the port assignment as shown.

It is worth noting that HCPNs embody a number of notions of binding: a variable of a transition, such as $a2$ of fig 2.2 is bound to a value in a binding element; a substitution transition such as $n1$ of fig 2.3(b) is bound to (or substituted by) an instance of the relevant subnet; a port node of the subnet such as *Incoming* is bound (or fused) to a corresponding socket node. HCPNs also cater for other styles of place fusion, which constitute yet another form of binding.

It is also worth noting that the semantics of HCPNs ignores the arcs and arc inscriptions incident on a substitution transition (node $n1$ of fig 2.3). The semantics of the construct is determined solely by the inscription. While gaining simplicity, this sacrifices a possibility for constraining the abstraction.

2.3 Electronic Data Interchange

Our next enhancement is to use the token ring of §2.2 as a network to support an Electronic Data Interchange (EDI) system. The added complexity being introduced is that the messages which circulate round the ring are complex objects (and not just records), and that each node now contains a number of subscribers which are also complex objects and which may interact with the message.

The fundamental issue here is that complex objects are best modelled by abstraction and encapsulation. Just as it was important for the token ring of §2.2 to encapsulate the logic of each node in a substitution transition, so it is important in this example to be able to encapsulate the data and activity of these complex objects. It is no longer desirable to have transitions like *Recv* and *Fwd* which treat the messages as passive data. The handling of a message at each node should now depend on some cooperative, synchronised activity between the message and the subscribers resident at the node.

A **document** in our EDI system consists of some standard information (including the address of sender and receiver and the identifier of the associated umbrella document) and a number of information fields. These fields are in the form of pairs, the first element of the pair being a field identifier and the second being the corresponding value, e.g. (cost, $125). The number of information fields varies during the lifetime of the document. For example, a bill of lading for some cargo may have a paper trail including a request for a letter of credit, a request for insurance cover, etc., each of which results in further information being added to the document. A document optionally has an associated reply.

An **umbrella document** encapsulates a document together with its paper trail, which consists of related subdocuments. Umbrella documents have some standard information (including the unique umbrella identifier and the original document fields) together with some variable information. The variable information includes the subdocuments which are yet to be posted, the subdocuments which have already been posted, and the replies expected to subdocuments which have been posted.

An **EDI subscriber** is identified by their address. Subscribers also have a number of mail folders – a spool area where documents which have been received are stored prior to examination, a mail box of documents which have been received and examined, and a folder of documents which have been posted (either originating with this subscriber or replies sent to received mail).

The **EDI system** consists of the token ring network (as in fig 2.1), the nodes of the network (each of which contains a number of subscribers), and the documents which circulate around the network. The EDI system provides the medium by which subscribers can generate documents, the documents are encapsulated in umbrellas, the subdocuments are delivered, and the subscribers respond to mail.

By encapsulating umbrellas and subscribers, the EDI system has a number of levels of activity and information containment. The system consists of a number of network nodes, each of which contains a variable number of subscribers. There is a set of umbrellas circulating round the network, each of which contains a dynamically varying set of documents. It is desirable to reflect the multiple levels directly in our OIIN model, since this direct use of abstraction becomes even more important in modelling more complex systems [6, 11]. Similarly, we choose to model synchronous interaction between subscribers and messages, since this is the most direct representation of the cooperative interaction (and is the style embodied in the original Z specification [30]).

In our OITN solution, each umbrella and subscriber is defined by an appropriately named class or subnet as in fig 2.4. Each of these classes has arcs incident on the class frame, which indicate that the instances of these classes can act as places by offering tokens to and accepting tokens from their environment. This transfer of tokens needs to be synchronised with actions removing tokens from or depositing tokens into the place. In this way, OITNs support synchronous interaction with subnets.

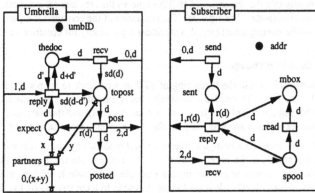

Fig 2.4: Umbrellas and subscribers as OPN pages

In order to understand the details of these classes, the following functions are assumed to be defined:

a(d), a(s):   the address given in document d and subscriber s

r(d):   the reply expected to document d, if any

sd(d):   the subdocuments given by the information fields of document d

                   ($sd(d-u)$ is the subdocuments given by d but not already in u)

The non-trivial guards for transitions are given below, with subscript s indicating subscribers:

recv$_s$: [a(d)=addr]   read$_s$: [r(d)=∅]   reply$_s$: [r(d)≠∅]

Each new umbrella is initialised by receiving a document via the *recv* transition. This saves the document in place *thedoc* and stores the subdocuments for posting in place *topost*. Subsequently, any of the subdocuments for posting can be posted by firing transition *post* which transfers the subdocument to place *posted*, and saves the expected reply in place *expect*. If an expected reply is received with the firing of transition *reply*, its information fields are used to augment the saved document, and possibly generate further subdocuments for posting. Transition *partners* with its compound test and inhibitor arcs is enabled only if variable x is bound to the expected replies and variable y is bound to the documents to be posted [15].

Each subscriber can spontaneously generate a document for posting with the *send* transition. It may also receive a document at the *recv* transition, and save it in place *spool* for later reading. If the document requires a response, the transition *reply* is fired to read and respond. Otherwise the transition *read* is fired. Both of these transitions serve to transfer the document from the user's spool area to their mailbox (the place *mbox*).

The nodes of the EDI system can be partly modelled as the OITN segments of fig 2.5 (cf. fig 2.3). Part (a) indicates the substitution of a token ring node by a subnet. In contrast to fig 2.3, the substitution is completely captured by the guard of the (substitution) transition. This indicates that the transition *n1* is bound to an instance *Site#11* of some subnet. The guard also indicates that the subnet place *Incoming* is bound to place *q4*, and similarly for the place *Outgoing* being bound to place *q1*. In this way, OITNs have a consistent notation for *binding* which captures the HCPN notions of transition substitution (and transition fusion), port assignment, and place fusion. This is possible because OITNs, like the π-calculus, recognise the pervasiveness of names. Consequently, the labels *q4*, *n1*, *q1* are not identifiers associated directly with places and transitions, but indirectly via names or object identifiers. It is the binding which determines which particular instance (given by an object identifier) an identifier is bound to, with the default being the local instance. The binding can vary within the usual constraints of polymorphism.

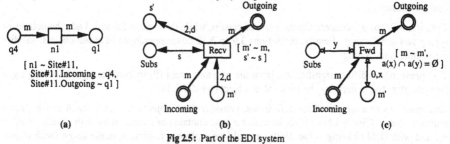

(a)   (b)   (c)

Fig 2.5: Part of the EDI system

Fig 2.5(b) indicates part of the subnet (of which one instance is *Site#11*) which processes incoming and outgoing EDI messages. Note that places *Incoming* and *Outgoing* are marked as exported by a double outline. The *Recv* transition indicates how an incoming message may interact with a subscriber: a message *m* is extracted from place *Incoming*; message *m'* (which is bound to *m*) offers a flagged subdocument *2,d*; a subscriber *s* is extracted from place *Subs*; subscriber *s'* (which is bound to *s*) accepts the flagged document *2,d*; the message *m* is forwarded for further processing. Note that binding finds a further use in allowing interaction across various levels of an object, as is necessary to unpack the levels of information containment. Note also that the binding is established dynamically or on a per-transition basis, in contrast to the place fusion of HCPNs which is static for the duration of the net. This means that OΠNs can cater for the dynamic communication topology and object mobility of the Actor model [1, 2] and the π-calculus [23, 24] (see §4.3).

Fig 2.5(c) indicates another part of the subnet (of which one instance is *Site#11*) which forwards messages provided they have no possible interaction with resident subscribers. Transition *partners* of *m'* supplies the messages *x* currently active in the umbrella, from which the addresses can be extracted as *a(x)*. This can be compared with the addresses of the resident subscribers, given by *a(y)*. This interaction uses compound test and inhibitor arcs [15].

Finally, it is worth noting that OΠN semantics does not ignore the arcs and arc inscriptions of substitution transitions (as in fig 2.5(a)). Instead, the interaction of the node *n1* with its environment (as specified by the incident arcs and arc inscriptions) must match the interaction of the subnet with the same environment. In this case, the arcs and arc inscriptions indicate that the same message object is forwarded, even though interaction in the subnet may have modified its contents.

3 Formal Definition of Object Petri Nets (OΠNs)

The formal definition of OΠNs is divided into a number of sections. Preliminary definitions, including the generic definitions for objects are given in §3.1. The static structure and dynamic behaviour of Object Petri Nets (OΠNs) are given in §3.2 and §3.3 respectively.

The following definitions and their associated explanatory notes should be read together.

3.1 Preliminary Definitions

Initially, we consider the structure of classes and objects (the instances of a class). A class is simply a set of values, but we distinguish between structured classes (which have composite values associating component values with variables) and primitive classes (which have values which are given directly, without the use of variables). Primitive classes include *integer*, *real*, *boolean*, etc. and multisets.

Definition 3.4 is central to this section because it defines objects (or structured class instances) as partial functions from object identifiers and variables to values. These functions are specified by listing the appropriate triples. Each object (and each component object) is associated with (and accessed via) an object identifier. It might be argued that a formal definition would be simpler if all classes were treated the same, but it turns out that primitive classes require special attention, whichever approach is taken. The difference is not that important, since each primitive class can be formed into a structured class by the use of a pseudo variable '_' (a process only required for multiset classes.)

Definitions 3.2 to 3.5 concern the structure of classes, while definitions 3.6 to 3.8 add the notions of typing and subclassing. Finally, definitions 3.9 to 3.12 cover operations between objects, validity constraints, and cloning.

This presentation differs significantly from the previous one [19] in introducing subclassing right from the start and in defining the *clone* of an object (see def 3.11).

Example: As an example of the definitions, consider three objects x, y and z each of which is a multiset object. Object x has object identifier o_1 and contains one component with object identifer o_2, and with field f having value *1995*. x could be a place p holding a single token (with object identifier o_2). Object y is identical to object x. It could be an indication of a token to be removed from place p. Object z is also a multiset with object identifier o_1. It has one component with object identifier o_3 with field f having value *1996*. z could be an indication of a token to be added to place p. Thus, we have:

$$x = (o_1, \_, o_2), (o_2, f, 1995)$$
$$y = (o_1, \_, o_2), (o_2, f, 1995)$$
$$z = (o_1, \_, o_3), (o_3, f, 1996)$$

Note that we abbreviate the multiset value $1\,{}^{\backprime}c$ as c. The difference and sum of x, y, z are given by:

$$x - y + z = (o_1, \_, o_3), (o_3, f, 1996)$$

This could be the result of removing y and adding z to the place p, identified by x.

Definition 3.1: The following are assumed to be well-defined:
(a) the **set of elements of type T** , which can be denoted by T.
(b) the **multiset of elements of type T**, which is denoted by T*.
(c) **multiset addition, subtraction, scalar multiplication, comparison** operations, denoted in the usual way. (The multiset difference $m_1 - m_2$ is assumed to be defined analogously to set difference, even when $m_1 \geq m_2$ does *not* hold.)
(d) a **relation** $R \subseteq A \times B$ has domain $\mathbf{dom(R)} = \{a \mid (a,b) \in R\}$ and range $\mathbf{rng(R)} = \{b \mid (a,b) \in R\}$.
(e) a **function** f: $A \rightarrow B$ is a relation on $A \times B$ with $\forall\, a \in A$: $|\{ b \mid (a,b) \in f\}| = 1$.
(f) a **partial function** f: $A \rightarrowtail B$ has $\mathrm{dom(f)} \subseteq A$.

Definition 3.2: We assume the availability of a set of **classes** Σ, a set of **primitive classes** Σ^o, sets of **object identifiers OID**, and $s|_X$ denotes the **restriction** of set s to $X \subseteq \Sigma$ where:
(a) $\mathrm{OID} \subseteq \Sigma^o \subseteq \Sigma$
(b) $O_C \in \mathrm{OID} \Rightarrow C \in \Sigma$
(c) $C^* \in \Sigma \Rightarrow C \in \Sigma$
(d) $C \in \Sigma \wedge O \in \mathrm{OID} \wedge C \cap O \neq \emptyset \Rightarrow C \in \mathrm{OID}$
(e) $s|_X = s \cap (\cup X)$

Notes:
(a) The available classes are divided into primitive and structured classes. Structured classes have components, while primitive classes do not. (See definitions 3.3 and 3.4.) The sets of object identifiers are primitive classes.
(b) Each set of object identifiers (O_C) is associated with a class (C) and gives identity to instances of that class. Not every class has an associated set of object identifiers.
(c) Each multiset class (C*) has its element type (C) a class in Σ.
(d) Any class which overlaps a set of object identifiers *is* a set of object identifiers.
(e) The notation for restriction is simply a shorthand as defined.

Definition 3.3: A **structured class** C is specified by:
(a) a non-empty set of **variables** V_C
(b) a boolean **guard** G_C: $C \rightarrow \mathrm{bool}$
(c) a **default initialisation** $I_C \in C$, where $G_C(I_C)$

Note:
- (a) The set of variables determines the components of a structured class.
- (b) A class has a boolean guard which determines whether an instance is valid (see def 3.10).
- (c) Each class has a default initial value, which satisfies the guard.

The boolean guard associated with each class determines the validity constraint for objects of that class, and hence corresponds to the notion of a class invariant as in the language *Eiffel* [21]. This can be used to constrain the possible objects of the class, and also to constrain the relationship of a subclass to its parent (see def 3.7 and the discussion in §4.5).

Definition 3.4: A (set of) **partial objects** x is a partial function $x: (\cup OID) \times (\cup V_c) \nrightarrow \Sigma^o$ with **object identifiers** oids(x), **undefined components** undef(x) and **root elements** elt(x) where:
- (a) $oids(x) = dom(dom(x)) - rng(x)$
- (b) $undef(x) = rng(x)|_{OID} - dom(dom(x))$
- (c) $elt(x) = \{ ((o, v), c) \mid o \in oids(x) \}$
- (d) $c = x((o,v)) \lor c \in x((o,v)) \Leftrightarrow (o, v, c) \in x$

Notes:
- (a) The objects of x are those with object identifiers which are not referenced in x, i.e. which do not occur in the range of x.
- (b) The undefined components are those which are referenced but not defined, i.e. which do not appear in the domain. Partial object(s) x is said to be **complete** if $undef(x) = \emptyset$.
- (c) The root elements are those directly connected to the object identifiers of x.
- (d) The membership of elements of x is extended to its multiset components. Note that we write $(o, v, c) \in x$ instead of $((o, v), c) \in x$ where there is no confusion.

Note that variables are associated with primitive type values. This implies that all the structure of nested objects must be captured with the use of object identifiers. This approach is more regular than allowing anonymous objects of class C of the form $V_c \rightarrow \Sigma$.

Note also that a primitive class does not have a set of variables and hence its values are given directly, and not by component values associated with variables (as in def 3.4). However, it is possible to create a structured class out of a primitive class by using a pseudo variable, say '_'. So, we might specify a structured class, C, holding integer multisets with $V_c = \{ \_ \}$, $G_c(x) = true$, $I_c = \{ o_1, \_, \emptyset \}$. In the following presentation, this construction is only used for multiset classes.

Definition 3.5: The components of partial object(s) x can be accessed as follows:
- (a) the **O-components** x[O]: $x[O] = \{ ((o, v), c) \in x \mid o \in O \cup rng(x[O]) \}$
- (b) the **O-components** [O]: $[O] = x[O]$
- (c) the **o-components** x[o]: $x[o] = x[\{o\}]$
- (d) the **non-o-components** x[~o]: $x[\sim o] = x - x[o]$

Notes:
- (a) The O-components of x are the components contained directly or indirectly within the objects with object identifiers in O. Note that the definition allows O to contain elements which are *not* object identifiers (in which case they contribute nothing to the set of O-components).
- (b) Where the object context is understood, it can be omitted.
- (c) The o-components of x are the components contained directly or indirectly within the object with object identifier o. If o is not an object identifier, x[o] will be empty.
- (d) The non-o-components of x are all those *except* those in the object with object identifier o.

Having defined the structure of classes, we now turn to consider their typing and type constraints.

Definition 3.6: There is a **type function** τ defined over Σ, and a **subclass** partial order <: over Σ:
- (a) $\tau: (\cup \Sigma) \rightarrow \Sigma$
- (b) $\forall C', C \in \Sigma: c \in C' \land C' <: C \Rightarrow c \in C$
- (c) $\forall C \in \Sigma: \forall c \in C: \tau(c) = C \Leftrightarrow (\forall C' \in \Sigma: c \in C' \land C' <: C \Rightarrow C' = C)$
- (d) $\forall C'^*, C^* \in \Sigma: C' <: C \Leftrightarrow C'^* <: C^*$
- (e) $\forall O_{c'}, O_c \in OID: C' <: C \Leftrightarrow O_{c'} <: O_c$

Notes:

(a) For every value of every class in Σ, τ specifies the class to which it (primarily) belongs.

(b) The subclass relationship is reflected in the membership of objects in classes. An object is a member not only of its primary class, but also of all its parent classes. This significantly simplifies the following presentation, since it is not necessary to state explicitly every situation where a subclass instance can appear in a superclass context.

(c) Following on from part (b), the type of a value is the smallest subclass containing that value.

(d) The subclass relationship between classes is paralleled by the subclass relationship between the corresponding multiset classes.

(e) The subclass relationship between classes is paralleled by the subclass relationship between the corresponding sets of object identifiers

Definition 3.7: Each structured class C has a **type function** τ_C: $(\cup \Sigma) \cup V_C \to \Sigma$ and the **subclass** partial order $<$: for structured classes satisfies the additional properties:

(a) $\forall\, C' \in \Sigma$: $\forall\, c \in C'$: $\tau_C(c) = \tau(c)$ and $\forall\, v \in V_C$: $\tau_C(v) \in \Sigma^\circ$

(b) $\forall\, C, C' \in \Sigma - \Sigma^\circ$: $C' <: C \Rightarrow \forall\, v \in V_C$: $v \in V_{C'} \wedge \tau_{C'}(v) <: \tau_C(v))$

(c) $\forall\, C, C' \in \Sigma - \Sigma^\circ$: $C' <: C \Rightarrow (G_{C'} \Rightarrow G_C)$

Notes:

(a) The type function for a class (τ_C) extends the function τ (of def 3.6) to the class variables. The type functions can be extended to expressions in the usual way.

(b) All the variables of the parent appear in the subclass, and their types are compatible.

(c) The guard for a subclass is stronger than that for the parent (as in [21, 22]).

Definition 3.8: A (set of) partial objects x: $(\cup \text{OID}) \times (\cup V_C) \to \Sigma^\circ$ is **properly typed** if:

(a) $\forall\, (o,v) \in \text{dom}(x)$: $\tau(o) = O_C \Rightarrow x((o,v)) \in \tau_C(v)$

(b) $\forall\, o \in \text{dom}(\text{dom}((x)))$: $\tau(o) = O_C \Rightarrow \text{dom}(x) \cap \{o\} \times (\cup V_{C'}) = \{o\} \times V_C$

Notes:

(a) The values associated with variables are of the appropriate type, with provision for polymorphism (as in def 3.6). We adopt the usual indexing notation for class instances, i.e. we write x(o.v) in place of x((o,v)), or simply o.v if the object context is self-evident.

(b) If x contains a value for one class field, it contains a value exactly for all the variables of the class. This simplifies the identification of undefined components, as in def 3.4(b).

Definition 3.9: For partial object(s) x, y with $(o, v) \in \text{dom}(x) \cap \text{dom}(y) \Rightarrow v = \_$, the operations $x \geq y$, $x + y$, $x - y$ are defined in terms of multisets.

Note: Each of x and y can be considered as a set and hence a multiset (def 3.1). Comparison, addition and subtraction then follow with these operations applied to the multiset components. Since the domains only overlap for multiset components, the result will also be a partial object.

Apart from the partial results during computation, OTINs are only concerned with objects that satisfy additional properties, in which case we refer to them as *valid*.

Definition 3.10: The partial object(s) x is **valid** iff:

(a) $\forall\, o \in \text{dom}(\text{dom}(x))$: $o' \in \text{rng}(x[o]) \Rightarrow o \notin \text{rng}(x[o'])$

(b) $\forall\, o \in \text{dom}(\text{dom}(x))$: $o \in O_{C*} \Rightarrow \text{rng}(x[\sim o])|_{\text{OID}} \cap \text{rng}(x[o])|_{\text{OID}} = \emptyset$

(c) $\forall\, o \in \text{dom}(\text{dom}(x))$: $\tau(o) = O_C \Rightarrow G_C(x[o])$

Notes:

(a) Object containment is not reflexive.

(b) The elements of multisets are self-contained.

(c) The appropriate guard is satisfied by each object in x.

Within object systems, the specific object identifiers associated with an object are not significant. Accordingly, we define how they can be consistently relabelled to form a clone object.

Definition 3.11: A **clone** h(x) of partial object(s) x is derived from an isomorphism h: $\cup \Sigma \to \cup \Sigma$ which consistently relabels object identifiers, where:

(a) $\forall c \in \cup \Sigma$: $\tau(h(c)) = \tau(c)$

(b) $\forall c \in \cup \Sigma - \cup OID$: $h(c) = c$

(c) $h(x) = \{ ((h(o), v), h(c)) \mid ((o, v), c) \in x \}$

Notes:

(a) The isomorphism maintains type-consistency.

(b) The isomorphism only modifies object identifiers, leaving other values invariant.

(c) The relabelled object has the object identifiers suitably modified.

Definition 3.12: The **completion** of a partial object x in the context of partial object y is given by:

(a) $\text{closure}_y(x) = x$ if $\text{undef}(x) = \emptyset$

(b) $\text{closure}_y(x) = \text{closure}_\emptyset(x + (y\,[\text{undef}(x)] - x))$ if $\text{undef}(x) \neq \emptyset$ and $y \neq \emptyset$

(c) $\text{closure}_\emptyset(x) = \text{closure}_\emptyset(x + h(I_C))$ if $o \in \text{undef}(x)$, $\tau(o) = O_C$, $h(I_C)$ is a clone of I_C, and $h(\text{oid}(I_C)) = o$.

Notes:

(a) An object with no undefined components is complete (see definition 3.4(b)).

(b) The undefined components of x, which are defined in y, are added (if not already present).

(c) Any remaining undefined components are defined by cloning the appropriate class initialisation. The cloning is normally required to introduce distinct object identifiers.

3.2 The Static Structure of Object Petri Nets

We now consider the definition of Object Petri Nets. The following definitions assume the reader is familiar with the style of definition found in [13].

While we introduce a single definition of OΠN classes, we distinguish between *state classes* and *transition classes*. State classes can contain fields for constants, places and transitions, but not arcs. They are instantiated to give the state or marking of the net. Transition classes also contain fields for arcs. They are instantiated to give a step of the net – each instance of a transition class thus corresponds to a transition binding of a CPN.

Unlike the previous presentation [19], test and inhibitor arcs are included from the start. Partly this is in the interests of brevity, and partly it is because the definitions have a certain symmetry which is obscured by separating the presentation into two parts. The primary motivation, however, is the belief that the ability to test the state of an object without modifying that state is essential to building clean interfaces for object models. Test and inhibitor arcs provide this possibility [15].

Given this paper's emphasis on the consistent use of polymorphism, and the consequent possibility of binding a simple place to a super place, OΠNs must define the semantics of test and inhibitor arcs incident on super places. The most direct interpretation is that such test and inhibitor arcs must match against the possible token offerings of the super place, by considering *all* possible steps involving internal transitions of the super place. It is anticipated that this would be computationally unmanageable. Instead, we require a super place to specify the possible token offerings explicitly with test and inhibitor arcs incident on the super place boundary. These internal arcs will then be matched against corresponding external arcs. See the example of figs 2.4 and 2.5.

Definition 3.13: An **Object Petri Net class** s is a tuple $s = (\Sigma_s, \tau_s, K_s, P_s, T_s, A_s, G_s, I_s)$. A **state class** is one with $A_s = \emptyset$, and a **transition class** is one with $A_s \neq \emptyset$, where:

(a) $\Sigma_s \subseteq \Sigma$ is a subset of the available classes

(b) $OID_s = \Sigma_s \cap OID$ is the relevant sets of object identifiers

(c) $V_s \subseteq V$ is the set of **variables** associated with class s, where $V_s = K_s \cup P_s \cup T_s \cup A_s$

(d) τ_s: $(\cup \Sigma_s) \cup V_s \to \Sigma_s$ is the **type function** for variables and values of the classes Σ_s

(e) $K_s = V_s - (P_s \cup T_s \cup A_s)$ is a set of **constants**

(f) $P_s = V_s - (K_s \cup T_s \cup A_s)$ is a set of **places** where $p \in P_s \Rightarrow \tau_s(p) = O_{C^*} \in OID_s$ and where C is a state class.

(g) $T_s = V_s - (K_s \cup P_s \cup A_s)$ is a set of **transitions** where $t \in T_s \Rightarrow \tau_s(t) = O_{C^*} \in OID_s$ and where C is a transition class.

(h) $A_s = V_s - (K_s \cup P_s \cup T_s) \subseteq P_s \times \{+, -, <, >\}$ is a set of **arcs** with $a \in A_s \Rightarrow \tau_s(a) = C^*$ where $\tau_s(p(a)) = O_{C^*}$, where $p(a)$ is the place of a.

(i) G_s is a boolean **guard expression**, i.e. $\tau_s(G_s) = \text{bool}$.

(j) I_s is a complete default **initial value** for class s.

Notes:

(a) The set of classes Σ_s relevant to a particular class s is a subset of the available classes.

(b) The set of sets of object identifiers OID_s relevant to a particular class s is a subset of the available set of sets of object identifiers.

(c) Variables are associated with the constants, places, transitions and arcs of the class s. The sets of variables for each kind of component do not overlap (as required by parts (e)-(h)).

(d) The type function τ_s gives a type for each value and variable used in the class (see def 3.3).

(e) Constants are components which do *not* have incident arcs and hence cannot be modified by the firing of a transition in the class s.

(f) Places are components which hold multisets of tokens as is the case with traditional Petri Net formalisms. Here, however, a place is an object with an associated object identifier and it is the object identifier which is associated with the multiset of tokens.

(g) Transitions are components which serve to modify the adjacent places. In a step, each transition is associated with a multiset of bindings. Again, the traditional Petri Net definition is modified by associating an object identifier with each transition.

(h) Arcs connect places to transitions and transitions to places. Since arcs only occur as components of transition classes, the adjacent transition can be determined by the context and need not be specified. Without loss of generality, there is at most one input arc from a place to a transition, at most one output arc from a transition to a place, and at most one test and one inhibitor arc for each place and transition pair. Hence, arcs can be identified by the place paired with an indicator of whether the arc is input (–), output (+), inhibitor (<) or test (>). Again without loss of generality, each arc is inscribed not by an expression, but by a variable of type matching the incident place. The context will indicate whether we are referring to the arc or the variable associated with the arc.

(i) The guard is a boolean expression dependent on the variables of the class. For a transition class, the guard is the transition guard of CPNs. For a state class, it is a state invariant.

(j) The default initial value for the class is given by a complete object.

Definition 3.14: We distinguish the **place object identifiers PLC**, the **transition object identifiers TRN**, and the simple place and transition components of an object x by:

(a) $PLC = \{ O_{C^*} \in OID \mid C \text{ is a state class } \}$

(b) $TRN = \{ O_{C^*} \in OID \mid C \text{ is a transition class } \}$

(c) the **simple places** plc(x): $\qquad plc(x) = mg(x)|_{PLC}$

(d) the **simple transitions** trn(x): $\qquad trn(x) = mg(x)|_{TRN}$

Notes:

(a) The place object identifiers are the sets of object identifiers for multisets of state classes.

(b) The transition object identifiers are the sets of object identifiers for multisets of transition classes.

(c) The simple places are those components which have type a multiset of state classes.

(d) The simple transitions are those components which have type a multiset of transition classes.

Definition 3.15: An **Object Petri Net** is a tuple $O\Pi N = (\Sigma, s_0, o_0)$ where:

(a) Σ is the set of available classes.

(b) $s_0 \in \Sigma$ is the **root class** which is instantiated to establish the net

(c) o_0 is the unique object identifier of the **root instance**, i.e. $O_{s_0} = \{ o_0 \}$

Note:

(a) The set of classes may include primitive classes (see def 3.2) as well as the $O\Pi N$ classes of definition 3.13.

(b) Even though the number of components of the net can vary as the net evolves, all its components can be derived indirectly from the single instance of the root class.

(c) Without loss of generality, we can assume that there *can* only ever be one instance of the root class which we call the root instance.

3.3 The Dynamic Behaviour of Object Petri Nets

As in the definition of CPNs [13], we define token elements, markings, binding elements and steps.

Definition 3.16: A **token** is a valid instance of a state class. A **token element** is a partial object $(p, \_, c)$ where p and c are the object identifiers of a place and a token, respectively. The **marking** of a place p is the completion of a multiset of token elements for p. The **marking M** of the Object Petri Net OΠN is a complete valid instance of the root class s_0 with object identifier o_0, which includes the markings of all the places of OΠN. The **initial marking M_0** is given by I_{s_0}.

Definition 3.17: A **binding** of a transition is a valid instance of a transition class. A **binding element** is a partial object $(t, \_, b)$, where t and b are the object identifiers for a transition and a binding, respectively. A **step** Y is a valid set of partial objects with root elements consisting of a multiset of binding elements. (A step is not necessarily complete.)

Example: Consider the segment of the net from fig 2.2(a) with places *q4* and *q1*. A marking *M* with the token for *1995* in place *q4* might be:

$$M = (o_0, q1, o_1), (o_0, q4, o_4), (o_0, n2, o_2), (o_1, \_, \emptyset), (o_2, \_, \emptyset), (o_4, \_, o_5), (o_5, year, 1995)$$

Here, places *q1* and *q4* have object identifiers o_1 and o_4 (respectively), while transition *n2* has object identifier o_2, with no associated binding elements. Place *q1* is empty, while place *q4* contains a single token with object identifier o_5, which is an object with field *year* with value *1995*. A step *Y* might be:

$$Y = (o_2, \_, o_3), (o_3, q1, o_1), (o_3, q4, o_4), (o_3, (q4,-), o_5), (o_3, (q1,+), o_6), (o_6, year, 1996)$$

Thus, the step has one binding element for transition *n2* (with object identifier o_2). The binding is an object with object identifier o_3. It indicates that the neighbouring places *q1* and *q4* are bound to object identifiers o_1 and o_4. (This is the same as in the marking, but it could be different if the places were fused to other places for the purpose of this binding element.) The binding also specifies that the input arc from place *q4* will be bound to the single token with object identifier o_5, and that the output arc incident to place *q1* will be bound to the single token given by o_6.

Definition 3.18: Given a step Y of Object Petri Net OΠN, the changes made by Y, and the associated markings are given by:

(a) $\displaystyle del_Y = \sum_{(t,\_,b)\in Y} \left(\sum_{(b,(p,-),o)\in Y} o - \sum_{(b,(p,+),o)\in Y} o \right)$

(b) $\displaystyle new_Y = \sum_{(t,\_,b)\in Y} \left(\sum_{(b,(p,+),o)\in Y} o - \sum_{(b,(p,-),o)\in Y} o \right)$

(c) $\displaystyle get_Y(O_P, O_T) = \sum_{\substack{(t,\_,b)\in Y \\ t\in O_T}} \sum_{\substack{(b,(p,-),o)\in Y \\ b.p\in O_P}} (b.p, \_, o)$

(d) $\displaystyle put_Y(O_P, O_T) = \sum_{\substack{(t,\_,b)\in Y \\ t\in O_T}} \sum_{\substack{(b,(p,+),o)\in Y \\ b.p\in O_P}} (b.p, \_, o)$

(e) $\displaystyle lo_Y(O_P, O_T) = \sum_{o\in O_P} \max_{\substack{(t,\_,b)\in Y \\ t\in O_T}} \max_{\substack{(p,>)\in \tau(t) \\ b.p=o}} (b.p, \_, b.(p,>))$

(f) $\displaystyle hi_Y(O_P, O_T) = \sum_{o\in O_P} \min_{\substack{(t,\_,b)\in Y \\ t\in O_T}} \min_{\substack{(p,<)\in \tau(t) \\ b.p=o}} (b.p, \_, b.(p,<))$

(g) $M^+ = M + [new_Y]$

(h) $M^{+'} = M^+ - get_Y(plc(M^+), trn(M^+)) + put_Y(plc(M^+), trn(M^+))$

(i) $M' = M^{+'} - [del_Y]$

Notes:

(a) del_Y is the multiset of object identifiers explicitly deleted by the step. It does *not* include object identifiers which are removed and replaced by the *same* transition. Such object identifiers are classified as *reserved*. It will be a requirement that object identifiers deleted by one transition (of the step) cannot be removed by another. Note that this is a multiset sum so that object identifiers deleted more than once will occur more than once in del_Y.

(b) new_Y is the multiset of object identifiers explicitly generated by the step. It does *not* include object identifiers which are removed and replaced by the same transition, i.e. the *reserved* object identifiers. It will be a requirement that object identifiers added by one transition (of the step) cannot be removed by another. Note that this is a multiset sum so that object identifiers generated more than once will occur more than once in new_Y.

(c) $get_Y(O_P, O_T)$ is the multiset of token elements explicitly removed from places with object identifiers in O_P by transitions with object identifiers in O_T. This does *not* specify the nested components, since these may change as a result of the step.

(d) $put_Y(O_P, O_T)$ is the multiset of token elements explicitly added to places with object identifiers in O_P by transitions with object identifiers in O_T. This does *not* specify the nested components, since these may change as a result of the step.

(e) $lo_Y(O_P, O_T)$ is the maximum over test arcs of tokens visible at places with object identifiers in O_P by transitions with object identifiers in O_T.

(f) $hi_Y(O_P, O_T)$ is the minimum over inhibitor arcs of the bounds on tokens at places with object identifiers in O_P and transitions with object identifiers in O_T.

(g) M^+ is the **augmented marking** and is obtained by extending the marking with the objects which are introduced by the step. ($[new_Y]$ is evaluated in the context of Y.) These objects may be introduced by cloning some existing object or by initialising a new instance. Some of the components of M^+ (including some of those of new_Y) may be discarded by the step.

(h) $M^{+'}$ is the **augmented follower marking** and is obtained by removing the get_Y components and adding the put_Y components for the simple places and the simple transitions. Note that this computation is properly defined, whether or not the step is enabled, on the basis of definition 3.1(c).

(i) M' is the **follower marking** and is obtained by removing the objects deleted by the step. ($[del_Y]$ is evaluated in the context of $M^{+'}$.) This may be considered as a garbage collection phase, or removing components which are no longer linked into the marking. Note that the follower augmented marking and the follower marking are introduced now rather than later in order to demonstrate the symmetry of the definitions.

Example: Continuing with the example above, step Y in marking M would have:

$del_Y = o_5$

$new_Y = o_6$

$get_Y(plc(M^+), trn(M^+)) = (o_4, \_, o_5)$

$put_Y(plc(M^+), trn(M^+)) = (o_1, \_, o_6)$

$M^+ = ..., (o_1, \_, \varnothing), (o_2, \_, \varnothing), (o_4, \_, o_5), (o_5, year, 1995), (o_6, year, 1996)$

$M^{+'} = ..., (o_1, \_, o_6), (o_2, \_, \varnothing), (o_4, \_, \varnothing), (o_5, year, 1995), (o_6, year, 1996)$

$M' = ..., (o_1, \_, o_6), (o_2, \_, \varnothing), (o_4, \_, \varnothing), (o_6, year, 1996)$

In other words, the step deletes the object with object identifier o_5, and generates the object with object identifier o_6. The token element specifying o_5 is removed from the place o_4 and the token element specifying o_6 is added to the place o_1. The augmented marking M^+ includes the token for o_6 but *not* its connection to place o_1. The augmented follower marking includes the token for o_5 but *not* its connection to place o_4. The follower marking removes the token o_5. Note that the same would apply irrespective of the complexity or levels of information containment of tokens o_5 and o_6.

Note that if we have the equivalent of a CPN, the only objects in the net would be the places and transitions. All tokens would be values from primitive classes. Consequently, we would have $del_Y = new_Y = \varnothing$, and get_Y, put_Y would be the complete multiset of token elements removed and added by the step (respectively). The augmented marking M^+ would be the same as M, and the augmented follower marking $M^{+'}$ would be the same as M'.

Definition 3.19: A step Y of Object Petri Net OΠN in marking M is **enabled** if:

(a) $del_Y \leq dom(dom(M^+)) \land new_Y \leq (\cup OID) - dom(dom(M))$

(b) $\forall o \in dom(dom(M'))$: $\tau(o) = O_C \Rightarrow G_C(M'[o])$

(c) $\forall o \in plc(M^+)$: $elt(M^+[o]) \geq get_Y(\{o\}, trn(M^+[\sim o])) + lo_Y(\{o\}, trn(M^+[\sim o]))$

$\land hi_Y(\{o\}, trn(M^+[\sim o])) \geq elt(M^+[o]) + put_Y(\{o\}, trn(M^+[\sim o]))$

(d) $\forall\, o \in mg(M^+) - plc(M^+)$: $get_Y(\{o\}, trn(M^+[\sim o])) = put_Y(\{o\}, trn(M^+[o]))$

$\land\ put_Y(\{o\}, trn(M^+[\sim o])) = get_Y(\{o\}, trn(M^+[o]))$

$\land\ lo_Y(\{o\}, trn(M^+[\sim o])) \le lo_Y(\{o\}, trn(M^+[o]))$

$\land\ hi_Y(\{o\}, trn(M^+[\sim o])) \ge hi_Y(\{o\}, trn(M^+[o]))$

$\land\ hi_Y(\{o\}, trn(M^+[o])) \ge$

$\qquad get_Y(\{o\}, trn(M^+[o])) + put_Y(\{o\}, trn(M^+[o])) + lo_Y(\{o\}, trn(M^+[o]))$

(e) $\forall\, o \in mg(M^+) - trn(M^+)$: $get_Y(plc(M^+), \{o\}) = get_Y(plc(M^+[\sim o]), trn(M^+[o]))$

$\land\ put_Y(plc(M^+), \{o\}) = put_Y(plc(M^+[\sim o]), trn(M^+[o]))$

$\land\ lo_Y(plc(M^+), \{o\}) = lo_Y(plc(M^+[\sim o]), trn(M^+[o]))$

$\land\ hi_Y(plc(M^+), \{o\}) = hi_Y(plc(M^+[\sim o]), trn(M^+[o]))$

Notes:

(a) The multiset of deleted object identifiers must occur in the augmented marking, and the multiset of newly-generated object identifiers must *not* occur in the existing marking.

(b) The guard (or invariant) must be satisfied for each component of the follower marking. Note that the definition of a binding (def 3.17) requires that the guard must be satisfied, and the definition of initialisation (def 3.10) insists that the guard holds for the initial marking.

(c) Each simple place must have sufficient token elements (see def 3.4) for all the input and test arc demands made, and must not exceed the inhibitor arc limits (even when the added tokens are taken into account). See [15].

(d) For super places, the tokens extracted externally must match the tokens offered internally, and vice versa. Because of this, the follower marking of definition 3.18 only considers the changes to simple places. Similarly, the internal and external test and inhibitor arcs must be compatible, and finally, the internal inhibitor arcs must encompass all token movement, analagously to part (c) for simple places.

(e) For super transitions, the token movement indicated by the external incident arcs must match the token movement from the internal transitions. The same applies to test and inhibitor arcs. Because of this, the follower marking of definition 3.18 only considers the changes due to simple transitions. Since the token movement of the external incident arcs encompasses *all* simple places, while the token movement for internal transitions only encompasses the external simple places, this implies that the external incident arcs cannot (through place fusion) affect any of the internal places. This would be contrary to good encapsulation practice.

Definition 3.20: If step Y of Object Petri Net OΠN is enabled in marking M, then it can **occur** leading to the follower marking M' (of def 3.18), in which case we write M [Y> M'.

4 Discussion

This section highlights a number of aspects of the formal definition. These include design choices, implementation considerations, and possible simplifications.

4.1 Steps

One design choice concerns the possible limitations on the binding elements which may occur in a step. The above definitions allow for objects deleted and generated by a step to have components modified by the same step. This possibility could be prohibited. Such a choice could be reasonable given the more flexible initialisation provisions given here (as compared to those of [19]). In this case, the components of objects [del_Y] and [new_Y] (def 3.18) would *not* be modified by the step and it would be sufficient to consider the effect of the step on the marking M instead of on the augmented marking M^+. However, the current approach was adopted because it was considered helpful to be able to extract a component of an object which is being discarded, or to add a component to a newly-generated object.

Another design issue related to steps is what we might call the *conflict of interest* between objects and object identifiers. The question is whether objects ought to be specified by giving all their components or simply by giving their object identifiers. So, when a place contains objects, should a step specify the addition or removal of all the token components or simply the object identifiers? The former approach was adopted in the original formulation of OPNs [17]. This was completely general, but also awkward in practice, especially if a transition simultaneously added or removed both an object and some of its components. The alternative approach was adopted in [19] with arc inscriptions only specifying the object identifiers. This was much simpler, but did not readily support cloning of objects, which is commonly desired in object-oriented languages. The current definition steers a middle course. Arc inscriptions specify the object identifiers of objects to be added or removed, with the result that a step serves to connect or disconnect the associated objects with the marking. At the same time, the components of newly-generated objects can be defined by cloning as part of the step since it is defined to be a set of partial objects (see [new$_Y$] in def 3.18(g)).

4.2 Bindings

The definition of token movement achieved by a step (def 3.18(c), (d)) indicates that the token elements are related to the place determined by the binding. In this way, a binding can dynamically modify the affected place, thus supporting arbitrary place fusion. Polymorphism allows the same mechanism to be used to bind a (simple) place to a compatibly typed super place. In a similar way, the fusion of two transitions is achieved with a binding element specifying one of the transitions, and a binding which is an instance of *both* the transition classes, i.e. a transition class formed by multiple inheritance. This guarantees that the binding has the variables from both transition classes and satisfies both the guards.

Another issue related to bindings is the support for arbitrary references between objects. Since markings must be valid objects (see defs 3.10, 3.16) the references between objects are significantly constrained, so much so that one might query the ability to model object systems with arbitrary referencing. These constraints have been applied so that the removal of an object identifier is known to imply the removal of all the components of the associated object (see def 3.18(i)). This eliminates the possibility of dangling references, thus simplifying implementation and invariant analysis. Bindings are also required to be valid, but the augmented marking (def 3.18(g)) is not. In other words, a binding can refer to arbitrary objects in the marking, and validity is only compromised when the augmented marking is formed, and is reestablished in the follower marking.

In other words, arbitrary references between objects are possible in the context of a binding. This is normally a transient value, but the establishment of these bindings is an implementation issue. For example, the choice of the bindings in a step to satisfy the guard of fig 2.5(a) could be rather involved (and is a matter for further study). A particular implementation may provide additional clues as to how a binding is to be chosen. For example, there may be away of indicating that a binding will be used until explicitly changed. In that case, it is the implementation which must protect against the possibility of dangling references.

4.3 Object mobility

Given that the consistent use of names in the π-calculus provided one of the key motivations for the OITN formalism, it is appropriate to examine the extent to which the object mobility of the π-calculus is also supported. A cursory examination of the example of §2.3 indicates that it is. The umbrella documents are objects which circulate around the ring. The ability to bind local places dynamically to these circulating objects (as in fig 2.5(b), (c)) makes it possible to interact with them.

One example of the π-calculus [23] is that of a car communicating via a mobile radio to one of a number of transmission stations. The car has two communication channels – an output channel for sending the words of a message, and an input channel for receiving notification of alternative channels (as the car moves out of range of one station and into the range of another). Communication channels are not provided directly in Petri Net formalisms, but only indirectly via transitions. As a result, one might conceive two possible OITN solutions to this problem, as in fig 4.1.

In part (a), the approach of §2.3 is adopted: the cars and the stations are collected together into places and a transition *xmit* serves to enable the communication, by selecting a car and a station and passing a message between them. The selection of the station can be constrained by a suitable attribute stored in the car.

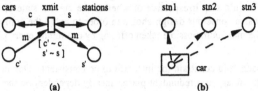

Fig 4.1: Possible approaches to object mobility

Part (b) is closer to the style of the π-calculus solution. Here, the car has a local place which is bound to one of the available stations. As discussed in §4.2, the formal definition does not allow this binding to be part of the marking unless the relevant station is a component of the car (which is somewhat unrealistic). Instead, the binding must be established each time the relevant transition (for communicating with the station) is enabled. As we have noted, the implementation may provide a way to simplify this, but it is not covered by the formalism.

4.4 Superplaces

The interaction with superplaces depends on synchronous channels [10]. Communication enabling, like the condition for enabling of superplace interaction (def 3.19(d)), can quickly become computationally prohibitive to test, since all possible combinations of binding elements for the communicating transitions need to be considered. In practice, this can be considerably simplified. In the implementation of LOOPN++ [16], each internal transition of a superplace which transfers tokens to the environment can do so only one token at a time. This means that each token is handled independently. This substantially reduces the backtracking required in the determination of a step, and at the same time reflects the idea that the tokens in a place are essentially independent.

Another aspect of superplaces which is open to simplification, particularly with respect to implementation efficiency, is the consistency of their interfaces. The formal definition does not place any constraints on the inscription of internal arcs incident on the superplace boundary. This is extremely general, but can lead to anomalous situations, such as that of fig 4.2. (Expressions are used as arc inscriptions to simplify the presentation, instead of using distinct variables which must be constrained in the guards.)

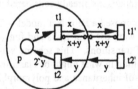

Fig 4.2: Inconsistent interface for a super place

Note that interface transitions are inconsistent – $t1$ offers one token to the environment based on having one token available in the the internal place p, while transition $t2$ stores two tokens in p based on the acceptance of one token from the environment. This means that the appropriate inscription on the inhibitor arc incident on transition $t1$ is uncertain. If the intention is to allow the enabling of a step including both $t1$ and $t2$ (provided p initially holds a token matching x), should the inscription be $x+y$ or $x+2'y$? The inscription $x+y$ satisfies the enabling condition for the inhibitor arc (def 3.19(d)). However, if an inhibitor arc is added between p and $t1$ with an inscription $x+y$ (which seems to be the appropriate choice given the inscriptions on the other arcs of $t1$), then the step would not be enabled (by virtue of def 3.19(c)).

This anomalous situation is the result of allowing superplaces with inconsistent interfaces. The above example is inconsistent because the transitions $t1$ and $t2$ have different relationships between the tokens transferred via the superplace boundary and the tokens stored in place p. It is also inconsistent because there is no constraint on the inhibitor arc relating it to the internal state of the superplace. In other words, such an inhibitor arc gives an upper bound on the tokens which a superplace might offer in a step irrespective of whether the current state of the superplace could supply more. This was a conscious design choice in order to avoid the computationally excessive task of determining the maximum possible token offering by examining every possible step involving internal transitions.

An alternative approach is to constrain the interface to be consistent. One possible approach is to treat the superplace boundary as a redundant place, linearly dependent on the internal places of the superplace [12]. For the example of fig 4.2, this would mean that the relationship between the tokens transferred via the boundary and those stored in place p must be the same for both $t1$ and $t2$. Further, the inhibitor arc incident on transition $t1$ would require an inhibitor arc between $t1$ and p with an inscription in the same relationship. Once such a linear dependency is specified, the arcs incident on the superplace boundary, together with their inscriptions, can be deduced from the internal arcs.

4.5 Guards

The formal definition includes a guard in all structured classes. This design choice is attractive for a number of reasons. Firstly, it allows a class to be defined with the range of possible values being constrained, and not just the product of all the possible values for all the possible fields. Secondly, object-oriented systems commonly support integrity constraints or class invariants. For example, the *Eiffel* language [22] uses invariants to guarantee robustness and to constrain the relationship between a subclass and its parents (as in def 3.7). This has been applied in a wide range of practical applications. Other, more rigid, approaches to constraining the relationship between a subclass and its parent do not necessarily work in practice [20], and hence support for invariants in OTINs makes the practical solution accessible. Thirdly, there is no need to make special provision for transition guards and there is an attractive duality between the constraints on bindings and state components.

5 Other Styles of Object-Oriented Petri Nets

Having given the formal definition of OTINs and shown how they can be transformed into behaviourally equivalent CPNs, we are now in a position to compare them with some other approaches to object-oriented Petri Nets described in the literature. This review only considers more recent comparable work and does not reproduce the comparisons already considered in [19].

Though not object-oriented, Valk proposed a Petri Net model – the Task Flow EN systems – with two levels of activity [31]. These were used for tasks and functional units, with synchronisation constrained by a static labelling of transitions. This scheme was subsequently extended to more than two levels of activity [32] resulting in what might be classified as Object-Based Petri Nets [19, 33], since there is no consideration of inheritance and polymorphism. This work strives to anchor multiple levels of activity in elementary net theory. This is an important concern, given the variety of object-oriented net formalisms being proposed. However, the ability to model multiple levels of activity arises more naturally in the context of complex systems, for which elementary net systems lack descriptive power (or expressive comfort), as noted in §2.

Another approach is that of the language CLOWN [4], which has emerged out of the work on OBJSA nets [5]. These have arisen from the enhancement of algebraic data types (from the data model of OBJ) with concurrency and synchronisation primitives (derived from Petri Nets). The lifecycle of each object is represented by a state machine, with methods being attached to transitions. Distinct objects can operate concurrently, and are synchronised by transition fusion, which thus corresponds to method activation. This implies that inter-object concurrency is the natural form of concurrency in CLOWN. More recent work [3] allows for intra-object concurrency through the use of compound classes. CLOWN objects can refer to other objects and can therefore model multi-level objects, but dangling references are possible. Inheritance is supported in CLOWN – a subclass may add attributes, redefine methods and specify a new state machine net. There is a requirement that an

ST-preorder relationship should exist between the state machine nets of parent and subclass. This requirement is a promising area of further work since it imposes some form of behavioural similarity (which can be automatically checked) between a class and its subclass. However, it is too early to decide whether such a requirement will satisfy the needs of practical application [20].

Buchs and Guelfi propose a formalism called CO-OPN which is similar to the approach above, but extends the use of algebraic data types to include the communication structures [8, 9]. Again, objects are defined by algebraic data types with the methods being associated with transitions. The net structure of an object is no longer restricted to a state machine but may have multiple internal transitions firing concurrently. The firing of internal transitions depends only on the internal state of the object, while the firing of external transitions requires synchronisation with the transitions of other objects. Synchronisation may follow a simultaneous, sequential or alternative regime. Objects can be parameterised, thus supporting generic object definitions. More recently, a revised formalism called CO-OPN/2 has been proposed [7], which introduces the notions of class, inheritance and sub-typing. It takes seriously the debate in the object-oriented research community concerning the difference between subtyping and subclassing, and consequently provides both notions in the formalism. It is questionable whether this makes the formalism excessively complex. The formalism also distinguishes weak and strong subtyping, the former relating only to type compatibility, while the latter encompassing behaviour compatibility [34]. The latter is considered appropriate for refinement, but it again remains to be seen whether this will be appropriate in practice [20].

Sibertin-Blanc and others have considered the use of Coloured Petri Nets to build complex systems out of components based on the client-server paradigm [27]. Here, a request for a service is made by a client passing an appropriate token to a server and waiting for a response before continuing. In order to guarantee suitable behaviour, a server will not accept a request unless it can guarantee to fulfil it, and a client will not make a request unless it is prepared to wait for the response. This has been extended into a Petri Net formalism called *Cooperative Nets* [28], where each object has an object identity or name, and may store the names of other objects. Each object has a number of interface places where requests for services (from clients) are submitted as tokens. When a transition fires, it may generate such a request token for another object, by suitably identifying the object and its interface place. This is akin to a dynamic form of place fusion (as discussed in §2). More recently, a textual version of Cooperative Nets, called SYROCO [29] has been proposed, which includes synchronous interaction and is intended to supply an open software development environment.

In summary, the above proposals have approached the inclusion of object-oriented concepts into Petri Net formalisms in a variety of ways. Some have concentrated on direct support for multiple levels of activity, while others have catered for this by allowing objects to contain references to others. We believe that it is important to support multiple levels of activity directly so that the deletion of an instance of a class can be tied to the simultaneous deletion of all its components, a property which is important for invariant analysis of such nets. Some approaches have exclusively supported synchronous interaction, as the appropriate mechanism for method calls in object-oriented languages. We believe that it is appropriate to support both synchronous and asynchronous interaction for the convenience of the modeller, even if both are mapped to some common underlying implementation. Finally, we believe that most of these approaches have not attempted to rationalise their definitions by following through the implications of object-orientation as we have attempted to do in this paper.

6 Conclusions and Further Work

This paper has presented a formal definition of Object Petri Nets which is claimed to be more general and elegant than previous definitions. It is more general because it caters for transition fusion as well as place fusion. It is more elegant because it captures all the notions of place substitution, transition substitution, place fusion and transition fusion under the single notion of binding. This is achieved by explicitly supporting names in the formalism, in line with the approach of the π-calculus. Thus, the binding of place and transition identifiers to names (or object identifiers) can be varied in a transition binding, thus resulting in the above rationalisation.

The definition in this paper is also more consistent in its use of inheritance and polymorphism. Thus every net or subnet is defined in terms of simple places and simple transitions. It is the use of polymorphism which allows these components to be substituted or bound to compatible subnets, and hence super places and super transitions.

The elegance of this definition is also enhanced by a more obvious duality between state and change of state, one of the key aspects of Petri Net theory [25]. Thus, both tokens and bindings are objects, and place and transition fusion are both seen as binding one object to a compatible object.

Object Petri Nets support the encapsulation, inheritance, polymorphism and dynamic binding which are common in object-oriented languages. The use of super places and super transitions makes it possible to model both synchronous and asynchronous interaction between objects. The flexibility of place fusion allows the partners of a communication to be dynamically chosen, thus supporting the dynamic interaction topologies proposed for Actor systems [2] and the π-calculus [23].

OPNs directly support the modelling of complex systems with multiple levels of activity. As well as being desirable for the modeller, this will be important for synchronising the actions of an object and its components, whether creation, deletion or migration. This is particularly significant in the development of appropriate net invariants.

The above properties should allow OPNs to reap the practical benefits of object-orientation including clean interfaces, reusable software components, and extensible component libraries [21]. Furthermore, despite their increased descriptive power, OPNs can be transformed into behaviourally equivalent CPNs [18], thus providing a basis for adapting CPN analysis techniques for use with OPNs.

The work on Object Petri Nets is an ongoing project. A graphical editor for OPNs is nearing completion, and the implementation of OPNs in textual form [16] is being refined. It will be important to consider how various aspects of the formal definition, such as the form of state invariants, ought to be constrained in order to facilitate an efficient implementation. There is also great scope for further work in tailoring analysis techniques to OPNs. Perhaps the most pressing need is to gain practical experience in the application of the OPN formalisms to realistic examples. Only in this way will it be possible to assess the appropriateness of this and other proposals, particularly in regard to the appropriate relationship between a class and its subclasses.

Acknowledgements

The author is pleased to acknowledge the helpful discussions on this and earlier work held with Alfredo Chizzoni, Søren Christensen, Fiorella De Cindio, Kurt Jensen, Chris Keen, Daniel Moldt, Kjeld Mortensen, Rüdiger Valk, Kees van Hee, David Wright, and the Networking Research Group of the University of Tasmania.

References

[1] G. Agha, S. Frølund, W.Y. Kim, R. Panwar, A. Patterson, and D. Sturman *Abstraction and Modularity Mechanisms for Concurrent Computing* Research Directions in Concurrent Object-Oriented Programming, G. Agha, P. Wegner, and A. Yonezawa (eds.), pp 3-21, MIT Press (1993).

[2] G.A. Agha *Actors: A Model of Concurrent Computation in Distributed Systems* The MIT Press series in artificial intelligence, MIT Press (1986).

[3] E. Battiston, A. Chizzoni, and F.D. Cindio *Inheritance and Concurrency in CLOWN* Proceedings of Workshop on Object-Oriented Programming and Models of Concurrency, Torino, Italy (1995).

[4] E. Battiston and F. de Cindio *Class Orientation and Inheritance in Modular Algebraic Nets* Proceedings of IEEE International Conference on Systems, Man and Cybernetics2, pp 717-723, Le Touquet, France, IEEE (1993).

[5] E. Battiston, F. de Cindio, and G. Mauri *OBJSA Nets: A Class of High-level Nets having Objects as Domains* Advances in Petri Nets 1988, G. Rozenberg (ed.), Lecture Notes in Computer Science 340, pp 20–43, Springer-Verlag (1988).

[6] M. Bever, K. Geihs, L. Heuser, M. Mühlhäuser, and A. Schill *Distributed Systems, OSF DCE, and Beyond* Proceedings of International DCE Workshop: DCE - the OSF distributed computing environment : Client/Server Model and Beyond, Lecture notes in computer science 731, Karlsruhe, Germany, Springer-Verlag (1993).

[7] O. Biberstein and D. Buchs *Structured Algebraic Nets with Object-Orientation* Proceedings of Workshop on Object-Oriented Programming and Models of Concurrency, Torino, Italy (1995).

[8] D. Buchs and N. Guelfi *CO-OPN: A Concurrent Object Oriented Petri Net Approach* Proceedings of 12th International Conference on the Application and Theory of Petri Nets, Gjern, Denmark (1991).

[9] D. Buchs and N. Guelfi *Open Distributed Programming Using the Object Oriented Specification Formalism CO-OPN* Proceedings of 5th International Conference on Putting into Practice Methods and Tools for Information System Design, Nantes (1992).

[10] S. Christensen and N.D. Hansen *Coloured Petri Nets Extended with Channels for Synchronous Communication* Proceedings of 15th International Conference on the Application and Theory of Petri Nets, Lecture Notes in Computer Science 815, pp 159-178, Zaragoza, Springer-Verlag (1994).

[11] P.A. Fishwick *Computer Simulation: Growth Through Extension* Proceedings of Modelling and Simulation (European Simulation Multiconference), pp 3-20, Barcelona, Society for Computer Simulation (1994).

[12] S. Haddad *A Reduction Theory for Coloured Nets.* Advances in Petri Nets 1989, G. Rozenberg (ed.), Lecture Notes in Computer Science 424, pp 209–235, Springer-Verlag (1990).

[13] K. Jensen *Coloured Petri Nets: Basic Concepts, Analysis Methods and Practical Use – Volume 1: Basic Concepts* EATCS Monographs in Computer Science, Vol. 26, Springer-Verlag (1992).

[14] K. Jensen, S. Christensen, P. Huber, and M. Holla *Design/CPN™: A Reference Manual* MetaSoftware Corporation (1992).

[15] C. Lakos and S. Christensen *A General Systematic Approach to Arc Extensions for Coloured Petri Nets* Proceedings of 15th International Conference on the Application and Theory of Petri Nets, Lecture Notes in Computer Science 815, pp 338-357, Zaragoza, Springer-Verlag (1994).

[16] C. Lakos and C. Keen *LOOPN++: A New Language for Object-Oriented Petri Nets* Proceedings of Modelling and Simulation (European Simulation Multiconference), pp 369-374, Barcelona, Society for Computer Simulation (1994).

[17] C.A. Lakos *Object Petri Nets – Definition and Relationship to Coloured Nets* Technical Report TR94-3, Computer Science Department, University of Tasmania (1994).

[18] C.A. Lakos *The Consistent Use of Names and Polymorphism to Achieve an Elegant Definition of Object Petri Nets* Technical Report R95-12, Computer Science Department, University of Tasmania (1995).

[19] C.A. Lakos *From Coloured Petri Nets to Object Petri Nets* Proceedings of 16th International Conference on the Application and Theory of Petri Nets, Lecture Notes in Computer Science 935, pp 278-297, Torino, Italy, Springer-Verlag (1995).

[20] C.A. Lakos *Pragmatic Inheritance Issues for Object Petri Nets* Proceedings of TOOLS Pacific 1995, Melbourne, Australia, Prentice-Hall (1995).

[21] B. Meyer *Object-Oriented Software Construction* Prentice Hall (1988).

[22] B. Meyer *Eiffel: The Language* Prentice Hall (1992).

[23] R. Milner *Elements of Interaction* Communications of the ACM, **36**, 1, pp 78-89 (1993).

[24] R. Milner, J. Parrow, and D. Walker *A Calculus of Mobile Processes, I* Information and Computation, **100**, 1, pp 1-40 (1992).

[25] C.A. Petri *"Forgotten" Topics of Net Theory* Advances in Petri Nets 1986 Lecture Notes in Computer Science, Springer-Verlag (1986).

[26] W. Reisig *Petri nets : An Introduction* EATCS Monographs on Theoretical Computer Science, Vol. 4, Springer-Verlag (1985).

[27] C. Sibertin-Blanc *A Client-Server Protocol for the Composition of Petri Nets* Proceedings of 14th International Conference on the Application and Theory of Petri Nets, Lecture Notes in Computer Science 691, pp 377-396, Chicago, Springer-Verlag (1993).

[28] C. Sibertin-Blanc *Cooperative Nets* Proceedings of 15th International Conference on the Application and Theory of Petri Nets, Lecture Notes in Computer Science 815, pp 471-490, Zaragoza, Spain, Springer-Verlag (1994).

[29] C. Sibertin-Blanc, N. Hameurlain, and P. Touzeau *SYROCO: A C++ Implementation of Cooperative Objects* Proceedings of Workshop on Object-Oriented Programming and Models of Concurrency, Torino, Italy (1995).

[30] P.A. Swatman, P.M.C. Swatman, and R. Duke *Electronic Data Interchange: A High-level Formal Specification in Object-Z* Proceedings of 6th Australian Software Engineering Conference, pp 341-354, Sydney, Australia, Springer-Verlag (1991).

[31] R. Valk *Modelling Concurrency by Task/Flow EN Systems* 3rd Workshop on Concurrency and Compositionality, 1991, E. Best, et al. (ed.), GMD-Studien 191, pp 207–215 (1991).

[32] R. Valk *Petri Nets as Dynamical Objects* Proceedings of Workshop on Object-Oriented Programming and Models of Concurrency, Torino, Italy (1995).

[33] P. Wegner *Dimensions of Object-Based Language Design* Proceedings of OOPSLA 87, pp 168-182, Orlando, Florida, ACM (1987).

[34] P. Wegner *Inheritance as an Incremental Modification Mechanism, or What Like Is and Isn't Like* Proceedings of ECOOP '88 - European Conference on Object Oriented Programming, Lecture Notes in Computer Science 322, pp 55-77, Oslo, Norway, Springer Verlag (1988).

Designing a Security System
by Means of Coloured Petri Nets

Jens Linneberg Rasmussen and Mejar Singh

Computer Science Department, Aarhus University
Ny Munkegade, Bldg 540, DK–8000 Aarhus C, Denmark
E-mail: {kajenika,singh}@daimi.aau.dk

Abstract. In this paper, we present an industrial use of Coloured Petri Nets (CP-nets) in designing a security system. An animation utility was developed which made it possible to perform user-friendly CP-net simulations. Furthermore, occurrence graphs (also known as reachability graphs and state spaces) were used for debugging the CP-net. In this way, a series of errors in the model were found and corrected. The CP-net design is used as a specification of the implementation of the security system. Therefore, finding errors by means of simulations and occurrence graph analysis reduces the amount of errors in the final implementation – making the software quality higher, which is the goal of the project.

1 Introduction

In the process of developing a new version of a security system at Dalcotech A/S[1], CP-nets were used for designing the software of the system. From the initial sketches of the system to the final system specification, CP-nets were applied. An animation utility was developed which made it possible to perform user-friendly CP-net simulations. Occurrence graphs were generated from the CP-net. They were used for investigating the dynamic properties of the CP-net and a series of errors in the model were located. The project demonstrates how the simulation and analysis methods associated with CP-nets are valuable when designing industrial systems.

We begin with introducing the security system and the project organisation. After this, we focus on the process of modelling the system and describe the final model. Additionally, we describe the use of the animation utility in simulations and the experiences gathered through occurrence graph analysis. Finally, we look at the pros and cons of using CP-nets for designing an industrial system.

Security System

This presentation of the PRISMA C–96 security system which is being developed by Dalcotech is based on [Dal95]. The security system will be the successor

---

[1] Dalcotech A/S is an engineering company which produces electronic systems, e.g. security systems. The company resides at: Bouet Møllevej 16, 9400 Nørresundby, Denmark, tlf: +45 98191799, fax: +45 98190799.

of PRISMA C–91 (described in [Dal94]), but the software is constructed from scratch and the functionality of the system is extended. The security system is an intruder alarm system with advanced functionality and with the capability of handling large installations. In Fig. 1, a small example of an installation is given [Dal95].

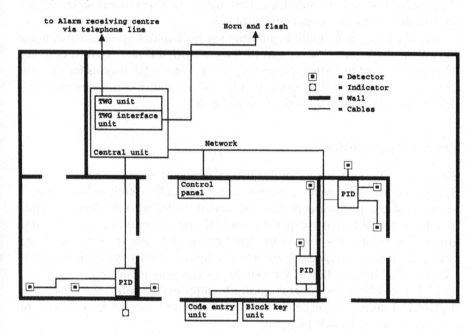

Fig. 1. PRISMA C–96 in a small installation

The *central unit* controls the security system and is connected to an alarm receiving centre via the public telephone system. Normally, an installation needs only one central unit. The PIDs (PRISMA Interface Devices) handle physical *input* (e.g. detectors) and *output* (e.g. light-indicators and horns). There can be several PIDs in an installation, and these are connected to the central unit by a network. The system is operated primarily from the *control panels* located inside the building(s). The control panels are equipped with a keypad and an alphanumeric display. They can be used for retrieving information on the condition of the system and for controlling it, e.g. by acknowledging alarms. Access to a building (and the control panels inside) is granted by presenting a valid code, key, or magnetic card (or combinations of these) at an *entry unit*. Generally, entry units are located on the exterior wall of the supervised building(s), close to the entrance. These units can be code entry units, block key units, or magnetic card readers.

The security system divides the building(s) into areas, which may overlap each other. Generally speaking, there are two states for an area: It can be *set*, meaning that the activation of a detector in the area will lead to an alarm. The alarm will be reported locally, by means of horns and light-indicators, and externally, by sending a message to the alarm receiving centre. Alternatively, an area can be *unset*. In this state, activation of a detector in the area will not cause an alarm. However, special input (e.g. glass-break detectors) can cause an alarm even if the corresponding area is unset. The states of the areas are controlled by the users via control panels and entry units. The most important function of the security system is to report an alarm whenever a detector is activated in a set area or under other illegal circumstances.

Installations can be quite large – the system handles up to a maximum of 100 areas, 1000 input, 1000 output, 1000 users (with different access codes), etc. The reliability of the system is essential, which is one of the reasons for applying CP-nets. The aim of the project is to increase the quality of software and make the new system more flexible than the old system.

Project Organisation

The project of using CP-nets for designing the security system was conducted by 8 people. Along with 3 security system engineers from Dalcotech we[2] participated in the *work group* which performed the actual modelling of the system. Additionally, 3 *consultants* participated in monthly project meetings with the work group. The consultants revised the work group's CP-nets, constructed design proposals, and were responsible for introducing the security system engineers to CP-nets. The introduction of CP-nets in the company is financially supported by ESSI – an EU-funded program meant to improve the quality of software production in Europe. The expenses for consultants, new machinery, CP-net tool, etc. were financed by this program.

2 Initial Models

During the first two months of the project, the three employees at Dalcotech became acquainted with CP-nets [Jen92] and with Design/CPN [CPN93] which was used as CP-net tool. Firstly, they attended an introductory course about developing CP-nets. After the course, the security system engineers continued developing CP-nets in connection with some small projects. Meanwhile, the system design was considered.

At this stage, it was discussed how CP-nets were to be applied in the project. What was the purpose of the model? Which level of detail should the model describe? Which parts of the system should be modelled? The work group found that the task was to design the software of the essential part of the security system: the central unit. Communication protocols, hardware failures, or the individual low-level behaviour of each control panel and entry unit were not to be modelled. On the other hand, the model should be sufficiently detailed. For instance, it should describe that some of the actions in the system are based on

---

[2] In this paper, *we* refers to us (the authors) alone. The work was done in connection with our master's thesis.

time-events. As an example, a user has to insert his key in the block key unit within 60 seconds after entering his code at the code entry unit. At this stage, the work group did not know how to model this; it was just recognised that time-events were an important factor of the system.

During the startup period the security system engineers realised that CP-nets were well suited for making detailed system designs. This was reflected by the use of CP-nets in connection with two other projects at Dalcotech. With respect to the security system model, the startup phase did not yield many tangible results but provided an increased understanding of how CP-nets were to be used. The startup phase lasted 2-3 months, partly because it takes a while to become familiar with the use of CP-nets, and partly because in this phase, the work group reflected upon the system's functionality. The work group had to set some goals for what to achieve with the CP-net. A *requirements specification* was initiated by the security system engineers. This was a textual description of the future system and was to serve as a guideline for developing the CP-net.

The development of the CP-net was definitely going to be part of a process that would lead to an implementation of the system. One of the consultants suggested investigating whether it could be possible to automatically generate code from the CP-net. The feasibility of this new idea was examined in connection with the project. However, at this stage it was uncertain whether it was possible to perform automatic code generation, or whether the CP-net was to serve as a template for a C/C++ implementation of the system.

Client/Server Model

With a possible C/C++ implementation in mind, the work group started modelling the system with a *client/server* approach, as suggested by Dalcotech. The work group found that it would ease implementation if the model described a client/server system. Also, the client/server approach is very flexible, allowing expansion of the system with few changes to old parts.

In this CP-net, a single place (containing message queues) models the process communication between clients and servers, each of which is modelled by a subnet. Data in one of the subnets of the client/server model can only be modified in the same subnet, which inhibits situations in which two processes try to modify the same data at the same time. In this way, a high degree of data control and modularity is obtained. On the other hand, every subnet (process) will need to communicate through the place containing the message queues in order to access data which is situated in other subnets. This induces a complex message structure used for communication.

During the startup phase, one of the consultants constructed an alternative CP-net, describing the security system. This model was also developed further while the client/server model was being constructed. The purpose of this model was to try to capture the essence of the system as described in the requirements specification of that time. The implementation of the model was not considered and therefore, we refer to this model as the *implementation independent* model. This model had a clearer structure which did not force all communication through one place. The main drawback of this model was that the final system would be less flexible (harder to expand). It seemed that the client/server approach made it easier to add an extra process or change a specific process later on. Furthermore, the security system engineers felt that the client/server model would be easier to implement because it consisted of well defined modules and message structures.

In April '95 it was decided that *one* approach had to be chosen and used for the rest of the project. The choice fell on the implementation independent approach. The loss of clarity and the focus on implementation inherent in the client/server approach were too serious drawbacks. If automatic code generation was to be applied, the implementation benefits from a client/server model would be insignificant. The bottom line is that *it was a question of modelling the security system concept instead of modelling an implementation of a security system*. It was found that a more comprehensible and general model would be desirable in the design of the system. Also, the employees at Dalcotech were becoming more acquainted with CP-nets and beginning to get an idea of how an implementation independent model could be implemented – even if automatic code generation was not to be applied. As a result of this decision, a new model was constructed. Parts of the previous models could be reused, but as a matter of fact this new model was developed almost from scratch, with major revisions of all data structures and functions previously declared.

3 Final Model

Having already made two (incomplete) models of the system, it was much easier to start on the new model. The work group was very careful with designing the fundamental structure of the CP-net. Quite some time was spent on identifying candidates of the system's processes and defining colour sets which were as comprehensible as possible, based on the requirements specification of that time. The work group used a *top-down* approach and focussed on determining the basic structure of the CP-net.

In Fig. 2, the top page of the CP-net is depicted. The main system components are the three substitution transitions[3] – each representing a subnet of the model. The Environment models the peripheral devices; the Central unit forms the main part of the model, it manages the global data and models the functionality of the security system; and the Configuration initialises the global data, by loading data from configuration-files.

On the places in the left column, the communication between the Central unit and the external devices present in the Environment takes place. These places will be referred to as interface places, as they constitute the I/O interface to the Environment. For instance, the digital input/output is modelled by two places. The Changed logical input place holds a list of the changed logical input (e.g. detectors) reported by PIDs in the Environment. The Changed output

---

[3] Substitution transitions are marked with an HS-tag.

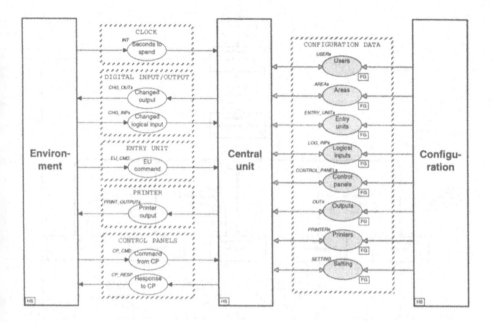

Fig. 2. Top level of the CP-net

place holds a list of changed output (e.g. light-indicators), which is sent by the Central unit to PIDs.

The shaded places in the right column represent global data initialised from the Configuration. Their colour sets are lists of records; each record has between 10 and 20 entries describing both static and dynamic data. The shaded places, marked by FG-tags, are all fusion places. A fusion place is a member of a set of places (called a fusion set), which conceptually represents one single place. The interface places are sockets. Sockets define the interface to substitution transitions; each socket is related to a port place on the subpage.

Splitting the model into Environment, Central unit, and Configuration subnets turned out to be a good design decision, as it isolates the software of the central unit from the peripheral hardware units and the configuration. It also allows three different versions of the environment and configuration: one set of subnets used during simulations, where the environment used our graphic animation utility described in Sect. 4; another set of subnets, used during the analysis phase; finally, a third set of subnets was used in case of automatic code generation, where the environment used serial port communication.

Figure 3 shows the subnet of the Central unit. The work group identified the five following processes: Input event handler receives changed input (from PIDs) and determines whether alarms are to be generated. Time handler controls time dependent data and handles time-out events. Control panel handler takes care of the user interaction to/from the control panels by sending menus to the control panel and executing user commands. Log handler logs events to the printers and the internal logs, e.g. when a detector triggers an

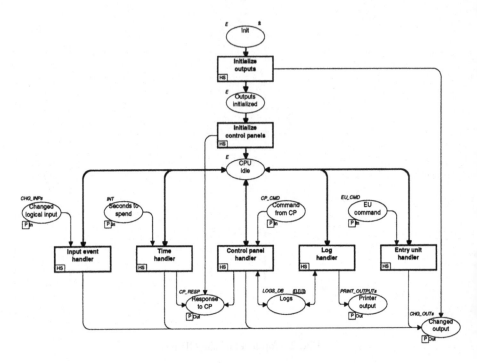

Fig. 3. Central unit

alarm, its name and alarm type will be logged. **Entry unit handler** handles entry unit commands, e.g. set/unset requests and user-code ratification. The processes can be activated if there are tokens on their input ports, representing environment changes, e.g. commands entered by the user on the control panel.

Central unit starts by initialising the outputs (e.g. switching on the control panel power-light) and sending start-menus to the control panel. It can then execute *one* of the five processes. The **CPU idle** place ensures that no more than one of the processes is chosen and finished before another can proceed. This is done in order to prevent the processes from operating on shared data simultaneously, which might lead to inconsistencies. Moreover, it reflects that only one processor exists in the hardware which is to be used for implementation.

Figure 4 shows the subnet of the **Input event handler**, describing its intrinsic behaviour. The sequential process-flow is indicated by thick arcs. White arrow-heads are used on one (the less important) of the two arcs that are needed in order to update a list. The %-operator used in the arc expressions returns the value to the right if the list of boolean expressions to the left all evaluate to true, but if one of the boolean expressions evaluates to false an empty multi-set is returned.

On the leftmost top place, changed input is given by the **Environment** (see Fig. 2), e.g. representing an opened infrared detector. An input can be either closed, opened, or sabotaged. The **Receive changed inputs** transition updates the input state on the **Logical inputs** fusion place. Subsequently, the **Get**

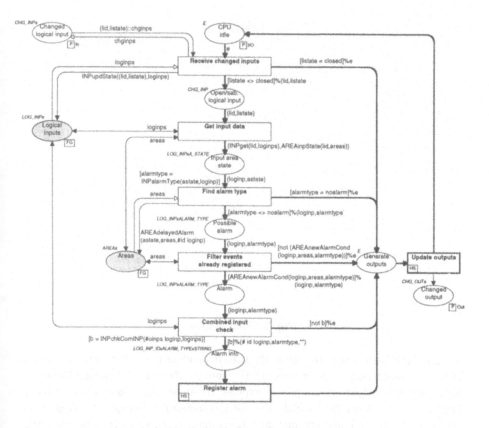

Fig. 4. Input event handler

input data transition retrieves the state of the area in which the input is situated. **Find alarm type** determines what alarm type is to be generated, e.g. a glass-break detector will generate an intrusion alarm type. Additionally, **Find alarm type** delays input changes belonging to areas temporarily unset by entry units. If an unset-command is not received from a control panel within 45 seconds, the delayed inputs will activate alarms (this is controlled by the time handler). Input not causing alarms will be filtered out. Then **Filter events already registered** ignores alarm conditions already registered. In order to reduce the number of false alarms input may be combined to depend on other input, which must all be open before triggering an alarm – this is controlled by **Combined input check**. Finally, an alarm condition is registered by the **Register alarm** substitution transition. The input event handler always ends by "calling" the rightmost **Update outputs** substitution transition, which re-evaluates the output expressions, as they might depend on the newly detected input change. A user-defined output expression is attached to each output in order to determine whether it is to be on or off. For example, the output expression for the horn could be: ALARM_COND(1) OR ALARM_COND(2), meaning that the horn is activated if there is an alarm condition either in area one or two.

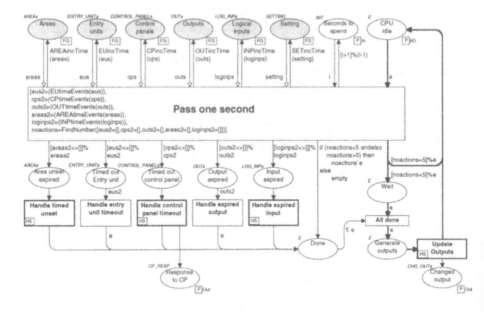

Fig. 5. Time handler

The `Time handler` manages all the time dependent data-structures, as shown in Fig. 5. All six fusion places have relative integer attributes, denoting the number of seconds before a time period expires. As an example, each area on place `Areas` has an integer-field which denotes the remaining seconds the area stays temporarily unset. The `Pass one second` transition increases time by updating the time attributes on all time dependent fusion places, and if any time-outs occur, they will be handled by the five `Handle...` transitions. For example, when a temporary unset period expires, the `Handle timed unset` subpage makes sure that the area-state is returned to set and that any delayed alarms are registered. After time-outs have been handled the `Update Outputs` substitution transition is called again, as on Fig. 4.

Timed CP-nets were not considered, as the CP-net was not intended for performance evaluations; instead, the CP-net should be able to explicitly cope with "real" clock-ticks from the environment. Therefore, the time has been integrated in the model by equipping the data structures with integer attributes, operating as local timers.

Having described two of the central unit processes, we remark that the user-interaction between the control panel and entry unit also constitutes a great portion of the model. All together, the model has 38 pages with 95 transitions and 325 places. In order to reuse identical net-structures, seven subpages have between two and four instances (copies). Furthermore, the model makes use of approximately 4,000 lines of CPN ML declarations and functions. CPN ML is the inscription language of Design/CPN and is based on the functional language Standard ML (SML) described in, e.g. [Ull94].

4 Simulations

Simulation is an important instrument for debugging and verification of CP-nets. It gives the developer an improved understanding of the dynamics of the model.

In order to enhance the user-interface during simulations, we constructed a graphic animation utility. This utility supports two-way communication between the user and the simulator. The CP-net animates graphic objects while the user can feed the CP-net with input, based on selections. Hereby, a more user-friendly simulation is obtained. In fact, the user does not have to inspect the CP-net, but can concentrate on how the net responds graphically to his actions. This is particularly useful when presenting the CP-net to users without any knowledge of CP-nets. But it also became useful when testing the CP-net in the simulator. The animation utility allows the user to build his own graphic "scenes" composed by objects such as: buttons, arcs, bitmap objects, all of which can be manipulated during simulation to mimic/imitate their real-world counterparts. In this way, the user can visualise the CP-net's response to different scenarios and display interesting system-states.

We have constructed a library for Design/CPN using the built-in routines for manipulation of graphic objects [OAF93]. It is called *Mimic/CPN* [RS95a] and its main primitives are to: hide or show objects, move or align objects, hide or show connectors, select an object/connector by clicking on it with the mouse, combine multiple user-selections, and save and restore states of graphic objects. All library functions are to be called from transition code segments, which are pieces of SML code, executed whenever the corresponding transition occurs. The graphic objects are built within the editor, and their graphic appearance is defined by the user. This allows the user to customise the graphic layout for his own specific needs.

The "mimic" name is actually inherited from the security system, in which an optional *mimic-board* is included, showing a drawing of the covered building with lights mounted underneath. The lights indicate various system-states such as: area states, input states, alarms. In order to model the mimic-board, we use a scanned picture of a building with detectors represented as circles, output (e.g. indicators) as squares, and icons for various devices: clock, horn, entry units, control panels, and printers (see Fig. 6). When simulating, the environment sub-net updates the graphic states of light-output, which are shown as either filled or empty squares – depending on whether the state is on or off. If, for instance, an output changes to on, the old square will be hidden, and a previously hidden filled square appears at the same point. Horn and flash output are shown as icons. Printer text, sent by the central unit, is appended in text-boxes on separate pages; which opens when the log icons are clicked. All graphic updates are done from transition code segments by means of Mimic/CPN functions.

The animation utility also allows the user to select from graphic objects, thereby making it possible for the user to control simulations. For example, when making an input change the user clicks on a detector and then selects its new state from the closed/opened/sabotaged symbols on the mimic-board. We also made entry unit commands (set and unset) and user-codes on the code

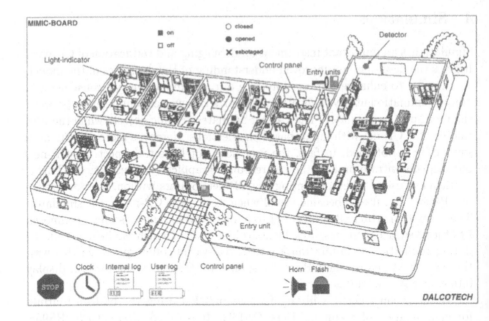

Fig. 6. Mimic-board

entry unit by letting the user click on the keypad – as he would do in the actual security system. When the user clicks the clock icon, a dialogue box will appear, and the entered amount of clock-ticks will be reported to the central unit. The entire control panel user-dialog, where the user browses in menus, received from the central unit, and selects commands has been implemented using the animation utility (see Fig. 7). It should be noted that the animation utility and the simulator are synchronised, i.e. when a select function is used, the simulator actually waits until an object has been chosen by the user.

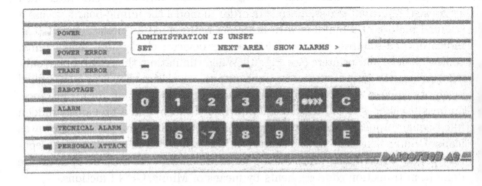

Fig. 7. Control panel

The work group used an iterative approach of modelling and simulation cycles. The first prototype was gradually refined, and eventually, it constituted the final model. Simulation was used to test and investigate the net behaviour – after having augmented or changed the CP-net. The animation utility was particularly useful for attracting the user's attention visually, if the graphic objects did not correspond to the user's expectations. Typically, simulations of 300-400 steps were performed. In these trial-and-error experiments, the animation utility was used for guiding and monitoring the simulations.

5 Occurrence Graph Analysis

The CP-net is used as a specification of the final program. Therefore, finding an error in the model by means of simulation or analysis can inhibit an error in the final program. Test simulations of the CP-net revealed many flaws in the model and provided feedback for improving the model during the entire design phase. At the end of the design phase, we applied occurrence graph analysis (also known as reachability graph analysis and state space analysis) in order to locate as many errors as possible before the Dalcotech engineers started implementing the system. It should be stressed that at the same time test simulations were performed by the security system engineers. The simulation experiments also continued after we finished analysing the CP-net.

In this section, we briefly illustrate the changes made to the CP-net in order to generate occurrence graphs. Furthermore, we give examples of how we investigated dynamic properties of the CP-net, and how we located errors in the model. For the analysis of the CP-net we used a prototype of the Design/CPN Occurrence Graph Tool [CJ95] – we refer to this tool as the OG tool. The OG tool makes generation of full occurrence graphs[4] (O-graphs) possible. After generating an O-graph, a set of standard queries can investigate dynamic CP-net properties, such as boundedness, liveness, etc. Additionally, queries which are customised by the user allow checking of system specific properties. The O-graphs of the OG tool are integrated with the simulator in the sense that it is possible to switch from a node in an O-graph to the state in the simulator which it represents – and vice versa.

Preparing the Model for Occurrence Graph Analysis

In advance, we had little idea about whether O-graph analysis would yield noteworthy results when used on the large security system model. The only thing that seemed certain was that plenty of reductions of the model and its configuration would be necessary. Some of the colour sets in the CP-net are records with between ten and twenty fields which either change dynamically during simulation or are determined by the configuration of the system. Even if "infinite" colour sets (e.g. integers) were reduced to small finite sets, the number of different reachable markings would be enormous. Hence, we did not expect to prove

---

[4] A full occurrence graph has a node for each reachable marking and an arc for each occuring binding element.

the correctness of the CP-net in all kinds of different configurations. Instead, we wanted to look for errors in the model and get an increased understanding of the behaviour of the CP-net by examining its dynamic properties. In this way, the analysis would become an "extended simulation" which investigated all possible occurrence sequences of small examples. This would supplement normal simulations which investigate "few" occurrence sequences of larger examples.

With the purpose of analysing the CP-net by generating O-graphs, we replaced the environment subnet used for simulations (which applied the animation utility) with "test environments". An example of such a test environment is displayed in Fig. 8. The environment page is a subpage of the top page from Fig. 2.

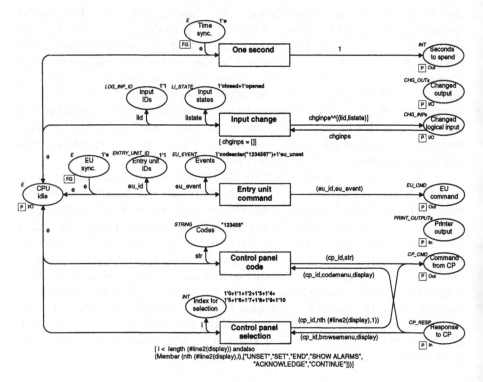

Fig. 8. Environment page for analysis

The displayed test environment produces clock-ticks (the One second transition), random input changes (Input change), random entry unit commands (Entry unit command), and random control panel commands (Control panel code and Control panel selection). The range of different input changes is decided by the tokens on the places Input IDs and Input states. Consequently, the displayed environment can only change the state of input number one to opened or closed. In similar ways, the range of different entry unit and control panel commands is controlled. However, when a browse menu is situated on Response to CP, a list of submenus can be chosen from. Here, a random submenu is chosen. The guard of Control panel selection ensures that only "interesting" selections are made, i.e. selections which are necessary in order to test the important functions of the control panel.

In order to be able to fully generate O-graphs, we applied a minimal configuration with only one area, one input, one control panel, etc. Furthermore, some of the data fields of the elements have been modified. For instance, time-out periods are all set to expired after one second. Of course, this was a serious reduction of the security system. For instance, we were not able to examine conflicts between two control panels attempting to operate on the same area. However, the number of states exploded if, e.g. two areas and two inputs were included, and therefore, we focussed on testing the minimal configuration.

In addition to changing the environment and configuration of the security system we had to modify the model in order to reduce the state space. For example, we excluded the log-facility from the model, and we limited the list of changed output (on place **Changed output** of Fig. 8), so that it held no more than one element. Furthermore, we made the **CPU idle** place accessible from the environment page – thereby only allowing environment transitions to occur when the CPU was idle. The fusion places **Time sync.** and **EU sync.** were introduced for synchronising the transitions **One second** and **Entry unit command** with the central unit. For instance, the **One second** transition cannot send a second clock-tick until the time handler in the central unit has finished treating the first clock-tick. These modifications were not important for the behaviour of the model. However, they were necessary in order to obtain finite O-graphs of reasonable sizes.

In spite of all these reductions of the CP-net, fully generated O-graphs could only be obtained on small test examples. We could not achieve a fully generated O-graph for the environment displayed in Fig. 8 using the minimal configuration. Instead, we made O-graphs for selected parts of the system by, e.g. omitting entry unit commands from being generated in the environment. The largest O-graph generated consisted of 150,000 nodes and 250,000 arcs, for which the OG tool demanded 230 Mbytes of RAM. For practical reasons, we generally worked with smaller O-graphs which had up to 50,000 nodes.

Investigating Dynamic Properties of the CP-net

Having described the problems connected with constructing O-graphs from the CP-net we now look at the results we achieved from the reduced CP-net. Despite the reductions, investigations of the dynamic properties of the CP-net helped improve the confidence in the model. Furthermore, we located approximately 15 non-trivial errors, of which we will describe a few.

In order to check dynamic properties of the CP-net we made *queries* using the OG tool. A set of standard queries makes it possible to investigate the reachability, boundedness, liveness, fairness, and home properties corresponding to the propositions of Sect. 1.4 in [Jen94]. As an example, a single function-call returns a list of all transition instances which are live.

The test environment allows infinite occurrence sequences. As we were able to fully generate O-graphs, the infinite occurrence sequences must be *cyclic* in some way. This made it interesting to investigate the set of strongly connected components (SCC) of the O-graph. Having generated an O-graph, the OG tool allows generation of an SCC-graph which has a node for each strongly connected component and thereby essentially identifies the cyclic structures of the O-graph. A terminal component of an SCC-graph represents a set of nodes in the O-graph from which no simulation can escape (while a terminal node of an O-graph represents a dead marking). When using test environments as shown in Fig. 8, we found that only one terminal strongly connected component existed. Thus, the individual nodes of the terminal component are home markings. This agrees with the intentions for the security system model. If alarms are generated, they can be acknowledged (and thereby removed), and if an area is unset, it can be set again, and so forth.

In general, the standard queries increased our confidence in the model. For instance, we established bounds on places and determined which transitions were live. However, some properties could not be verified by means of standard queries. For this purpose, the OG tool provides functions for traversing nodes, arcs, or strongly connected components of the O-graph, using user-defined functions to get the desired information. As an example, we made a query which searched all nodes of the O-graph and checked that only one process token existed.

The most important property of the security system is to report an alarm to the alarm receiving centre whenever a detector is triggered under illegal conditions. We applied the functions of the OG tool to make queries that found all nodes where "illegal" input change had arrived at the Open/sab. logical input place of Fig. 4. For all these nodes, we found the first successor node in which the input change had been treated by the input event handler. On this set of nodes in which the input change had been treated, we verified that the system was in an alarm condition and that the horn had been turned on. Thereby, we had proved that in a minimal configuration, illegal input changes to the security system always causes alarms to be generated.

As described above, we investigated standard as well as system-specific dynamic properties of the security system model. However, we also found a series of errors in the model by means of O-graph analysis. The functions used in the model all raise SML exceptions if something goes wrong. If an exception is encountered during O-graph generation, the generation halts. The OG tool makes it possible to switch a node in the O-graph into its state in the simulator. In this way, the marking which causes problems can be examined in the simulator. Furthermore, the OG tool makes it easy to find predecessors (and successors) of nodes in the O-graph. By examining predecessors, an occurrence sequence leading to the critical node can be found.

As an example, we discovered that the O-graph generation halted when the control panel timed out (which normally happens after 30 seconds of idle time). The error appeared when a user had been acknowledging alarms, a time-out had occurred, and the user had chosen to acknowledge alarms again. It turned out to be caused by an index error in the control panel data-structure. It is unlikely, that the occurrence sequence leading to the error would have been used

– revealing the error – in further test simulations. Therefore, the advantage of calculating all occurrence sequences in the O-graph was valuable in this case.

The OG tool makes it possible to display selected parts of the marking of nodes and the binding element of arcs. By examining these markings and binding elements, we also found a number of errors. For instance, we discovered that the list of area-data situated on the **Areas** place (see Fig. 2) was expanded (its elements were duplicated) during certain occurrence sequences. As the configuration of the system ought not to change during execution, this was an obvious error in the model. However, we had not encountered the problem during the previous test simulations.

The main drawback of making full O-graphs for the complex CP-net is that only greatly reduced configurations can be used if a state space explosion is to be avoided. We *did* make some O-graphs for larger configurations (with, e.g. two control panels) and a wider range of test environments. However, in order to reduce the state space we had to execute the central unit only a fixed number of times (normally one time). Thereby, the O-graphs became totally acyclic, and most of the standard dynamic properties of the O-graphs became uninteresting. For instance, no transitions were live, as no cycles existed in the O-graph. However, we also found errors in the model by this approach. As an example, we discovered that inconsistencies could occur when users at separate control panels were acknowledging alarms. A user could attempt to acknowledge an alarm which was already acknowledged by another user. As the error occurred when two users made certain selections at the same time from two different control panels, it is uncertain whether the error would have been located by further test simulations (which cannot test all such combinations).

All in all, O-graph analysis turned out to be a very efficient way of debugging the reduced CP-net. We found approximately 15 non-trivial errors in the model by this approach. Some of these errors would have existed in the final implementation of the security system, if they had not been located by O-graph analysis. Furthermore, investigations of the dynamic properties improved our understanding of the model behaviour. For these reasons, preparing the model for O-graph analysis proved worthwhile despite the difficulties it involved. Consequently, we can recommend using O-graphs for analysing CP-nets – even if the CP-nets are complex and only parts of them can be examined.

6 Evaluation

As the design of the system evolved throughout the modelling process, the work group frequently had to reject or radically change parts of the model. The most serious changes (apart from rejecting the client/server approach) were made in May '95 after employees from Dalcotech had learned more about the requirements that a security system needs to fulfil in order to be approved by the German standard for security systems. Furthermore, monthly meetings with the consultants induced a number of changes to the model. In this way, the modelling process was conducted in a *prototyping* fashion.

In parallel with constructing the CP-net, a written requirements specification was worked out. For those of us without knowledge of security systems, this was an important aid in the modelling process, as we could see a detailed description of what we were supposed to model. The detailed CP-net design made it possible to elaborate the requirements specification. The correlation between the requirements specification and the model was useful and also resulted in *two* system descriptions instead of one. This was useful later on, as both were consulted in order to retrieve information about the system.

In Fig. 9, an overview of the activities in the project is displayed. It illustrates how the requirements specification, the animation utility, and the implementation independent model were developed in parallel with the final model. The

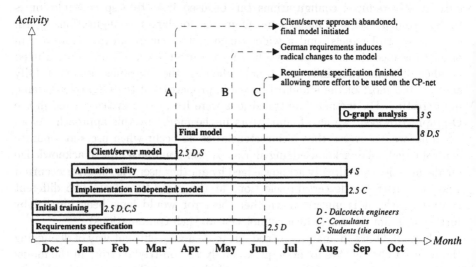

Fig. 9. Overview of project activities

number after each activity is an *estimate* on the amount of man months used on the activity. After each estimate, it is indicated who performed (the main part of) the respective activity. As an example, the figure illustrates that the Dalcotech engineers did not work on the implementation independent model, the animation utility, or the O-graph analysis. However, they did approximately half the work on the client/server model and the final model. The A, B, and C lines represent important events in the modelling process. Totally, two man years were used on the displayed activities. Fortunately, the development of the animation utility is only needed once, and in future CP-net projects at Dalcotech, the initial training of the employees is not necessary.

Apart from the activities displayed in Fig. 9, investigations of the feasibility of automatic code generation were conducted by the consultants and Dalcotech employees throughout most of the project. The SML code of the simulation-engine was extracted for automatic code generation, and was isolated by ignoring, e.g. all code used for graphic updating. The SML code was then ported to the Moscow ML [RS95b] environment, which is based on the Caml Light [Ler95] runtime system. The generated code is small in size, but the interpreted programs run about ten times slower than compiler generated code. The stand-alone executable of the SML code used approximately 450 Kbytes of RAM, and the data also took up 450 Kbytes on the 80386 PC-card which is used in the central units of the security system. The results of these tests were promising, as it turned out to be possible to execute the SML code in its new environment. However, the automatically generated code was too slow and optimisations of the code were needed, if it was to be used in the final system. At this point Dalcotech decided not to proceed with automatic code generation. A tight schedule, lack of resources, and uncertainty about the success of this new approach were the main reasons for this decision.

The implementation of the security system began in October '95 after having finally decided not to proceed with automatic code generation. The CPN ML functions and colour sets are translated into C/C++ functions and types. Furthermore, the structure of the CP-net is used as a template for the C/C++ program. In order to implement the CP-net, some changes of the design have been necessary. However, the Dalcotech employees introduce such changes by updating the CP-net before implementing the changes. Thereby, the CP-net is constantly up-to-date with the actual implementation of the system. It should be noted that no automatic verification tool is used for checking that the C/C++ program is consistent with the CP-net. It is of course possible that the changes to the CP-net and the C/C++ code in itself will contain new errors. With respect to the adjusted CP-net, it was considered to apply O-graph analysis again in order to locate such errors. However, due to lack of resources this was not done. Instead, systematic test runs are performed.

Using CP-nets for Designing the Security System

The use of CP-nets in the Dalcotech project proved successful for the following reasons:

1. CP-nets provided a graphic as well as a formal description of the system. It is valuable to have a formal model which describes the main components and actions of the system in a graphic way. It is easy to find out technical details about the systems behaviour by looking at the CP-net.
2. Being able to execute CP-nets is a very powerful feature. By means of simulations, the members of the work group gained confidence in the existing CP-net. Using the animation utility described in Sect. 4, simulations became realistic tests of the functionality of the system, even though the hardware etc. was not yet developed at the time. The interface to the simulations resembles that of the final system. Thereby, the security system model can be demonstrated to customers.
3. O-graph analysis of the CP-net was a valuable aid for debugging the model. Some of the located errors could have emerged in the implementation of the system. Even though the analysis was limited by the complexity of the model, the errors which were found justify spending resources on O-graph analysis.

Finding and correcting errors in the final system can be very expensive. Therefore, the analysis methods associated with CP-nets are very valuable.

4. CP-nets have computer tools supporting their drawing, simulation and analysis. In our case Design/CPN provided almost all the functionality that we wished for. The tool has good drawing facilities for making CP-nets and makes it easy to syntax check and simulate the CP-nets. We did not meet barriers with respect to what we could model by means of the tool. Using the OG tool, we verified many dynamic properties of the CP-net, and additionally, we were able to locate a series of errors in the CP-net.

5. The use of CP-nets has improved the quality of software development at Dalcotech. The Dalcotech employees recognised CP-nets as "a new way of thinking" which takes time to learn, but which is worth the effort. CP-nets were used in all phases: from the initial sketches of the system to the final technical model which was used for coding the actual implementation.

6. CP-nets were used successfully by Dalcotech in connection with two other projects on the Dalcotech employees' own initiative. Additionally, CP-nets are used for designing the low-level system which underlies the model described in this paper. Having become familiar with CP-nets, the security system engineers chose CP-nets for a rapid design of the low-level system. The security system engineers expect to use CP-nets in future projects, too.

The project also revealed problems with respect to using CP-nets in design of industrial systems:

1. CP-nets were unknown to the security system company as they are to most industrial companies. Consequently, training of the security system engineers was necessary. In spite of an intensive training course, it took some time before the engineers learned how to transfer their ideas into applicable CP-nets. Dalcotech would not have introduced CP-nets without the financial support from ESSI which covered expenses in connection with training, consultants, CP-net tool etc.

2. Using Design/CPN, it took a while to check the syntax, simulate etc. when the model began to grow larger. In general, it became somewhat cumbersome to add to the model and simulate the extensions. The Dalcotech employees also meant that better editing/debugging facilities of SML code in Design/CPN could have been very useful. A new version of the tool is presently being developed. The simulator is many times faster [HH94] and a new approach for declaring colours, etc. is included. Thereby, the problems concerning turn-around time are greatly reduced.

3. In order to apply O-graph analysis to the large CP-net many reductions of the model and its configuration were necessary. Therefore, analysing the CP-net was not a straightforward task. In some cases, using O-graphs with symmetrical or equivalent markings would decrease the need of prior reductions to the CP-net. However, the security system model is an example of a CP-net which does not have many symmetrical markings and which therefore is not well-suited for such an approach.

Having discussed the positive and negative aspects of using CP-nets for designing the security system, we conclude that CP-nets were very applicable in this industrial design project. The project demonstrates the value of the interactive simulations and formal analysis methods associated with CP-nets. The employees at Dalcotech recognised CP-nets as a useful design instrument and will continue applying CP-nets in future projects.

Acknowledgements

We would like to thank the Dalcotech company for letting us participate in the project. Furthermore, we thank Torben Andersen, Klaus L. Nielsen, Søren V. Hansen from Dalcotech and John Mølgaard from DELTA for fruitful cooperation and constructive ideas. Finally, we are grateful for the many valuable comments and ideas for this paper and our work in general from our advisors Kurt Jensen and Søren Christensen.

References

[CJ95]   Søren Christensen and Kurt Jensen. *Design/CPN Occurrence Graph Tool, User's Manual (vers. 1.0)*. Computer Science Department, Aarhus University, 1995.

[CPN93] Meta Software Corporation, Cambridge, MA, USA. *Design/CPN Reference Manual for the Macintosh (vers. 2.0)*, 1993.

[Dal94] Dalcotech A/S. *PRISMA C-91 System Manual*, 1994.

[Dal95] Dalcotech A/S. *PRISMA C-96 Requirements Specification*, 1 edition, 1995.

[HH94]   Torben B. Haagh and Tommy R. Hansen. *Optimising a Coloured Petri Net Simulator*. Master's thesis, Computer Science Department, Aarhus University, 1994.

[Jen92] Kurt Jensen. *Coloured Petri Nets – Basic Concepts, Analysis Methods and Practical Use, Volume 1*. Monographs in Theoretical Computer Science. Springer-Verlag, 1992.

[Jen94] Kurt Jensen. *Coloured Petri Nets – Basic Concepts, Analysis Methods and Practical Use, Volume 2*. Monographs in Theoretical Computer Science. Springer-Verlag, 1994.

[Ler95] Xavier Leroy. *The Caml Light System, Release 0.7. Documentation and User's Manual*. INRIA, France. Available at ftp.inria.fr in directory lang/caml-light, 1995.

[OAF93] Meta Software Corporation, Cambridge, MA, USA. *Design/CPN Internal Functions Programmer's Reference (vers. 2.0)*, 1993.

[RS95a] Jens L. Rasmussen and Mejar Singh. *Mimic/CPN, A Graphical Animation Utility for Design/CPN (vers. 1.5)*. Computer Science Department, Aarhus University, 1995.

[RS95b] Sergei Romanenko and Peter Sestoft. *Moscow ML Owner's Manual (vers. 1.31)*. Available at ftp.dina.kvl.dk in directory pub/mosml, 1995.

[Ull94] Jeffrey D. Ullman. *Elements of ML Programming*. Prentice-Hall International Editions, 1994.

Modeling and Analysing DART Systems Through High-Level Petri Nets

Libero Nigro and Francesco Pupo

Dipartimento di Elettronica, Informatica e Sistemistica,
Università della Calabria, I-87036 Rende (CS) - Italy,
Email: l.nigro@unical.it, pupo@apollo.deis.unical.it

Abstract. The work described in this paper is concerned with modeling and analysing distributed object-oriented real time (RT) systems, developed according to DART - Distributed Architecture for Real Time - which provides an object-oriented life cycle for RT systems. The paper first gives an overview of DART, then discusses the usefulness of mapping DART systems onto high level Petri nets as part of an iterative design process. The resultant framework is a modular operational model, based on a powerful Petri nets tool which delivers both the user-interface utilised for animation purposes and a set of mechanisms suited for the analysis of time-dependent behaviour.

Key words: real time, scheduling analysis, distributed systems, object-orientation, Petri nets, animation, executable specifications

1 Introduction

The development of a real-time system is a well-known complex task in that it must fulfill both the *logical correctness* and, most importantly, the *timing correctness* of the target system. Logical correctness is a shared issue with conventional concurrent systems and is concerned with the achievement of the required system functions. Timing correctness, on the other hand, is related to guaranteeing that a given logical behaviour is provided at the *right time*, not before nor after a *due* time. In order to achieve logical correctness an RT system design can be based on software engineering techniques such as structured design or object-orientation. However, the challenge remains in coping with the timing correctness too. For instance, object-oriented languages generally lack abstractions that are directly related to real-time activities, e.g., for ensuring an explicit and tight control of the external environment, maintaining the stability of which is the ultimate goal of an RT system. Timing control is highly dependent on both language assumptions *and* the features of the runtime system used. The Distributed Architecture for Real Time - DART - [13, 15], represents a simple yet powerful model and programming paradigm suited to the development of distributed, object-oriented RT systems. DART was directly derived from Shlaer and Mellor object-oriented analysis method [21]. It focuses on the concept of "Operating Software" [23], by offering a set of mechanisms for communication, synchronisation and flexible scheduling control which are not extensions of hidden OS mechanisms. DART emphasises a language-centered rather than a classical system-centered approach. DART's main contribution rests on specifying and then meeting the systems timing constraints using the concept of *time predictability* [22, 7, 17]. Moreover, it can easily be integrated to work with different network and protocols suited for RT operation. The effective use of DART strongly depends on accurate analysis of RT system behaviour and performance. The transformation process from analysis down through to design and implementation in popular object-oriented languages like C++ and

Oberon-2 has been demonstrated and is straightforward. This paper considers the possibility of mapping DART on to high level Petri nets with the goal of allowing both animation and simulation of a chosen system. The paper focuses on a powerful Petri net based tool, Cabernet [19], which provides the user-interface for animation purposes and a set of mechanisms useful for an in depth analysis of both functional and timing behaviours through prototyping and execution of a modelled system.

The structure of the paper is the following. First a description of DART is given together with its application to a realistic process control system. The paper goes on by showing a transformation of DART into the terms of the high level Petri nets supported by Cabernet as a part of the overall design process. Usefulness of the mapping process is illustrated by analysing the modelled RT system example leading to an implementation which will exhibit the required logical *and* timing behaviour. Finally, some directions of further work are summarised in the conclusions.

2 Distributed Architecture for Real Time

DART is a minimal software architecture and programming model suited to the development of distributed, object-oriented real-time systems. Its programming in-the-large and programming in-the-small levels have been adapted from Shlaer-Mellor Object Oriented Analysis (OOA) method from which are inherited the basic models. Key points of DART are *timing predictability* and a *seamless systems development life cycle*, where the same concepts and entities (i.e., objects, event interactions, timers ...) propagate unchanged from analysis down through to design and implementation. A whole system is decomposed into a set of cooperating subsystems. Each subsystem is associated with an application domain and is responsible for providing selected functions and real-time behaviour. Communications among subsystems is ensured through the services of a deterministic communications network. A subsystem hosts a community of active objects which implement the static structure and dynamic behaviour of the subsystem in a decentralised way.

2.1 OOA Models

The following provides a summary of the fundamental OOA models. An example of the application of these models to a problem domain is given later in this paper.

- *Information Model* (IM). This is the crucial starting point of a project, it shows the objects involved in a system and the relationships between partner objects. Each object specification includes a list of internal attributes.
- *Object lifecycle.* Captures the dynamic behaviour of an active object. It is an extended, timed finite state machine diagram or equivalent State Transition Table (STT). An object lifecycle is composed of *states*, *events* and *actions*. Events cause state transitions, an event may also occur as a consequence of firing a timer. Actions are associated with states (Moore model). An action is executed each time the corresponding state is entered.
- *Object Communication Model (OCM).* Summarises the dynamic interactions between object state models by showing the various exchanged events and for each event its sender and receiver and name of the message which encapsulates it.
- *Action Data Flow Diagram* (ADFD). Can be used to formalise the elementary processes occurring in the execution of an action.

2.2 Subsystem Architecture

This section summarises the basic concepts of DART programming in-the-small, further details can be found in the references [13, 15].

- DART relies upon a *concurrent view* of the world. Objects are the basic building blocks of a (sub)system, they operate concurrently and synchronise their execution.
- *Events* are the basic form of communication. They are asynchronous in character and are responsible for causing state transitions. *Event data* can be transmitted by value only. A receiver object responds to an event on the basis of its current state. Once an event is acted upon it has been consumed and cannot be re-used. Consequently, the handling of an event which cannot be served immediately can be postponed by storing it in data or states. Generally the arrival of an event causes a state transition and the execution of an action associated with the target state.
- *Action execution* is atomic. At most one action may be executing at any point in time within a given object. Furthermore events which arrive at the object boundary when an action is in progress remain outstanding until the action is completed. Actions within separate state machines can be executed concurrently.
- *Concurrency semantics* depend on the *time interpretation*. In "simultaneous" time interpretation, action concurrency means simultaneous execution, i.e., time overlapped. In "interleaved" time interpretation, only one action can be running at any point in time. The action is the unit of interleaving. In DART, simultaneous execution is possible among different subsystems which are allocated on different physical processors. Interleaved execution is adopted within a subsystem.
- *Synchronisation* among concurrent objects rests on communication and object lifecycle. No recourse to conventional mechanisms (e.g., semaphores and their equivalents) is used.
- In any interpretation of time, concurrency leads to the existence of *competition situations* when different objects (clients) compete for the use of common resources (servers). In general, a "monitor object" (*service assigner*) can be introduced whose task is to filter client requests. Concurrent requests are sequenced by the monitor. Moreover, as a server becomes idle, it is assigned to a single client according to a monitor policy.
- An external event is said to trigger a *thread of control* (i.e., a *transaction*). Such a thread is the chained execution of the actions involved in the overall processing of the event. One action may cause further action executions by raising further events. A thread can thus split itself into two or more paths as a result of events generated from within actions. A thread may terminate internally within the system in an action which does not generate more events, or externally in an output terminator. *Terminators* are objects which interface the system to its environment. They are responsible for converting external messages into equivalent DART events and vice versa.
- A state action can in general make read/write requests upon attributes belonging to itself or to other objects in its local subsystem. It is the responsibility of the developer to assure that an action is always provided of consistent data. Coping with data consistency problems is normally achieved by the exchange of suitable events which guarantee that accesses to an attribute occur whilst maintaining its data integrity.
- Besides active, concurrent objects, DART supports *passive objects* too, these do not have a lifecycle. A passive object is a usual data abstraction accessed via a set of synchronous accessor methods.

- The behaviour of a whole subsystem is influenced by a notion of "real time", supported by a hardware clock system. Based on this time concept, timers can be set up to program periods of time during which, for instance, a physical process is allowed to perform a meaningful activity. A *timer* is equipped both of a *time interval* and an *event* (*timeout*) which is sent to the target object at the fire time. It is assumed that objects can ask for the current real time, and for the time remaining on a specific timer before it fires.

2.3 System Architecture

The key points of DART programming in-the-large level are as follows.

- Since the (apparent) concurrency normally existing in a DART subsystem, introduced by the action interleaving schema of scheduler, there is no need to allocate more than a subsystem onto a same physical processor. As a consequence, the structure of a typical DART system consists of a collection of subsystems interconnected by a communications network. From this point of view, DART is flexible enough to easily be integrated with different protocols and communication means (e.g., CAN bus, Token Ring, and so forth). As a matter of simplification, DART can conveniently make use of the services of a transport layer.
- Inter-subsystem messages conform to a *system-wide format* for expressing the *destination subsystem*, the *destination object* and the *event* of the destination object which is requested. Event data, if there are any, are packed in a system-independent way which facilitates *interoperability*. A system-wide message may need *translation* at the subsystem receiver site into a format compatible with the receiver address and data space. An object which can be target of system-wide messages is said a *network object*.
- The design of the large scale programming model normally makes use of a coordinator or administrator object [8] on a per subsystem basis. The administrator *represents* the subsystem and it is the only network object of the subsystem. Inter-subsystems events are received by the administrator and then delegated and routed to the relevant local objects. The administrator lifecycle can embody a *control algorithm* assigned to the subsystem. A two level coordination structure among administrators is often required by a real-time system, with local administrators acting as autonomous and independent computing agents cooperating one with another, peer-to-peer, by exchanging inter-subsystem messages, but which behave as *slaves* of a *master* subsystem which is responsible of issuing global coordination events directed to all or part of the slave administrators. From this perspective, slave administrators can recognise a limited set of *control messages* which implement *configuration operations*. The master subsystem normally hosts the human computer interface (HCI) used for the operator console and to initialise and configure the system.

2.4 Scheduling Issues

The evolution of the active objects of a subsystem is regulated by a *scheduler* object which can be customised by programming. Many scheduler alternatives are possible, for example:

1. A basic scheduler handles a *timer-queue* and an *event-queue*. The timer-queue holds a ranked list of all the timers which have been set to fire, with the most imminent at the head. The event-queue buffers all the events which

have been sent through the basic *non-blocking send* operation. The scheduler performs an *event-loop* where, at each iteration, an event is selected from the event-queue or from the *most recently fired timer* and dispatched to its relevant destination object. As a consequence, an action may be triggered into execution. On completion of the action, the scheduler event-loop is resumed. The basic scheduler just outlined ensures that messages get dispatched "as soon as possible". Fire times of timers express only *lower time bounds*. Such a simple scheduler structure is adequate for non hard real- time systems and relies on a combination of asynchronous messages (*event-driven model*) and timers utilised to schedule activities according to a time line (*time-driven model*).

2. When time has to be managed in a more precise way, a *cyclic executive* structure can be implemented [14]. First, a *time frame* is identified, possibly by a preliminary harmonisation of periodic object frequencies. The time frame is then utilised to structure event handling on the basis of a *major cycle* consisting of one or more *minor cycles*. It is required that a thread of control triggered by an external event must complete before the frame end. To fulfill timing requirements it may be necessary to split the controlling process among different computing resources.

3. A more interesting scheduler can be achieved by refining the basic scheduler at point 1. with a *time-driven* schema similar to that of the pure cyclic executive at point 2. The external environment is sensed by periodic objects awakened by timer events whose frequencies are not necessarily harmonised. Every timer is paired with a *timethread*, i.e., the event chain generated by handling the external condition parsed by the periodic object. Each timethread has two attributes which define it, its *period* and *deadline*. The deadline can coincide with the period end or it can possibly occur before period completion. The *event-driven* component of the scheduling structure is responsible for the necessary interleaving among timethreads. The scheduler selects the next event to be dispatched from the timethreads on the basis of the *earliest-deadline-first* (EDF) strategy.

More specialised control structures can be designed by taking into account, for instance, a priority notion among objects or messages and so forth. Generally speaking, scheduling analysis of a DART subsystem is simplified by the following factors.

- *Avoiding interrupts.* Interrupts *are not* used for controlling the external environment at present. Instead, *periodic terminator* objects implemented through timers are introduced. These objects poll at certain frequencies the environment in order to sense for situations which raise external events. Sporadic objects are converted into periodic ones by considering the statistical average of changes of the associated environment condition. In reality, one interrupt only is retained: the *hardware clock tick*, which is essential to adjust the real time clock of a subsystem. The timer concept of DART is based entirely on this interrupt.

- *Atomic actions.* Atomic actions imply that the overhead cost for context switching operations of a pre-emptive scheduling schema is avoided. Scheduling analysis requires knowledge of periodic object frequencies and *worst case action execution times*. Having knowledge of the action duration is critical for guaranteeing timing constraints are met. This in turn requires adequate resources to be available (e.g., multiple processors), and/or for a design of object lifecycles with an acceptable action granularity. The following rules should be observed during the development of actions.

- No iteration construct should be used in an action unless it has a time-bounded execution
- Table look-up operations should be preferred to complex and time consuming algorithms
- The use of dynamic storage should be minimised.

- *Time-bounded communications network.* Subsystems of a whole DART application should be connected through a network with *predictable* transmission times. For instance, a CAN (Controller Area Network) bus could be used. A real time protocol based on CAN requires, at design time, an identification of the importance (urgency, priority) of system-wide and inter-subsystem events. To every event a *unique identifier* is then assigned which acts as its priority when the message is up to be transmitted through the net. Simultaneous transmissions are resolved by the CAN arbitration mechanism which automatically favours the message which has the highest priority message identifier. CAN makes it possible to predict event transmission times both for high priority and low priority messages [24].

2.5 Programming Issues

DART programming model [13, 15] closely adheres to the "spirit of Oberon" [25]: *make it as simple as possible but not simpler.* It makes possible to deliver small sized subsystem code which would fit well in micro-controllers with limited resources. DART programming in-the-small level has been implemented in the popular languages Oberon-2 and C++ according to the *orthogonality principle.* DART mechanisms and concepts are supported by a few base classes thus preserving the preexisting syntax and semantics of the host language. The realisations fully exploit the object-oriented mechanisms (e.g., inheritance can be used to organise reusable active object hierarchies). Active objects are provided by *event handlers* [20] which ensure type-safe communications. The transformation from modeling descriptions (e.g., object lifecycles) to elegant programming notation is straightforward [15]. DART programming in-the-large has been experimented with AT&T TLI on a Token Ring network of Dos/Win PCs [16]. It has been also interfaced to CAN bus [13, 12]. For more general applications (parallel computing and distributed discrete-event simulation using Time Warp [18, 2] DART has been implemented on top of UNIX/PVM to exploit the services of heterogeneous distributed systems of workstations interconnected by standard communications devices (e.g., Ethernet).

2.6 An Example

The application of DART is illustrated by an example commonly utilised in literature (e.g., [4]) which possesses many of the characteristics which typify embedded hard real-time systems. It is a simple pump control system for a mining environment. The system is used to pump mine-water, which collects in a sump at the bottom of the shaft, to the surface (see Figure 1). The relationship between the control system and the environment devices is depicted in Figure 2. Dashed arrows indicate an interaction with sporadic objects whereas normal arrows express an interaction with periodic objects.

Functional Requirements Can be divided into four components: the pump operation, the environment monitoring, the operator interaction and the system monitoring.

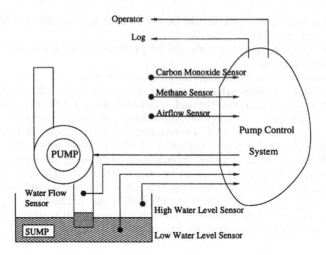

Fig. 1. A mine drainage control system

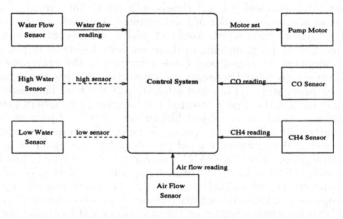

Fig. 2. Interface between the control system and the environment devices

- *Pump operation.* When the water reaches a high level (or when requested by the operator) the pump is turned on and the sump is drained until the water reaches the low level. At this point (or when requested by the operator) the pump is turned off. A flow of water in the pipes can be detected if required. The pump can only be allowed to operate if the methane level in the mine is below a critical level.

- *Environment monitoring.* The environment must be controlled in order to detect a critically high methane level in the air. The monitoring also measures the level of carbon monoxide in the mine and detects if there is an adequate flow of air. Alarms must be activated if gas levels become critical.

- *Operator interaction.* The system is controlled from the surface by the operator's console. The operator must be informed of all critical events.

- *System monitoring.* All the system events are to be stored in an archival database.

Timing Requirements These can be divided into three components: the monitoring periods, the shutdown deadline and the operation information deadline.

- *Monitoring periods.* For the purpose of this example, all the periods are assumed the same for all sensors: 60 seconds. For the methane there is a more stringent requirement due to the proximity of the pump and the need to ensure that it never operates when the methane level exceeds a critical threshold.
- *Shutdown deadline.* To avoid explosions there is a deadline within which the pump must be switched off once the methane is detected to be beyond a critical level. This deadline is related to the methane sampling period, to the rate at which the methane accumulate, and the safety margin between the level of methane assumed critical and the level at which it explodes. The relationship is expressed by the inequality:

$$R * (P + D) < P$$

  where
  - R is the rate at which the methane can accumulate
  - P is the methane sampling period
  - D is the shutdown deadline
  - M is the safety margin.

  P and D can be traded off against each other. In the example, a sampling period P of 5 seconds and a shutdown D of 1 second will be assumed.
- *Operator information deadline.* The operator must be informed within 1 second of the detection of methane or carbon monoxide readings, within 2 seconds of a critical low air-flow reading, and within 3 seconds of a failure in the operation of the pump.

The Table 1 summarises periods, deadlines and terminator attributes. As one can see, the operator is periodically polled without a deadline. The response is intended to be provided "as soon as possible".

	Periodic/Sporadic	arrival times	deadline
CH4 sensor	P	5	1
CO sensor	P	60	1
Water flow control	P	60	3
Airflow sensor	P	60	2
High/Low water sensor	S	100	20
Operator	P	1	-

Table 1. Timing attributes

System modeling An Information Model for the pump control system is shown in Figure 3, where the relationships between objects are named according to the viewpoint of the involved partner objects. The subtype/supertype relationship among sensors should be noted. The dynamic behaviour of the pump monitor and environment monitor active objects is portrayed respectively in Figure 4 and 5, where, for simplicity, logging actions are omitted. A transition from one state to another is flagged by the triggering *event*. The action associated with a given state is drawn inside the state box. The remaining details about object lifecycles should be self-explanatory. A summary of the events exchanged among the object lifecycles is captured in the Object Communication Model shown in Figure 6, where object responsibility decreases down through the diagram, with slave objects (*terminators*) reported at the bottom. It is worthy of note that the adopted OOA models are well revealing and help in making a project *visible*. For instance, the object lifecycles and the Object Communication Model permit an object design and implementation without introducing any distortions [15]. Of course, before embarking into the details of design/implementation phases, it is essential to proceed with a thorough analysis and verification of the modelled system from the functional and timing point of view. In particular, the scheduling analysis represents a fundamental step. The analysis activities can greatly benefit from prototyping and executing the system specifications. These issues are covered in the next section of the paper. The chosen control system could be partitioned into two DART subsystems: one running on a processor on the surface (OperatorHCI) and the other running on a processor in the mine (PumpCS). The OperatorHCI could host the asynchronous human-computer-interface of the operator and the data logger object. The PumpCS subsystem could allocate the pump monitor and the environment monitor objects. The two subsystems should be allowed to communicate one with another through the services of a predictable interconnection device (e.g., CAN bus). However, for the sake of simplicity, in the rest of the paper a single subsystem uniprocessor implementation will be assumed.

3 Modeling DART through Petri nets

In the following, a possible mapping of DART concepts on to Petri nets is presented. The fundamental motivation for using such a representation rests on the exploitation of a well-known formal notation used as the standard user-interface for animation purposes and for the analysis of timing properties. Petri nets, *per se*, only provide a powerful way to specify control flow in complex systems. However, they have been extended in the past in several different directions with the goal of making them suitable to specifying also other aspects such as data, functionality and timing (e.g., [3]). Timed Environment/Relationship (TER) nets are an example of extended nets [9] which are suited to verifying properties of hard real-time systems. TER nets were chosen as a basis for modeling and analisying DART systems for the following reasons:

- tokens are *environments*, i.e., they carry data values
- relations are associated with every transition to specify
  1. an *action*, i.e., the transformation of input token value tuples into output token value tuples and
  2. a *predicate* that the tuples of token values must satisfy in order to be chosen to fire the transition.
- timing information are provided by associating a *timestamp* (chronos) as a particular value item to each token. The timestamp of a token represents the time when the token has been produced. Timestamps are updated by the relations associated with the transitions
- a pair of functions $(t_{\min}, t_{\max})$, called *time interval*, is associated with each transition t. $t_{\min}$ determines the minimum time at which transition t *can* fire; $t_{\max}$ determines the maximum time at which transition t *must* fire unless disabled by the firing of another transition at a time not greater than the value determined by $t_{\max}$. This is a *strong time model* according to [9]. Both functions $t_{\min}$ and $t_{\max}$ are relative to the enabling time at which transition

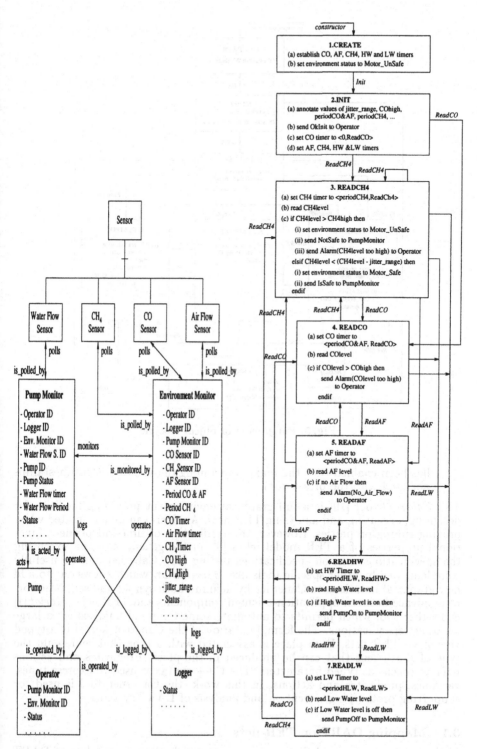

Fig. 3. An Information Model for the pump control system

Fig. 4. Environment monitor object life-cycle

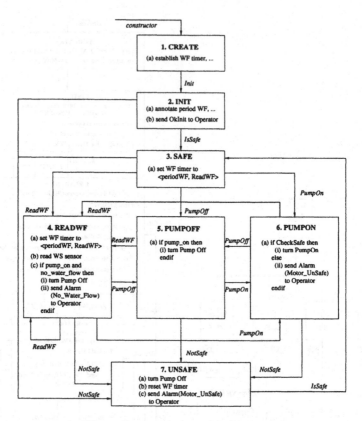

Fig. 5. Pump monitor object lifecycle

t has been enabled (i.e., the maximum among the values of the timestamps associated with the preset of t).

The *Cabernet* tool [19] is a full graphical environment providing facilities for designing, executing and analysing TER nets. Analysis can be automated when proving timing properties. However, the verification of time-independent functional properties of a TER model relies [9], in the general case, on executing the specifications, i.e., on specifications testing [11]. Cabernet can be usefully complemented by other specific tools like *Meta-Editor*, which allows one to add new graphical notations to Cabernet by defining both syntax and semantic correspondence between application-oriented components and TER nets. *Merlot* [1] is another tool which permits to automate some temporal analysis for a large set of net specifications. The formal notation of the Cabernet kernel is a typed version of TER nets. Each place is associated with a type that represents the type of the tokens that can be produced in the place. Predicates and actions refer to tokens as typed parameters. The C++ syntax is used to express types, variables, predicates and actions. In this work, the Cabernet tool is used for supporting prototyping, execution and analysis of DART systems.

3.1 Mapping DART on TER nets

In this section some guidelines are proposed which allow a modeling of DART in TER net terms. The approach is highly modular and facilitates reusability.

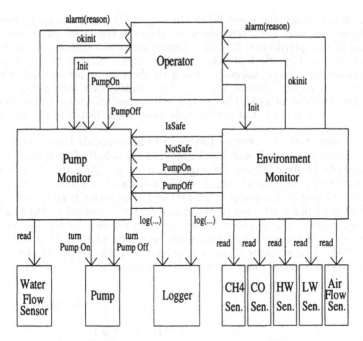

Fig. 6. Object Communication Model

Distinct subnets are associated with an active object, a scheduler object, a RT communications network and protocol and so forth. In the following, attention is focused on basic object and scheduler subnets together with a specification of functional and timing behaviour. Object subnets are not aware of the strategy embodied in a scheduler structure. Results from execution and analysis of a TER model can directly be interpreted into DART terms and in this way the overall design can be analysed and improved step by step.

Object subnet DART objects are timed finite state machines which can be mapped on to TER (sub)nets according to the general schema depicted in Figure 7. There is an input transition which selects an arrived event directed to this object. A recognised event is copied in the input place. A single output place exists where all the events generated by the object are stored. The object subnet admits a few levels of transitions and places. The first level of transitions is represented by the *state transitions* (T12, T21). The second level is composed by the *reject transitions* (T'12, T'21) and the last one consists of the *action transitions* (TA1, TA2). The first level of places is associated with the possible current states (S1, S2), whereas the second level refers to *temporary states* (S'1, S'2) controlling the action transitions. A state transition is enabled by the corresponding current state and by the availability of an expected event. For instance, T12 is only enabled when the current state is S1 and the received event is E1. The exact enabling condition is expressed by the predicate component of the state transition. Reject transitions are useful for throwing away events unexpected by the current state. For instance, T'12 gets enabled when the current state is S1 and the arrived event was *not* E1. The unexpected event is rejected by T'12 which generates an empty valid token in the output place. However, the current status remains unchanged. An *event token* is an environment holding identification information (the event label, the object ID, the belonging timethread) and, if there are any, the event data. A *status token* is an environment capturing the object attributes. It gets modified as a consequence of executing an action transition. The effect of executing the action relevant to a reached state is summarised in a single token produced in the output place. Such a token is an environment which contains a queue of events together with a flag indicating if the token is valid or not. The action component of every action transition contains, in C++ syntax, the action code early identified into the Information Model.

The object subnet can be extended in order to accomodate the existence of *synchronous accessor methods* which are executed on behalf of an action transition and contribute to its duration. In Figure 8 is exemplified an action transition TAh of an object O1 (*client*) which needs a data item produced on the basis of the internal status of an object O2 (*server*). The latter is supposed equipped of an additional place wherein is maintained, at any instant in time, a copy of the internal status. The accessor method is then realised by an *accessor transition* enabled by a *request* token generated by a request transition of O1, which precedes TAh. The accessor transition produces a token holding the requested item, directly in a *result* place in the preset of transition TAh of O1.

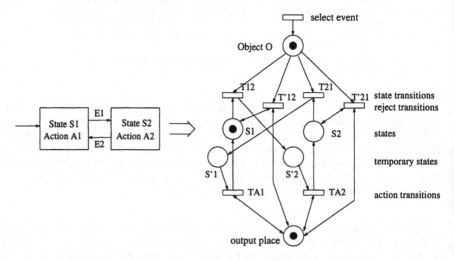

Fig. 7. An object TER subnet

Scheduler subnet An example of an EDF based scheduler is shown in Figure 9. For simplicity, only two timers are considered. The scheduler subnet possesses a single input transition (Schedule) and a single output place (Dispatch). The Dispatch place should be linked to the select event transition of all the object subnets (see Figure 7). The Schedule transition must be linked to the output place of all the object subnets. To each timer is associated a timethread, i.e., a place which hosts a token environment utilised for queuing all the events belonging to the thread of control triggered by the external situation monitored by the timer. The first event which initiates a timethread is a timer event. As the thread of control develops, e.g., splitting itself into two or more paths, all the associated events get queued on the same timethread. The identification of the

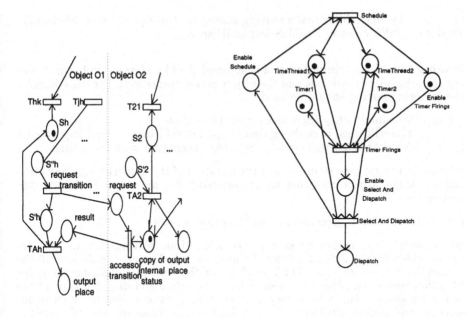

Fig. 8. A synchronous accessor transition

Fig. 9. An EDF based scheduler

belonging thread propagates from the event just processed by an object to the events generated by an object action transition. A timer consists of a token environment holding the fire time, the timeout event, the period and a status which can be active or inactive. Active timers are set automatically upon firing. An object action transition can synchronously activate/deactivate a timer. The token environment of a timethread carries, besides a queue of scheduled events, the deadline information, measured from the beginning of the timer period, within which the thread is expected to complete.

The Schedule transition is responsible for selecting a timethread upon which must be scheduled the events generated by an object. The action component of the Schedule transition determines the actual sent events on the basis of the validity attribute of the token in the output place of the various objects. The TimerFirings transition allows timers to fire. Timer firings occur when all the timethreads are empty or when one or more timers have been expired, i.e., the current time is greater than or equal to the fire time. More precisely, the action component of TimerFirings transition sets again each active expired timer and puts the relevant timeout event into the corresponding timethread. The SelectAndDispatch transition implements the fundamental event selection and dispatch process of the scheduler. From this point of view timethreads are ordered by their deadlines which act as priorities: a lower deadline is a greater priority. In particular, the action component of SelectAndDispatch transition is in charge of extracting the first event from the highest priority non empty timethread and of producing it into the Dispatch place. The activities of scheduling sent events by Schedule, permitting timers to fire by TimerFirings and selecting and dispatching the next event into the Dispatch place by SelectAndDispatch, are strictly sequenced by an enabling token moving among the three places EnableSchedule, EnableTimerFirings and EnableSelectAndDispatch.

Timing functions Timing control is accomplished by appropriately constraining the scheduler transitions and the object action transitions. For the scheduler the following setting should be used:

(1) t_{min}(Schedule) = t_{max}(Schedule) = enabling time
(2) t_{min}(TimerFirings) = enabling time; t_{max}(TimerFirings) = enabling time+δ
(3) t_{min}(SelectAndDispatch) = t_{max}(SelectAndDispatch) = enabling time

where δ is a worst case estimation of the duration of the TimerFirings transition (all the timer firings occurring simultaneously). For an action transition, the setting should be

(4) $t_{min}(action)$ = $t_{max}(action)$ = enabling time + wcet

where wcet is the action worst case execution time. Settings (1) and (3) are yet provided by the default setting of Cabernet, which is also applied to all the remaining transitions of a DART model. For the sake of simplicity, the value of action wcets is assumed charged of both the select and dispatch time of the scheduler and the times to be payed for scheduling events into the timethreads queues. The timing attributes of an enabled action transition will advance the current time of the action worst case execution time.

3.2 Analysis of a DART-TER model

The soundness of the DART approach is exemplified in the following by analysing a TER net translation of the pump control system discussed in section 2.6. The modelling process enhances the separation of concerns. At the programming in-the-small level the emphasis is on objects, schedulers, interconnections devices and so forth, modelled as subnets according to standard guidelines. At the programming in-the-large level a system topology is established by linking subnets one with another. The resultant DART-TER net can be executed and analysed to verify that functional and timing properties of the whole system are met. The modelling and verification processes can be iteratively applied in order to fix problems discovered during DART-TER operational analysis.

Figure 10 shows a TER net for the pump control system. For simplicity, the logger has been omitted. A first concern in analysis is the functional behaviour test. Toward this, the net can be animated without timing constraints. Separately, for each external unsolicited event (e.g., a CH4 reading, a CO reading, an Operator command, ...) the corresponding thread of control can be observed by tracing the event exchanges and state transitions in the involved objects. In particular, the net can be executed starting from an initial marking corresponding to a system state and an external event and controlling, at the user-interface, the sequence of firings associated with the timethread, possibly by a step-by-step firing. In addition, the information about threads of control can be recorded in a data base by exploiting the C++ interface of Cabernet, thus allowing off-line analysis. Following the functional behaviour analysis, attention is turned to timing behaviour. First the DART-TER net is annotated with the timing constrains by appending to each transition the functions t_{min} and t_{max} according to the guidelines described in the section 3.2. Obviously, an estimation of the worst case execution time wcet for every action transition is required. Then a detailed testing of temporal behaviour can be conducted by executing the DART-TER net starting from an initial marking achieved by setting:

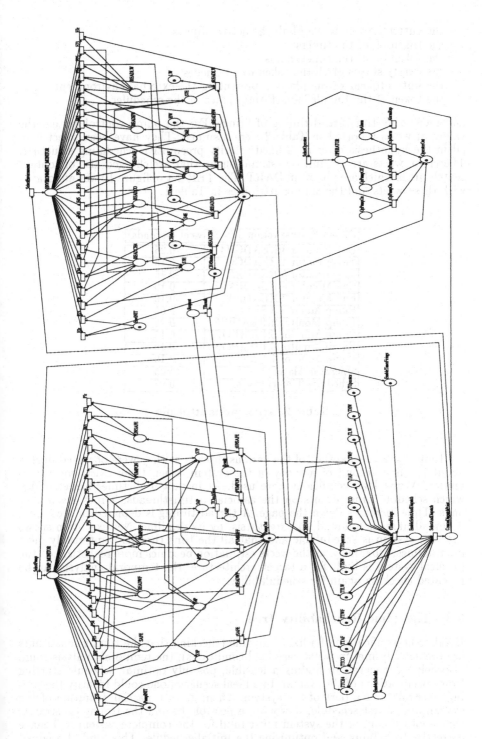

Fig. 10. Pump Control System DART-TER net

- the current status token of all the active objects
- the token of all the timers
- the token of all the timethreads
- the empty status of all the token event queues
- the initial tokens of the places representing the external environment
- the token of the EnableTimerFirings place of the scheduler

and registering timethread timing of firings. For the pump control system the values of wcets reported in Table 2 were assumed. They have been selected in [4] in the hypothesis of using a fixed-priority preemptive scheduling structure. Therefore, wcets are pessimistic estimations in that are charged also of context switch times which are absent in DART. The TER net of figure 10 was annotated with these wcets and the timing attributes in Table 1.

Object.Status:Action	wcet (seconds)
Env Monitor.3:READCH4	0.25
Env Monitor.4:READCO	0.15
Env Monitor.5:READAF	0.15
Env Monitor.6:READHW	0.10
Env Monitor.7:READLW	0.10
Pump Monitor.3:SAFE	0.05
Pump Monitor.4:READWF	0.15
Pump Monitor.5:PUMPOFF	0.10
Pump Monitor.6:PUMPON	0.10
Pump Monitor.7:UNSAFE	0.10
Operator:Alarm	0.05
Scheduler:TimerFirings	0.07

Table 2. Action wcets attributes

Table 3 gives a snapshot of the various threads as they were observed during about the first second of activity of the system, when there exists most contention. Worst load conditions were used. At instant 0 all timer firings occur, which schedule timer events on the associated timethreads. Times are in seconds. Time increases down the page. A thread timing is depicted in a column. A thread name is followed, in parenthesis, by its deadline. Net execution shows that the system is schedulable and that all the time threads comfortably meet its deadlines. Obviously, as the wcets estimation becomes more realistic, i.e., the physical system is built for a target architecture, the temporal analysis can be repeated checking again for schedulability.

3.3 The Time Reachability Tree

Besides the verification of a fixed schedule (checking that the timing constraints are predictably met in worst case load conditions), the temporal analysis could also help, in general, in finding a feasible, possibly optimal, schedule starting from specifications of the system. In a ideal scenario, one could identify the timing constraints of a part of the system. Then, given a partial definition of the system and a first schedule, it would be possible to find the timing constraints that another part of the system must fulfil for the complete system to behave correctly, so refining and optimising the initial schedule. This kind of analysis can be based on the construction of a tree of symbolic states called Time Reachability Tree (TRT) [10], of which there is an implementation in Cabernet. A

Timer firings	CH4(1)	CO(1)	AF(2)	WF(3)	HW(20)	LW(20)	OP(-)
0.07							
	0.25						
		0.15					
	0.10						
		0.05					
	0.05						
			0.15				
			0.05				
				0.15			
				0.05			
					0.10		
						0.10	
					0.05		
						0.05	
							0.05

Table 3. A snapshot of the threads of control

symbolic state consists of a set of (numeric) markings. Such a set is obtained by associating places with symbolic values, representing the timestamps, suitably constrained. The technique depends on the possibility of identifying symbolic enablings that characterise all the enablings in a given symbolic state. Once the symbolic enablings in a symbolic initial state S have been identified, it is possible to derive a new set of symbolic states that represent the set of markings that can be reached from the markings depicted in S by a single firing. In general, the construction of the TRT is not fully automatable since the computation of the successor state is undecidable and the TRT can be infinite. However, in many cases, the complexity of the time functions is restricted to linear functions and one can exploit existing algorithms for proving bounded invariance (behaviours that the system does not exhibit within a given temporal interval) and bounded response (behaviours expected from the system in a given temporal interval) properties. These properties can be examined by building a finite portion of the TRT, corresponding to the behaviour of the modelled system up to a given finite time limit. Properties can refer to states of the net (particular values or conditions of token-environments in certain places) and events (firings). Properties can also refer to sequences of states and events, e.g., it would be possible to check for the existence of consecutive firings of two transitions. The following gives a flavour of the use of TRT by applying it to a verification of the shutdown deadline of the pump after a high level of CH4 has been detected. In this case the property to be checked is the firing of two subsequent transitions in a given time interval (1 second deadline) and the test should end by stating if the property is satisfied or not, according to the timing constraints of the system. The initial symbolic state S0, marked with the same symbolic value T0=0, coincides with the initial status of the system (see initial marking in Figure 10) in the presence of worst load conditions. The property to be verified is expressed by:

$$[\#t\text{READCH4}, ?t\text{UNSAFE}] \ : \ T\#2 > T\#1 \ \&\& \ T\#2 < T\#1 + 1 \ :$$

where the operator $\#$ means for each, the operator $?$ means exists, and $T\#1$ and $T\#2$ are the firing times of respectively the first and second transitions of the formula in square parenthesis. In other words, considering the first second of activity, it is asked to verify if from a firing of tREADCH4 with a critical methane level, it follows a transition of tUNSAFE in the same time interval. A TRT was derived using a bread-first exploration strategy and terminates by confirming that the property is satisfied. It is worth noting that, according to Figure 10, non deterministic behaviour, e.g., choosing the next event from CH4 or CO timethreads, which have the same deadline, is resolved within the action component of the SelectAndDispatch transition. Cabernet enhances the TRT analysis power by showing, for each symbolic state in the TRT, information like last fired transition and corresponding firing time, and depicting an instance of the net which satisfies the selected state.

4 Conclusions

DART is a development methodology for hard real-time systems, centered on a minimal set of abstractions: light-weight objects modelled as finite state machines, event-driven multitasking which can be made time-sensitive by using timers and deadlines, programmer-defined non-preemptive scheduling control structures which can be tuned at the application at hand and easy integration with different networks and protocols suited for RT systems. DART mechanisms can directly be embodied in a high-level object oriented language like C++ and Oberon-2. Other languages are also possible. Several test-beds (e.g., [16, 2] confirmed that DART makes possible an effective and systematic approach to the building of time-dependent critical applications. This paper tried to demonstrate the usefulness of mapping DART on to Timed/Environment Relationship (TER) nets [9], an example of high level Petri nets. The Cabernet tool [19] provides a full graphical environment which offers powerful facilities both for the modeling process and the analysis phase based on TER nets. The mapping process preserves modularity. Analysis results, e.g., confirmation of the fulfilment of deadlines, derived from a DART-TER model can directly be interpreted into DART terms. If timing dependency or deadline problems are discovered then the DART model can be refined and the Cabernet tool used iteratively to find a practical design which meets the constraints. The ongoing activity is geared toward applying modeling and analysis to industrial case studies in the presence of RT interconnections devices (e.g., CAN bus, Token bus, ...). The possibility of using the MetaEditor tool which comes with Cabernet, to develop the abstract user-interface of DART appears very attractive. Another research issue which deserves further investigation involves a better realisation of sporadic objects with associated interrupts. Sporadic objects would eliminate unnecessary polls but should be naturally integrated in the DART framework, by absolutely avoiding the undesirable effects of interrupts, i.e., unpredictable action durations. An interesting direction advocated in ([5, 6]) consists in decoupling the processing unit into two parts: the Object Processor, which hosts a DART subsystem and is devoted to executing object actions in response to events dispatched by a scheduler; and the Scheduler Processor, which is involved in the scheduling activities by first converting interrupts into DART-like events. The two processors could be linked by a dual-port memory. This direction too would greatly take advantage from the modeling and analysis methods based on TER nets.

Acknowledgement

The authors are grateful to Francesco Tisato and Brian Kirk for their helpful comments on preliminary drafts of this paper, to Carlo Bellettini for his support during the experimental work with Cabernet and to the anonymous referees whose suggestions improved the presentation of the paper.

References

1. Bellettini C, Felder M, Pezzè M. MERLOT: A Tool for Analysis of Real-Time Specifications, *Proc. of 7th International Workshop on Software Specification and Design*, Redondo Beach, CA, pp 110-119, 1993.
2. Beraldi R, Nigro L. Distributed Object oriented Simulation Environment: An implementation of Time Warp using PVM, *Proc. of EUROSIM 95, Software Tools and Products Volume*, Vienna, September, 1995.
3. Bruno G, Marchetto G. Process-Translatable Petri Nets for the Rapid Prototyping of Process Control Systems, *IEEE Transactions on Software Engineering*, vol. SE-12(2), 346-357, 1986.
4. Burns A, Wellings A J. HRT-HOOD: A Structured Design Method for Hard Real-Time Systems, *Real Time Systems*, 6(1):73-114, 1994.
5. Colnaric M. Real-Time System Architecture Design Guidelines, *Proc. of Real Time '95*, Ostrava (Cz Rep), 103-111, 1995.
6. Colnaric M, Verber D, Gumzei R, Halang W. Implementation of Hard Real-Time Embedded Control Systems, to appear on *Real Time Systems*.
7. Faulk S R, Parnas D L. On Synchronisation in Hard-Real-Time Systems, *Communications of ACM*, 31(3):274-287, 1988.
8. Gentleman W M. Message Passing Between Sequential Processes: the Reply Primitive and the Administrator Concept, *Software-Practice and Experience*, 11, 435-466, 1981.
9. Ghezzi C, Mandrioli D, Morasca S, Pezzè M. A Unified High-Level Petri Net Model for Time Critical Systems, *IEEE Transactions on Software Engineering*, 17(2):160-172, 1991.
10. Ghezzi C, Morasca S, Pezzè M. Timing analysis of time basic nets, *Journal of Systems and Software*, 27(7):97-117, 1994.
11. Kemmerer R A. Testing formal specifications to detect design errors, *IEEE Transactions on Software Engineering*, vol. SE-11, 1985.
12. Kirk B. Real Time Protocol Design for Control Area Networks, *Proc. of Real Time '95*, Ostrava (Cz Rep), 251-268, 1995.
13. Kirk B, Nigro L. *Distributed Architecture for Real Time, in Oberon-2*, in Advances in Modular Languages, P. Shulthess (ed.), Universitatsverlag Ulm GmbH, Germany, 325-366, 1994.
14. Locke C D. Software architecture for hard real-time applications: cyclic executive vs. fixed priority executives, *Real Time Systems*, 4(1):37-53, 1992.
15. Nigro L. A real time architecture based on Shlaer Mellor object lifecycles, *Journal of Object Oriented Programming*, 8(1):20-31, 1995.
16. Nigro L. Distributed Architecture for Real Time: an example using distributed measurement control, *Proc. of Real Time '95*, Ostrava (Cz Rep), 1-14, 1995.
17. Nigro L, Tisato F. RTO++: a framework for building hard real-time systems, *Journal of Object Oriented Programming*, 6(2):35-47, 1993.
18. Nigro L, Veneziano G. Control abstractions in Modula-2: a case study using advanced backtracking, *Informatica*, 18(2):229-243, 1994.
19. Pezzè M. Cabernet: A Customizable Environment for the Specification and Analysis of Real-Time Systems, *Tech. Rep. 54-94*, Polytechnic of Milan, 1994.
20. Reiser M, Wirth N. *Programming in Oberon*, Addison Wesley, ACM Press, 1992.
21. Shlaer S, Mellor S J. *Object Lifecycles - Modeling the world in states*, Yourdon Press, 1992.
22. Stankovic J A. Misconceptions about real-time computing, *IEEE Computer*, 21(10):10-19, 1988.
23. Stankovic J A, Ramamritham K. Real-time computing systems: The next generation, *Tutorial on Hard Real-Time Systems*, Stankovic J A (ed.), IEEE Press, 14-37, 1988.
24. Tindell K, Burns A, Wellings A J. Analysis of Hard Real-Time Communications, *Real Time Systems*, 9(2):147-171, 1995.
25. Wirth N. A plea for lean software, *IEEE Computer*, 28(2):64-68, 1995.

{SC}*ECS:
A Class of Modular and Hierarchical Cooperating Systems *

Laura Recalde, Enrique Teruel, and Manuel Silva

Departamento de Informática e Ingeniería de Sistemas
Centro Politécnico Superior de Ingenieros de la Universidad de Zaragoza
María de Luna 3, E-50015 Zaragoza (Spain)

Abstract. We introduce a new class of Place/Transition net systems, {SC}*ECS, a generalisation of Equal Conflict systems [13] and Deterministic Systems of Sequential Processes [7, 11, 9, 10, 6]. {SC}*ECS are modular and hierarchical by definition: An {SC}*ECS is made up of a set of {SC}*ECS modules that asynchronously communicate through buffers in a restricted way.

After defining the class, we take advantage of its modular and hierarchical structure to analyse it, i.e. the study will not be made on the flat net, but taking into account its global architecture. We present a polynomial time method to synthetise {SC}*ECS nets which are well-formed, i.e. structurally bounded and live.

1 Introduction

Modular construction is a widely used approach in systems design. DSSP (Deterministic Systems of Sequential Processes), for instance, are defined as a set of modules, the Sequential Processes, that communicate by message passing through destination-private buffers that do not affect modules' internal choices. The definition of DSSP has evolved since their introduction, with a different name, in [7]. First, weights on arcs adjacent to buffers were allowed [11, 9, 10] and recently the restriction that the buffers be source-private was removed [6]. The imposed restrictions allow cooperation but not competition between modules.

In this work we define a new class of systems, that we call {SC}*ECS, which generalises DSSP retaining some of their most interesting properties. The generalisation comes in two ways. First, the cooperation between modules is not restricted to happen at a single flat level but in a hierarchical fashion: the cooperating modules at some level are themselves {SC}*ECS. Second, the lowest

---

* This work has been partially supported by the projects CICYT TIC-94-0242, Esprit BRA Project 7269 (QMIPS) and contract CHRX-CT94-0452 (MATCH) within the HCM Programme of the EU. L. Recalde was granted by the Dpto. de Educación y Cultura del Gobierno de Navarra.

level modules are allowed to be EC (Equal Conflict) systems instead of Sequential Processes. EC systems [13] are a weighted generalisation of FC (Free Choice) [3, 2] thus they admit non sequential behaviour presenting both conflicts and synchronisations. In turn, these basic modules could be seen as the composition of simpler ones since for EC systems decomposition results exist [13] and even sound and complete bottom-up and top-down synthesis procedures are available for FC systems [2]. Therefore we could speak of a "loose coupling" (through buffers) between modules and a "tight coupling" (e.g. transition merging, place fusion, etc.) between the components of a lowest level module, although we concentrate here in the former.

In summary, the class is *recursively* defined starting from EC systems as elementary {SC}*ECS and composing {SC}*ECS by asynchronous message passing through a set of destination-private buffers that do not disturb internal conflict resolutions. Actually, this recursive definition is the reason for the class' name: Systems of Cooperating Systems of... Cooperating Equal Conflict Systems.

It is important to note that, although there is only cooperation *between modules* at a given level, there can be a sort of *competence* between transitions within a module for the tokens in a shared buffer (differently from DSSP). This is especially clear when these transitions belong to different submodules. Note that, despite the destination-private restriction, a buffer can be input to two diferent submodules *when they have been previously interconnected*. This is a first example of the importance of the *building process*, and not only the flat net, in this class.

Although {SC}*ECS are a strict superclass of EC and DSSP they fortunately keep (part of) their tractability. In particular we prove in this paper a polynomial-time rank-based characterisation of well-formedness (Th. 20) extending similar results in [13, 6]. Compared to these previous works, it is important to notice that to achieve this goal the hierarchical and compositional definition of {SC}*ECS nets is extensively taken into account.

The remainder of the paper is organised as follows: In Sect. 2 basic notations and definitions are presented. Section 3 contains more specialised preliminary material, including structural conflicts, EC systems, and the EC rank theorem. The formal definition of the class of {SC}*ECS and some of its basic properties constitute Sect. 4. In Sect. 5 structural properties mainly concerned with T-semiflows of these nets are studied. Besides providing a deeper understanding of the class, they will be used to derive, in Sect. 6, a polynomial time characterisation of well-formedness.

2 Preliminaries

The reader is assumed to be familiar with Petri net theory (see [4]for a survey). Nevertheless in this section we recall the basic concepts and introduce the notation to be used. For the sake of readability whenever a net or system is defined it "inherits" the definition of all the characteristic sets, functions, parameters... with names conveniently marked.

Place/Transition net and related concepts. A P/T *net* is a triple $\mathcal{N} = (P, T, W)$ where P and T are disjoint finite sets of *places* and *transitions*, and $W : (P \times T) \cup (T \times P) \to \mathbb{N}$ defines the *weighted flow relation*: if $W(u, v) > 0$, then we say that there is an *arc* from u to v, with *weight* or *multiplicity* $W(u, v)$. Since a P/T net can be seen, and drawn, as a bipartite weighted directed graph, several graph concepts, like connectedness (without loss of generality, nets are assumed to be connected), strong connectedness, etc. can be extended to nets. In particular, let $v \in P \cup T$; its *preset* and *postset* are given by: ${}^\bullet v = \{u \mid W(u, v) > 0\}$, and $v^\bullet = \{u \mid W(v, u) > 0\}$. The preset (postset) of a set of nodes is the union of presets (postsets) of its elements. The weighted flow relation can be alternatively defined by: $Pre(p, t) = W(p, t)$, $Post(p, t) = W(t, p)$. These functions can be represented by matrices[2]. The *incidence matrix* is defined as $C = Post - Pre$. A net $\mathcal{N}'$ is *subnet* of $\mathcal{N}$ $(\mathcal{N}' \subseteq \mathcal{N})$ iff $P' \subseteq P$, $T' \subseteq T$ and W' is the restriction of W to P' and T'. Subnets are *generated by* subsets of nodes of both kinds. A subnet generated by a subset V of nodes of a single kind is assumed to be that generated by $V \cup {}^\bullet V \cup V^\bullet$. Subnets generated by a subset of places (transitions) are called *P-(T-)subnets*.

Place/Transition system and related concepts. A function $M : P \to \mathbb{N}$ is called *marking*, and can be represented by a vector. A P/T *system* is a pair $(\mathcal{N}, M_0)$ where $\mathcal{N}$ is a P/T net and M_0 is the *initial* marking. A transition t is *enabled* at M iff $M \geq Pre[P, t]$. Being enabled, t may *occur* (or *fire*) yielding a new marking $M' = M + C[P, t]$, and this is denoted by $M \xrightarrow{t} M'$. An *occurrence sequence* from M is a sequence of transitions $\sigma = t_1 \cdots t_k \cdots$ such that $M \xrightarrow{t_1} M_1 \cdots M_{k-1} \xrightarrow{t_k} \cdots$. If the firing of sequence σ yields the marking M', this is denoted by $M \xrightarrow{\sigma} M'$; denoting by $\vec{\sigma}$ the *firing count vector* of σ, clearly $M' = M + C \cdot \vec{\sigma}$.

Boundedness and liveness of P/T Systems. A P/T system is *bounded* when every place is bounded, i.e., its token content is less than some bound at every reachable marking. It is *live* when every transition is live, i.e., it can ultimately occur from every reachable marking. Boundedness is necessary whenever the system is to be implemented, while liveness is often required, specially in reactive systems. A net $\mathcal{N}$ is *structurally bounded* when $(\mathcal{N}, M_0)$ is bounded for *every* M_0, and it is *structurally live* when *there exists* an M_0 such that $(\mathcal{N}, M_0)$ is live. Consequently, if a net $\mathcal{N}$ is structurally bounded and structurally live there exists some marking M_0 such that $(\mathcal{N}, M_0)$ is bounded and live (B&L). In such case, non B&L is exclusively imputable to the marking, and we say that the net

---

[2] Places and transitions are supposed to be arbitrarily, but fixedly, ordered. Therefore rows and columns can be indexed by the sets P and T. For instance, the submatrix of m corresponding to row $p \in P$ and columns in $\tau \subseteq T$ is denoted by $m[p, \tau]$. Similarly for vectors. The usual multiplication of scalars, vectors and/or matrices a and b is denoted by $a \cdot b$. The (componentwise) comparisons of a and b are denoted by $a \geq b$ and $a > b$, while $a \gneq b$ denotes $a \geq b$ but $a \neq b$, not to be confused with $a \not\geq b$, meaning that $a \geq b$ is false.

is *well-formed*. Notice that, with this definition, in general, well-formedness is *not* necessary for B&L, although it happens to be in some selected subclasses, as EC systems [13].

Flows and Semiflows. Flows (semiflows) are integer (natural) annullers of C. Right and left annullers are called T- and P-(semi)flows, respectively. We call a semiflow *minimal* when its support[3] is not a proper superset of the support of any other, and the greatest common divisor of its elements is one. Flows are important because they induce certain invariant relations which are useful for reasoning on the behaviour. Actually, several structural properties are defined in terms of the existence of certain annullers, or similar vectors:

- $\mathcal{N}$ is *consistent* iff $X > 0$ exists such that $C \cdot X = 0$.
- $\mathcal{N}$ is *conservative* iff $Y > 0$ exists such that $Y \cdot C = 0$.

A couple of basic properties of T-semiflows that we will use are:

Proposition 1. *Let $\mathcal{N}$ be a P/T net and let X be a T-semiflow of $\mathcal{N}$.*

1. *If $t \in \|X\|$, then for all $p \in t^\bullet$, $p^\bullet \cap \|X\| \neq \emptyset$, and for all $p' \in {}^\bullet t$, ${}^\bullet p' \cap \|X\| \neq \emptyset$.*
2. *If X is minimal, then there is no other X' minimal T-semiflow of $\mathcal{N}$ such that $\|X\| = \|X'\|$.*

The next proposition can be easily deduced from Th. 3.1 in [8]. It states than in a bounded system any sequence can seen as a linear combination of T-semiflows plus a rest that is bounded.

Proposition 2 (Boundedness of non repetitive subsequences).
Let $(\mathcal{N}, M_0)$ be a bounded system. There exists $\kappa \in \mathbb{N}$ such that for every firing sequence σ, its Parikh vector can be decomposed as $\vec{\sigma} = \sum_{i=1}^{r} \alpha_i X_i + R$, with $\alpha_i \in \mathbb{N}$ and $R \leq \kappa \cdot \mathbb{1}$; where $\{X_1, \ldots, X_r\}$ is the set of the minimal T-semiflows of the net.

3 Conflicts, Net Subclasses, and Rank Properties

The basic module of the class we consider here are EC nets [13], whose definition and some results we recall next. Among these results we highlight the well-formedness characterisation in terms of the rank of the incidence matrix (Th. 8). A basic concept underlying EC nets, the rank theorem, etc. are structural conflicts. We consider a conflict as the situation where not everything that is enabled can be fired:

Definition 3. *Let $(\mathcal{N}, M_0)$ be a P/T system. Two transitions $t, t' \in T$ are in Conflict relation at marking M iff there exist $k, k' \in \mathbb{N}$ such that $M \geq k \cdot Pre[P, t]$ and $M \geq k' \cdot Pre[P, t']$, but $M \not\geq k \cdot Pre[P, t] + k' \cdot Pre[P, t']$. This notion is extended to sets "pairwisely".*

---

[3] The set $\|v\|$ of the non-zero components of vector v.

Some relations between transitions of the net, which are related to conflicts, can be defined:

Definition 4. Let $\mathcal{N}$ be a P/T net.

1. Two transitions, $t, t' \in T$, are in *Choice (or Structural Conflict) relation* iff $t = t'$ or ${}^\bullet t \cap {}^\bullet t' \neq \emptyset$. This relation is not transitive.
2. Two transitions, $t, t' \in T$, are in *Coupled Conflict relation* iff there exist $t_0, \ldots, t_k \in T$ such that $t = t_0$, $t' = t_k$ and, for $1 \leq i \leq k$, t_{i-1} and t_i are in Choice relation. It is an equivalence relation on the set of transitions, and each equivalence class is a *Coupled Conflict set*. We denote by $\mathcal{C}$ the set of Coupled Conflict sets.
3. Two transitions, $t, t' \in T$, are in *Equal Conflict relation* iff $t = t'$ or $Pre[P, t] = Pre[P, t'] \neq 0$. This is also an equivalence relation on the set of transitions, and each equivalence class is an *Equal Conflict set*. We denote by $\mathcal{E}$ the set of Equal Conflict sets, and by $\bar{t}$ the equivalence class of t. This notation is extended to sets: $\bar{T} = \bigcup_{t \in T} \bar{t}$.

Choices (places with more than one output transition) are the "topological construct" making possible the existence of conflicts. The Coupled Conflict relation is the transitive closure of the Choice relation. An Equal Conflict set is such that whenever any transition belonging to it is enabled, then all of them are. Sometimes it will be important to know that the transitions we are working with belong to different Equal Conflict sets. To provide a simple way to state this concept we introduce the following definition:

Definition 5. Let $\mathcal{N}$ be a P/T net. A set of transitions, $T' \subseteq T$, is EC-free in $\mathcal{N}$ iff there is no pair of different transitions in Equal Conflict relation.

A well-known, polynomial time, necessary condition for well-formedness is conservativeness and consistency. This condition has been improved by adding an upper bound for the rank of the incidence matrix. This necessary condition, together with a sufficient one of the same kind, is recalled next:

Theorem 6 [13, 6] (A Nec. and a Suf. Cond. for Well-formedness).
Let $\mathcal{N}$ be a P/T net.

1. *If $\mathcal{N}$ is well-formed, then it is conservative, consistent, and $\mathrm{rank}(C) < |\mathcal{E}|$.*
2. *If $\mathcal{N}$ is conservative, consistent, and $\mathrm{rank}(C) = |\mathcal{C}|-1$, then it is well-formed.*

In some subclasses these general conditions are improved, as we shall show to be the case of the subclass we study here. For the purpose of this work, let us first recall here the definition of other syntactical subclasses of P/T nets:

Definition 7. Let $\mathcal{N}$ be a P/T net.

1. $\mathcal{N}$ is *Choice-free* iff $\forall p \in P : |p^\bullet| \leq 1$.
2. $\mathcal{N}$ is *Join-free* iff $\forall t \in T : |{}^\bullet t| \leq 1$.

3. $\mathcal{N}$ is *Equal Conflict (EC)* iff ${}^\bullet t \cap {}^\bullet t' \neq \emptyset \Rightarrow Pre[P,t] = Pre[P,t']$.

Equal Conflict systems generalise (extended) FC ones. Among other properties for EC systems (and FC) both rank bounds in Th. 6 collapse, thus leading to a well-formedness characterisation [1, 13]. This is known as the *rank theorem*. We state it as in [1]. (Although the statement in [13] is slightly different, it can easily be deduced from Th. 20 and 28 of that work.)

Theorem 8 [13] (EC Rank Theorem).
  Let $\mathcal{N}$ be an EC net. $\mathcal{N}$ is well-formed iff it is strongly connected, conservative, and $\mathrm{rank}(C) = |\mathcal{E}| - 1$.

4 {SC}*ECS, a Modular and Hierarchical Class

4.1 Definition of the Class. Levels of an {SC}*ECS

Definition 9. $\mathcal{N} = (P,T,W)$ is an {SC}*ECS net iff it is an EC net or P is the disjoint union of $P_1, \ldots, P_n$, and B; T is the disjoint union of $T_1, \ldots, T_n$; and:

1. For every $i \in \{1,\ldots,n\}$, the module $\mathcal{N}_i = (P_i, T_i, W_i)$ is an {SC}*ECS net, where W_i is the restriction of W to P_i and T_i.
2. For each buffer $b \in B$:
   (a) There exists dest$(b) \in \{1,\ldots,n\}$ such that $b^\bullet \subseteq T_{\mathrm{dest}(b)}$
   (b) Every Equal Conflict set of every module is preserved by the buffers, i.e. for every $e \in \mathcal{E}_i$, $t,t' \in e \Rightarrow W(b,t) = W(b,t')$.

An {SC}*ECS is an {SC}*ECS net with an initial marking.

The incidence matrix of a non trivial {SC}*ECS net can be written in a structured way reflecting modularity:

$$C = \left(\begin{array}{cccc} C_1 & 0 & \cdots & 0 \\ \hline 0 & C_2 & \cdots & 0 \\ \hline \vdots & \vdots & \ddots & \vdots \\ \hline 0 & 0 & \cdots & C_n \\ \hline B_1 & B_2 & \cdots & B_n \end{array} \right) = \left(\begin{array}{c} \mathrm{diag}\{C_i\} \\ \hline C_B \end{array} \right) \qquad (1)$$

where C_i is the incidence matrix of $\mathcal{N}_i$ and B_i is the matrix that represents its connections to the buffers.

Figure 1 shows an example of an {SC}*ECS. This {SC}*ECS net consists of two modules $\mathcal{N}_1$ and $\mathcal{N}_2$ (the nets enclosed by the dashed lines), which are also {SC}*ECS nets, connected by two buffers: b_1 and b_2. Each module is composed of three submodules, all of them EC nets, interconnected by four and three buffers respectively. There is no restriction about the number of modules a buffer receives tokens from (e.g. the buffer b_{22} is output place for two submodules, $\mathcal{N}_{22}$ and $\mathcal{N}_{23}$) but they must be *destination-private*. (e.g. although the destination

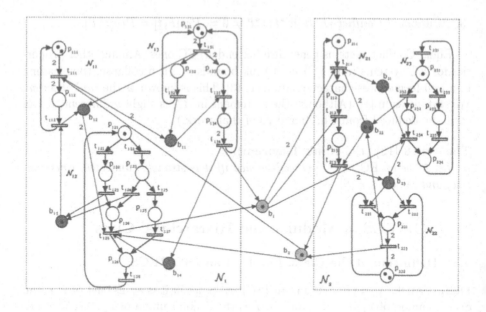

Fig. 1. A two levels {SC}*ECS.

of b_2 are two submodules, $\mathcal{N}_{11}$ and $\mathcal{N}_{12}$, at the level at which b_2 has been added both submodules belong to the same module $\mathcal{N}_1$).

This shows clearly the *hierarchical* nature of this class. Not only the obtained *flat net* is important, but also the way it has been built, i.e. the *relation between its components*. Now, we could use this system as a module and connect it to other {SC}*ECS getting more and more complex systems each time.

Three kinds of communication can be modelled. First, within the basic EC modules, as a generalisation of FC systems, we may have both syncronisations and conflicts. If the modules are seen as a composition of submodules (e.g. state machine decomposition of live and bounded FC systems) then we could speak of *"tight coupling"*. Second, between the modules at a given level there is only *cooperation*: (1) different modules cannot compete for the tokens (resources) contained in one buffer due to the destination-private restriction, and (2) arbiter modules cannot be modelled because buffers do not affect the resolution of internal conflicts. In this sense it can be said that the behaviour of one module cannot be conditioned by decisions of the others but only, possibly, delayed. Third, since buffers can be input of several transitions in a module, there is a sort of *competition* between these transitions for the tokens in the buffers, specially if they belong to different submodules. It is said to be a weak competition because the "competing" transitions are *somehow synchronised* by the interconnection within their module.

The hierarchical definition of the class induces the notion of "levels" of a net, as the number of layers that have been used to build it. This concept is formalised in the following definition:

Definition 10. Let $\mathcal{N}$ be an {SC}*ECS net.

If $\mathcal{N}$ has been built as $\mathcal{N} = (P_1 \cup \cdots \cup P_n \cup B, T_1 \cup \cdots \cup T_n, W)$, then levels($\mathcal{N}$) = $\max_{1 \leq i \leq n}$ {levels($\mathcal{N}_i$)} + 1, where $\mathcal{N}_i = (P_i, T_i, W_i)$ and W_i is the restriction of W to P_i and T_i.

Otherwise $\mathcal{N}$ is an EC net and levels($\mathcal{N}$) = 0.

A given {SC}*ECS net may have different number of levels depending on the way it has been constructed. Take for instance the net of the system in Fig. 1. The left module, $\mathcal{N}_1$, is a one level {SC}*ECS net because it has been built connecting $\mathcal{N}_{11}$, $\mathcal{N}_{12}$, and $\mathcal{N}_{13}$ which where zero levels. Nevertheless in this particular case each buffer is input of a single transition and therefore $\mathcal{N}_1$ happens to be EC, so it could have been rated as zero levels. It would be possible to define the number of levels of a net as the minimum of all the possible values, but this might change the role of some places from buffers to modules' places and viceversa. We will rather preserve the designer point of view so we will assume that every {SC}*ECS has its building process attached (except when the construction process is not important — in some proofs — where the minimum can be used for the sake of simplicity).

Regarding the membership problem we want to emphasise that although to verify whether a given net is in the class is combinatorial this is not a problem because we *assume* that we have got a net that has been built to be an {SC}*ECS net. (The appropriate syntax checks can easily be made at each step during the construction.)

The definition of DSSP [6] can be restated in terms of {SC}*ECS: A DSSP is a one level {SC}*ECS where the modules are monomarked State Machines.

4.2 Local and Global Fairness

In [13] it is proved that the infinite firing sequences of a bounded strongly connected EC system where every transition occurs infinitely often are characterised as those where every solution of a conflict that is effective infinitely often is taken infinitely often, generalising the results in [14] for FC systems. Here we generalise this property to bounded strongly connected {SC}*ECS. To do so we use the notion of EC-local fairness. It is identical to local fairness except that only Equal Conflicts, and not Coupled Conflicts, are taken into account. Clearly both definitions coincide for EC nets, where $\mathcal{E} = \mathcal{C}$. The new definition looks quite natural if we think on DSSP, whose modules are monomarked State Machines: since each State Machine has got just one token, two transitions in the same module cannot be in Conflict relation unless they are in Equal Conflict relation (thus, the only effective conflicts are Equal Conflicts).

Definition 11. Let $(\mathcal{N}, M_0)$ be a P/T system.

1. A firing sequence σ is *EC-locally fair* iff it satisfies: If for some $e \in \mathcal{E}$, σ can be written $M_0 \xrightarrow{\sigma_{i_0}} M_{i_1} \xrightarrow{t_{i_1}\sigma_{i_1}} M_{i_2} \xrightarrow{t_{i_2}\sigma_{i_2}} \cdots$ with $\sigma_{i_0}, \sigma_{i_1}, \sigma_{i_2}, \ldots$ finite sequences,

such that for every $k > 0$, $t_{i_k} \in e$, then for each $t \in e$ the sequence σ can be written $M_0 \xrightarrow{\sigma_{j_0}} M_{j_1} \xrightarrow{t\sigma_{j_1}} M_{j_2} \xrightarrow{t\sigma_{j_2}} \cdots$ with $\sigma_{j_0}, \sigma_{j_1}, \sigma_{j_2}, \ldots$ finite sequences.

2. A firing sequence σ is *globally fair (or impartial)* iff it is finite or every $t \in T$ appears infinitely often.

In other words, we could say that a sequence is EC-locally fair if for any Equal Conflict that is enabled infinitely often, each possible outcome of the conflict is fired infinitely often.

Theorem 12 (EC-local and global fairness).

Let $(\mathcal{N}, M_0)$ be a bounded strongly connected {SC}*ECS. A firing sequence σ is globally fair iff it is EC-locally fair.

Proof. For "$\Rightarrow$", if σ is finite it is obvious. Otherwise, since every transition is fired infinitely often, for any EC set every transition is fired infinitely often.

For "$\Leftarrow$" the same proof given in [13] for EC systems is also valid here changing the Coupled Conflict sets for EC sets. ◇

So, global fairness can be achieved by local control of EC choices. If we take care that, for every EC set, all the transitions are fired with a fair policy, we can assure that the sequence we will obtain will be globally fair.

An important consequence of this result is the equivalence of liveness and deadlock-freeness for bounded and strongly connected {SC}*ECS. The proof is analogous to that in [13] for EC systems.

Corollary 13 (Liveness and deadlock-freeness).

Let $(\mathcal{N}, M_0)$ be a bounded strongly connected {SC}*ECS. Then $(\mathcal{N}, M_0)$ is live iff it is deadlock-free.

4.3 Properties and Hierarchy

Our aim is to make use of the modular and hierarchical nature of {SC}*ECS for their analysis. This means applying the knowledge we may have of the modules' properties to obtain information about the whole net. To simplify the notation, and turning again to the recursive definition of the class, we introduce the following definitions:

Definition 14. Let $\mathcal{N}$ be an {SC}*ECS net and let Π be a property (e.g. strong connectedness, well-formedness, ...).

1. $\mathcal{N}$ is *recursively Π*, r-Π, iff $\mathcal{N}$ fulfills Π and, if levels($\mathcal{N}$) > 0 with $\mathcal{N} = (P_1 \cup \cdots \cup P_n \cup B, T_1 \cup \cdots \cup T_n, W)$, then every $\mathcal{N}_i$ is r-Π.
2. $\mathcal{N}$ is *quasi-recursively Π*, qr-Π, iff levels($\mathcal{N}$) $= 0$ or $\mathcal{N} = (P_1 \cup \cdots \cup P_n \cup B, T_1 \cup \cdots \cup T_n, W)$ with every $\mathcal{N}_i$ being r-Π.

Notice that there exist properties for which r-Π is equivalent to Π, though it is not so in general. For instance, using the incidence matrix of an {SC}*ECS net in the structured form presented in (1) it is not difficult to see that consistency is equivalent to r-consistency. Nevertheless, paradoxical as it may seem, conservativeness does not imply r-conservativeness: the reader may check that the net in Fig. 2 is conservative although its modules are not.

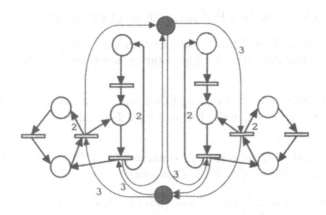

Fig. 2. A conservative, but non r-conservative {SC}*ECS net.

With this notation in mind, our main goal is to obtain a polynomial time characterisation of r-well-formedness for qr-well-formed {SC}*ECS nets. In other words, a polynomial time procedure to build r-well-formed nets.

5 On Allocations and T-semiflows

Allocations have been an important tool in the development of the FC and EC theory. Essentially they are functions that select transitions from conflict sets. Allocations and allocatability were defined for FC nets in [3]. Here we take the extensions given in [13] particularised to EC sets: we take one transition per EC set. The aim of this technical section is to study, by means of allocations, some properties of the T-semiflows of an {SC}*ECS that are needed later on to prove the rank theorem. The organisation of the section is as follows:

After the definition, we present Algorithm 1 (a recursive generalisation of a similar one from [13]). This algorithm provides a way to generate EC-allocations with "good properties", as stated by Lemma 16. Quite informally, *given a subset of transitions* or *seed*, what the algorithm does is orderly visiting the EC sets that are not represented in the seed, selecting from them one appropriate transition *to direct tokens towards the input places of the transitions in the seed*.

In Th. 17 we describe minimal T-semiflows of a net in terms of their projections on the modules. Obviously, a T-semiflow of an {SC}*ECS net is also a T-semiflow when restricted to a module. The theorem proves a kind of converse:

under some assumptions we can choose a minimal T-semiflow per module and combine them to build a T-semiflow of the whole net. This emphasises the modularity of the class. Moreover, a couple of methods to build allocations containing the support of exactly one minimal T-semiflow are provided, and a polynomial time sufficient condition for EC-allocatability is obtained (later on, in Th. 20, we will find that r-well-formed {SC}*ECS satisfy this condition, thus they are EC-allocatable).

Definition 15. Let $\mathcal{N}$ be a P/T net, with set of EC sets $\mathcal{E}$.

1. A mapping $\alpha : \mathcal{E} \to T$ that assigns to each $e \in \mathcal{E}$ one of its transitions $t \in e$ is an EC-*allocation* over $\mathcal{N}$. The notation is extended to sets: $\alpha(\gamma)$ denotes $\bigcup_{e \in \gamma} \alpha(e)$.

2. $\mathcal{N}$ is EC-*allocatable* iff for every EC-allocation over $\mathcal{N}$ the T-subnet generated by the allocated nodes has at least a T-semiflow.

Lemma 16. *Let $\mathcal{N}$ be an r-strongly connected {SC}*ECS net and T' the result of applying Algorithm 1 with $T^0 \subseteq T$ as seed:*

1. *The algorithm is well-defined and ends in a finite number of steps.*
2. *There exists an EC-allocation, α, with $\alpha(\mathcal{E}) \subseteq T'$. Moreover, if $t \in T' - \alpha(\mathcal{E})$ then $t \in T^0$. Thus, in case T^0 is EC-free in $\mathcal{N}$, the algorithm defines an EC-allocation.*
3. *If X' is a T-semiflow of the T-subnet generated by T' then $\|X'\| \cap T^0 \neq \emptyset$.*
4. *If $\mathcal{N}$ is r-EC-allocatable, then for every transition $t \in T$ there exists an EC-allocation such that t is in the support of every T-semiflow of the T-subnet generated by the allocated transitions. Thus, $\mathcal{N}$ is consistent.*

Proof. Since there is recursion on the levels of the net the algorithm is applied to, the lemma is proven by induction on the number of levels of $\mathcal{N}$. For levels$(\mathcal{N}) = 0$ the result is proven in [13]. Otherwise assume it holds for levels$(\mathcal{N}) < k$ and let levels$(\mathcal{N}) = k$.

Part 1. By induction hypothesis, we know we can apply the algorithm to the modules of the net. The only tricky thing is the way the propagation through the modules is done.

Assume we have already gone through some modules and let T^j be the set of selected transitions. Since $\mathcal{N}$ is strongly connected, if $\overline{T^j} \neq T$ there exists $1 \leq i \leq n$ such that $T_i \cap \overline{T^j} = \emptyset$ and $T_i^{\bullet\bullet} \cap \overline{T^j} \neq \emptyset$ (i.e. there exists an output buffer of T_i which is also an input buffer of T^j). Take $t \in T_i$ verifying $t^{\bullet\bullet} \cap \overline{T^j} \neq \emptyset$. Since the buffers preserve the EC sets, $t^{\bullet\bullet} \cap T^j \neq \emptyset$. This way we can extend the algorithm to one more module. Each step returns an EC-allocation over one module of the net, so the algorithm ends in a finite number of steps.

Part 2. For every $1 \leq i \leq n$ we know there exists an EC-allocation α_i such that $\alpha_i(\mathcal{E}_i) \subseteq T_i'$ (induction hypothesis). Merging all these EC-allocations we get an EC-allocation, α, with $\alpha(\mathcal{E}) \subseteq T'$. If $i \notin S$ then $\alpha_i(\mathcal{E}_i) = T_i'$ because,

Algorithm 1.

    Input: $\mathcal{N}$, an r-strongly connected {SC}*ECS net, and
                      T^0, a non empty subset of transitions.

    Output: T', a subset of transitions that contains T^0 and whose most
                      important characteristics are stated in Lemma 16

Begin

    If levels($\mathcal{N}$) = 0 **do**

        Let $j := 0$

        While $\overline{T^j} \subset T$ **do**

            $T^{j+1} := T^j \cup \{t\}$, where $\overline{t} \cap \overline{T^j} = \emptyset$ and $t^{\bullet\bullet} \cap T^j \neq \emptyset$

            $j := j + 1$

        od

    od

    Else do { Rem: $\mathcal{N} = (P_1 \cup \cdots \cup P_n \cup B, T_1 \cup \cdots \cup T_n, W)$ }

        Let $S := \{i \mid T_i \cap T^0 \neq \emptyset\}$

        For every $i \in S$ **do**

            $T_i^0 := T^0 \cap T_i$

            $T_i' := $ Algorithm 1($\mathcal{N}_i, T_i^0$)

        od

        Let $j := 1$; $T^j := \bigcup_{i \in S} T_i'$

        While $\overline{T^j} \neq T$ **do**

            Take $t \in T_i$ such that $T_i \cap \overline{T^j} = \emptyset$ and $t^{\bullet\bullet} \cap T^j \neq \emptyset$

            $T_i' := $ Algorithm 1($\mathcal{N}_i, \{t\}$)

            $T^{j+1} := T^j \cup T_i'$

            $j := j + 1$

        od

    od

    $T' := T^j$

End

---

since the seed is one transition, it is EC-free in N_i . Furthermore, for any $i \in S$, if there exists $t \in T_i' - \alpha_i(\mathcal{E}_i)$, then $t \in T_i^0 \subseteq T^0$ (induction hypothesis again). So, it is proven that if there exists $t \in T' - \alpha(\mathcal{E})$, then $t \in T^0$.

Part 3. If we could prove that $i \in S$ exists such that the restriction of X' to T_i is not null then $X'[T_i']$ would be a T-semiflow of $\mathcal{N}_i'$ (the T-subnet generated by T_i'). Then, by induction hypothesis, $\|X'[T_i']\| \cap T_i^0 \neq \emptyset$, what clearly leads to $\|X'\| \cap T^0 \neq \emptyset$.

Let $1 \leq i \leq n$ be such that the restriction of X' to T_i is not null and assume $i \notin S$. Then $X'[T_i']$ is a T-semiflow of $\mathcal{N}_i'$ and thus t, the seed when applying the algorithm to $\mathcal{N}_i$, is in $\|X'[T_i']\|$ (induction hypothesis). By Prop. 1.1 every output buffer of t is input buffer of a transition in $\|X'\|$. Let $j > 1$ be such that $T^j = T^{j-1} \cup T_i'$. Taking into account the way T^j has been built, and the destination-private hypothesis on the buffers, we know $T^{j-1} \cap \|X'\| \neq \emptyset$.

Repeating the reasoning, we go through the modules in the reverse way than we did with the algorithm. We finally reach T^1 and thus i must be in S.

Part 4. Apply Algorithm 1 with $T^0 = \{t\}$. By Part 3 the EC-allocation it induces fulfills the conditions of the statement. $\diamond$

Theorem 17. *Let $\mathcal{N}$ be a strongly connected and conservative qr-well-formed {SC}*ECS net with* $\text{rank}(C) \leq |\mathcal{E}| - 1$.

1. *For every $\mathcal{N}_i$, let X_i be a given minimal T-semiflow. Then there exists a T-semiflow of $\mathcal{N}$, X, such that $X = \sum_{i=1}^{n} k_i \cdot \widehat{X_i}$, where $\widehat{X_i}[T_i] = X_i$, $\widehat{X_i}[T - T_i] = 0$ and $k_i \in \mathbb{N}$.*
2. *$\mathcal{N}$ is EC-allocatable.*
3. *Let X be a minimal T-semiflow of $\mathcal{N}$. Then $\|X\|$ is EC-free and, in case levels$(\mathcal{N}) > 0$, for every $1 \leq i \leq n$ $X[T_i]$ is proportional to a minimal T-semiflow of $\mathcal{N}_i$. Furthermore, if $\mathcal{N}'$ is the T-subnet obtained applying Algorithm 1 with $T^0 = \|X\|$, then every T-flow of $\mathcal{N}'$ is multiple of $X[T']$.*
4. *For any $t \in T$, if we apply Algorithm 1 with $T^0 = \{t\}$, then the T-subnet its image generates, $\mathcal{N}'$, has got a unique minimal T-semiflow and every T-flow is multiple of it.*

Proof. (For the sake of readability not every detail is given here. See [5] for a complete proof.) By induction on the number of levels of a net. If levels$(N) = 0$ Part 1 does not apply and Part 2, Part 3 and Part 4 are proven in [13]. Assume the theorem is proven for levels$(\mathcal{N}) < k$ and let levels$(\mathcal{N}) = k$.

Part 1. For every module, $\mathcal{N}_i$, take $T_i^0 = \|X_i\|$ and apply Algorithm 1. This will give an EC-allocation over $\mathcal{N}_i$ such that for the T-subnet generated by its image, $\mathcal{N}'_i$, $X_i[T'_i]$ is its unique minimal T-semiflow and every T-flow is multiple of it (induction hypothesis, Part 3). Merging all these partial EC-allocations an EC-allocation over $\mathcal{N}$ is obtained. Let $\mathcal{N}'$ be the T-subnet generated by its image. If we prove that $\mathcal{N}'$ has a T-semiflow it will be a T-semiflow of $\mathcal{N}$ that is proportional to the chosen T-semiflows of the modules. Since rank$(C') \leq$ rank$(C) < |\mathcal{E}| = |T'|$ at least a T-flow of $\mathcal{N}'$ exists. Breaking it up into a positive and a negative part, and using the induction hypothesis, they can be shown to be T-semiflows.

Part 2. Let α be an EC-allocation over $\mathcal{N}$ and $\mathcal{N}'$ the T-subnet its image generates. Let α_i be the restriction of α to $\mathcal{E}_i$, which is clearly an EC-allocation over $\mathcal{N}_i$. Since $\mathcal{N}_i$ is EC-allocatable (Th. 6.1 and induction hypothesis) a minimal T-semiflow of $\mathcal{N}_i$ with its support contained in the image of the allocation exists. Apply Part 1 using these T-semiflows and a T-semiflow of $\mathcal{N}$ with its support contained in T' will be obtained.

Part 3. First we prove that $\|X\|$ is EC-free. Let $A = \{i \mid X[T_i] \neq 0\}$. For every $i \in A$, $X[T_i]$ is a T-semiflow of $\mathcal{N}_i$, so a minimal T-semiflow of $\mathcal{N}_i$, $\widetilde{X_i}$, exists such that $\|\widetilde{X_i}\| \subseteq \|X\|$. In other words, from every module whose intersection with $\|X\|$ is not null, take a minimal T-semiflow with its support contained in the support of X. Now, define $T^0 = \bigcup_{i \in A} \|\widetilde{X_i}\|$ and use it as seed for Algorithm 1. It can be proven that $\|X\|$ contains the support of any minimal T-semiflow of the T-subnet generated by the output of the algorithm. On the other hand, $\mathcal{N}$ is EC-allocatable (Part 2) so at least a T-semiflow exists whose support is contained in the image of the allocation. By minimality of X, $\|X\|$ must be contained in the image of the EC-allocation. Therefore there are no EC-related transitions in $\|X\|$. $X[T']$ is the unique minimal T-semiflow of $\mathcal{N}'$ because two minimal T-semiflows cannot have the same support (Prop. 1.2). To see that any T-flow is multiple of it the idea is to break the T-flow up in two vectors, one with the positive entries and the other with the negative ones, and prove that one of them must be null.

Part 4. Assume there are two minimal T-semiflows: X' and $\widetilde{X}'$. It can be proven that any module for which the restriction of the support of X' is not null, verifies that the support of both X' and $\widetilde{X}'$ coincide on it. Then $\|X'\| \subseteq \|\widetilde{X}'\|$ and since both are minimal T-semiflows, $X' = \widetilde{X}'$ (Prop. 1.2). To prove that every T-flow is multiple of this T-semiflow, just notice that the same allocation can be obtained using $\|X'\|$ instead of $\{t\}$ as seed for Algorithm 1, and then the result can be deduced from Part 3. $\Diamond$

6 Well-formedness of {SC}*ECS

We are ready now to obtain the sought characterisation of well-formedness for {SC}*ECS nets. The first step is proving that when the modules are live it is always possible to find a marking for the buffers so that the whole system is live. Since liveness is equivalent to deadlock-freeness (Cor. 13) we just need to find an initial marking from which every reachable marking enables at least one transition.

Theorem 18 (Compositional liveness).
*Let $\mathcal{N}$ be a strongly connected and conservative qr-well-formed {SC}*ECS net with $\text{rank}(C) \leq |\mathcal{E}| - 1$ and $\text{levels}(\mathcal{N}) > 0$. For every $\mathcal{N}_i$ let M_{0_i} be such that $(\mathcal{N}_i, M_{0_i})$ is live. Then M_0 exists such that $M_0[P_i] = M_{0_i}$ for every $1 \leq i \leq n$ and the system $(\mathcal{N}, M_0)$ is live.*

Proof. Constructive: we define an M_0 that is later proven sufficient (usually not necessary!) to make $(\mathcal{N}, M_0)$ a live {SC}*ECS.

Let $\{X_1, \ldots, X_r\}$ be the set of minimal T-semiflows of $\mathcal{N}$. For every $1 \leq l \leq r$, $X_l[T_i]$ is multiple of a minimal T-semiflow of $\mathcal{N}_i$ (Th. 17.3), so $X_l[T_i] = \sum_{j=1}^{r_i} \lambda_{i,j}^l \cdot X_{i,j}$, where $\{X_{i,1}, \ldots X_{i,r_i}\}$ is the set of minimal T-semiflows of $\mathcal{N}_i$ and $\lambda_{i,j}^l \in \mathbb{N}$ (every $\lambda_{i,j}^l = 0$ but one). Let $\mu(i,j)$ be the maximum number of times $X_{i,j}$ appears in the T-semiflows of $\mathcal{N}$, $\mu_{i,j} = \max_{1 \leq l \leq r} \{\lambda_{i,j}^l\}$.

For each $b \in B$ define the number of tokens taken from b when $X_{\text{dest}(b),j}$ occurs:

$$k(b,j) = Pre[b, T_{\text{dest}(b)}] \cdot X_{\text{dest}(b),j} \geq -C[b, T_{\text{dest}(b)}] \cdot X_{\text{dest}(b),j} \qquad (2)$$

By Prop. 2, for every $\mathcal{N}_i$ there exists $\kappa_i \in \mathbb{N}$ such that for every firing sequence σ_i, $\vec{\sigma}_i = \sum_{j=1}^{r_i} \alpha_{i,j} \cdot X_{i,j} + R_i$, with $\alpha_{i,j} \in \mathbb{N}$, $R_i \leq \kappa_i \cdot \mathbb{1}$.

Define $M_0[P_i] = M_{0_i}$ for every $1 \leq i \leq n$ and, for every $b \in B$:

$$M_0[b] = \sum_{j=1}^{r_{\text{dest}(b)}} \mu_{\text{dest}(b),j} \cdot k(b,j) + \kappa_{\text{dest}(b)} \cdot Pre[b, T_{\text{dest}(b)}] \cdot \mathbb{1} \qquad (3)$$

that is, enough tokens to enable at once any minimal T-semiflow involving transitions of dest(b).

Assume $(\mathcal{N}, M_0)$ is not live. Then it can reach a deadlock marking M_d firing some sequence σ (Cor. 13). We can write $\vec{\sigma} = \sum_{l=1}^{r} \lambda_l \cdot X_l + X_R$, for some $\lambda_l \in \mathbb{N}$ and $X_R \not\geq X_l$ for every $1 \leq l \leq r$. Then, for every $b \in B$:

$$M_d[b] - M_0[b] = C[b,T] \cdot \vec{\sigma} = C[b,T] \cdot X_R \geq C[b, T_{\text{dest}(b)}] \cdot X_R[T_{\text{dest}(b)}] \qquad (4)$$

because if $t \notin T_{\text{dest}(b)}$ then $W(b,t) = 0$, so $C[b,t] \geq 0$. Fix a module $\mathcal{N}_i$. Clearly:

$$X_R[T_i] = \sum_{j=1}^{r_i} \lambda_{i,j} \cdot X_{i,j} + X_{R_i} , \qquad (5)$$

for some $\lambda_{i,j} \in \mathbb{N}$ and $X_{R_i} \not\geq X_{i,j}$ for every $1 \leq j \leq r_i$. Moreover, by Prop. 2

$$X_{R_i} \leq \kappa_i \cdot \mathbb{1} . \qquad (6)$$

Let $b \in B \cap {}^\bullet T_i$.

$$
\begin{aligned}
M_d[b] &\overset{(4)}{\geq} M_0[b] + C[b, T_i] \cdot X_R[T_i] \\
&\overset{(5)}{=} M_0[b] + \sum_{j=1}^{r_i} \lambda_{i,j} \cdot C[b, T_i] \cdot X_{i,j} + C[b, T_i] \cdot X_{R_i} \\
&\overset{(2)}{\geq} M_0[b] - \sum_{j=1}^{r_i} \lambda_{i,j} \cdot k(b,j) - Pre[b, T_i] \cdot X_{R_i} \\
&\overset{(6)}{\geq} M_0[b] - \sum_{j=1}^{r_i} \lambda_{i,j} \cdot k(b,j) - \kappa_i \cdot Pre[b, T_i] \cdot \mathbb{1} \\
&\overset{(3)}{=} \sum_{j=1}^{r_i} (\mu_{i,j} - \lambda_{i,j}) \cdot k(b,j).
\end{aligned}
$$

If $\mu_{i,j} > \lambda_{i,j}$ for every $1 \leq j \leq r_i$, then $M_d[b] \geq \sum_{j=1}^{r_i} k(b,j)$ for every $b \in B \cap {}^\bullet T_i$, so the buffers would have enough tokens to fire every T-semiflow in $\mathcal{N}_i$. In such case M_d could not be a deadlock, against the hypothesis. Therefore, there exists $1 \leq J(i) \leq r_i$, with $\mu_{i,J(i)} \leq \lambda_{i,J(i)}$. We can repeat this for every $\mathcal{N}_i$. By Th. 17.1, a T-semiflow of $\mathcal{N}$, $X = \sum_{i=1}^{n} \beta_i \cdot \widehat{X_{i,J(i)}}$ (can be assumed to be minimal) exists such that $\widehat{X_{i,J(i)}}[T_i] = X_{i,J(i)}$ and $\widehat{X_{i,J(i)}}[T - T_i] = 0$. By the way $\mu_{i,j}$ was defined, $\beta_i \leq \mu_{i,J(i)}$, so $\lambda_{i,J(i)} \geq \beta_i$. Let $X_R^i[T_i] = X_R[T_i]$ and $X_R^i[T - T_i] = 0$. Then,

$$X_R = \sum_{i=1}^{n} X_R^i \overset{(5)}{\geq} \sum_{i=1}^{n} \lambda_{i,J(i)} \cdot \widehat{X_{i,J(i)}} \geq \sum_{i=1}^{n} \beta_i \cdot \widehat{X_{i,J(i)}} = X \; ,$$

contradiction.   ◊

Note that every initial marking greater than that obtained from (3) makes the system live too. But this does not mean that liveness of r-well-formed {SC}*ECS nets is monotonic wrt. the marking of the buffers, as the example of Fig. 3 shows: although the depicted system is live, it can be deadlocked if a token is added to b_1 (fire t_{21}, t_{121}, and then twice t_{111}). Liveness monotonicity wrt. the marking of the buffers holds for DSSP [9, 10], where the monomarkedness of the modules guarantees that the only effective conflicts are equal, thus avoiding "situations" as the one appearing in Fig. 3.

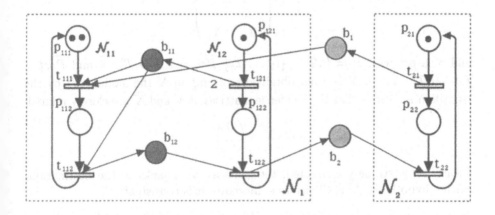

Fig. 3. A live system that becomes non live when adding a token to the marking of b_1.

We have proven that we can put enough tokens in the buffers to make the system live provided the modules are live. What about the converse, i.e. given a live system are the modules live when considered in isolation? The system of Fig. 3 is a counterexample for this too: the module on the left is not live by itself (fire t_{121} and then twice t_{111}).

After Th. 18 clearly strong connectedness, conservativeness, and rank$(C) \leq |\mathcal{E}| - 1$ is a sufficient condition for well-formedness of qr-well-formed {SC}*ECS nets. Together with the general necessary condition (Th. 6.1), this yields a complete characterisation of well-formedness. However, it can be improved by adding a lower bound to the rank of the incidence matrix as is proved in the next theorem:

Theorem 19 (A Lower Bound for the Rank).

*Let $\mathcal{N}$ be a r-strongly connected and consistent {SC}*ECS net. Then,*

$$\text{rank}(C) \geq \sum_{i=1}^{n} \text{rank}(C_i) + n - 1 \geq |\mathcal{E}| - 1 .$$

Proof. Let us apply induction on the levels of $\mathcal{N}$.

If levels($\mathcal{N}$) = 0, assume rank(C) < $|\mathcal{E}| - 1$. From Th. 20 of [13] it can be deduced that $\mathcal{N}^{rd}$ is EC-allocatable. But then, applying Th. 28 of [13], rank(C) = $|\mathcal{E}| - 1$, contradiction.

Assume the inequality holds for levels($\mathcal{N}$) < k and let levels($\mathcal{N}$) = k. Let

$$X = \begin{pmatrix} X_1 \\ \vdots \\ X_n \end{pmatrix} > 0$$

be such that $C \cdot X = 0$. Define

$$\tilde{X} = \begin{pmatrix} X_1 & \cdots & 0 \\ \vdots & \ddots & \vdots \\ 0 & \cdots & X_n \end{pmatrix}$$

and $\tilde{\mathcal{N}}$ a net with $\tilde{P} = B, \tilde{T} = \{\tilde{t}_1, \ldots, \tilde{t}_n\}, \widetilde{Pre} = Pre[B,T] \cdot \tilde{X}$ and $\widetilde{Post} = Post[B,T] \cdot \tilde{X}$, i.e. $\tilde{\mathcal{N}}$ is a net obtained replacing in $\mathcal{N}$ the module $\mathcal{N}_i$ by the transition $\tilde{t}_i$. Notice that the incidence matrices of $\mathcal{N}$ and $\tilde{\mathcal{N}}$ are closely related:

$$C \cdot \tilde{X} = \begin{pmatrix} 0 \\ \tilde{C} \end{pmatrix} .$$

Clearly $\tilde{\mathcal{N}}$ is strongly connected, the same as $\mathcal{N}$. Thanks to the destination-private hypothesis, $\tilde{\mathcal{N}}$ is Choice-free. Moreover it is consistent:

$$0 = C \cdot X = C \cdot \tilde{X} \cdot \begin{pmatrix} 1 \\ \vdots \\ 1 \end{pmatrix} = \begin{pmatrix} 0 \\ \tilde{C} \end{pmatrix} \cdot \begin{pmatrix} 1 \\ \vdots \\ 1 \end{pmatrix} ,$$

thus rank($\tilde{C}$) = $n - 1$ [12]. Since $\tilde{C}$ is obtained linearly combining columns of C,

$$\text{rank}(C) = \text{rank}\left(\begin{array}{c|c} \text{diag}\{C_i\} & 0 \\ \hline C_B & \tilde{C} \end{array} \right) \geq \sum_{i=1}^{n} \text{rank}(C_i) + \text{rank}(\tilde{C}) .$$

By induction hypothesis rank(C_i) $\geq |\mathcal{E}_i| - 1$, therefore

$$\text{rank}(C) \geq \sum_{i=1}^{n} \text{rank}(C_i) + \text{rank}(\tilde{C}) \geq \sum_{i=1}^{n} (|\mathcal{E}_i| - 1) + n - 1 = |\mathcal{E}| - 1 .$$

$\diamondsuit$

We note that the consistency condition cannot be removed or even replaced by conservativeness and the modules satisfying the rank condition. The net in Fig. 2 is r-strongly connected, conservative, and the rank of each module equals the number of its EC sets minus one; however rank$(C) = |\mathcal{E}| - 2$. We note also that the general sufficient condition given by Th. 6.2 is not suitable for {SC}*ECS nets because if a net is not EC then $|C| < |\mathcal{E}|$ and so it is impossible that rank$(C) = |C| - 1$.

Linking together Th. 8, Th. 18, Th. 6.1, and Th. 19 we get the announced polynomial-time characterisation of well-formedness for the class:

Theorem 20 ({SC}*ECS Rank Theorem).
*Let $\mathcal{N}$ be a qr-well-formed {SC}*ECS net. $\mathcal{N}$ is well-formed iff it is strongly connected, conservative, and* rank$(C) = |\mathcal{E}| - 1$.

Proof. If levels$(\mathcal{N}) = 0$, see Th. 8.

Otherwise, for the '$\Leftarrow$', since the modules are r-well-formed they can be lively marked. Applying Th. 18 the result follows. The converse can be deduced from Th. 6.1 and Th. 19. $\diamond$

The above result induces a method to obtain well-formed {SC}*ECS nets, in fact r-well-formed ones, by construction: The designer builds modules and verifies strong-connectedness, conservativeness, and rank$(C) = |\mathcal{E}| - 1$. Then, these modules are interconnected, and a qr-well-formed {SC}*ECS net is obtained. The same properties are verified in the composite net. This process can be repeated once and again yielding increasingly complex r-well-formed {SC}*ECS nets. Since the three properties can be verified in polynomial time it follows that:

Corollary 21. *Determining r-well-formedness of an {SC}*ECS net by construction is a polynomial time problem.*

Regarding the computation of the rank at a given level, it should be noticed that we can take advantage of the previous computations at the lower levels, thanks to the structure of the matrix reflected in (1).

Since "good modules interconnection" produces the minimum rank (Th. 19 and Th. 20), a different synthesis procedure can be envisaged: we do not check the rank at each step, but only strong connectedness and conservativeness. Only at the end we check consistency and the rank property of the complete net. If they are fulfilled, then we know at once that every module at every level fulfilled them. Otherwise we can detect the fault checking the last level modules: if they are all consistent and verify the rank property, then the problem was in the last level buffers, otherwise we can investigate similarly every faulty module that we have found.

7 Conclusion and Future Work

The class of {SC}*ECS is capable of modelling structured systems where cooperation between modules through buffers is introduced in a hierarchical fashion.

Also a limited form of competition can be modelled, since several transitions within a module can be output of the same buffer, even if they belong to different submodules. From this modular bottom-up point of view, the class of {SC}*ECS is a generalisation of DSSP in two senses: (1) more general modules are allowed and (2) a weak kind a competition can be modeled that in DSSP is impossible due to the marking restrictions.

From a top-down point of view the kind of cooperation that is allowed can be seen as a broad generalisation of Choice-free nets [12], where the individual transitions are the "agents" while the places play the role of buffers through which agents pass their "results" to one another. In both cases the syntactical restrictions aim at preventing or limiting competition.

Naturally, compared to Choice-free, {SC}*ECS allow far richer "agents" to be modelled. Nevertheless, in [5] we have introduced an abstract view of {SC}*ECS nets, where the details of the modules are conveniently hidden by substituting them by coarse transitions. This abstract (fluid in fact) view, that we call *coarse net*, is a Choice-free net with parametric weights. It contains major information on the structural properties of the net that it represents. We believe that many open points in the structure theory of {SC}*ECS will be solved through the further investigation of this coarse net.

We summarise next the most important properties that we have proved for {SC}*ECS, the ones we have counterproved, and announce some results comparing it all with similar results for DSSP and EC nets and systems.

Theorem 12 characterises globally fair sequences of bounded strongly connected {SC}*ECS as those that are locally fair, generalising a similar result for EC nets (see [13]). Corollary 13 proves that in bounded strongly connected {SC}*ECS liveness is equivalent to deadlock-freeness, as happens for EC systems [13] and DSSP [6]. The modularity of minimal T-semiflows of r-well-formed {SC}*ECS nets, together with other properties concerning allocations and semiflows are contained in Th. 17, closely related to results for EC nets [13].

Liveness monotonicity wrt. the marking of the buffers does not hold for {SC}*ECS (Fig. 3), although it does for bounded EC systems [13] and DSSP [9, 10].

The main result in this paper is the polynomial time characterisation of well-formedness for qr-well-formed {SC}*ECS nets: the rank theorem (Th. 20). It generalises analogous rank theorems for DSSP [6] and EC nets [13]. As part of our ongoing work we have recently proved that this characterisation can be extended to qr-strongly connected {SC}*ECS and also that, as it happens for EC nets [13], well-formedness is necessary for well-behavedness. Thus, we can decide in polynomial time whether a qr-strongly connected {SC}*ECS net can be lively and boundedly marked.

References

1. J. Campos, G. Chiola, and M. Silva. Properties and performance bounds for closed Free Choice synchronized monoclass queueing networks. *IEEE Trans. on Automatic Control*, 36(12):1368–1382, 1991.

2. J. Esparza and M. Silva. On the analysis and synthesis of Free Choice systems. In G. Rozenberg, editor, *Advances in Petri Nets 1990*, volume 483 of *Lecture Notes in Computer Science*, pages 243–286. Springer Verlag, 1991.

3. M. H. T. Hack. Analysis of production schemata by Petri nets. Master's thesis, MIT, 1972. (Corrections in *Computation Structures Note 17*, 1974).

4. T. Murata. Petri nets: Properties, analysis and applications. *Proceedings of the IEEE*, 77(4):541–580, 1989.

5. L. Recalde. A class of modular and recursively defined models of concurrent systems: {SC}*ECS. Master's thesis, DIIS. Univ. Zaragoza, September 1995.

6. L. Recalde, E. Teruel, and M. Silva. On well-formedness analysis: The case of deterministic systems of sequential processes. In J. Desel, editor, *Proc. of the Int. Workshop on Structures in Concurrency Theory (STRICT)*, Workshops in Computing, pages 279–293. Springer, 1995.

7. W. Reisig. Deterministic buffer synchronization of sequential processes. *Acta Informatica*, 18:117–134, 1982.

8. M. Silva. Towards a synchrony theory for P/T nets. In K. Voss et al., editors, *Concurrency and Nets*, pages 435–460. Springer Verlag, 1987.

9. M. Y. Souissi. *Une Etude de la Préservation de Propriétés par Composition de Réseaux de Pétri. Quelques Extensions aux Réseaux à Files. Application à la Validation de Protocoles de Communication*. PhD thesis, Univ. Pierre et Marie Curie (Paris VI), February 1990.

10. Y. Souissi. Deterministic systems of sequential processes: A class of structured Petri nets. In G. Rozenberg, editor, *Advances in Petri Nets 1993*, volume 674 of *Lecture Notes in Computer Science*, pages 406–426. Springer Verlag, 1993.

11. Y. Souissi and N. Beldiceanu. Deterministic systems of sequential processes: Theory and tools. In *Concurrency 88*, volume 335 of *Lecture Notes in Computer Science*, pages 380–400. Springer Verlag, 1988.

12. E. Teruel, J. M. Colom, and M. Silva. Choice-free Petri nets: A model for deterministic concurrent systems with bulk services and arrivals. *IEEE Trans. on Systems, Man, and Cybernetics*. To appear.

13. E. Teruel and M. Silva. Structure theory of Equal Conflict systems. *Theoretical Computer Science*, 153(1-2):271–300, 1996.

14. P. S. Thiagarajan and K. Voss. A fresh look at Free Choice nets. *Information and Control*, 61(2):85–113, 1984.

Behavioural Equivalence for Infinite Systems – Partially Decidable!

Kim Sunesen and Mogens Nielsen

BRICS[1], Department of Computer Science, University of Aarhus
Ny Munkegade, DK-8000 Aarhus C.,{ksunesen,mnielsen}@daimi.aau.dk

Abstract. For finite-state systems non-interleaving equivalences are computationally at least as hard as interleaving equivalences. In this paper we show that when moving to infinite-state systems, this situation may change dramatically.

We compare standard language equivalence for process description languages with two generalizations based on traditional approaches capturing non-interleaving behaviour, *pomsets* representing global causal dependency, and *locality* representing spatial distribution of events.

We first study equivalences on Basic Parallel Processes, BPP, a process calculus equivalent to communication free Petri nets. For this simple process language our two notions of non-interleaving equivalences agree. More interestingly, we show that they are decidable, contrasting a result of Hirshfeld that standard interleaving language equivalence is undecidable. Our result is inspired by a recent result of Esparza and Kiehn, showing the same phenomenon in the setting of model checking.

We follow up investigating to which extent the result extends to larger subsets of CCS and TCSP. We discover a significant difference between our non-interleaving equivalences. We show that for a certain non-trivial subclass of processes between BPP and TCSP, not only are the two equivalences different, but one (*locality*) is decidable whereas the other (*pomsets*) is not. The decidability result for *locality* is proved by a reduction to the reachability problem for Petri nets.

Keywords: Process Calculi, Petri Nets, Behavioural Equivalence, Partial Order Methods, Decidability.

1 Introduction

This paper is concerned with decidability issues for behavioural equivalences of concurrent systems, notably linear-time equivalences focusing on global/local causal dependency between events. Our results may be seen as a contribution to the search for useful verification problems which will be decidable/tractable when moving from the standard view of interleaving to more intentional non-interleaving views of behaviour.

All known behavioural equivalences are decidable for finite-state systems, but undecidable for most general formalisms generating infinite-state systems,

---

[1] Basic Research in Computer Science,
Centre of the Danish National Research Foundation.

including process calculi, like CCS and TCSP, and labelled Petri nets. To study systems in between various infinite-state process algebras have been suggested, see [5] for a survey. One of the most interesting suggestions is *Basic Parallel Processes*, BPP, introduced by Christensen [4]. BPPs are recursive expressions constructed from inaction, action, variables, and the standard operators prefixing, choice and parallel compositions. By removing the parallel operator one obtains a calculus with exactly the same expressive power as finite automata. BPPs can hence be seen as arising from a minimal concurrent extension of finite automata and therefore a natural starting point of exploring infinite-state systems. Another reason for studying BPP is its close connection to communication-free nets, a natural subclass of labelled Petri nets [4, 10]. It was hence shown in [4] that any BPP process can be effectively transformed into an equivalent BPP process in full standard form while preserving bisimilarity. Moreover, there is an obvious isomorphism between the transition system of a BPP process in full standard form and the labelled reachability graph of a communication-free net, for details see [8]. Hence, BPP is formally equivalent to the class of communication-free Petri nets with respect to any interleaving equivalence coarser than or equal to bisimilarity. BPPs were first suggested in [4, 6] and accompanied by a positive result stating that (strong) bisimulation is decidable on BPP. Later Hirshfeld showed that in contrast language (trace) equivalence is undecidable [10] for BPP. The picture has since been completed by a result showing that in the branching-time/linear-time spectrum of [30] only bisimulation is decidable, see [12]. For a survey on results for infinite-state systems see [5].

Also, various generalizations of behavioural equivalences to deal with non-interleaving behaviour have been studied, see for instance [31]. The basic idea is to include in your notion of equivalence some information of causal dependency between events, following the ideas from the theory of Petri nets and Mazurkiewicz traces [26, 19]. For finite-state systems non-interleaving equivalences are computationally at least as hard as interleaving equivalences, see [14, 18]. In this paper we show that when moving to infinite-state systems, this situation may change dramatically. For infinite-state systems a number of non-interleaving bisimulation equivalences have been proven decidable on BPP, e.g. causal bisimulation, location equivalence, ST-bisimulation and distributed bisimulation [15, 4]. In this paper we concentrate on non-interleaving generalizations of language equivalence.

More precisely, we compare standard language equivalence for process description languages with two generalizations based on traditional approaches to deal with non-interleaving behaviour. The first is based on *pomsets* representing global causal dependency, [24], and the second on *locality* [3] representing spatial distribution of events.

We first study the equivalences on Basic Parallel Processes, BPP. For this simple process language our two notions of non-interleaving equivalences agree, and furthermore they are decidable, contrasting the result of Hirshfeld [10] that language equivalence is undecidable. This result is inspired by a recent result of Esparza and Kiehn [9] showing the same phenomenon in the setting of model checking.

We follow up investigating to which extent the result extends to larger subsets of CCS and TCSP. We discover here a significant difference between our two non-interleaving equivalences. We show that for a certain non-trivial subclass of processes between BPP and TCSP, BPP_S, not only are the two equivalences different, but one (*locality*) is decidable whereas the other (*pomset*) is not. The decidability result for *locality* is proved by a reduction to the reachability result for Petri nets.

Finally, we show that there is also a difference between the power of the parallel operators of CCS and TCSP. Adding the parallel operator of Milner's CCS to BPP, BPP_M, we keep the decidability of *locality* and *causal dependency* equivalences, whereas by adding the parallel operator of Hoare's TCSP, BPP_H, both become undecidable.

Our results are summarized in the following table where *yes* indicates decidability and *no* undecidability. The results of the first column are all direct consequences of Hirshfeld's result on BPP [10]. The second and third show the results of this paper.

	Language equiv.	Pomset equiv.	Location equiv.
BPP	no	yes	yes
BPP_S	no	no	yes
BPP_H	no	no	no
BPP_M	no	yes	yes
TCSP& CCS	no	no	no

The operational semantics from which our pomsets are derived is based on an enrichment of the standard semantics of CCS [20] and TCSP [28] decorating each transition with some extra information allowing an observer to observe the location of the action involved. The location information we use to decorate transitions is derived directly from the concrete syntax tree of the process involved. We have chosen here to follow the technical *static* setup from [22], but could equally easy have presented an operational semantics in the *dynamic* style of [3]. The decidability results are based on the theory of finite tree automata and a new kind of synchronous automata working on tuples of finite trees. For this latter model we show decidability of the emptiness problem using a reduction to the zero reachability problem for Petri nets.

In Sections 2 and 3 we present the syntax and operational semantics of the CCS/TCSP-style language used throughout the paper, and define formally the equivalences to be studied. The next three sections are used to establish our results for BPP, TCSP and CCS respectively. First, in Section 4 we show that both non-interleaving equivalences are decidable for BPP processes. TCSP-style subsets are considered in Section 5, where we show that all our equivalences are undecidable on BPP_H and that for BPP_S locality equivalence is decidable, whereas pomset equivalence is not. In Section 6 we deal with the CCS-style subsets We show that the result of Section 4 extend to BPP_M, and no further. All proofs are sketched only, for details we refer to [27].

Acknowledgements: We would like to thank the anonymous referees for their comments and remarks.

2 A TCSP-style language

We start by defining the abstract syntax and semantics of a language, BPP_H, including a large subset of TCSP [11, 23]. The definition is fairly standard. As usual, we fix a countably infinite set of *actions* $Act = \{\alpha, \beta, \ldots\}$. Also, fix a countably infinite set of *variables* $Var = \{X, Y, Z, \ldots\}$. The set of process expressions *Proc* of BPP_H is defined by the abstract syntax

$$E ::= 0 \mid X \mid \sigma.E \mid E + E \mid E \parallel_A E$$

where X is in Var, σ in *Act* and A a subset of *Act*. All constructs are standard. 0 denotes inaction, X a process variable, σ. prefixing, $+$ non-deterministic choice, and $\parallel_A$ TCSP parallel composition of processes executing independently with forced synchronization on actions in the *synchronization set*, A. For convenience, we shall write $\parallel$ for $\parallel_\emptyset$.

A *process family* is a family of recursive equations $\Delta = \{X_i \stackrel{\text{def}}{=} E_i \mid i = 1, 2, \ldots, n\}$, where $X_i \in Var$ are distinct variables and $E_i \in Proc$ are process expressions containing at most variables in $Var(\Delta) = \{X_1, \ldots, X_n\}$.

A *process* E is a process expression of *Proc* with a process family Δ such that all variables occurring in E, $Var(E)$, are contained in $Var(\Delta)$. We shall often assume the family of a process to be defined implicitly. Dually, a process family denotes the process defined by its *leading variable*, X_1, if not mentioned explicitly. Let $Act(E)$ denote the set of actions occurring in process E and its associated family. A process expression E is *guarded* if each variable in E occurs within some subexpression $\sigma.F$ of E. A process family is guarded if for each equation the right side is guarded. A process E with family Δ is guarded if E and Δ are guarded. Throughout the paper we shall only consider guarded processes and process families.

We enrich the standard operational semantics of TCSP [28] by adding information to the transitions allowing us to observe an action together with its location. More precisely, the location of an action in a process P is the path from the root to the action in the concrete syntax tree represented by a string over $\{0, 1\}$ labelling left and right branches of $\parallel_A$-nodes with 0 and 1, respectively, and all other branches with the empty string ϵ.

Let $\mathcal{L} = \mathcal{P}(\{0, 1\}^*)$, i.e. finite subsets of strings over $\{0, 1\}^*$, and let l range over elements of $\mathcal{L}$. We interpret prefixing a symbol to $\mathcal{L}$ as prefixing elementwise, i.e. $0l = \{0s \mid s \in l\}$. With this convention, any process determines a $(Act \times \mathcal{L})$-labelled transition system with states the set of process expressions reachable from the leading variable and transitions given by the transitions rules of table 1. The set of *computations* of a process, E, is defined now as usual as sequences of transitions, decorated by action and locality information:

$$c: E = E_0 \xrightarrow[l_1]{\sigma_1} E_1 \ldots \xrightarrow[l_n]{\sigma_n} E_n$$

We let $loc(c)$ denote the set of locations occurring in c, i.e. $loc(c) = \bigcup_{1 \leq i \leq n} l_i$.

Example 1. Consider the process

$$p_1 = a.b.c.0 \parallel_{\{b\}} b.0.$$

The following is an example of an associated computation (representing the unique maximal run)

$$p_1 \xrightarrow[\{0\}]{a} b.c.0 \parallel_{\{b\}} b.0 \xrightarrow[\{0,1\}]{b} c.0 \parallel_{\{b\}} 0 \xrightarrow[\{0\}]{c} 0 \parallel_{\{b\}} 0$$

Consider alternatively the process

$$p_2 = a.b.0 \parallel_{\{b\}} b.c.0$$

with computation

$$p_2 \xrightarrow[\{0\}]{a} b.0 \parallel_{\{b\}} b.c.0 \xrightarrow[\{0,1\}]{b} 0 \parallel_{\{b\}} c.0 \xrightarrow[\{1\}]{c} 0 \parallel_{\{b\}} 0$$

$\square$

$$\sigma.E \xrightarrow[\{\epsilon\}]{\sigma} E \quad (prefix) \qquad\qquad \frac{E \xrightarrow[l]{\sigma} E'}{X \xrightarrow[l]{\sigma} E'}, (X \stackrel{\text{def}}{=} E) \in \Delta \ (unfold)$$

$$\frac{E \xrightarrow[l]{\sigma} E'}{E + F \xrightarrow[l]{\sigma} E'} \ (sum_l) \qquad\qquad \frac{F \xrightarrow[l]{\sigma} F'}{E + F \xrightarrow[l]{\sigma} F'} \qquad\qquad (sum_r)$$

$$\frac{E \xrightarrow[l]{\sigma} E'}{E \| _A F \xrightarrow[0l]{\sigma} E' \| _A F}, \sigma \notin A \ (com_l) \qquad\qquad \frac{F \xrightarrow[l]{\sigma} F'}{E \| _A F \xrightarrow[1l]{\sigma} E \| _A F'}, \sigma \notin A \ (com_r)$$

$$\frac{E \xrightarrow[l_0]{\sigma} E' \quad F \xrightarrow[l_1]{\sigma} F'}{E \| _A F \xrightarrow[0l_0 \cup 1l_1]{\sigma} E' \| _A F'}, \sigma \in A \ (com)$$

Table 1. Transition rules for BPP$_H$

3 Language, pomset, and location equivalence

Let $\sqsubseteq$ be the prefix ordering on $\{0,1\}^*$, extended to sets, i.e. for $l, l' \in \mathcal{L}$

$$l \sqsubseteq l' \iff \exists s \in l, s' \in l'. s \sqsubseteq s'.$$

For a given computation

$$c : E_0 \xrightarrow[l_1]{\sigma_1} E_1 \ldots \xrightarrow[l_n]{\sigma_n} E_n,$$

we define the location dependency ordering over $\{1, 2, \ldots, n\}$ as follows:

$$i \leq_c j \iff l_i \sqsubseteq l_j \wedge i \leq j.$$

Finally, we let $\leq_c^*$ denote the transitive closure of $\leq_c$.

Definition 1. *Behavioural Equivalences.*
Processes E and E' are said to be *language equivalent*, $E \sim_{lan} E'$, iff for every computation of E

$$c : E \xrightarrow[l_1]{\sigma_1} E_1 \ldots \xrightarrow[l_n]{\sigma_n} E_n$$

there exists a computation of E'

$$c' : E' \xrightarrow[l_1']{\sigma_1} E_1' \ldots \xrightarrow[l_n']{\sigma_n} E_n'$$

and vice versa.
E and E' are said to be *pomset equivalent*, $E \sim_{pom} E'$, iff the above condition for language equivalence is satisfied, and c' is further required to satisfy $i \leq_c^* j \iff i \leq_{c'}^* j$.
E and E' are said to be *location equivalent*, $E \sim_{loc} E'$, iff the above condition for language equivalence is satisfied, and c' is further required to satisfy that there exists a relation $\mathcal{R} \subseteq loc(c) \times loc(c')$ satisfying that for each $1 \leq i \leq n$, $\mathcal{R}$ restricts to a bijection on $l_i \times l_i'$, and for each $i \leq j$, $s_0(\mathcal{R} \cap l_i \times l_i')s_0'$ and $s_1(\mathcal{R} \cap l_j \times l_j')s_1'$, $s_0 \sqsubseteq s_1 \iff s_0' \sqsubseteq s_1'$. $\square$

Notice that the condition in the definition of pomset equivalence requires identical global causal relationship between the events of c and c', whereas the condition in the definition of location equivalence requires the same set of local causal relationships (up to renaming of locations). Also, notice that our notion of pomset equivalence is consistent with formal definitions from e.g. [14], and that location equivalence is a natural application of the concepts from [3] to the setting of language equivalence.

Example 2. It follows immediate from the definition that for our process language considered so far, location equivalence is included in pomset equivalence, which in turn is included in language equivalence. The standard example of processes a.0 ∥ b.0 and a.b.0 + b.a.0 shows that the inclusion in language equivalence is strict. The different intuitions behind our two non-interleaving equivalences may be illustrated by the two processes from Example 1. Formally, the reader may verify that p_1 and p_2 are pomset equivalent but not location equivalent. Intuitively, both processes may perform actions a,b, and c in sequence, i.e. same set pomsets, but in p_1 one location is responsible for both a and c, whereas in p_2 two different locations are responsible for these actions. □

4 BPP

In this section we investigate the calculus known as Basic Parallel Processes [4], BPP – a syntactic subset of CCS and TCSP which can be seen as the largest common subset of these. The abstract syntax of BPP expressions is

$$E ::= 0 \mid X \mid \sigma.E \mid E + E \mid E \parallel E$$

and the semantics is just as presented in the previous section. A BPP *process* is a process only involving BPP expressions. Note that BPP is nothing but our previous language restricted to parallel compositions without communication.

Theorem 2. *For BPP, $\sim_{loc} = \sim_{pom} \subset \sim_{lan}$.*

Proof. From the fact that all observed locations are singletons it easily follows $\sim_{loc}$ and $\sim_{pom}$ coincide on BPP. The strict inclusion follows from Example 2. □

Definition 3. A Σ-*labelled net* is a four-tuple (S, T, F, l) where S (the *places*) and T (the *transitions*) are non-empty finite disjoint sets, F (the *flow relations*) is a subset of $(S \times T) \cup (T \times S)$ and l is a *labelling function* from T to Σ. A *marking* of a net is a multiset of places. Finally, a *Petri net* is a pair (N, M_0) where N is a labelled net and M_0 is an (*initial*) marking. The *preset* of a transition $t \in T$ is the set $^\bullet t = \{s \mid (s, t) \in F\}$. A Petri net is *communication-free* iff for every $t \in T$, $\mid ^\bullet t \mid = 1$. □

As mentioned in the introduction BPP is formally equivalent to the class of communication-free Petri nets with respect to any interleaving equivalence coarser than or equal to bisimilarity. With the normal form result below it is straightforward to obtain a similar result for location and pomset equivalence.

An important property of BPP is the fact that due to the lack of communication the location/pomset ordering of computations have a particularly simple form.

Proposition 4. *For any BPP process E, and any computation*

$$c : E = E_0 \xrightarrow[l_1]{\sigma_1} E_1 \ldots \xrightarrow[l_n]{\sigma_n} E_n$$

the ordering $\leq_c^*$ is a tree ordering.

Proof. Every observed location is a singleton and hence every location has at most one predecessor. □

We use the rest of this section to prove that $\sim_{loc}$ and $\sim_{pom}$ are decidable on BPP processes. The proof relies on the proposition above, and a reduction to the equivalence problem for recognizable tree languages which is well-known to be decidable, see e.g. [7] or for a brief treatment [29].

4.1 Normal form

We present a definition of normal form for BPP processes and a normal form result. The normal form we use is based on the following structural congruence.

Definition 5. Let $\equiv \subseteq Proc \times Proc$ be the least congruence with respect to all operators such that the following laws hold.

Abelian monoid laws for $+$:

$$E + F \equiv F + E$$
$$E + (F + G) \equiv (E + F) + G$$
$$E + 0 \equiv E$$

Abelian monoid laws for $\|$:

$$E \parallel F \equiv F \parallel E$$
$$E \parallel (F \parallel G) \equiv (F \parallel E) \parallel G$$
$$E \parallel 0 \equiv E$$

Idempotence law for $+$:

$$E + E \equiv E$$

Linear-time laws:

$$(E + F) \parallel G \equiv (E \parallel G) + (F \parallel G)$$
$$\sigma.(E + F) \equiv \sigma.E + \sigma.F$$

$\square$

Proposition 6. $\equiv$ *is sound in the sense that if $E \equiv F$ then $E \sim_{pom} F$.*

Proof. Induction in the structure of the proof of $E \equiv F$. $\square$

As parallel composition is commutative and associative, it is convenient to represent a parallel composition of distinct variables $X_0 \parallel \ldots \parallel X_k$ by the set $\{X_0, \ldots, X_k\}$, and 0 by the empty set. We denote by $\mathcal{P}(M)$ the set of all finite subsets of M.

Definition 7. A BPP family $\Delta = \{X_i \stackrel{\text{def}}{=} E_i \mid i = 1, 2 \ldots, n\}$ is defined to be in *normal form* if and only if every expression E_i is of the form

$$E_i \equiv \sum_{j=1}^{n_i} \sigma_{ij} \alpha_{ij}$$

where $\sigma_{ij} \in Act$ and $\alpha_{ij} \in \mathcal{P}(Var(\Delta))$. $\qquad \square$

BPP processes in normal form and communication-free nets are closely related. Hence, mapping variables to places and actions to action labelled transitions, induces an obvious isomorphism between the computations of a BPP process in normal form and the firing sequences of a communication-free net.

From the soundness of $\equiv$ it is fairly straightforward to prove the following normal form result.

Proposition 8. *Let Δ be a BPP family with leading variable X_1. Then a BPP family in normal form Δ' can be effectively constructed such that $\Delta'' \sim_{pom} \Delta'$, where Δ'' is Δ extended with a new leading variable $X_1' = s.X_1$, for some $s \notin Act(\Delta)$ and $X_1' \notin Var(\Delta)$.*

Proof. Due to the guardedness and the fact that the leading equation of Δ'' has no unguarded choice any unguarded choice can be made guarded by propagating it to its nearest preceding prefixing. $\qquad \square$

Note that for example the process $(a \parallel b) + c$ can not be brought on normal form while preserving pomset equivalence whereas the process $s.((a \parallel b) + c)$ can. Hence, the point of the slightly technical normal form result is that prefixing the leading equation of two BPP processes by the same action respects and reflects pomset equivalence.

4.2 Finite tree automata

In this section we show how to effectively construct a finite tree automaton $\mathcal{A}_\Delta$ from a BPP family Δ in normal form in such a way that pomset equivalence reduces to equivalence of recognizable tree languages.

Let $\Sigma = \Sigma_0 \cup \ldots \cup \Sigma_n$ be a ranked finite alphabet. The set of all trees over Σ, T_Σ is the free term algebra over Σ, that is, T_Σ, is the least set such that $\Sigma_0 \subseteq T_\Sigma$ and such that if $a \in \Sigma_k$ and for $i = 1, \ldots, k$, $t_i \in T_\Sigma$, then $a[t_1, \ldots, t_k] \in T_\Sigma$. For convenience, we use a and $a[]$ interchangeably to denote members of Σ_0.

Definition 9. A non-deterministic top-down finite tree automaton, *NTA*, is a four-tuple $\mathcal{A} = (\Sigma, Q, S, \delta)$, where Σ is a ranked finite alphabet, Q a finite set of states, $S \subseteq Q$ is a set of initial states, and δ is a ranked finite family of labelled transition relations associating with each $k \geq 0$, a relation $\delta_k \subseteq Q \times \Sigma_k \times Q^k$. $\qquad \square$

Definition 10. Let $\mathcal{A} = (\Sigma, Q, S, \delta)$ be a NTA and let $t \in T_\Sigma$. A *configuration* of $\mathcal{A}$, is a multiset of pairs from $Q \times T_\Sigma$. Denote by $conf_\mathcal{A}$ the set of all configurations of $\mathcal{A}$. For $\sigma \in \Sigma$, let $\xrightarrow{\sigma} \subseteq conf_\mathcal{A} \times conf_\mathcal{A}$ be the labelled transition relation between configurations defined by

$$\{|(q, t)|\} \cup c \xrightarrow{\sigma} \{|(q_1, t_1), \ldots, (q_k, t_k)|\} \cup c,$$

if $\sigma \in \Sigma_k$, $t = \sigma[t_1, \ldots, t_k]$, $(q, \sigma, q_1, \ldots, q_k) \in \delta_k$ and $c \in conf_\mathcal{A}$. We write $\rightarrow$ for the union over all $\sigma \in \Sigma$ of $\xrightarrow{\sigma}$, and $\rightarrow^*$ for the reflexive and transitive closure of $\rightarrow$. A (*successful*) *run* of $\mathcal{A}$ on input t is a derivation $\{|(q_0, t)|\} \rightarrow^* \emptyset$, where $q_0 \in S$. The tree language, $L(\mathcal{A})$, *recognized* by $\mathcal{A}$ consists of all trees t, for which there is a successful run of $\mathcal{A}$ on t.

A transition relation δ is *permutation closed* if for all $q, q_1, \ldots, q_k \in Q$, $k \geq 0$, and permutations π on $\{1, \ldots, k\}$

$$(q, \sigma, q_1, \ldots, q_k) \in \delta_k \iff (q, \sigma, q_{\pi(1)}, \ldots, q_{\pi(k)}) \in \delta_k$$

A *NTA* is *permutation closed* if its transition relation is permutation closed. $\square$

Construction 11. Given a BPP family Δ in normal form with leading variable X_1, define a permutation closed *NTA* $\mathcal{A}_\Delta = (Act(\Delta), Var(\Delta), \{X_1\}, \delta)$ such that for every $(X \stackrel{\text{def}}{=} \sum_{i=1}^n \sigma_i \alpha_i) \in \Delta$, every index $1 \leq j \leq n$ and for every $\{Y_1, \ldots, Y_k\} \subseteq \alpha_j$,

$$(X, \sigma_j, Y_1, \ldots, Y_k) \in \delta_k$$

The ranking of the alphabet $Act(\Delta)$ is induced by the definition of δ. $\square$

The crucial property of this construction is formulated in the following proposition.

Proposition 12. *Given BPP families Δ and Δ' in normal form and with leading variables X and X', respectively. Then*

$$X \sim_{pom} X' \iff L(\mathcal{A}_\Delta) = L(\mathcal{A}_{\Delta'})$$

Proof. The proof is by induction in the length of computations/runs and uses Proposition 4 and the fact that the constructed *NTAs* are permutation closed. $\square$

Theorem 13. *For BPP, $\sim_{pom}$ and $\sim_{loc}$ are decidable, whereas $\sim_{lan}$ is undecidable.*

Proof. The undecidability result was proved in [10]. The decidability results follow from Theorem 2, Propositions 8 and 12 and the fact that the equivalence problem for *NTA* recognizable tree languages is decidable, see e.g. [7]. $\square$

5 Extending towards full TCSP

We now return to the TCSP subset, BPP_H, defined in Section 2. In contrast to BPP, BPP_H allows communication. In this section we show that this extension right away leads to undecidability of both $\sim_{pom}$ and $\sim_{loc}$. In proving the undecidability, an interesting difference in the complexity of the reductions used appears. To show that $\sim_{pom}$ is undecidable we only need one static occurrence of the parallel operator with a non-empty synchronization set, whereas to show that $\sim_{loc}$ is undecidable we seem to need much more sophisticated techniques. We end the section with a study of a non-trivial subset of BPP_H, called static BPP_H or just BPP_S, that makes the difference between $\sim_{pom}$ and $\sim_{loc}$ explicit: for BPP_S, $\sim_{loc}$ is decidable whereas $\sim_{pom}$ remains undecidable.

5.1 BPP_H and TCSP

When allowing non-empty synchronization sets $\sim_{pom}$ and $\sim_{loc}$ become different.

Theorem 14. *For BPP_H, $\sim_{loc} \subset \sim_{pom} \subset \sim_{lan}$*

Proof. Follows from Definition 1 and Examples 1 and 2. $\square$

Theorem 15. *For BPP_H, $\sim_{pom}$ and $\sim_{loc}$ are undecidable.*

Proof. It is well-known that BPP_H is Turing powerful, see e.g. [4] where it is shown how to simulate Minsky counter machines [21] in BPP_H. Given the encoding of Minsky counter machines there is a standard way of reducing the halting problem for Minsky counter machines to an equivalence problem. Given a Minsky counter machine N construct first a BPP_H process E_N that simulates N and then another process F_N that is an exact copy of E_N except for F_N having a distinguished action, say h, not in E_N such that h is enabled if and only if N halts. Hence E_N is equivalent to F_N if and only if N does not halt, and the undecidability of the equivalence follows. Hence in particular pomset and location equivalence are undecidable. $\square$

For $\sim_{pom}$ the following stronger result shows that even for a very restricted subset of BPP_H pomset equivalence remains undecidable.

Proposition 16. *Let E and F be BPP processes with identical alphabets Σ and let S be the BPP process $S \stackrel{\text{def}}{=} \sum_{a \in \Sigma} a.S$.*

$$E \sim_{lan} F \iff E \parallel_{\Sigma} S \sim_{pom} F \parallel_{\Sigma} S$$

Proof. The intuition is that the process S works as a sequentializer. The proof essentially consists of transforming computations of E into computations of $E \parallel_{\Sigma} S$ and vice versa. $\square$

We do not know of any way to prove the undecidability for $\sim_{loc}$ without referring to the full Turing power of BPP_H.

5.2 BPP$_S$

A natural restriction when dealing with non-interleaving behaviours is to allow only parallel composition in a fixed static setup, see e.g. [2, 1]. This of course leads to finite-state systems. We generalize the idea to possibly infinite-state systems.

Let BPP$_S$ be the syntactic subset of BPP$_H$ obtained by allowing only synchronization, i.e. the $\|_A$ operator with $A \neq \emptyset$, at top level and restricting the synchronization sets to be the set of all actions possible in either of the components. A BPP$_S$ process can hence be seen as a fixed set of BPP processes synchronizing on every action. Formally, a BPP$_S$ expression is given by the abstract syntax

$$E ::= E_1 \|_\Sigma \ldots \|_\Sigma E_l,$$

where each E_i is a BPP expression and $\Sigma = \bigcup_{i=1,\ldots,l} Act(E_i)$. A BPP$_S$ process E is a BPP$_S$ expression with a BPP family Δ such that all variables occurring in E are contained in $Var(\Delta)$. A BPP$_S$ family is a process family with the property that the leading variable defines a BPP$_S$ expression, all other variables define BPP expression and that the leading variable does not occur on any rightside. A BPP$_S$ family $\Delta = \{X \stackrel{\text{def}}{=} X_1 \|_\Sigma \ldots \|_\Sigma X_l\} \cup \Delta'$ is in *normal form* if the BPP family Δ' is in normal form. We call l the *arity* of Δ.

We leave it to the reader to check that also for BPP$_S$ we have the following relationships.

Theorem 17. *For BPP$_S$, $\sim_{loc} \subset \sim_{pom} \subset \sim_{lan}$.*

Proof. Follows from Definition 1 and Examples 1 and 2. $\qquad\square$

Theorem 18. *For BPP$_S$, $\sim_{loc}$ is decidable whereas $\sim_{pom}$ is not.*

Proof. From Proposition 16 it follows that $\sim_{pom}$ is undecidable for BPP$_S$ processes. We use the rest of this section to prove that $\sim_{loc}$ is decidable. $\qquad\square$

5.3 Synchronous automata on tuples of finite trees

The synchronous automata on tuples of finite trees (*SATT*s) we define below consists of a tuple of non-deterministic top-down tree-automata and work on tuples of finite trees such that each *NTA* works on its component of a tuple while synchronizing with the others. *SATT*s are closely related to communicating finite automata, see e.g. [32], and may be seen as communicating finite tree-automata. Let $\hat{T}_\Sigma = T_\Sigma \times \ldots \times T_\Sigma$ denote the set of l-tuples of finite trees over the alphabet Σ.

Definition 19. For $i = 1, \ldots, l$ let $\mathcal{A}_i = (\Sigma, Q_i, S_i, \delta_i)$ be permutation closed *NTA*s. A synchronous automaton on tuples of finite trees, *SATT*, is a pair $\mathcal{A}^\otimes = ((\mathcal{A}_1, \ldots, \mathcal{A}_l), S_A)$, where $S_A \subseteq S_1 \times \ldots \times S_l$. $\qquad\square$

Definition 20. Let $\mathcal{A}^{\otimes} = ((\mathcal{A}_1, \ldots, \mathcal{A}_l), S_{\mathcal{A}})$ be a *SATT*. A *configuration* of $\mathcal{A}^{\otimes}$ is a tuple in $conf_{\mathcal{A}_1} \times \ldots \times conf_{\mathcal{A}_l}$. The set of configurations of $\mathcal{A}^{\otimes}$ is denoted by $conf_{\mathcal{A}^{\otimes}}$. Let $\Rightarrow_{\mathcal{A}} \subseteq conf_{\mathcal{A}^{\otimes}} \times conf_{\mathcal{A}^{\otimes}}$ be the transition relation between configurations defined by

$$(c_1, \ldots, c_l) \Rightarrow_{\mathcal{A}} (c'_1, \ldots, c'_l)$$

if and only if for some $\sigma \in \Sigma$, $c_i \xrightarrow{\sigma} c_i'$ for all $i = 1, \ldots, l$. We denote by $\Rightarrow_{\mathcal{A}}^{*}$ the reflexive and transitive closure of $\Rightarrow_{\mathcal{A}}$. A (*successful*) run of $\mathcal{A}^{\otimes}$ on input $(t_1, \ldots, t_l) \in \hat{T}_{\Sigma}$ is a derivation $(\{|(q_1, t_1)|\}, \ldots, \{|(q_l, t_l)|\}) \Rightarrow_{\mathcal{A}}^{*} (\emptyset, \ldots, \emptyset)$, where $(q_1, \ldots, q_l) \in S_{\mathcal{A}}$. The tree-tuple language, $L(\mathcal{A}^{\otimes})$, *recognized* by $\mathcal{A}^{\otimes}$ consists of all tree-tuples $\hat{t}$, for which there is a run of $\mathcal{A}^{\otimes}$ on $\hat{t}$. □

A tree-tuple is said to be *well-synchronized* if it belongs to the language of some *SATT*. Let $\hat{T}_{\Sigma}^{\otimes}$ denote the set of well-synchronized tree-tuples. Next, we show that the class of tree-tuple languages over $\hat{T}_{\Sigma}^{\otimes}$ recognized by *SATT*s is closed under Boolean operations. But first, a property which is convenient for defining complement.

Definition 21. A *SATT* $\mathcal{A}^{\otimes} = ((\mathcal{A}_1, \ldots, \mathcal{A}_l), S_{\mathcal{A}})$ is in *standard form* if for every $\hat{t} = (t_1, \ldots, t_l) \in L(\mathcal{A}^{\otimes})$ there is exactly one tuple $(q_1, \ldots, q_l) \in S_1 \times \ldots \times S_l$ such that $(\{|(q_1, t_1)|\}, \ldots, \{|(q_l, t_l)|\}) \Rightarrow_{\mathcal{A}}^{*} (\emptyset, \ldots, \emptyset)$. □

Proposition 22. *Any SATT can effectively be transformed into a SATT in standard form recognizing the same language.*

Proof. It is not hard to see that in general the set of initial state tuples of a *SATT* can not be reduced to a singleton. Hence, the transformation consists of first rebuilding each *NTA* $\mathcal{A}_i$ using the effective boolean closure such that for each $t \in L(\mathcal{A}_i)$ there is exactly one successful run on t and then redefining $S_{\mathcal{A}}$ appropriately replacing each tuple by possibly more than one tuple. □

Let $\mathcal{A}_{\Sigma}$ be the *NTA* that recognizes T_{Σ}. Given *NTA*s $\mathcal{A}$ and $\mathcal{B}$ let $\mathcal{A} \cup \mathcal{B}$ and $\bar{\mathcal{A}}$ denote the *effectively* constructible *NTA*s recognizing the union of the languages recognized by $\mathcal{A}$ and $\mathcal{B}$ and the complement of the language recognized by $\mathcal{A}$, respectively, see [7] for the detailed constructions. In the following we also use the fact that due to the non-determinism any *NTA* can be *effectively* transformed into a *NTA* with only one initial state recognizing the same language.

Definition 23. Let $\mathcal{A}^{\otimes} = ((\mathcal{A}_1, \ldots, \mathcal{A}_l), S_{\mathcal{A}})$ and $\mathcal{B}^{\otimes} = ((\mathcal{B}_1, \ldots, \mathcal{B}_l), S_{\mathcal{B}})$ be *SATT*s. Define

1. $\mathcal{A}^{\otimes} \cup \mathcal{B}^{\otimes} = ((\mathcal{A}_1 \cup \mathcal{B}_1, \ldots, \mathcal{A}_l \cup \mathcal{B}_l), S_{\mathcal{A}} \cup S_{\mathcal{B}})$
2. Let furthermore $\mathcal{A}^{\otimes}$ be in standard form. Define

$$\bar{\mathcal{A}}^{\otimes} = (\bigcup_{i \in \{1, \ldots, l\}} ((\mathcal{C}_1^i, \ldots, \mathcal{C}_l^i), S^i)) \cup ((\mathcal{A}_1, \ldots, \mathcal{A}_l), S)$$

where $C_j^i = (\Sigma, Q_j^i, \{p_j^i\}, \delta_j^i)$ such that $C_i^i = \bar{A}_i$ and for $j \neq i$, $C_j^i = A_\Sigma$, $S^i = \{(p_1^i, \ldots, p_l^i)\}$ and $S = S_1 \times \ldots \times S_l \setminus S_A$. $\quad\square$

Clearly, $A^\otimes \cup B^\otimes$ and $\bar{A}^\otimes$ are *SATT*s. The following proposition states that they have in fact the expected properties.

Proposition 24.

1.) $L(A^\otimes \cup B^\otimes) = L(A^\otimes) \cup L(B^\otimes)$
2.) $L(\bar{A}^\otimes) = \hat{T}_\Sigma^\otimes - L(A^\otimes)$

Proof. The proof of 1.) is straightforward, and 2.) relies on the fact that $A^\otimes$ is in standard form. $\quad\square$

Another important property is the decidability of the emptiness problem for *SATT*. We establish this by a reduction to the zero reachability problem for Petri nets, as defined in Section 4, which is decidable [17, 16]. The representation of finite tree automatas as Petri nets was studied in [25]. Here, we translate *NTA*s into communication-free nets and *SATT*s into synchronized products of communication-free nets. As our Petri nets are not weighted we use the easily shown fact that any *NTA* can be *effectively* transformed into another *NTA* recognizing the same language but with the property that its transition relation δ satisfies that for all $(q, \sigma, q_1, \ldots, q_k) \in \delta_k$, $q_i = q_j \Rightarrow i = j$. The construction below translates at one swoop a *SATT* into a Petri net.

Before we give the general construction of Petri nets from *SATT*s we consider an example.

Example 3. Given the *SATT* $A^\otimes = ((A_1, A_2), \{(p_1, q_1)\})$ where
$A_1 = (\{a, b, c\}, \{p_1, p_2, p_3\}, \{p_1\}, \delta_1)$, $A_2 = (\{a, b, c\}, \{q_1, q_2, q_3\}, \{q_1\}, \delta_2)$,

$$\delta_1 = \{(p_1, a, p_2, p_3), (p_1, a, p_3, p_2), (p_2, c), (p_3, b)\} \text{ and}$$
$$\delta_2 = \{(q_1, a, q_2), (q_2, b, q_3), (q_3, c)\}$$

we construct the Petri net in figure 1. $\quad\square$

For notational ease let $\delta^\sigma = \bigcup_{0 \leq k} \{(q, \sigma, q_1, \ldots, q_k) \in \delta_k\}$.

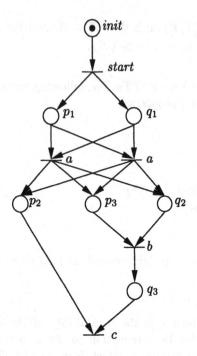

Fig. 1. The Petri net associated with the *SATT* of example 3

Construction 25. Given a *SATT* $\mathcal{A}^{\otimes} = ((\mathcal{A}_1, \ldots, \mathcal{A}_l), S_{\mathcal{A}})$ such that the Q_is are disjoint and a action *start* $\notin \Sigma$. Let i range over $1, \ldots, l$, let $\delta = \bigcup_i \delta_i$. Define the labelled Petri net $P_{\mathcal{A}} = ((S, T, F, l), M_0)$ with places $S = \bigcup Q_i \cup \{init\}$, transitions

$$T = \{(\sigma, \eta_1, \ldots, \eta_l) \mid \eta_i \in \delta_i^{\sigma} \text{ for } \sigma \in \Sigma\} \cup S_{\mathcal{A}},$$

flow relation $F = F_1 \cup F_2 \cup F_3 \cup F_4$, where

$F_1 = \{(q, t) \mid t = (\sigma, \eta_1, \ldots, \eta_l) \in T \wedge \eta_i = (q, \sigma, q_1, \ldots, q_k)\},$
$F_2 = \{(t, q_j) \mid t = (\sigma, \eta_1, \ldots, \eta_l) \in T \wedge \eta_i = (q, \sigma, q_1, \ldots, q_k) \wedge 1 \leq j \leq k\},$
$F_3 = \{(t, q_i) \mid t = (q_1, \ldots, q_l) \in S_{\mathcal{A}}\},$ and
$F_4 = \{(init, t) \mid t \in S_{\mathcal{A}}\},$

labelling function $l : T \to (\Sigma \cup \{start\})$ given by $l(t) = \sigma$, if $t = (\sigma, \eta_1, \ldots, \eta_l)$ and *start* otherwise, and initial marking $M_0 = \{|init|\}$. □

Note that for each transition not in $S_{\mathcal{A}}$ the cardinality of the preset is exactly l. Also, since we are interested mainly in the following reduction the labelling of $P_{\mathcal{A}}$ is irrelevant.

Proposition 26. $L(\mathcal{A}^{\otimes}) \neq \emptyset \iff$ *the zero-marking is reachable in* $P_{\mathcal{A}}$

Proof. Induction in the length of runs/firing sequences showing that each successful run of $\mathcal{A}^{\otimes}$ corresponds to a firing sequence of P_A reaching the zero-marking, and vice versa. □

From the proposition above and the Boolean closure we get

Proposition 27. *The emptiness and the equivalence problem for SATT is decidable.*

Proof. The decidability of the emptiness problem follows immediately from Proposition 26 and the decidability of the zero reachability problem [17, 16]. The decidability of the equivalence problem then follows immediately due to the closure under Boolean operations. □

Let Perm_l denote the set of all permutations on $\{1, \ldots, l\}$.

Construction 28. Given a BPP_S family Δ in normal form with leading equation $X = X_1 \parallel_{\Sigma} \ldots \parallel_{\Sigma} X_l$ and corresponding BPP families $\Delta_1, \ldots, \Delta_l$ with leading variables $X_1, \ldots, X_l$, respectively. Define

$$A_{\Delta}^{\otimes} = \bigcup_{\pi \in \mathrm{Perm}_l} ((A_{\Delta_{\pi(1)}}, \ldots, A_{\Delta_{\pi(l)}}), S_{A_\pi}),$$

where $S_{A_\pi} = S_{\pi(1)} \times \ldots \times S_{\pi(l)}$ and S_i the set of initial states of A_{Δ_i} □

The essential property of the construction is expressed by the following proposition.

Proposition 29. *Let Δ and Δ' be BPP_S families in normal form of the same arity, and with leading variables X and Y, respectively. Then*

$$X \sim_{loc} Y \iff L(A_{\Delta}^{\otimes}) = L(A_{\Delta'}^{\otimes})$$

Proof. Induction in the length of runs/computations showing that each run of A_Δ and $A_{\Delta'}$ corresponds to a computation of respectively Δ and Δ', and vice versa. □

Theorem 30. *For BPP_S, $\sim_{loc}$ is decidable.*

Proof. Its is not hard to see from Proposition 8 that we can assume that the families Δ and Δ' are in normal forms. Moreover, since arity checking is syntactically easy to check, the result follows from Construction 28 and Propositions 29 and 27. □

6 Extending towards full CCS

In this section we study the extensions of BPP obtained by adding first CCS-synchronization and then CCS-restriction. To avoid confusion we begin by explaining the syntax and semantics of both. Let Act and Var be as in section 2 and let $\overline{Act} = \{\bar{\alpha}, \bar{\beta}, \ldots\}$ such that $\bar{\ }$ is a bijection between Act and $\overline{Act}$, mapping $\bar{\bar{\alpha}}$ to α. Let $Act_\tau = Act \cup \overline{Act} \cup \{\tau\}$ be the set of *actions*, where τ is a distinguished action not in Act or $\overline{Act}$. τ is known as the *invisible* action. Any other action is *visible*. The set of CCS process expressions is defined by the abstract syntax

$$E ::= 0 \mid X \mid \sigma.E \mid E + E \mid E \parallel E \mid E \backslash L$$

where X is in Var, σ in Act_τ and L a subset of Act. 0, X, $\sigma.$, and $+$ are as for BPP. $\parallel$ is CCS parallel composition of processes executing independently with the possibility of pairwise CCS-synchronization and $\backslash L$ is CCS-restriction. The semantics is given by the transition rules of BPP together with the rules of Table 2. Following [20] we restrict ourselves to guarded processes, in the sense that every variable occurs within a prefix $\sigma.F$, where $\sigma \neq \tau$.

$$\frac{E \xrightarrow[l_0]{\sigma} E' \qquad F \xrightarrow[l_1]{\bar{\sigma}} F'}{E \parallel F \xrightarrow[0 l_0 \cup 1 l_1]{\tau} E' \parallel F'} \ (\tau - com)$$

$$\frac{E \xrightarrow[l]{\sigma} F}{E\backslash L \xrightarrow[l]{\sigma} F\backslash L}, \ \sigma, \bar{\sigma} \notin L \ (res)$$

Table 2. Transition rules for CCS communication and restriction

6.1 BPP$_M$

BPP$_M$, is the subset of CCS obtained by adding the transition rule τ-*com* of table 2 to BPP with actions from Act_τ and hence introducing CCS-synchronization. Since there is no restriction operator in BPP$_M$ communication cannot be forced. Whenever a communication occurs in a computation, also the computation with the communicating actions occurring separately is possible. Conversely, if there is a computation in which two complementing actions occur independently then the same computation except from the two actions now communicating exists. The proof of the following proposition relies on this observation.

Proposition 31. *The BPP$_M$ processes E and F are pomset (location) equivalent if and only if the BPP processes E and F are pomset (location) equivalent.*

Proof. The only if direction is obvious. The if direction is shown by induction in the number of communications in computations and based on first undoing and then doing communications. □

From this proposition we immediately get the following results.

Theorem 32. *For BPP_M, $\sim_{loc} = \sim_{pom} \subset \sim_{lan}$.*

Proof. From Proposition 31 and Theorem 2. □

Theorem 33. *For BPP_M, $\sim_{pom}$ and $\sim_{loc}$ are decidable whereas $\sim_{lan}$ is undecidable.*

Proof. From Proposition 31 and Theorem 13. □

Comparing this result with the earlier results on BPP_H shows a clear difference between adding CCS- and TCSP-communication to BPP. In the former case $\sim_{pom}$ and $\sim_{loc}$ still coincide and remain decidable whereas in the latter $\sim_{loc}$ is strictly finer than $\sim_{pom}$ and they both become undecidable.

6.2 CCS

CCS is BPP_M extended with the CCS-restriction operator. For CCS, $\sim_{pom}$ and $\sim_{loc}$ no longer coincide.

Theorem 34. *For CCS, $\sim_{loc} \subset \sim_{pom} \subset \sim_{lan}$.*

Proof. The inclusions follow from Definition 2. The strictness of the inclusions follow from Example 2 and the following examples.

$$q_1 = (a.b.c.0 \parallel \bar{b}.0)\backslash\{b\}, \ q_2 = (a.b.0 \parallel \bar{b}.c.0)\backslash\{b\}$$

Clearly, $q_1 \sim_{pom} q_2$ but $q_1 \not\sim_{loc} q_2$. □

Theorem 35. *For CCS $\sim_{pom}$ and $\sim_{loc}$ are undecidable.*

Proof. Due to the well-known Turing power of CCS, see e.g. [28], both $\sim_{pom}$ and $\sim_{loc}$ are undecidable for CCS processes. The reduction is similar to the one sketched in Section 5.1. □

7 Conclusion

We have presented results illuminating the delicate bounds between the decidable and the undecidable in the setting of behavioural equivalences for infinite-state concurrent systems. We would like to see our results as a contribution to the search for useful verification problems which will be decidable/tractable when moving from the standard view of interleaving to more intentional non-interleaving views of behaviour.

Our results raise many open questions to be addressed. We have concentrated on the question of decidability of certain equivalences for process calculi. However, there are immediate links to other questions, like *regularity* of processes, see [13] for a recent result showing that regularity is in fact decidable for Petri nets with respect to (even!) language equivalence. Secondly, we have focussed on various process calculi extensions of BPP, and although these, of course, imply results for the corresponding Petri net extensions of communication-free nets, it would be interesting to look for independent extensions in terms of net subclasses with decidable non-interleaving equivalences. Also, many other non-interleaving equivalences exist besides our chosen pomset and location equivalences, and which deserve to be explored. In particular, we do not claim that our notion of location equivalence is the only natural formalization of local causality, other possibilities exist.

References

1. S. Abramsky, Eliminating Local Non-determinism: a New Semantics for CCS, Computer Systems Laboratory, Queen Mary College, Report no. 290 (1981).
2. L. Aceto, A Static View of Localities, Formal Aspects of Computing, 6 (2), 202–222 (1994).
3. G. Boudol, I. Castellani, M. Hennessy and A. Kiehn, Observing Localities, Theoretical Computer Science, 114, 31–61, 1993.
4. S. Christensen. *Decidability and Decomposition in Process Algebras* Ph.D. Thesis, University of Edinburgh, CST-105-93, 1993.
5. S. Christensen, H. Hüttel, Decidability Issues for Infinite-State Processes - a Survey, EATCS Bulletin 51, 156–166 (1993).
6. S. Christensen, Y. Hirshfeld and F. Moller, Bisimulation equivalence is decidable for basic parallel processes, CONCUR '93, Springer LNCS 715, 143–157 (1993).
7. J. Engelfriet, Tree automata and tree grammars, University of Aarhus, DAIMI FN-10 (1975).
8. J. Esparza, Petri nets, commutative context-free grammars, and Basic Parallel Processes, in Proceedings of Fundamentals of Computation Theory, (FCT'95), LNCS 965, Springer Verlag 1995.
9. J. Esparza and A. Kiehn, On the Model Checking Problem for Branching Logics and Basic Parallel Processes, CAV '95, Springer LNCS 939, 353–366 (1995).
10. Y. Hirshfeld, Petri Nets and the Equivalence Problem, CSL '93, Springer LNCS 882, 165–174 (1994).
11. C. A. R. Hoare, *Communicating Sequential Processes*, Prentice Hall, International Series in Computer Science (1985).
12. H. Hüttel, Undecidable Equivalences for Basic Parallel Processes, TACS '94, Springer LNCS 789, 454–464 (1994).
13. P. Jančar and F. Moller, Checking Regular Properties of Petri Nets, CONCUR '95, Springer LNCS 962, 348–362 (1995).
14. L. Jategaonkar and A. Meyer, Deciding true concurrency equivalences on finite safe nets. ICALP '93, Springer LNCS 700, 519–531 (1993).

15. A. Kiehn and M. Hennessy, On the Decidability on Non-interleaving Equivalences, CONCUR '94, Springer LNCS 836, 18–33 (1994)

16. S.R. Kosaraju. Decidability of Reachability in Vector Addition Systems. 14th Annual ACM Symposium on Theory of Computing, San Francisco, 267–281 (1982).

17. E.W. Mayr, Persistence of Vector Replacement Systems is Decidable. Acta Informatica 15, 309–318 (1981).

18. E.W. Mayr and A.R. Meyer, The Complexity of the Finite Containment Problem for Petri Nets, Journal of the ACM, 28(3), 561–576 (1981).

19. A. Mazurkiewicz, Basic notions of trace theory, in de Bakker, de Roever and Rozenberg (eds.), *Linear Time, Branching Time and Partial Orders in Logics and Models for Concurrency*, Springer LNCS 354, 285–363 (1988).

20. A.R.G. Milner, *Communication and concurrency*, Prentice Hall (1989).

21. M.L. Minsky, *Computation - Finite and Infinite Machines*, Prentice Hall (1967).

22. M. Mukund and M. Nielsen, CCS, Locations and Asynchronous Transition Systems, FST & TCS '92, Springer LNCS 652, 328–341 (1992).

23. E.R. Olderog and C.A.R. Hoare, Specification- Oriented Semantics for Communicating Processes, Acta Informatica, 23, 9–66 (1986).

24. V.R. Pratt, Modeling concurrency with partial orders, International Journal of Parallel Programming, 15(1), 33–71 (1986).

25. W. Reisig, A note on the representation of finite tree automata. *Information Processing Letters*, 8(5):239-240, June 1979

26. W. Reisig, *Petri Nets - an Introduction*, EATCS Monograph in Computer Science, Springer (1985).

27. K. Sunesen and M. Nielsen, Behavioural equivalence for infinite systems - partially decidable!, Technical Report RS-95-55, BRICS, Aarhus University (1995).

28. D. Taubner, *Finite Representations of CCS and TCSP Programs by Automata and Petri Nets*, Springer LNCS 369 (1989).

29. W. Thomas, Automata on Infinite Objects, in *Handbook of Theoretical Computer Science*, vol B, ed. J. van Leeuwen, Elsevier, 133–192 (1990).

30. R.J. van Glabbeek, *Comparative concurrency semantics and refinement of actions*, PhD thesis, CWI Amsterdam (1990).

31. R.J. van Glabbeek and U. Goltz, Equivalence Notions for Concurrent Systems and Refinement of Actions, MFCS '89, Springer LNCS 379, 237–248 (1989).

32. P. Wolper and P. Godefroid, Partial Order Methods for Temporal Verification, Concur '93, Springer LNCS 715, 233–246 (1993).

Topological Aspects of Traces

Jaap van Oosten

BRICS*

Department of Computer Science

University of Aarhus, Denmark[†]

17 March 1996

Abstract

Traces play a major role in several models of concurrency. They arise out of "independence structures" which are sets with a symmetric, irreflexive relation.

In this paper, independence structures are characterized as certain topological spaces. We show that these spaces are a universal construction known as "soberification", a topological generalization of the ideal completion construction in domain theory. We also show that there is a group action connected to this construction.

Finally, generalizing the constructions in the first part of the paper, we define a new category of "labelled systems of posets". This category includes labelled event structures as a full reflective subcategory, and has moreover a very straightforward notion of bisimulation which restricts on event structures to strong history-preserving bisimulation.

Key words and phrases: Partial order, independence relations, traces, topology, event structures, bisimulation

Introduction. This paper is a little mathematical study of some models of concurrency. The most elementary one is the concept of an independence structure, which is nothing but a set L with a binary, irreflexive and symmetric rlation on it, the independence relation. This leads to the notion of a *trace*: a string of elements of L, modulo the equivalence generated by swapping adjacent, independent elements of the string.

There are two aspects of finite traces: they form an order, hence a topology; on the other hand they form a monoid, a quotient of the free monoid on L. Unfortunately, these two points of view are hard to bring together, since the monoid structure can never be continuous or even order-preserving. It is therefore not surprising that many papers on trace theory consist of two, disjoint, parts. In this paper I concentrate on the order-theoretic and topological aspects.

---

*Basic Research in Computer Science,
  Centre of the Danish National Research Foundation.
†Current address: Dept. of Mathematics, Utrecht University, The Netherlands

In the first two sections, the set of infinite traces is considered from a topological point of view. It is well-known (e.g. [Kwi]) that taking both finite and infinite traces, yields a Scott domain. There are reasons to consider also the set of just the infinite traces, arguing that important processes are always ongoing, potentially infinite things.

The infinite traces in themselves do not constitute a Scott domain for lack of finite elements, but as a topological space they arise as what can be seen as a generalization of the ideal completion construction in domain theory: namely, *soberification*. This is done in section 2. Since the points of the space are traces, that is labelled partial orders, it is also shown how many important topological properties of these points can be expressed entirely in terms of this labelled poset structure.

Section 3 treats a group action on the set of all infinite L-words, depending on the independence relation, such that the orbits of the action are the equivalence classes under: finitely many swappings of independent pairs. One has a nice characterization of these orbits in terms of certain "action graphs", called "good". One observes that inclusion of good graphs corresponds to order-*reflecting* bijections between the corresponding partial orders; which is similar to the order-reflecting bijection that a map of event structures induces on a configuration by restriction.

Therefore, in section 4, we look at a category of systems of posets and systems of order-reflecting bijections as maps. We see that labelled event structures are a full reflective subcategory of this, as are independence structures.

1 Characterization of Independence Structures

Let X be a set. The purpose of this section and the next one is to represent independence relations on X as topological spaces. The points of these spaces are X-labelled partial orders, and one can express topological properties of these points in terms of this labelled poset structure.

Some definitions:

In a partial order $(P, \leq)$ we write $x <^d y$ for "x is directly below y", i.e. $x < y \land \forall z (x \leq z \leq y \Rightarrow x = z \lor y = z)$.

$(P, \leq)$ is *locally finite* if for every $p \in P$, the set $\downarrow p = \{q \in P \mid q \leq p\}$ is finite.

An X-*labelled* partial order is a partial order $(P, \leq)$ together with a function $l : P \to X$.

Lemma 1.1 *A countable partial order is locally finite if and only if it has a linear extension which can be embedded in ω.*

Proof. Trivial. ∎

Definition 1.2 *Let $(P, \leq)$ be a countable, locally finite, infinite poset with labelling $l : P \to X$, such that for every $x \in X$ the set $l^{-1}(x)$ is a linearly ordered subset of P. For $\alpha \in X^\omega$ we say that α extends $(P, \leq)$ if there is a bijective function $f : P \to \omega$ which is order preserving and such that $\alpha \circ f = l$.*

Note that by the requirements on $(P, \leq)$, such f is unique if it exists. For two such posets $(P, \leq, l)$ and $(Q, \leq', l')$ we say that $(P, \leq, l)$ extends $(Q, \leq', l')$ if every α which extends $(P, \leq, l)$, also extends $(Q, \leq', l')$. This is equivalent to: there is a bijective function $Q \to P$ which is order preserving, i.e. the order on Q is a subset of the order on P.

Definition 1.3 *Let $\mathcal{P}$ be a set of countably infinite, locally finite X-labelled partial orders. Such a set is called a* complete system *if the following conditions hold:*

i) *For all $(P, \leq, l)$ in $\mathcal{P}$ and all $x \in X$, $l^{-1}(x)$ is a linearly ordered subset of P;*

ii) *Every $\alpha \in X^\omega$ extends exactly one $(P, \leq, l)$ in $\mathcal{P}$;*

iii) *If $p <^d q$ in some $(P, \leq, l)$ in $\mathcal{P}$ then for all $(P', \leq', l')$ in $\mathcal{P}$ and $r, s \in P'$ with $l'(r) = l(p), l'(s) = l(q)$, either $r \leq' s$ or $s \leq' r$.*

Given two complete systems $\mathcal{P}$, $\mathcal{Q}$ on X, say $\mathcal{P} \preceq \mathcal{Q}$ if for every $(P, \leq, l)$ in $\mathcal{P}$ there is a unique $(Q, \leq', l')$ in $\mathcal{Q}$ such that $(P, \leq, l)$ extends $(Q, \leq', l')$. This defines a preorder on the class of complete systems on X.

Definition 1.4 *Given an independence relation I on X we define an equivalence relation $\sim_I$ on X^ω as the smallest which contains (α, β) whenever for some $i \in \omega$, $(\alpha(i), \alpha(i+1)) \in I$, $\beta(j) = \alpha(j)$ for $j \neq i, i+1$, and $\beta(i) = \alpha(i+1)$ and $\beta(i+1) = \alpha(i)$.*

Furthermore we put $\alpha \approx_I \beta$ if for all $n \in \omega$ there is both a $\beta' \sim_I \beta$ with $\forall i < n.\alpha(i) = \beta'(i)$, and an $\alpha' \sim_I \alpha$ with $\forall i < n.\beta(i) = \alpha'(i)$.

Theorem 1.5 *There is an order-isomorphism between isomorphism classes of complete systems on X and independence relations on X, ordered by inclusion.*

Proof. Any $\approx_I$-equivalence class C determines a countably infinite, locally finite poset $\bigcap C$ since there is a set P such that all $\alpha \in C$ can be regarded as linear orders on P: writing for $x \in X$ and $\alpha \in X^\omega$, x_α for the number of times x appears in α (may be ω), $x_\alpha = x_\beta$ whenever $\alpha \approx_I \beta$, and $P = \{(x, i) \mid x \in X \wedge i < x_\alpha\}$ for some $\alpha \in C$. So the order on P is the intersection of all the linear orders on P, determined by the elements of C.

We put $l(x, i) = x$, and it is clear that $(P, \leq, l)$ is a locally finite labelled poset such that $l^{-1}(x)$ is a linearly ordered subset. Put $\mathcal{P}_I = \{\bigcap C \mid C \in X^\omega / \approx_I\}$.

The operation $I \mapsto \mathcal{P}_I$ is obviously injective since if $(x, y) \in I$, $(x, y) \notin J$ there will be P in $\mathcal{P}_I$, $p, q \in P$ with labels x, y respectively, and p, q incomparable in P; but there will be no such in $\mathcal{P}_J$.

Conversely, given a complete system $\mathcal{P}$ on X, by requirement ii) of definition 1.3, we can define an equivalence relation $\approx_\mathcal{P}$ on X^ω by: $\alpha \approx_\mathcal{P} \beta$ if α and β extend the same element of $\mathcal{P}$. On the other hand we put:

$$I = \{(x, y) \mid \text{for some } (P, \leq, l) \in \mathcal{P}, \text{ there is } p, q \in P \text{ with}$$
$$l(p) = x, l(q) = y \text{ and } p, q \text{ unrelated w.r.t. } \leq\}$$

By i) of definition 1.3, I is an independence relation. To show that $\mathcal{P}$ is isomorphic to $\mathcal{P}_I$ it clearly suffices to show that $\approx_{\mathcal{P}}$ is the equivalence relation $\approx_I$ induced by I as in the first part of the proof.

One direction is clear: if $\alpha \approx_{\mathcal{P}} \beta$ then certainly $\alpha \approx_I \beta$. For the converse, assume $\alpha \approx_I \beta$. Let $(P, \leq, l)$ the element of $\mathcal{P}$ that α extends. I show that also β extends $(P, \leq, l)$.

Since $\alpha \approx_I \beta$ there is a bijection $f : P \to \omega$ such that $\beta \circ f = l$ and which is order-preserving when restricted to subsets of the form $l^{-1}(x)$.

If β does not extend $(P, \leq, l)$ there are $p < q$ in P with $f(q) < f(p)$. Since P is locally finite, we then also have for some $p <^d q$ in P that $f(q) < f(p)$. If $g : P \to \omega$ is the unique order preserving bijection with $\alpha \circ g = l$, then since α extends $(P, \leq, l)$, we must have $g(p) < g(q)$. But from iii) in definition 1.3 it follows that $(l(p), l(q))$ cannot be in I. Therefore, contrary to our assumption, it cannot be that $\alpha \approx_I \beta$.

So the operation $I \mapsto \mathcal{P}_I$ is also, up to isomorphism, surjective; and that the 1-1 correspondence is monotone both ways, is obvious. ∎

Corollary 1.6 *Up to isomorphism, a complete system on X, corresponding to the independence relation I, can always be taken as the set of all locally finite partial orders $(P, \leq)$ such that:*

1. *$P \subseteq X \times \omega$ such that $(x, j) \in P$ and $i \leq j$ implies $(x, i) \in P$ and $(x, i) \leq (x, j)$ in P;*

2. *if (x, i) and (y, j) are incomparable in $(P, \leq)$ then $(x, y) \in I$;*

3. *if $(x, i) <^d (y, j)$ in P then $(x, y) \notin I$.*

The labelling function is always the first projection.

In the sequel we shall always assume this representation, which despenses with the need to mention the labelling.

The elements of $\mathcal{P}_I$ will be called the (infinite) *traces* w.r.t. (X, I). An element p of $(P, \leq, l) \in \mathcal{P}_I$ (up to a suitable equivalence) is called an *event*. Note that, despite our representation of P as subset of $X \times \omega$, it is not sufficient to denote an event by (x, i); to specify an event, it is also necessary to give the partial order it is considered to be an element of. We return to this matter in section 4.

2 Topology on Infinite Traces

Convention In this chapter, the set X is assumed to be *countable*.

Theorem 1.5 displays every $\mathcal{P}_I$ as a quotient of X^ω by the equivalence relation $\approx_I$. X^ω has a natural topology, with basic opens determined by finite initial segments $s \in X^*$:

$$\mathcal{U}_s = \{\alpha \in X^\omega \mid \alpha = s \star \beta \text{ for some } \beta\}$$

where $\star$ denotes concatenation of sequences.

It seems therefore straightforward to give $\mathcal{P}_I$ the *quotient topology*, that is the largest topology making the quotient map $\pi_I : X^\omega \to \mathcal{P}_I$ continuous, and is defined by: $\mathcal{U}$ is open in $\mathcal{P}_I$ if and only if $\pi_I^{-1}(\mathcal{U})$ is open in X^ω.

Proposition 2.1 *i)* *With the topologies given,* $\pi_I : X^\omega \to \mathcal{P}_I$ *is an open map;*

ii) *the space* $\mathcal{P}_I$ *is not* T_1, *unless* $I = \emptyset$;

iii) *a set* $\mathcal{U} \subseteq \mathcal{P}_I$ *is open if and only if for each* $P \in \mathcal{U}$ *there is a finite initial segment* ($\downarrow$-*closed subset*) P' *of* P *such that for all* Q *which contain* P' *as initial segment,* $Q \in \mathcal{U}$.

Proof. As in the proof of theorem 1.5, we use the equivalence relations $\sim_I$ and $\approx_I$ of definition 1.4.

i) This means that for every open $\mathcal{U}$ in X^ω, $\pi_I[\mathcal{U}]$ is an open subset of $\mathcal{P}_I$. This is equivalent to: if $\mathcal{U}$ is open in X^ω then $\{\alpha \mid \exists \beta \in \mathcal{U}.\alpha \approx_I \beta\}$ is open in X^ω. It suffices to check this for basic opens $\mathcal{U}_s$. But if $\alpha \approx_I \beta$ and $\beta \in \mathcal{U}_s$, there is $\beta' \in \mathcal{U}_s$ such that $\alpha \sim_I \beta'$; since this $\sim_I$-equivalence only involves an initial segment of α, there is an open neighborhood V of α all of whose members are $\sim_I$-equivalent, hence $\approx_I$-equivalent, to some $\beta'' \in \mathcal{U}_s$.

ii) The T_1 property means: for any two distinct points in the space, there is for either point an open set containing that point, but not the other point. This is equivalent to: for any point x, the set $\{x\}$ is closed.

If $I \neq \emptyset$, say $(x, y) \in I$, then the strings x^ω and yx^ω are not $\approx_I$-equivalent yet every basic open neighborhood of x^ω contains an element $\approx_I$-equivalent to yx^ω; so every open neighborhood of $\pi_I(x^\omega)$ contains $\pi_I(yx^\omega)$.

iii) Every finite initial segment of α gives a finite $\downarrow$-closed subset of $\pi_I(\alpha)$, and conversely for every finite $\downarrow$-closed subset Q of $\pi_I(\alpha)$ there is $\alpha' \approx_I \alpha$ which starts by enumerating Q. ∎

Since $\mathcal{P}_I$ in general is not T_1 it makes sense to look at its *specialization ordering*: $x \leq y$ if for every open set $\mathcal{U}$, $x \in \mathcal{U}$ implies $y \in \mathcal{U}$ (Note that the T_1 property is equivalent to the property that this ordering is discrete). To be precise, as defined it is only a preorder, but in the case of $\mathcal{P}_I$ it is a partial order ($\mathcal{P}_I$ is T_0). Given the topology on $\mathcal{P}_I$, $x \leq y$ in the specialization order means that x embeds as initial segment ($\downarrow$-closed subset) in y.

It is worth remarking here that both the topology on $\mathcal{P}_I$ and its specialization order are known in the literature on traces. [Kwi] and others (e.g. [KPP], [D]) consider the set $\mathcal{O}_I$ of both finite and infinite traces. This set is naturally ordered, and has the structure of a Scott domain.

The topology on $\mathcal{P}_I$ just described, is the subspace topology w.r.t. the inclusion of $\mathcal{P}_I$ in $\mathcal{O}_I$, hence the specialization ordering on $\mathcal{P}_I$ is the restriction to $\mathcal{P}_I$ of the order on $\mathcal{O}_I$.

However, the topology on $\mathcal{P}_I$ is *not* in general the Scott topology for its specialization order. This is easily seen at the case $I = \emptyset$: then $\mathcal{P}_I = X^\omega$ and the specialization order is discrete, so every $\{x\}$ is Scott-open; but X^ω is not a discrete space.

Moreover, the order on $\mathcal{P}_I$ turns out to be a non-algebraic, directed complete partial order.

First, let us characterize compatible subsets of $\mathcal{P}_I$ and establish that $\mathcal{P}_I$ has joins of compatible subsets as well as directed joins.

Lemma 2.2 *A set $\{(P_k, \leq_k)| \, k \in K\}$ of elements of $\mathcal{P}_I$ is compatible (i.e., has an upper bound) if and only if for each $k, l \in K$:*

- $P_k \cap P_l$ *is an initial segment of both P_k and P_l, and the orders $\leq_k$ and $\leq_l$ coincide on $P_k \cap P_l$;*

- *for each $(x, i) \in P_k \setminus P_l$ and $(y, j) \in P_l \setminus P_k$, $(x, y) \in I$.*

If these conditions hold, the set $\{(P_k, \leq_k)| \, k \in K\}$ has a join $\bigvee_{k \in K}(P_k, \leq_k)$ in $\mathcal{P}_I$, given by the order

$$\left(\bigcup_{k \in K} P_k, \bigcup_{k \in K} \leq_k\right)$$

Proof. The necessity of the conditions is easy to verify. For example, if P_k and P_l are both initial segments of another element, and $(x, i) \in P_k \setminus P_l$, $(y, j) \in P_l \setminus P_k$, these elements must be incomparable in the larger poset, hence $(x, y) \in I$.

Conversely, if the conditions are met it is trivial to check that the given union is indeed a poset with the required properties, and the join of the P_k. ∎

From the lemma, it is straightforward to deduce that two elements x and y of $\mathcal{P}_I$ are incompatible if and only if there are disjoint open sets U and V with $x \in U$ and $y \in V$.

Now we will show that although, through lack of finite elements, $\mathcal{P}_I$ is not a Scott domain (nor is it continuous as a poset), it arises out of a subspace by a topological construction which is a direct generalization of the *ideal completion* construction in the theory of Scott domains. The generalization I mean is *soberification*. We need some definitions.(A good reference to these matters is [Joh])

In a topological space, we say that a closed set F is *irreducible* if whenever $F \subseteq G_1 \cup G_2$ for closed G_1, G_2, then $F \subseteq G_1$ or $F \subseteq G_2$.

A space X is called *sober* if every nonempty irreducible closed set is the closure of $\{x\}$ for a unique $x \in X$, denoted $\overline{\{x\}}$.

The category of sober spaces is reflective in the category of all topological spaces. The reflector is called *soberification*. The soberification $Sob(X)$ of a space X is the set of all nonempty irreducible closed subsets of X, topologized by: for each open $a \subseteq X$, there is an open

$$\mathcal{U}_a = \{F \in Sob(X)| \, F \cap a \neq \emptyset\}$$

of $Sob(X)$.

Every continuous poset with the Scott topology is sober, because a nonempty closed (that is, $\downarrow$-closed and closed under directed joins) subset F is irreducible if and only if $\{x \mid \exists f \in F.x \ll f\}$ is directed ($\ll$ is the *way below*-relation). Since directed joins exist and F is closed under them, it follows that $F = \downarrow\{\bigvee F\}$, i.e. F is the closure of $\{\bigvee F\}$.

In the case of a Scott domain D, consider D_{fin}, the set of finite elements of D, as a subspace of D with the Scott topology. A nonempty closed subset F of D_{fin} is irreducible if and only if it is an ideal, and therefore $D = Sob(D_{\text{fin}})$, which is just the ideal completion construction.

To clarify the point of showing that a space X is of the form $Sob(X')$ for a subspace X' of X, compare with Scott domains. Given Scott domains D and E, a continuous map $D \to E$ is completely determined by its action on the finite elements of D, and this property is very useful in semantics. Likewise, if $X = Sob(X')$ then for any sober space Y, any continuous map $X' \to Y$ has a unique extension to a continuous map $X \to Y$.

With respect to the Scott topology, the finite elements x have the property that $\uparrow\{x\}$ is open; in particular, $\{x\}$ is open in $\overline{\{x\}}$. In fact one can show (see [Joh], exercise II.1.7) that if a sober space X is the soberification of a subspace X', then X' has to contain at least the subset

$$X_D = \{x \in X \mid \{x\} \text{ is open in } \overline{\{x\}}\}$$

In our case, let us start by identifying $(\mathcal{P}_I)_D$.

Recall that, in a partial order, an *antichain* is a set of pairwise incomparable elements. A *chain* is (for us) a set of the form $\{c_n \mid n \in \omega\}$ with $c_0 < c_1 < c_2 < \dots$

Lemma 2.3 *Any infinite, locally finite partial order P contains a chain or an infinite antichain.*

Proof. Let B be the tree of sequences $\langle s_0, \dots, s_n \rangle$ of elements of P with s_0 a minimal element and $s_0 <^d s_1 <^d \dots <^d s_n$, as well as the empty sequence. Just mapping the empty sequence to any point of P, and $\langle s_0, \dots, s_n \rangle$ to s_n defines, by locally finiteness of P, a surjection of B onto P.

If P does not contain an infinite antichain, the tree B is finitely branching, and since P hence B is infinite, by König's Lemma B must have an infinite branch, which gives a chain in P. ∎

Proposition 2.4 *For $x \in \mathcal{P}_I$, the following are equivalent:*

1. $\{x\}$ *is open in* $\overline{\{x\}}$;

2. *for every chain C in x, $x \backslash \downarrow C$ is finite, and x contains no infinite antichain;*

3. *there is no directed set $Y \subseteq \{x' \mid x' < x\}$ such that $\bigvee Y = x$*

Proof. We prove 1) $\Rightarrow$ 3) $\Rightarrow$ 2) $\Rightarrow$ 1).

1) $\Rightarrow$ 3): it is obvious, by the characterization of the topology on $\mathcal{P}_I$, that this topology is contained in the Scott topology for its specialization order. So opens are inaccessible for directed joins, and 3) is immediate from 1).

3) $\Rightarrow$ 2): if x contains a chain $C = c_0 < c_1 < c_2 < \ldots$ with $x \setminus \downarrow C$ infinite, then $\{\downarrow (x \setminus \downarrow C) \cup \downarrow c_n \,|\, n \in \omega\}$ is a chain of elements of $\mathcal{P}_I$ below x with join x; and similarly, if x contains an infinite antichain it is easy to write x as a nontrivial join of a chain of proper infinite initial segments.

2) $\Rightarrow$ 1): by 2) and lemma 2.3, x contains an infinite chain C, and $x \setminus \downarrow C$ is finite. This holds for any chain, so for every other chain D, $\downarrow C \subseteq \downarrow D$. So if P is the finite initial segment $\downarrow (x \setminus \downarrow C)$, then every $y \leq x$ which contains P will be x, since also y will contain a chain. In other words, $\{x\} = \downarrow x \cap \mathcal{U}_P$, and $\{x\}$ is open in $\downarrow x = \overline{\{x\}}$. $\blacksquare$

In our analogue of the construction of a Scott domain as soberification, we need more than just the subspace $(\mathcal{P}_I)_D$. At least when X is infinite and contains infinite I-cliques, there will be points x with the property that $\overline{\{x\}} \cap (\mathcal{P}_I)_D = \emptyset$. These are, as is readily seen, those x that satisfy: for every chain C in x, $\downarrow C$ contains an infinite antichain.

Let us first see that $\mathcal{P}_I$ is sober.

Theorem 2.5 $\mathcal{P}_I$ *is a sober space.*

Proof. Let $F \subset \mathcal{P}_I$ be irreducible, closed. Then if $x, y \in F$, x and y must be compatible, otherwise there exist disjoint opens U, V with $x \in U$, $y \in V$; and $F = (F \setminus U) \cup (F \setminus V)$.

Now the join $x \vee y$ of x and y must also be in F for if not, there is a finite initial segment P of $x \vee y$ such that $\mathcal{U}_P \cap F = \emptyset$; letting $Q = P \cap x$ and $R = P \cap y$, we find that $\mathcal{U}_Q \cap \mathcal{U}_R = \mathcal{U}_P$, so $F = (F \setminus \mathcal{U}_Q) \cup (F \setminus \mathcal{U}_R)$; contradiction.

So if F is nonempty, F is directed and since F is closed under directed joins, F is the closure of $\{\bigvee F\}$. $\blacksquare$

Here it is good to insert a remark on the assumption for this chapter that X is countable. Theorem 2.5 fails in general for uncountable X, for if I is the maximal independence relation on X, then $\mathcal{P}_I$ itself is irreducible, but not the closure of a point (points are countable orders).

Theorem 2.6 *For every* $x \in \mathcal{P}_I$, $\uparrow x$ *is a Scott domain.*

Proof. The finite elements in $\uparrow x$ are the $x \vee P$, P a finite X-labelled poset compatible with x. Every $y \geq x$ is a directed join of these. $\blacksquare$

Now we prove our soberification result, stated in abstract topological form. The application to $\mathcal{P}_I$ is obtained by letting the subspace X' be

$$(\mathcal{P}_I)_D \cup \{x \in \mathcal{P}_I \,|\, \overline{\{x\}} \cap (\mathcal{P}_I)_D = \emptyset\}$$

It is easy to see, using theorems 2.5 and 2.6, that the conditions of the next theorem are satisfied in this case.

Theorem 2.7 *Let* X *be a sober space and* $X' \subset X$ *a subspace such that for all* $x \in X$ *the following two conditions hold w.r.t. the specialization ordering:*

1. x is a directed join of elements of X';

2. *For every* $y, y' \in \downarrow x \cap X'$ *we have: either there is a directed set* $X'' \subset \uparrow y \cap X'$ *such that* $y' \leq \bigvee X''$, *or there is a directed* $X'' \subset \uparrow y' \cap X'$ *with* $y \leq \bigvee X''$.

Then X *is the soberification of* X'.

Proof. Since X is sober, X' is a T_0-space and we know that $Sob(X')$ will be a subspace of $Sob(X) \cong X$. It suffices therefore to establish a bijection between the nonempty, irreducible closed subsets of X and those of X'.

We define $Sob(X) \to Sob(X')$ by $F \mapsto F \cap X'$, and $Sob(X') \to Sob(X)$ by $A \mapsto \overline{A}$ (closure in X). We check that these maps are well-defined and each other's inverses.

If $F \in Sob(X)$ then by sobriety of X, $F = \overline{\{x\}}$ for some $x \in X$. $F \cap X'$ is certainly nonempty since $\downarrow x \cap X'$ contains a directed subset with join x; suppose it is reducible in X', then $F \cap X' = (F_1 \cap X') \cup (F_2 \cap X')$ for F_1, F_2 closed in X. Let $y_1 \in (F_1 \cap X') \setminus F_2$, $y_2 \in (F_2 \cap X') \setminus F_1$. By symmetry we can assume there is directed $X'' \subseteq \uparrow y_1 \cap X'$ with $y_2 \leq \bigvee X''$. Since the F_i are downward closed and X'' is directed, either $X'' \subseteq F_1$ or $X'' \subseteq F_2$; a contradiction in both cases. So $F \cap X'$ is irreducible in X'.

Conversely, if $A \in Sob(X')$ then $\overline{A}$ is always irreducible closed in X. And always, for A closed in X', $A = \overline{A} \cap X'$.

Furthermore, if F closed in X then $\overline{F \cap X'} \subseteq F$, and if moreover F is irreducible nonempty, so $F = \overline{\{x\}} = \downarrow x$, then since $x = \bigvee(\downarrow x \cap X')$, $x \in \overline{\downarrow x \cap X'}$ so $F \subseteq \overline{F \cap X'}$. ∎

In particular, if the set of labels X is finite (this is often imposed on independence structures, see [BT]), no $x \in \mathcal{P}_I$ can contain an infinite antichain because that implies the existence of an infinite I-clique. So then $\mathcal{P}_I = Sob((\mathcal{P}_I)_D)$. We formulate this as a corollary:

Corollary 2.8 *Suppose X does not contain infinite I-cliques. Let Y be a sober space. Then every continuous map: $(\mathcal{P}_I)_D \to Y$ has a unique extension to a continuous map $\mathcal{P}_I \to Y$.*

In trace semantics ([BT]) and other considerations on traces, e.g. fairness (see [Kwi]), one often considers the *maximal* elements of $\mathcal{P}_I$. Again, maximality of $x \in \mathcal{P}_I$ can be expressed entirely in terms of the partial order x. The condition given here is similar to the one given in [Kwi].

First, we say that the *level* of $\xi \in x$ is the maximal length of a sequence $\xi_1 <^d \xi_2 <^d \ldots <^d \xi_n = \xi$. So the level of a minimal element is 1, and if the level of ξ is $n > 1$, there is η of level $n - 1$ with $\eta <^d \xi$. By locally finiteness, levels are well-defined; we write x_i for the set of elements of x of level i.

Again let $\pi : x \to X$ the labelling function.

Proposition 2.9 x *is maximal in* $\mathcal{P}_I$ *if and only if the following conditions hold:*

i) *for every $i \geq 1$ and for every $a \in X$ such that $(a, b) \in I$ for all $b \in \pi[x_i]$, there is $j \geq i$ and $\xi \in x_j$ such that $(\pi(\xi), a) \notin I$;*

ii) *x contains no maximal element.*

Proof. If condition i) doesn't hold there is a minimal i with the property that for some $a \in X$, $\forall j \geq i \forall \xi \in x_j . (\pi(\xi), a) \in I$. Taking such a; now either $i = 1$ and we can simply add a minimal element of x with label a, unrelated to any other element of x, or $i > 1$ and we add likewise an element ξ with label a and put $\xi > \zeta$ for all ζ of level $i - 1$ for which $(\pi(\zeta), a) \notin I$ By minimality of i, such ζ exist.

If condition ii) doesn't hold, say $\xi = (a, i)$ is a maximal element, we can put an element $(a, i + 1)$ on top of ξ. In either case therefore, x was not maximal.

Conversely, if the conditions hold and $x < y$ in $\mathcal{P}_I$ then x is embedded as initial segment of y, so $x_i = y_i \cap x$ for $i \geq 1$. If $\xi \in y \setminus x$, $\xi \in y_i$ then $(\pi(\xi), \pi(\eta)) \in I$ for all $\eta \in x_i$ (and note that $x_i \neq \emptyset$ since x has no maximal elements) so there is $j > i$ and $\zeta \in x_j$ with $(\pi(\zeta), \pi(\xi)) \notin I$. But then $\xi < \zeta$ and x was not an initial segment of y; contradiction. ∎

3 Independence Relations and Group Actions on X^ω

In this section we see that the equivalence relation $\sim_I$ on X^ω, defined in the proof of theorem 1.5, is induced by the action of a group on that set. We study certain graphs which can be associated with this action, and connect these to traces.

The computational motivation for this is as follows. Traces – equivalence classes of elements of X^ω – are seen as processes, while the elements of X^ω themselves are *instances* of (or *runs* of) these processes. If two runs are equivalent i.e. runs of the same process, they are linked by a rewriting procedure. The group action codifies the finite parts of this rewriting procedure.

Definition 3.1 *The group T of twists is generated by the natural numbers, and the following relations:*

- $i^2 = \mathrm{id}$ *for* $i \in \mathbb{N}$;

- $i \cdot j = j \cdot i$ *whenever* $|i - j| > 1$;

- $(i \cdot (i + 1))^6 = \mathrm{id}$.

The name "twists" is mine; I am not aware of any appearance of the group T in the literature.

Proposition 3.2 *Every independence relation I on X induces a group action $\mu_I : T \times X^\omega \to X^\omega$ of T on X^ω, such that the orbits of the action are exactly the $\sim_I$-equivalence classes in the proof of 1.5.*

Proof. We define μ_I (writing $i \cdot \alpha$ for $\mu_I(i, \alpha)$ on generators by:

$$i \cdot \alpha = \begin{cases} (\alpha(0), \ldots, \alpha(i + 1), \alpha(i), \alpha(i + 2) \ldots) & \text{if } (\alpha(i), \alpha(i + 1)) \in I \\ \alpha & \text{otherwise} \end{cases}$$

By symmetry of I, $i \cdot (i \cdot \alpha) = \alpha$, so the first relation in the definition of T is respected, and similarly, for $|i-j| > 1$, one easily sees that $i \cdot (j \cdot \alpha) = j \cdot (i \cdot \alpha)$. As to the third relation, we distinguish cases. If all of $\{\alpha(i), \alpha(i+1), \alpha(i+2)\}$ are pairwise dependent or all are pairwise independent, then clearly $(i \cdot (i+1))^3 \cdot \alpha = \alpha$, and this also holds if two of the possible three independences hold. If only one of them is true, say $(\alpha(i), \alpha(i+1)) \in I$, then $(i \cdot (i+1))^2 \cdot \alpha = \alpha$. So the third relation is respected in all cases.

It is clear that the orbits of X^ω under the action μ_I are exactly the $\sim_I$-equivalence classes. ∎

Now we picture the action μ_I as an irreflexive, undirected graph, with as vertices the elements of X^ω, and edges labelled by the generators of T for nontrivial action: there is an edge $\alpha \overset{i}{\underline{\hspace{1cm}}} \beta$ if α and β are different and $\beta = i \cdot \alpha$ (which entails $\alpha = i \cdot \beta$).

Let $\mathcal{G}(X)$ stand for the action graph corresponding to the *largest* independence relation, viz. $X \times X \setminus \Delta$. Here there is an edge $\alpha \overset{i}{\underline{\hspace{1cm}}} i \cdot \alpha$ if and only if $\alpha(i) \neq \alpha(i+1)$. It is clear that two vertices α, β are connected by a path in this graph if and only if there are a γ and finite strings s, t of the same length such that $\alpha = s \star \gamma$, $\beta = t \star \gamma$ and s and t are permutations of each other.

We consider certain subgraphs of $\mathcal{G}(X)$.

Definition 3.3 *A subgraph G of $\mathcal{G}(X)$ is called* good *if it is connected, and for all configurations in $\mathcal{G}(X)$ of the form:*

if α and β are in G and are vertices of such a configuration, then any minimal path in the configuration joining α and β, belongs to G.

Example. The finite subgraph

$$\alpha \overset{i}{\underline{\hspace{1cm}}} i \cdot \alpha$$

is good, as is

$$\alpha \overset{i}{\underline{\hspace{1cm}}} i \cdot \alpha \overset{i+1}{\underline{\hspace{1cm}}} (i+1) \cdot i \cdot \alpha$$

but

$$\alpha \overset{i}{\underline{\hspace{1cm}}} i \cdot \alpha \overset{i+1}{\underline{\hspace{1cm}}} (i+1) \cdot i \cdot \alpha \overset{i}{\underline{\hspace{1cm}}} i \cdot (i+1) \cdot i \cdot \alpha$$

is not, because it should have contained also the other path in the hexagon.

Note, that any connected subgraph G of $\mathcal{G}(X)$ determines a subset $P \subset X \times \omega$ and (since all vertices of G are linear orders on P) a partial order on P which is locally finite and extends the order on $X \times \omega$, restricted to P, just as in theorem 1.5; we call this partial order $\bigcap G$.

Theorem 3.4 G *is good if and only if G is a connected component of the full subgraph of $\mathcal{G}(X)$ on $\{\alpha \,|\, \alpha$ extends $\bigcap G\}$.*

Proof. For any poset $(P, \leq)$ as in Corollary 1.6, any connected component of $\{\alpha \,|\, \alpha$ extends $(P, \leq)\}$ is good; this is easy. Conversely suppose G is good, $\alpha \in G$, β extends $\bigcap G$ and there is a path from α to β in $\mathcal{G}(X)$. I show that this path lies in G. By induction on the length of the path, it suffices to show this for paths of length 1.

So suppose $\alpha \xrightarrow{\quad i \quad} \beta = i \cdot \alpha$ in $\mathcal{G}(X)$. By the fact that α extends $(P, \leq) = \bigcap G$, let $\alpha(i)$ and $\alpha(i+1)$ correspond to (x, n) and (y, m) respectively. Since β also extends $(P, \leq)$ we have that (x, n) and (y, m) are incomparable in $(P, \leq)$, so there is $\gamma \in G$ which enumerates (y, m) before (x, n); and since G is connected there is a path σ in G joining α and γ. At some point along σ, the crucial swap takes place. The basic possibilities are:

$$(\vec{x}a\alpha(i)\alpha(i+1)\ldots) \xrightarrow{\quad N \quad} (\vec{x}\alpha(i)a\alpha(i+1)\ldots) \xrightarrow{\quad N+1 \quad} (\vec{x}\alpha(i)\alpha(i+1)a\ldots)$$

$$\Big\downarrow N$$

$$(\vec{x}\alpha(i+1)\alpha(i)a\ldots)$$

in which case, by goodness of G, we also had

$$(\vec{x}a\alpha(i)\alpha(i+1)\ldots) \xrightarrow{\quad N+1 \quad} (\vec{x}a\alpha(i+1)\alpha(i)\ldots)$$

in G; or,

$$(\vec{x}\alpha(i)\alpha(i+1)\ldots) \xrightarrow{\quad M \quad} (\vec{x'}\alpha(i)\alpha(i+1)\ldots) \xrightarrow{\quad N \quad} (\vec{x'}\alpha(i+1)\alpha(i))$$

with $|M - N| > 1$, where, similarly, we also had

$$(\vec{x}\alpha(i)\alpha(i+1)\ldots) \xrightarrow{\quad N \quad} (\vec{x}\alpha(i+1)\alpha(i)\ldots)$$

in G.

So unless σ was already the path $\alpha \xrightarrow{\quad i \quad} i \cdot \alpha$, we could move the crucial swap forward. ∎

Furthermore it follows from the proof of theorem 3.4 that for two good graphs G and G', if $G' \subset G$ then $\bigcap G'$ extends $\bigcap G$. So an inclusion of good subgraphs of $\mathcal{G}(X)$ corresponds to an order-*reflecting* map between partially ordered sets.

This will guide our reflections in the next section, where we consider the dynamics of independence structures and the connection with Winskel's Event Structures.

4 Independence Structures, Event Structures and an Enveloping Category

In [WN94], a mathematical analysis of various models of concurrency is carried out in terms of specific adjunctions between categories: reflections and coreflections. In this section, the results are in the same line: a category LSP is defined into which both independence structures (being sets with an independence relation, and independence-preserving maps) and event structures embed as full reflective subcategories.

The category in question is a category of *labelled systems of posets* (LSP's).

Definition 4.1 *A labelled system of posets (LSP) is a triple* $(X, \mathcal{P}, l)$ *where* X *is a set,* $l : X \to L$ *is a labelling function and* $\mathcal{P}$ *a collection of poset structures on subsets of* X, *satisfying:*

- *if* $(P, \leq) \in \mathcal{P}$ *then for all* $x \in L$, $l^{-1}(x) \cap P$ *is a linearly ordered subset of* P;

- *every* $(P, \leq) \in \mathcal{P}$ *is locally finite;*

- *if* $(P, \leq) \in \mathcal{P}$ *and* $Q \subset P$ *is a downwards closed subset, then* $(Q, \leq) \in \mathcal{P}$.

The set $\mathcal{P}$ is naturally ordered by: $P \preceq Q$ if P is an initial segment of Q (as partial orders; I write P instead of $(P, \leq)$ whereever possible).

Definition 4.2 *Let* $(X, \mathcal{P}, l)$ *and* $(Y, \mathcal{Q}, m)$ *be LSP's with* $l : X \to L$ *and* $m : Y \to M$. *A map from* $(X, \mathcal{P}, l)$ *to* $(Y, \mathcal{Q}, m)$ *is a triple* $(F, (f_P)_{P \in \mathcal{P}}, \kappa)$ *where* $\kappa : L \to M$ *is a function and* $F : \mathcal{P} \to \mathcal{Q}$ *a* $\preceq$-*preserving function, and* $(f_P : P \to F(P) | P \in \mathcal{P})$ *a system of order-reflecting bijective functions such that if* $p \in P \preceq Q$, *then* $f_P(p) = f_Q(p) \in F(Q)$, *and moreover* $m(f_P(p)) = \kappa(l(p))$.

This defines a category **LSP** of labelled systems of posets.

Definition 4.3 *Let* $(X, \mathcal{P}, l)$ *be an LSP. An event of* $(X, \mathcal{P}, l)$ *is an element of the colimit of* $\mathcal{P}$, *viewed as a system of posets and embeddings as initial segments. Concretely, an event is an equivalence class of pairs* (p, P) *with* $p \in P \in \mathcal{P}$, *under the equivalence relation which is generated by:* $(p, P) \sim (q, Q)$ *if there is* R *into which both* P *and* Q *embed as initial segments, and* $p = q$.

Given two events e_1, e_2 we say that e_1, e_2 are *consistent* if they have representatives (p, P) and (q, Q) such that P and Q have an upper bound in $\mathcal{P}$ w.r.t. $\preceq$. We write $e_1 \# e_2$ (and say e_1 and e_2 are *in conflict*) if they are inconsistent. We write $e_1 \leq e_2$ if there is $p \leq q$ in P, such that (p, P) and (q, P) are representatives of respectively e_1, e_2. Since the labelling map $l : X \to L$ obviously also gives a map on the set of events, we have a structure $(E, \leq, \#, l)$ where E is the set of events of $(X, \mathcal{P}, l)$. We call this structure $Ev(X, \mathcal{P}, l)$.

Proposition 4.4 $Ev(X, \mathcal{P}, l)$ *is a labelled event structure.*

Let us recall that a labelled event structure is a tuple $(E, \leq, \#, l)$ where $(E, \leq)$ is a poset, $\#$ is a binary conflict relation which is symmetric and irreflexive and satisfies $e \# e' \leq e'' \Rightarrow e \# e''$; and $l : E \to L$ is a labelling map. We require that if $l(e) = l(e')$ and $\neg(e \# e')$, then $e \leq e'$ or $e' \leq e$ holds.

A *map* of event structures $(E, \leq, \#, l : E \to L) \to (E', \leq', \#', l' : E' \to L')$ is a pair $(\kappa : L \to L', \eta : E \to E')$ such that $\kappa \circ l = l' \circ \eta$ and:

- $e' \leq' \eta(d) \Rightarrow \exists d' \leq d.e' = \eta(d')$

- if e_1, e_2 are consistent (i.e. not in the conflict relation) and not equal, then so are $\eta(e_1)$ and $\eta(e_2)$.

We have a category **LES** of labelled event structures. The only difference between the definition here and the one by Winskel and Nielsen is that they take the maps η partial. This is not a fundamental difference: see remark at the end of this section.

Proposition 4.5 LES *is a reflective subcategory of* **LSP**.

Proof. Given a labelled event structure $(E, \leq, \#, l)$, its *configurations* are the downwards closed subsets of E that are conflict-free. By $\mathcal{D}(E)$ we denote the set of configurations of $(E, \leq, \#)$. $\mathcal{D}(E)$ is ordered by inclusion which is, by definition of configuration, inclusion as initial segment.

Every $P \in \mathcal{D}(E)$ is a locally finite poset and $l^{-1}(x) \cap P$ is linearly ordered in P by the requirement we put on labelled event structures, so $(E, \mathcal{D}(E), l)$ is an LSP.

Any map $\eta : E \to E'$ of event structures (we suppress reference to labels wherever irrelevant) gives, for every $P \in \mathcal{D}(E)$, an order-reflecting bijection from P to $\eta[P]$, so we have in fact a functor $\psi : \textbf{LES} \to \textbf{LSP}$.

Conversely, there is a functor $\phi : \textbf{LSP} \to \textbf{LES}$. On objects, $\phi(X, \mathcal{P}, l)$ is the event structure $Ev(X, \mathcal{P}, l)$ described before. From the definition of events of an LSP, it is clear that any map of LSP's induces a map on events, which is readily seen to be a map of labelled event structures.

There is a natural 1-1 correspondence between maps $(X, \mathcal{P}, l) \to \psi(E, \leq, \#, m)$ of LSP's and maps $\phi(X, \mathcal{P}, l) \to (E, \leq, \#, m)$ of labelled event structures: every function $F : \mathcal{P} \to \mathcal{D}(E)$ satisfying the requirements for a map of LSP's, induces a unique function $Ev(X, \mathcal{P}, l) \to E$ which is a map of event structures, and vice versa; so ϕ is left adjoint to ψ.

Furthermore, there is a bijection $Ev(\psi(E, \leq, \#, l)) \to E$, given by the counit of the adjunction. This is, by well-known category theory, to say that ψ is full and faithful (a fact which can also easily be seen directly). ∎

Given the adjunction, it is easy to characterize the event structures among the LSP's:

Proposition 4.6 *An LSP* $(X, \mathcal{P}, l)$ *is a labelled event structure if and only if for every conflict-free subset A of $Ev(X, \mathcal{P}, l)$ there is a $P \in \mathcal{P}$ such that every $e \in A$ has a representative (p, P) in P.*

We now turn to independence structures. The category **IS** has as objects pairs (L, I) where L is a set and I an independence relation on L; maps $(L, I) \to (L', I')$ are functions $f : L \to L'$ such that $x I x'$ implies $f(x) I' f(x')$ for $x, x' \in L$. In other words, **IS** is a category of undirected, irreflexive graphs.

Proposition 4.7 IS *is a full reflective subcategory of* **LSP**.

Proof. The functor $\theta : \mathbf{IS} \to \mathbf{LSP}$ is defined on objects by $\theta(L, I) = (L \times \omega, \mathcal{A}_I, \pi : L \times \omega \to L)$ where $\mathcal{A}_I$ is the set of poset structures on parts of $L \times \omega$ which are initial segments of elements of $\mathcal{P}_I$, $\mathcal{P}_I$ is as in theorem 1.5, the posets regarded as subsets of $L \times \omega$. Every map $f : (L, I) \to (L', I')$ of independence structures gives a map $f^\omega : L^\omega \to L'^\omega$ such that whenever α and β are $\approx_I$-equivalent, $f^\omega(\alpha)$ and $f^\omega(\beta)$ are $\approx_{I'}$-equivalent; therefore there is a map $F : \mathcal{P}_I \to \mathcal{P}_{I'}$, continuous and therefore $\preceq$-preserving, which gives a map $\mathcal{A}_I \to \mathcal{A}_{I'}$.

$f^\omega : L^\omega \to L'^\omega$ gives also a homomorphism between the action graphs of the respective actions of T on L^ω and L'^ω, and this homomorphism is injective on connected components. This means that for any α, f gives an order-reflecting bijection from the poset represented by α to the one represented by $f^\omega(\alpha)$. I.e. we have a map of LSP's; and therefore a functor $\mathbf{IS} \xrightarrow{\theta} \mathbf{LSP}$.

In the other direction, we define $\zeta : \mathbf{LSP} \to \mathbf{IS}$ by $\zeta(X, \mathcal{P}, l : X \to L) = (\mathrm{im}(l), I)$ where the relation I is defined by $x I y$ if for some $P \in \mathcal{P}$ and $p, q \in P$: $l(p) = x$, $l(q) = y$ and p and q are incomparable in P. This is a symmetric and irreflexive relation (because $l^{-1}(x) \cap P$ is linearly ordered).

Given a map $(F, (f_P)_P, \kappa) : (X, \mathcal{P}, l) \to (Y, \mathcal{Q}, m)$, its ζ-image is the factorization of $\kappa \circ l$ through $\mathrm{im}(l)$.

Since the projection $L \times \omega \to L$ is surjective, clearly $\zeta\theta(L, I) = (L, I)$. The rest of the proof is left to the reader. Again, θ is full and faithful because the counit is an isomorphism. ∎

Now it follows directly from lemma 2.2 and its corollary that $\mathcal{P}_I$ has directed joins, that the image of every independence structure under the embedding in **LSP**, satisfies the condition in proposition 4.6; therefore:

Proposition 4.8 *The embedding* **IS** $\to$ **LSP** *factors through the embedding* **LES** $\to$ **LSP***; therefore,* **IS** *is a full reflective subcategory of* **LES**.

Proof. Only the last statement needs verification; however it is a general fact for two full reflective subcategories that if one is contained in the other, it is reflective in the other. ∎

We can also set up an adjunction between the category **LSP** and the category of Mazurkiewicz trace languages, as defined in [WN94], and establish that the latter is full reflective in the former. Again, the embedding factors through **LES**, and we obtain essentially the result of [WN94] that Mazurkiewicz trace languages are a full reflective subcategory of labelled event structures.

Remark about bisimulation. In [JNW], the authors observe that in many structures for concurrency, standard notions of bisimulation can be derived from so-called "open maps". These in turn can be defined in terms of a suitable *path lifting property*. One has to define a suitable "path category" with respect to which one then can define the open maps, and subsequently bisimulation as a span of open maps (for definitions, the reader is referred to the mentioned text).

The category **LSP** also has a notion of bisimulation, which restricts to the notion of strong history-preserving bisimulation for labelled event structures. There is, in this case, a very simple definition for open maps in **LSP**: define $(X, \mathcal{P}, l) \xrightarrow{(F,(f_P)_{P}, \kappa)} (Y, \mathcal{Q}, m)$ to be open iff $F : \mathcal{P} \to \mathcal{Q}$ is an open map of posets. A monotone map $\varphi : (P, \le) \to (Q, \le')$ is open if for $q_1 \le' \varphi(p) \le' q_2$ there are $p_1 \le p \le p_2$ with $\varphi(p_i) = q_i$ for $i = 1, 2$. We have:

Proposition 4.9 *If* $(X, \mathcal{P}, l) \xrightarrow{(F,(f_P)_{P}, \kappa)} (Y, \mathcal{Q}, m)$ *is an open map, then every* $f_P : P \to F(P)$ *is an isomorphism of labelled posets.*

Proposition 4.10 *The open maps of LSP's restrict on event structures to the open maps of [JNW].*

Remark about partiality. As remarked before, my definitions differ from those in [WN94] only in that they take (as maps of event structures and independence structures) *partial* functions. This is an inessential difference, in that the basic adjunctions also exist when the maps are partial. This is left to the reader.

References

[BT]   Diekert & Rozenburg (eds), *The Book of Traces*, World Scientific (Singapore) 1994

[D]   Diekert, *On the concatenation of infinite traces*, TCS **113** (1993), 35-54

[Joh]   P.T. Johnstone, *Stone Spaces*, Cambridge University Press 1982

[JNW]   A. Joyal, M. Nielsen & G. Winskel, *Bisimulation from open maps*, BRICS Report RS-94-7, Aarhus 1994. To appear in Information and Computation.

[KPP]   Kwiatkowska, Penczek & Peled, *A hierarchy of partial order temporal properties*, in: LNCS 827

[Kwi]   Marta Kwiatkowska, *Defining Process Fairness for Non-Interleaving Concurrency*, LNCS 472, 1991, and:
        On Topological characterization of behavioral properties, in: Reed, Roscoe & Wachter(eds), *Topology and Category Theory in Computer Science*, Clarendon (Oxford), 1991

[WN94]   G. Winskel & M. Nielsen, *Models of Concurrency*, BRICS Report RS-94-12, Aarhus 1994. To appear in: Abramsky, Gabbay & Maibaum (eds), *Handbook of Logic in Computer Science*, Oxford University Press

Asynchronous Control Device Design by Net Model Behavior Simulation

V.I.Varshavsky and V.B.Marakhovsky

The University of Aizu, Aizu-Wakamatsu, Fukushima 965-80, Japan

Abstract. We discuss the problem of designing asynchronous control devices for discrete event coordination specified by a Petri net model. The design is based on the compilation of standard circuit modules corresponding to PN fragments into a net modeling PN behavior and on the semantic interpretation of the modeling circuit. The impossibility of asynchronous implementation of the indivisible operation of marking change at the circuit level leads to the necessity of modeling PN with modified rules of marking change. Modifications of the known modules, a number of new module types, the rules of the module connections, and the procedures of minimization given that considerably improve the quality of the obtained solutions in terms of both speed and area. The design "reefs" are fixed. The minimization procedures are usually associated with a change of marking change rules producing the problems of providing the equivalence of the initial and modified PNs.

1 Introduction

Behavior specification of asynchronous and real-time systems using net models, such as Petri Nets (PN) [1], Signal Transition Graphs (STG) [2], Change Diagrams (CD) [3], etc., is efficient and popular, at least because it is so illustrative. A net specification, though like any other one, produces two classes of tasks: correction analysis of the specification and synthesis of the corresponding control devices. The tasks of analysis are associated with two aspects of correctness: the correctness in reference to the specified behavior and the correctness in reference to the applied method of hardware implementation. For example, CDs that are correct in terms of behavior specification can be incorrect in terms of the possibility of deriving guard functions (logical implementation) due to conflicting states or because one-gate-implementation of a particular guard function is impossible within the used technology (e.g. non-monotonic or too complicated guard functions) [3]. Self-Timing and Self-Timed Design Methodology are usually based on net models. The methods of direct or semantic translation of a net model into a control circuit were suggested and are being developed [3, 4, 5]. We will restrict our consideration to the problems of direct translation methods applied to PN specifications. The general ideas that form the base of direct translation methods are simple and apparent enough:

1. The behavior of an asynchronous system and, hence, the net model of this behavior are determined by the handshakes between the system blocks and/or

between the system blocks and the control structure; even without such a handshake in an explicit form, the behavior of each block can be treated as handshake interaction with the environment which is the rest of the system;

2. In terms of behavior control at the level of the net specification of block handshake signal interaction, the model of the block is a delay incorporated between the signals *Request* and *Acknowledgment*; the state of the block itself can be expressed by a Boolean variable b_j which has a stable value $\{0, 1\}$ when $Req = Ack$ and an excited value $\{0^*, 1^*\}$ otherwise (for example, $b_j = 0^*$ when $Req = 1$ and $Ack = 0$);

3. Thus, the behavior specification of the system can be transformed into the behavior specification of some speed-independent circuit with events $+b_j$ and $-b_j$; if a speed-independent circuit is synthesized from this specification, every variable b_j in it corresponds to one and only one gate;

4. A circuit built in this way models the system behavior in the sense that the partial order on the events in the controlled blocks corresponds to the partial order on the values of variables b_j; if the modeling circuit is speed-independent with respect to the outputs of the gates that represent variables b_j, then, according to the Muller – Bartkey theorem [6], the partial order relationships will not change if we incorporate any finite non-controlled delays in series with the outputs of these gates (controlled block is such delay with respect to its *Req* and *Ack* signals);

5. Thus, the modeling circuit is at the same time a control circuit and the task of control circuit synthesis is reduced to the task of synthesizing a circuit that models a net specification; note that direct translation itself creates only hardware implementation for event frame; semantic content requires labeled PN and additional procedures.

The general idea of the direct translation method was formulated fairly long ago [7]. It assumes that net specification fragments are in correspondence with standard circuit modules. Furthermore, sets of such fragments and sets of the corresponding modules are known [5]. The direct translation method does not guarantee minimum solutions in terms of the usual criteria of logical function minimization. However, it has a number of attractive properties as the simplicity of the synthesis procedure, guaranteed implementation complexity (that depends linearly from the complexity of the net specification), and, what is especially important, the simplicity of the wire connections which in its turn provides the simplicity of testing.

Note that different net models that have the same or close expressive power can specify the same task by nets of different complexity. Hence, the circuit built from them by the direct translation method can be of different complexity, too. Of course, it would be attractive to use hybrid models which in every particular case take into account the advantages of one or another approach to specify the behavior of one or another fragment. However, this way produces extra difficulties that are associated with the necessity of using different analysis methods and specification correctness criteria together. This is an independent task beyond the scope of this work.

In this work, we will fix our attention to a number of methodological problems that were beyond the scope of the preceding publications, consider several novel procedures of direct translation and show a number of new modules that extend the possibilities of the direct translation method and improve the quality of the obtained solutions.

2   Modeling events and marking change

In all net models, a change of markings is an indivisible operation. We will illustrate it in fig.1a by an example of PN. In a PN model, only events are asynchronous; according to the definition, the input marking α of an event S changes to its output marking β momentarily (synchronously) after the event S. Evidently, such a procedure cannot be implemented directly in speed-independent hardware. This difficulty can be overcome by at least three approaches.

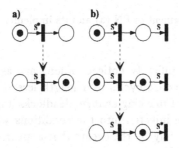

Fig. 1. Marking change procedure: common procedure (a), modified procedure(b).

First, the marking change can be synchronized by an external clock signal. This approach is justified when we work in a fully synchronous system and the external synchronization is not in conflict with the asynchronous behavior of its blocks;

Second, we can use the hypothesis about limited dispersion of gate delays (actually, this is true in modern technologies). We assume that erasing the input tokens is initiated by the completion of the event, performed concurrently with the recording of the output tokens and completed in a feasible time.

Finally (hereafter we will be oriented to this approach), the rules of PN functioning can be changed (fig.1b), namely, the change of tokens can be ordered: the completion of an event initiates the appearance of a tokens at the output conditions that, in its turn, initiates the erasing of the tokens at the input conditions. The rules for the event advent change: an event is initiated if there are tokens at all its input conditions and no tokens at all the input conditions of the events that are its immediate ancestors. As it is shown in [8], such a change in the rules of PN functioning does not change the general behavior semantics.

We will start our consideration from simple persistent safe PNs. A circuit that models a change of markings should contain a latch to keep the token and logic to provide the behavior discipline described above. A circuit from the simplest distributor cells, DC, [9] that models the change of tokens in a linear fragment of a PN (fig.2a) is given in fig.2b where two adjacent DCs interact as follows:

$$-S_{i-1} \rightarrow \underbrace{+d_i \rightarrow -c_i}_{\text{writing dot in } DC_i} \rightarrow \underbrace{+c_{i-1} \rightarrow -d_{i-1}}_{\text{deleting dot from } DC_{i-1}} \rightarrow +S_{i-1} \rightarrow -S_i \rightarrow \dots \, .$$

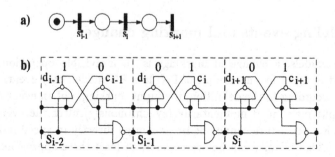

Fig. 2. Linear fragment of PN (a) and its implementation on DC (b).

However, there is a difficulty noticed as long ago as in [4]: using DCs, it is impossible to directly realize cycles of a length less than three because, with the accepted discipline of marking change, deadlocks appear. The cycles can be conservative (tokens are injected into the conditions within the cycle and are removed from the cycle only by the events that appear within the cycle) (fig.3) and non-conservative (tokens can be injected and removed by external events). Note that conservative cycles of the length 1 have no special substantial sense.

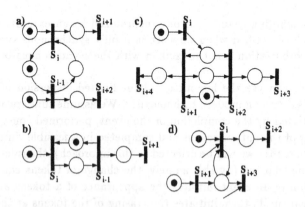

Fig. 3. Conservative cycles of length 2: general (a), pipeline (b) and toggle (c). Non-conservative cycle of length 1 (d).

Conservative cycles of the length 2 always have one of the two tokens. Hence, to save the marking within the cycle, it is enough to have one latch. In the simplest case (fig.3b), a conservative cycle of the length 2 specifies the pipelining of the events within the cycle. To model such a behavior, it is enough to slightly modify the DC by inserting an extra connection between the cells (fig4).

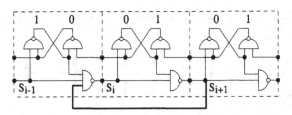

Fig. 4. Implementation for the cycle of length 2 depicted in fig.3b.

For more complicated conservative and non-conservative cycles of the length 2, the trivial solution is incorporating a virtual event that transforms a cycle of the length 2 to a cycle of the length 3. However, for special configurations, sufficiently simple solutions can be offered. An example is the fragment in fig.3c that specifies the behavior of a simple toggle: ... $\rightarrow S_i \rightarrow S_{i+1} \rightarrow ... \rightarrow S_i \rightarrow S_{i+2} \rightarrow ...$. Having modified the available circuit solutions, we obtain the standard implementation for this fragment (fig.5).

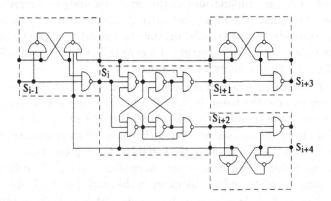

Fig. 5. Toggle implementation.

The implementations for join and fork fragments are known from [5]. For an event with n input conditions, n fictitious events are incorporated in [5] along with a complicated DC to represent a fictitious token and realized event. The example of join fragment implementation is shown in fig.6.

To simplify these fragments we can make some changes. Improved implementations for join and fork fragments are shown in fig.7.

In fig.7a, the example of a module is given that represents an event with two input conditions. The event is implemented as the output of a gate F with logical function $S_3 = \overline{S_1 d_{13} S_2 d_{23} + d_4(d_{13} + d_{23})}$. For a bigger number of input conditions, the event cannot be implemented on a single gate due to technological

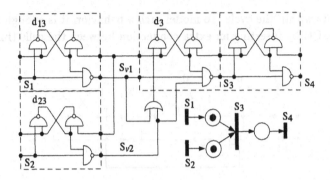

Fig. 6. Join fragment implementation.

and circuit restrictions on gate complexity. With n input conditions, the function $S_0 = \overline{\&_1^n S_j \&_1^n d_{j0} + d_{n+1}(\vee_i^n d_{j0})}$ where S_0 represents the event, S_j $(j = 1, n)$ represent the input events, d_{j0} represent the input tokens and d_{n+1} represents the output token can be realized by standard methods of checker implementation [5]. Note that the implementation in fig.7a is faster and simpler than that in fig.6.

Actually, the basic modules given here are enough to create a library of standard modules (as combinations of the input and output circuits of the basic modules that reflect the input and output configurations of the fragments). This library provides direct translation (by the substitution of the corresponding module instead of a PN fragment) of a simple, persistent, and safe PN with non-repeated events into an event frame, namely:

1. Every condition is in correspondence with a latch that saves the token; the condition to save the token is $S_j = 0$ (S_j is a variable that represents the preceding event); the condition to erase the token is writing tokens into all the output positions of the succeeding events. The conditions of writing and erasing tokens can be logical functions from the variables that represent several preceding events and the output conditions of the succeeding events. These functions can be immediately implemented on the gates in the latch (see fig.7c,d).

2. Every event S_j is in correspondence with either a single NAND-gate, in the simplest case, or a complicated date the output of which $S_j \to 0$ when there are tokens at all the input conditions of this event and no tokens at the input conditions of its immediately preceding events and $S_j \to 1$ when there are no tokens in all the input conditions. For example, in fig.7d, the gates F_2 and F_3 perform the checker function for events S_2 and S_3 depending on the additional conditions (bold input lines with arrows).

Note that the above consideration contains only technical changes that simplify the implementation and, in most cases of direct translation, and obviates using extra modules as compared to the preceding works [3,5,10]. We are going to demonstrate the possibility of a radically simplified implementation for several kinds of PN fragments. Let us look more attentively at the behavior of an event frame.

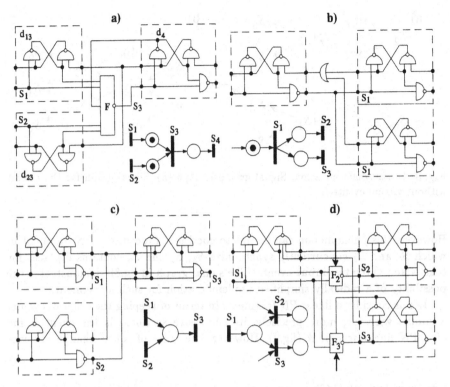

Fig. 7. Standard modules: event join (a), event fork (b), exclusive join (c), alternative fork (d).

First of all, let us agree about signal interpretation in an event frame. Every event S_j in a PN is in correspondence with two changes of signal S_j in the implementation of the event frame. An event itself will be considered to be modeled by the change $-S_j$ ($S_j = 1 \rightarrow S_j = 0$). The opposite change $+S_j$ ($S_j = 0 \rightarrow S_j = 1$) brings the information about erasing the tokens in the input conditions of event S_j.

Let us take a look at fig.6 and 7a. For the PN fragments given in these figures, latches to save tokens d_{13} and d_{23} are, generally speaking, superfluous. However, the implementation concerns only the event $S3$ and its input conditions. Since the procedure of event frame compilation is local, the behavior of this module should not depend on the configuration of the outputs of the events S_1 and S_2. Now let the events S_1 and S_2 of the PN fragment in fig.6 have output forks to tokens d_{15} and d_{26} respectively. The signal graph for this circuit is given in fig.8a and signal graph for the circuit with excluded DC for virtual events S_{v1} and S_{v2} is given in fig.8b. It is easy to see that, while in the first case $S_1 \parallel S_6$ and $S_2 \parallel S_5$, in the second case $S_1 \rightarrow S_6$ and $S_2 \rightarrow S_5$.

IT is natural to call this effect *cross-synchronization*. It can be caused by incorrect implementation and by the changing of marking change discipline.

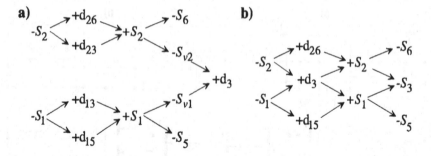

Fig. 8. Cross-synchronization. Signal graph for fig.6 (a). Signal graph for the circuit without virtual events (b).

However, this phenomenon can be put to good use. Indeed, the PN fragment in which S_1 and S_2 concurrently synchronize S_3, S_5 and S_6 contains 6 places for tokens and three two-input events. The signal graph in fig.8b demonstrates the possibility of a radically simplified implementation for such a fragment.

Let us consider a linear PN fragment. In terms of keeping the order of events, a simple wire (or, more precisely, its successive sections) is the event frame of a PN linear fragment (fig.9). However, the rules of token change are vio-

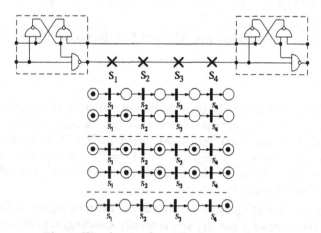

Fig. 9. Wired event frame for linear fragment.

lated: first, the tokens fill all the positions of the fragment and then they all are erased. In terms of the PN-specified behavior relative to the defined order on the events, nothing changes. What changes do take place in the event frame? For the implementation in fig.2, the signals in the linear fragment change as follows: $-S_1 \rightarrow +S_1 \rightarrow -S_2 \rightarrow +S_2 \rightarrow ... \rightarrow -S_k \rightarrow +S_k$; and for the implementation (fig.9): $-S_1 \rightarrow -S_2 \rightarrow ... \rightarrow -S_k \rightarrow +S_1 \rightarrow +S_2 \rightarrow ... \rightarrow +S_k$. The only result is that the changes of signals S_j have transposed while the total duration of the phases is the same. In doing so, the implementation has been considerably simplified and the time of DC firing has been eliminated. Note once more that everything we have said is true only for non-repeated PN.

Since for a linear fragment we have give up the original discipline of token changes[1], nothing can prevent us from doing the same for an arbitrary fragment. Multi-conditional events will be implemented by Muller C-elements (example in fig.10). Such an approach is apparently applicable only to the fragments without

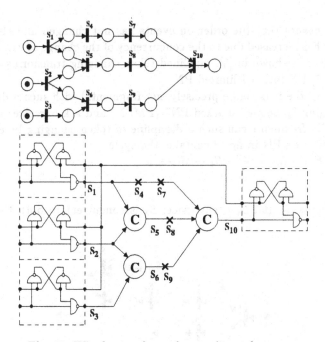

Fig. 10. Wired event frame for non-linear fragment.

internal tokens both in the initial and finite states. Besides, since the tokens within a fragment are erased just once, a token can be placed in a position inside the fragment only once (no cycles) and the outputs from the fragments should be synchronized (this requirement means that after every initiation there is one and only one output from the fragment). The restrictions are fairly strong; let us try to remove or, at least, to weaken them.

Let us return to the linear fragment and consider as its event frame a simple pipeline on C-elements (fig.11). In this case, tokens fill up the fragment distributing through it with the maximum speed, like in the case of a wire frame. However, the condition for erasing a token appears just after the event it has initiated. There are two waves propagating through the fragment: a wave of events (tokens) and a wave of erasing tokens. The signal interaction is as follows:

$$\ldots \rightarrow -S_1 \rightarrow -S_2 \rightarrow -S_3 \rightarrow \ldots \rightarrow -S_k \rightarrow \ldots$$
$$\downarrow \qquad \downarrow \qquad \downarrow$$
$$\ldots \rightarrow +S_1 \rightarrow +S_2 \rightarrow \ldots \rightarrow +S_{k-1} \rightarrow +S_k \rightarrow$$

---

[1] The first step toward giving up "purity of the idea" was made in Fig.1b.

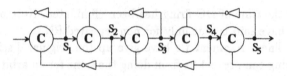

Fig. 11. Pipelined event frame for linear fragment.

We can ensure that the order on events has not changed and the speed of functioning has increased due to the concurrency of the change of variables. Note also that, as it is shown in [5], pipelined token changing transforms any unsafe (or k-limited) PN into a 1-limited PN.

Let us call the nets (more precisely, net fragments) with such a discipline of token changing "pipelined marked PN" (PMPN) and ask ourselves a question: "To what PN fragments can such a discipline of token changing be extended?" The fragment of a PN in fig.12 contains the cycle

$$S_2 \to S_4 \to S_6 \to S_2 \to S_4 \to S_5 \to \ldots$$
$$S_1 \to \updownarrow$$
$$S_3 \to \ldots$$

in which events S_2 and S_4 occur twice. S_3 is concurrent to both the first and

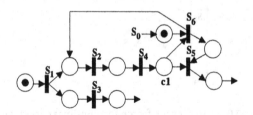

Fig. 12. PN fragment for pipelined implementation.

second firings of these events. Since a token in PMPN is represented by a state of the wire $(S_j = -1)$ and there is no separate memory for each of the output tokens of event S_1, the input token of S_2 can be erased only simultaneously with erasing the input token of S_3. This violates the partial order on the events. The second firing of S_2 is possible only after the firing of S_3. (Note that this property can be used for synchronization without additional tools. However, this is the case of design art rather than a regular synthesis procedure.) To recover the correct functioning, it is enough to incorporate a virtual event S_{v1} between S_1 and S_3.

The next problems cluster round the event S_6. Inserting tokens into its output conditions must be not later then erasing tokens from its input conditions. A similar problem for this fragment appears when using DC, too. The source of this problem is a modified discipline of marking changing. Indeed, $c1$ is an input condition both for S_5 and S_6 which is the immediate ancestor of S_5. So, S_5 can falsely fire from the first token in $c1$ if the token will not be erased before firing of S_6. To overcome this difficulty, it is also enough to incorporate a virtual event S_{v2} that checks the absence of tokens in the input conditions of S_6. S_{v2} is realized by an H-element with automaton equation $S_{v2} = S_6 \bar{S}_5 xy + S_{v2}(S_6 + \bar{S}_5)$. The event frame implementation for the PN fragment (fig.12) is given in fig.13.

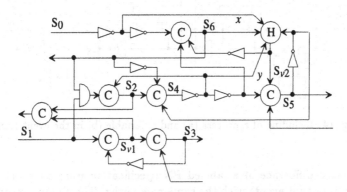

Fig. 13. Pipelined implementation for PN in fig.12.

The transition to PMPN simplifies the implementation of event frames (because the latches for saving tokens are removed) and speeds up the functioning (because the procedures of writing and erasing tokens are concurrent). However, this may violate both the specified partial order on the events and functioning correctness. Partial pipelining of the token changing procedure should be treated as a procedure of minimizing the hardware implementation of the event frame. Unfortunately, we do not have yet regular procedures of such minimization and formal procedure of checking the correctness of its results. What we have said above should be treated as setting the problem of simplifying the implementation of an event frame and accelerating its operation.

3   Semantic interpretation of an event frame (labeled PN)

A net specification of a control system defines a partial order on events that have a certain semantics. The contents of a specified event can be a four-phase protocol of process P_j initiation (full handshake): $-Req(P_j) \rightarrow -Ack(P_j) \rightarrow +Req(P_j) \rightarrow +Ack(P_j)$, or two-phase protocol of switching the state of a variable X_j: $Req(+X_j) \rightarrow +X_j \rightarrow Ack(+X_j)$ or $Req(-X_j) \rightarrow -X_j \rightarrow Ack(-X_j)$.

As we have already mentioned in the introduction, regardless of the internal semantics of the process, its model in reference to the signals $Req \rightarrow Ack$ is an arbitrary delay that can be incorporated into a gap of the wire that models the corresponding event. If the full handshake immediately corresponds to the protocol of event interaction in the event frame (fig.14a), the protocol of variable switching requires that a memory is inserted to keep the variable value and the protocol is supplemented up to four phases (fig.14b). Unlike [3, 10], the method of incorporating event $\pm X$ suggested here eliminates the danger of self-exciting on the fragment $\rightarrow +X \rightarrow -X \rightarrow$ and allow us to change the rules of marking change.

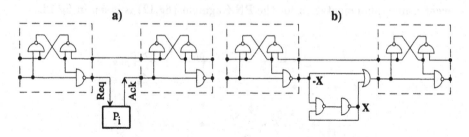

Fig. 14. Including of a process (a) and a variable (b) in the event frame.

The basic difference of a labeled PN specification from an event frame is the possibility that events with the same name exist. The circuits providing the incorporation of the same process (variable) into different places of the event frame are given in fig.15.

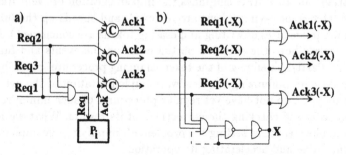

Fig. 15. Multiple including of a process (a) and a variable (b) in an event frame.

The C-elements in fig.15a have simplified functions: $AckJ = Ack + ReqJ \cdot AckJ$. We will also call attention to a very important property of the circuit in fig.15b. Let the events $-X_k, i$ and $-X_k, j$ be concurrent $(-X_k, i \parallel -X_k, j)$. (Here the designation $-X_k, j$ means the j-th use of the event $-X_k$ in the PN.) Then one of the events (any one) sets the new value of X_k and the other one just checks the fact of switching. Dotted wire in fig.15b is necessary only for this case to provide the property of speed-independence.

Note that such kind of concurrency is in correspondence with the non-distributivity of an algebraic structure formed by the infinite unfolding of an event partial order diagram [11,12]. To represent non-distributive specifications in the language of Change Diagrams [3] an OR-vertex is introduced. As far as we know, direct translation methods as applied to non-distributed specifications have been considered only in [10] in a similar but slightly different way.

If a process P_k is auto-concurrent with itself $(P_k, i \parallel P_k, j)$, the implementation requires more sophisticated solution than in the case of changing the value of a variable. A process can be auto-concurrent in two cases: 1) using common resource in concurrent branches of the PN (arbitration) and 2) single initiation of a process from any branch of the PN (non-distributivity).

Let us leave the case of arbitration beyond the scope of our consideration and discuss the variant of inserting an auto-concurrent process into a non-distributive event frame. For the PN fragment in fig.16a $(P_k, i \parallel P_k, j)$, the second initiation of event P_k, i (P_k, j) is possible only after the event P_k, j (P_k, i) is initiated, i.e. the process P_k is initiated according to the rule of "non-exclusive OR". In fig.16b, a semantic interpretation of the fragment in fig.16a is given using the language of Change Diagrams.

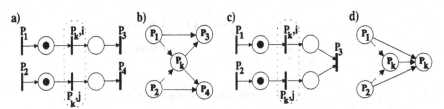

Fig. 16. Non-distributive semantics: PN fragment for $(P_k, i \parallel P_k, j)$ (a); CD with OR-vertex for the same behavior (b); PN fragment with arbitrary order of P_k, i and P_k, j (c); and corresponding CD specification (d).

A detailed PN interpretation of fig.16a fragment behavior is given in fig.17a. It is easy to see that the process P_k is initiated by the first token injected into its input condition of capacity 2 by any of the processes P_1 or P_2. The second token on this position is "eaten" by the virtual event P_v and a new initiation of P_1 and P_2 is possible only after P_v has fired.

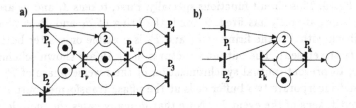

Fig. 17. PN behavior detalization: for fragments in fig.16a,b (a), for fragments in fig.16c,d (b).

The fragment in fig.16c can be semantically interpreted in a different way: events P_k, i and P_k, j are ordered but their order can be arbitrary, $(P_k, i \rightarrow P_k, j$ or $P_k, i \leftarrow P_k, j)$. A simplest substantial example of such a process can be a multi-input counter that changes the PN behavior when it produces the overflow signal. Note that the graphic facilities of Change Diagrams are more clear and illustrative (fig.16d). The PN interpretation of this fragment behavior is similar to fig.17b.

For the implementation of a standard module that provides auto-concurrent insertion of a process into an event frame (like that in fig.17a), let us first consider a circuit that transforms a single change of an input signal into a four-phase interaction protocol (fig.18). The behavior of the module in fig.18 is described by a signal graph:

$$(-x \text{ or } -y) \rightarrow +b \rightarrow \underbrace{-Req \rightarrow -Ack \rightarrow +c \rightarrow -d \rightarrow +Req \rightarrow +Ack}_{\text{four phase interaction with output PN fragment}} \rightarrow -a \rightarrow \dots$$

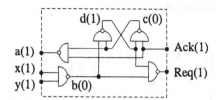

Fig. 18. Phase transforming module.

Let us call attention to the following fact: *tokens are injected into an output PN fragment and propagate through it without respect to the procedures of erasing tokens in the input PN fragment.* In other words, the insertion of such a module into an event frame once more changes the discipline of marking change. Erasing tokens in the input PN fragment is determined by the input protocol of the module. The signal graph of the protocol is the following:

$$\underbrace{-a \rightarrow (+x \text{ and } +y)}_{\text{deleting token}} \rightarrow -b \rightarrow +d \rightarrow -c \rightarrow +a \rightarrow \dots .$$

The general circuit of a standard module that provides auto-concurrent insertion of a process into an event frame modeling the specification in fig.17a is given in fig.19. This circuit functions specially. First, tokens d_1 and d_2 are erased only together, after P_v has fired. Second, the firing of P_1 and P_2 is blocked by the feedback with P_v (the firing of P_1 and P_2 is possible only after both tokens d_1 and d_2 are erased). The output structures in fig.19 are shown schematically. Actually, to prevent mutual synchronization of the reverse phase of P_3 and P_4 one should incorporate two buffer cells at the phase transformer output to save the output tokens of the event P_k. Note that in many cases the module of auto-concurrent incorporation of processes can be considerably simplified. However, this requires that the translation is not local.

A fragment of a PN can also be treated as a process that allows us to decompose the event frame using standard sub-frames (sub-PN) several times. New extra problems appear in the case of multi-pole sub-PN but this is a topic for a special consideration and we will not discuss it here.

For non-repeated PN fragments in which any event variable switches in both directions no more than once, the general approach to design a wired implemen-

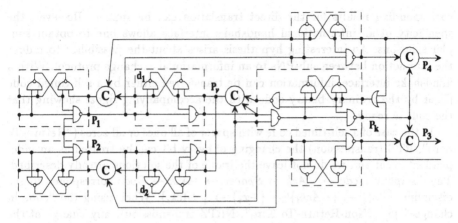

Fig. 19. Event frame for auto-parallel inclusion of the process P_k.

tation can be simplified as compared to the implementation in fig.10 by using latches to save the values of variables. An example of such an implementation is given in fig.20.

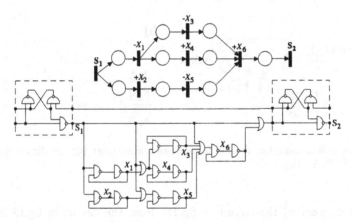

Fig. 20. Non-repeated PN fragment and its implementation.

4 External variables and predicate-transition PN

The behavior of a control unit specified by a PN can depend and usually depends on external variables. This dependence is determined by the way the control device interacts with the environment. We can recognize at least two types of interaction: 1) "request – acknowledgment" interaction and 2) "wait" interaction.

With "request – acknowledgment" interaction, the environment can in principle be treated as the output pole (poles) of the PN fragment and all the corresponding results for the direct translation can be applied. However, the specificity of an informational handshake interface allows one to obtain simpler solutions. An interesting hypothesis arises about the possibility to reduce the interaction between sub-PNs to an informational exchange protocol. With a handshake interface, information can be transferred either by a self-timed code [5] or by the ordinary binary code with the accompanying signal showing that the code is ready.

In the case of a self-timed code with spacer of all ones or all zeros ("Return-To Zero", RTZ transmission) the changing of every bit can be treated as an independent event and we do not leave the frame of the solutions already described. The discipline $Code \rightarrow -Ack \rightarrow Spacer \rightarrow +Ack \rightarrow Code$ corresponds to the discipline $\{-x_i\} \rightarrow \{-Ack_i\} \rightarrow \{+x_i\} \rightarrow \{+Ack_i\}$. When using the "code in changes" [5] ("Non-Return-To Zero", NRTZ transmission), any change of the interface bit y_i indicates the transmission of a bit $x_i = 0$ of the self-timed code. The spacer is artificially created when realizing the protocol of the interaction between the event frame with interface inputs and the environment. The module realizing this protocol should have the following signal graph:

$\rightarrow \{-y_i\} \rightarrow \{-x_i\} \rightarrow \{-Ack(x_i)\} \rightarrow \{+x_i\} \rightarrow \{+Ack(x_i)\} \rightarrow \{-Ack(y_i)\} \rightarrow$
$\{+y_i\} \rightarrow \{-x_i\} \rightarrow \{-Ack(x_i)\} \rightarrow \{+x_i\} \rightarrow \{+Ack(x_i)\} \rightarrow \{+Ack(y_i)\} \rightarrow ....$

Its implementation is given in fig.21a.

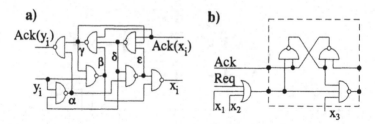

Fig. 21. Interface modules: double four-phase transformer (a); predicate module for the term $\bar{X}_1\bar{X}_2X_3$ (b).

The difference of the circuit in fig.21a from the circuit in fig.18 is that it generates four-phase interaction for each change of signal y_i and its behavior is described by the following signal graph:

$+y \rightarrow -\alpha \rightarrow +\varepsilon \rightarrow$
$\underbrace{-x \rightarrow -Ack(x) \rightarrow +\gamma \rightarrow -\beta \rightarrow +x \rightarrow +Ack(x) \rightarrow -\delta \rightarrow +\alpha \rightarrow -Ack(y) \rightarrow}_{\text{four phase interface for } +y}$
$-y \rightarrow +\beta \rightarrow$
$\underbrace{-x \rightarrow -Ack(x) \rightarrow +\delta \rightarrow -\varepsilon \rightarrow +x \rightarrow +Ack(x) \rightarrow -\gamma \rightarrow +Ack \rightarrow}_{\text{four phase interface for } -y}.$

In the case of an interface with an accompanying signal T, an important feature is the agreement of ordering the changes of the code and accompanying signal. If the signal T accompanies information signals $\pm x_i$ only by one its value (for example, $T = 0$) then direct interaction between PN fragment and external environment is possible with help of the interface protocol (RTZ):

$\{\pm x_i\} \rightarrow -T \rightarrow -Ack \rightarrow +T \rightarrow +Ack \rightarrow \{\pm x_i\} \rightarrow .$

If the signal T accompanies signals $\pm x_i$ by each of its switching, an intermediate interface module is needed (like that in fig.21a) realizing the interface protocol (NRTZ):

$\{\pm x_i\} \rightarrow -T \rightarrow -Req(PN) \rightarrow -Ack(PN) \rightarrow +Req(PN) \rightarrow +Ack(PN) \rightarrow -Ack \rightarrow ...\{\pm x_i\} \rightarrow +T \rightarrow -Req(PN) \rightarrow -Ack(PN) \rightarrow +Req(PN) \rightarrow +Ack(PN) \rightarrow +Ack \rightarrow ... .$

The variables x_i can be inserted into the event frame as predicates in the way represented in fig.22b. Note that if the accompanying signal represents an event internal to the corresponding PN, the same circuit provides the realization of "wait" predicate (obviously, only if the semantics of the specification provides the required discipline of predicate variables changing).

A special case is "internal" predicates which provide extra possibilities for minimization. If among the PN events there are events interpreted as switches of variables, the net states are determined not only by the current marking but also by the current values of these variables. For autonomous PN, the current marking uniquely determines the state of variables (the reverse, in general, is not true). For PN with external inputs, different states of variables may correspond to the same marking. Figuratively speaking, marking is a dynamically unstable state and a set of variable values is stable on a certain set of markings. As we use extra memory to save variable values, it seems attractive to provide that an event associated with a variable switch affects the immediately succeeding events not through the distribution of the corresponding tokens but in a predicate form. An example is shown in fig.22. Let the events in the PN fragment (fig.22a) satisfy

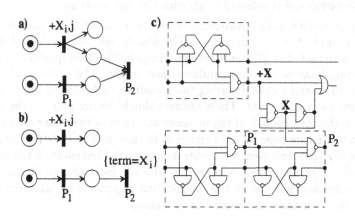

Fig. 22. Internal predicate application: PN fragment (a); PTPN fragment (b) and its implementation (c).

the condition: $[\forall k \neg \exists - X_i, k \parallel (P_1 \text{ or } P_2)] \& [\forall (k \neq j) \neg \exists + X_i, k \parallel (P_1 \text{ or } P_2)]$. Then this fragment can be represented as a fragment of a predicate-transition PN, PTPN,(fig.22b). Its implementation is given in fig.22c. Note also that when the cycle length is 1 (fig.3d), the state of the latch used to save the input token of event S_{i+3} can be used as a predicate variable for event S_i.

5 Conclusion

Design is an art and, apparently, it will remain an art for a long time. Although it sounds paradoxical, as the level of CAD tools grows, the place of invention in the process of obtaining new unexpected solutions becomes more and more distinct. Formal and automated procedures are permanently perfecting means of labour that take over more and more work liberating the designers from routine for pure creation.

Invention displays maximally at the level of specification which, in fact, already contains all the advantages and shortcomings of the future device. Human thinking is usually structural, that is why net models are so popular when specifying behavior. They are illustrative, structured and have rich compositional possibilities. Formal procedures and computer aided tools for net specification correctness analysis produce an excellent environment for man-machine design.

However, the conventional hardware implementation is associated with quite a different basis – logic gates and standard logic devices. The transition to a standard set of logic modules largely breaks the initially conceived structure of the device or, at least, significantly distorts it. When a good system designer constructs a specification, he always keeps in mind the possibilities of hardware implementation and is oriented to standard circuit solutions.

Very attractive is the possibility to create a circuit basis adequate to the general nature of net specifications. The idea is not new and, probably, was initially formulated in [7] and developed in [3-5,10]. Many problems remained out of consideration or were mentioned just briefly. The problem of changing the rules of marking change during the transition to hardware implementation seems especially important. These changes should be invisible for the designer, hidden in the construction of the modules and rules of their connection. On the other hand, and we hope we have managed to show it, a local extra change of the marking change rules opens up a wide avenue to implementation minimization. The problem of identifying subsets for which a local change of the marking change rules does not distort the specification appears to be interesting for the experts on net models and helpful for designers.

Here we have tried to look more thoroughly at the general problem of direct translation of a net specification into the corresponding hardware structure. Many of the circuit solutions are discussed here for the first time. But the major impression we want the readers to get after reading this work is that the problem is open. It is like the well-known "Mechano LEGO" for children – the set of modules and construction elements can permanently grow increasing the quality and variety of the obtained devices.

References

1. Petri, C.A.: Kommunikation mit Automaten. PhD thesis, Technical report Schriften des IIM Nr. 3, Institut für Instrumentelle Mathematik, Bonn (1962)
2. Chu, T.-A.: Synthesis of Self-Timed VLSI Circuits from Graph-Theoretic Specifications. PhD thesis, MIT/LCS/TR-393, MIT Laboratory for Computer Science, June (1987)
3. Kishinevsky, M., Kondratyev, A., Taubin, A. and Varshavsky, V.: Concurrent Hardware. The Theory and Practice of Self-timed Design, J.Wiley and Sons (1993)
4. Astanovsky, A.,Varshavsky, V., Marakhovsky, V., et al.: Aperiodical Automata. Ed. by V.Varshavsky, Moscow, Nauka, (1976) (in Russian)
5. Varshavsky, V., Kishinevsky, M., Marakhovsky, V., et al.: Self-timed Control of Concurrent Processes. Ed. by V.Varshavsky, Kluver Academic Publishers (1990) (Russian Edition - 1986).
6. Muller,D.E.,Bartky, W.S.: Theory of Asynchronous Circuit. Univ. of Illinois, Digital Comput. Lab., Rep. 75, 78, pt.1(1966), pt.2(1967)
7. Blanchard, M., Gillon, J., Cavarroc, J.C.: Concertion Modulaire à Automatismes Industriels Asynchronos. Jornée AFCET, Grenoble, France (1972)
8. Tiusanen, M.: Some unsolved Problems in Modeling Self-timed Circuits Using Petri Nets. Bulletin of the European Association for Theoretical Computer Sciences, 36, October (1988) 152-160
9. Varshavsky, V.I., Marakhovsky, V.B., et al.: A Cell of Asynchronous Signal Distributor. USSR Patent Certificate Number 758940, The Inventions Bulletin, 8 (1980)
10. Varshavsky, V.I., Tiusanen, M.: Hardware Support of Concurrent Processes Interaction and Synchronization: on the Principle of Auto-correct Implementation. Helsinki University of Technology, Digital System Laboratory. Series B: Technical Reports, 4, May (1988)
11. Muller, D.E.: Lecture Notes on Asynchronous Circuit Theory. Digital Computer Laboratory, University of Illinois, Urbana, Spring (1961)
12. Miller, R.E.: Switching Theory II. Sequential Circuits and Machines. John Wiley and Sons, New York (1965)

Covering Step Graph

François Vernadat, Pierre Azéma, François Michel
e-mail : {vernadat,azema,fmichel}@laas.fr

LAAS-CNRS
7 avenue du Colonel Roche F-31077 Toulouse cedex - France

Abstract. Within the framework of concurrent systems, several verification approaches require as a preliminary step the complete derivation of the state space. Partial-order methods are efficient for reducing the state explosion due to the representation of parallelism by interleaving. The covering step graphs are introduced as an alternative to labelled transition systems. A transition step consists of several possibly concurrent events. In a covering step graph, steps of independent transitions are substituted as much as possible to the subgraph which would result from the firing of the independent transitions. Attention must be paid to the case of conflict and confusion. An algorithm for the "on the fly" derivation of step graphs is proposed. This algorithm is then extended to behaviour analysis by means of observational equivalence. A performance evaluation is made with respect to other methods.

Keywords: concurrent systems, state space exploration, partial-order, verification methods

1 Introduction

The state space derivation represents the preliminary step of several verification methods for concurrent systems. This approach is made attractive by the existence of efficient and automatic verification techniques, such as bisimulation and model-checking. The combinational explosion is the main limitation of this approach.

Many studies are currently in progress for reducing this problem: on the fly bisimulation [FM 90], symbolic marking graph derivation through symmetrical folding [Jen 87], so-called partial order techniques which attempt to avoid the combinational explosion resulting from the concurrency interpretation by means of interleaving: persistent sets [Val 89], sleep sets [GW 91]. An other direction, followed in [Esp 93, McMil 95], avoids the state explosion by using model checker directly on the description of the system (unfolding of Petri Nets).

The partial order techniques (see [WG 93] for a general survey) represent the framework of the approach developed in this paper. The basic principle of these approaches consists into considering a single specific path among all the sequences which possess the same Mazurkiewicz's trace [Maz 87]. In the case of persistent or sleep sets, only a subset of enable transitions is examined, the derived graph is then a subgraph of the whole graph.

The proposed approach visit all the transitions, but some independent events are put together to build a single transition step, the firing of this transition step is then atomic. The resulting graph is referred as covering step graph (CSG). A notion of step graph appears in [Rei 85]: this graph is obtained in a standard way except in presence of independently enable transitions where a step gathering these transitions is considered as an additional event and put to the set of transitions. The standard LTS consequently is a sub-graph of the Reisig's step graph. The covering step graph proposed here, is radically different because steps of independent transitions are substituted as much as possible to the subgraph which would result from the firing of the independent transitions. The potential benefit of such a substitution may be exponential with respect to the number of "merged" independent transitions. Some examples, in the paper, show that this potential benefit may effectively be obtained.

The covering step graph supplies an alternative structure to the traditional reachable state graph. The CSG analysis allows the control of global reachability properties, such as liveness or presence of deadlock. Furthermore, step graphs may be used for checking behaviour properties such as observational equivalence. An algorithm for the "on the fly" derivation of CSG is proposed. This algorithm is then extended to behaviour analysis by means of observational equivalence. A performance evaluation is made with respect to other methods.

Section 2 presents the covering step graph definition and the general properties preserved by such structures. Section 3 shows how to modify a standard enumeration algorithm in order to obtain a step graph. This algorithm is adapted in section 4 for computing a step graph observationally equivalent to the regular state graph. A preliminary assessment of the proposed approach is developed in section 5, by comparing it with stubborn and symmetry methods in the case of a standard data base example. Current work and extensions are finally mentioned.

2 Covering Step Graphs

2.1 Basic Notions

Labelled Transition Systems (LTS) (definition)
A LTS is a quadruple $< S, s_o, T, \rightarrow >$ where : S is a set of states, s_0 a distinguished state in S , T is a set of transition labels, $\rightarrow$ is a set of transitions ($\rightarrow \subset S \times T \times S$).
Notations:
$s \overset{t}{\longrightarrow}$ iff $\exists s' \in S : (s, t, s') \in \rightarrow$, $s \overset{t}{\not\longrightarrow}$ iff not $s \overset{t}{\longrightarrow}$,
$s \overset{t}{\longrightarrow} s'$ to mean that $(s, t, s') \in \rightarrow$
$s_0 \overset{w}{\Rightarrow} s_n$ iff $w = t_1.t_2 \ldots t_n$ and $s_0 \overset{t_1}{\longrightarrow} s_1, s_1 \overset{t_2}{\longrightarrow} s_2, \ldots, s_{n-1} \overset{t_n}{\longrightarrow} s_n$

2.1.1 Independency Relation

Let $\Sigma = < S, s_o, T, \rightarrow >$ be a LTS and $\wr$ a binary relation over the set of transitions ($\wr \subset T \times T$)

definition: [WG 93] $\wr$ is an independent relation over Σ iff $\forall s, s_1, s_2 \in S, \forall t_1, t_2 \in T$:
$[t_1 \neq t_2, s \xrightarrow{t_1} s_1, s \xrightarrow{t_2} s_2$ and $t_1 \wr t_2] \Rightarrow s_1 \xrightarrow{t_2} s'$ and $s_2 \xrightarrow{t_1} s'$

if t_1 and t_2 are enabled in s then if $s \xrightarrow{t_1} s_1$ (resp $s \xrightarrow{t_2} s_2$) then t_2 is enabled in s_1 (resp t_1 is enabled in s_2), moreover, there exists a unique state s' such that both $s_1 \xrightarrow{t_2} s'$ $s_2 \xrightarrow{t_1} s'$ (commutativity of enabled independent transitions)

The complement relation of $\wr$ is called conflict, or dependency, relation. The conflict relation is denoted $\#$.

Independency relation in Petri Net [Rei 85]

Let $R = < P, T, Pre, Post >$ a (Place/Transition or Condition/Event) net, m_0 an initial marking and $G(R, m_0)$ the reachability graph associated with the marked net $< R, m_0 >$

For each transition $t \in T$ we define: $\bullet t =_{def} \{p \in P : Pre(p, t) \neq 0\}$,
$t^\bullet =_{def} \{p \in P : Post(p, t) \neq 0\}$ and $\bullet t^\bullet =_{def} \bullet t \cup t^\bullet$

Place/Transition Nets Let $\wr_{P/T} \subset T \times T$ defined as $t_1 \wr_{P/T} t_2$ iff $\bullet t_1 \cap \bullet t_2 = \emptyset$
For each Place/Transition marked net $< R, m_0 >$,
$\wr_{P/T}$ is an independency relation over $G(R, m_0)$.

Condition/Event Nets Let $\wr_{C/E} \subset T \times T$ defined as $t_1 \wr_{C/E} t_2$ iff $\bullet t_1^\bullet \cap \bullet t_2^\bullet = \emptyset$
For each Condition/Event marked net $< R, m_0 >$,
$\wr_{C/E}$ is an independency relation over $G(R, m_0)$.

2.1.2 Mazurckiewicz's Traces [Maz 87]

Concurrent Alphabet: A concurrent alphabet is a couple $\mathcal{E} = (\alpha, \#)$ where α is an alphabet and $\#$ the *dependency* in α ($\#$ is a reflexive and symmetric relation). The complementary relation $\#^C$ of relations[1] $\#$ in α is called independency relation in α

Mazurckiewicz's Traces:

Let $\mathcal{E}$ be concurrent alphabet $(\alpha, \#)$, the equivalence relation in α^*, $\equiv_{\mathcal{E}}$ is defined by: $W \equiv_{\mathcal{E}} W'$ iff there exists a finite sequence $(W_O, W_1, \ldots W_n)$ such that
$W_0 = W$ and $W_n = W'$ and
$\forall i, 1 \geq i \geq n : \exists u, v \in \alpha^*, \exists (a, b) \in \#^C$ such that $W_{i-1} = ubav$ et $W_i = uabv$
Relation $\equiv_{\mathcal{E}}$ defines the trace equivalence over $\mathcal{E}$. Equivalence classes of $\equiv_{\mathcal{E}}$ are called traces over $\mathcal{E}$. A trace generated by a string w is denoted by $[w]_{\mathcal{E}}$.

Trace properties:

$*$ $\equiv_{\mathcal{E}}$ is the least congruence in the monoïd $(\alpha, ., \epsilon)$ satisfying:
$(a, b) \in \#^C \Rightarrow a.b \equiv_{\mathcal{E}} b.a$

$*$ Let $\Sigma, \wr, \mathcal{E}$ be respectively a LTS $< S, s_o, T, \rightarrow >$, an independency relation over Σ and an associated concurrent alphabet $(T, \wr^C)$, then:
$\forall s \in S, w_1, w_2 \in T^*$ such that $s \xrightarrow{w_1} s_1$ & $s \xrightarrow{w_2} s_2 : [w_1]_{\mathcal{E}} = [w_2]_{\mathcal{E}} \Rightarrow s_1 = s_2$

2.2 Transition Steps

In what follows, Σ denotes a LTS and $\wr$ an independency relation over Σ.

---

[1] For a binary relation $R \subset U \times U$, R^C denotes the binary relation $(U \times U) \setminus R$

2.2.1 Transition Steps (definition)

The transition set E defines a **transition step** wrt $\wr$ iff $\forall t_1, t_2 \in E : t_1 \wr t_2$
In the sequel, $Step(T, \wr)$ denotes the set of all transition steps derived from T
wrt $\wr$. When no confusion is possible, we note $Step(T)$ instead of $Step(T, \wr)$.
Property: $E \in Step(T, \wr), F \subset E \Rightarrow F \in Step(T, \wr)$

2.2.2 Traces and transition steps

Set of sequences associated with a transition step
Let Seq be the mapping $\mathcal{P}(T) \longmapsto \mathcal{P}(T^*)$ defined by:
$Seq(\emptyset) = \{\epsilon\}$ and $Seq(E) = \bigcup_{e \in E} Seq(E \setminus \{e\}) \odot e$
where for $\Omega \in \mathcal{P}(T^*)$ and $w \in T^* : \Omega \odot w = \{w.\omega : \omega \in \Omega\}$
Property: $\forall E \in Step(T, \wr), \omega_1$ and $\omega_2 \in Seq(E) \Rightarrow [\omega_1]_{(T, \wr^C)} = [\omega_2]_{(T, \wr^C)}$

Trace of a transition step The former property allows to define the transition
step trace as the trace of a string associated with the step by means of function
Seq. For $E \in Step(T, \wr)$, $[E]_{(T, \wr^C)} =_{def} [\mathcal{E}]_{(T, \wr^C)}$ where $\mathcal{E} \in Seq(E)$

Trace of a sequence of transition steps is defined as the trace of the string
obtained by concatenation of the strings associated with each step of the sequence: For $E_1.E_2 \ldots E_n \in Step(T, \wr)^*$,
$[E_1.E_2 \ldots E_n]_{(T, \wr^C)} =_{def} [\mathcal{E}_1.\mathcal{E}_2 \ldots \mathcal{E}_n]_{(T, \wr^C)}$ where $\mathcal{E}_i \in Seq(E_i)$

2.2.3 Extension of the reachability relation to transition steps

Let $\rightarrow$, be the extension of $\rightarrow$ to transition steps defined by:
$$\forall s, s' \in S, \forall E \in Step(T) : s \overset{E}{\rightarrow} s' \text{ iff } \forall e \in E : s \overset{e}{\longrightarrow} \text{ and } \exists \omega \in Seq(E) : s \overset{\omega}{\longrightarrow} s'$$

2.2.4 Extensions of $\wr$ to string and transition steps

Support of a string: Let $\| \ \|$, be the mapping $T^* \longmapsto \mathcal{P}(T)$ defined by:
$\|\epsilon\| = \emptyset$ and $\|u.\omega\| = \{u\} \cup \|\omega\|$

Extensions of $\wr$
Let $\wr \subset Step(T) \times Step(T)$ defined by $T_1 \wr T_2$ iff $T_1 \times T_2 \subset \wr$
Let $\wr \subset Step(T) \times T^*$ defined by $T_1 \wr w$ iff $T_1 \times \|w\| \subset \wr$
Let $\wr \subset T^* \times T^*$ defined by $w_1 \wr w_2$ iff $\|w_1\| \times \|w_2\| \subset \wr$

2.2.5 Diamond properties

1. For $t \in T, w \in T^* :$ If $s \overset{t}{\longrightarrow}, s \overset{w}{\Rightarrow}$ and $t \wr w$ then $s \overset{t.w}{\Rightarrow} s'$ and $s \overset{w.t}{\Rightarrow} s'$
2. For $w_1, w_2 \in T^* :$ If $s \overset{w_1}{\Rightarrow}, s \overset{w_2}{\Rightarrow}$ and $w_1 \wr w_2$ then $s \overset{w_1.w_2}{\Rightarrow} s'$ and $s \overset{w_2.w_1}{\Rightarrow} s'$
3. For $w \in T^*, E \in Step(T) :$ If $s \overset{w}{\Rightarrow}, s \overset{E}{\rightarrow}$ and $E \wr w$ then $s \overset{\mathcal{E}.w}{\Rightarrow} s'$ and $s \overset{w.\mathcal{E}}{\Rightarrow} s'$
   $\forall \mathcal{E} \in Seq(E)$
4. For $E_1, E_2 \in Step(T) :$ If $s \overset{E_1}{\rightarrow}, s \overset{E_2}{\rightarrow}$ and $E_1 \wr E_2$ then $s \overset{E_1 \cup E_2}{\rightarrow}$
   And $s \overset{\mathcal{E}_1.\mathcal{E}_2}{\Rightarrow} s'$ and $s \overset{\mathcal{E}_2.\mathcal{E}_1}{\Rightarrow} s' \ \forall \mathcal{E}_i \in Seq(E_i)$

Sketches of proofs: (1) by recurrence on $|w|$. (2) by recurrence on $|w_i|$ and property (1). (3) is a consequence of (2) $(E \wr w \Rightarrow \mathcal{E} \wr w)$. (4) $E_1 \wr E_2 \Rightarrow E_1 \cup E_2 \in Step(T)$ and property 3.

2.3 Covering Step Graph

This section introduces the Covering Step Graph definition. The general reachability properties preserved by Step Graph are presented.

2.3.1 Covering Step Graph (definition) Let $\Sigma =< S, s_0, T, \rightarrow >$, be a LTS $\wr$ an independency relation over Σ (we let $\# = \wr^C$)

$\boldsymbol{\Sigma} =< S', s_0, \boldsymbol{T}, \twoheadrightarrow >$ is a Covering Step Graph for Σ, wrt $\wr$ iff

(1) $S' \subset S$ $\hspace{3cm}$ (2) $\boldsymbol{T} \subset Step(T, \wr)$

(3) $\forall s, s' \in S \cap S' : s \overset{E}{\twoheadrightarrow} s'$ implies $s \overset{E}{\rightarrow} s'$

(4) $\forall s \in S \cap S', \forall s' \in S' :$

$$s \overset{\omega}{\Rightarrow} s' \text{ implies} \begin{cases} \exists s" \in S', \exists \omega' \in T^*, \exists P_{\omega.\omega'} \in \boldsymbol{T}^* : \\ s' \overset{\omega'}{\Rightarrow} s", s \overset{P_{\omega.\omega'}}{\twoheadrightarrow} s" \text{ and } [w.w']_{(T,\#)} = [P_{\omega.\omega'}]_{(T,\#)} \end{cases}$$

note: any LTS may be viewed as a CSG by considering $\wr = \emptyset$

Milner's Scheduler This scheduler is described in [Mil 85]. n sites cyclically execute action a_i then action b_i. A scheduler constraints the execution of the whole system in such a way that the n sites alternatively perform actions $a_1, a_2, \ldots a_n$. Figure 1 depicts the net of the whole system (scheduler + n sites), then the corresponding LTS, and a derived CSG.

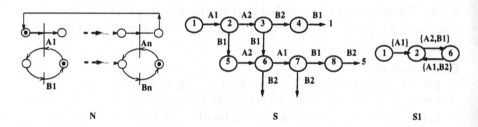

Fig. 1. Milner's Scheduler

Condition (1), each state of the step graph is a state of the standard LTS. For the Milner's scheduler, $t_1 \wr t_2$ iff $t_1 \neq t_2$, consequently each subset of T constitutes a step of transitions and (2) holds. Condition (3) means that each transition step in the CSG ($S1$) corresponds to a firing sequence in the standard LTS (S): for instance :$2 \overset{\{A2,B1\}}{\rightsquigarrow} 6$ corresponds to $2 \overset{A2.B1}{\Rightarrow} 6$ (or $2 \overset{B1.A2}{\Rightarrow} 6$).

Finally, condition (4) expresses a "covering condition" between the firing sequences of the standard LTS and the step sequences of the CSG.

In S, $1 \overset{a_1.a_2.b_2.b_1}{\Longrightarrow} 1$, condition 4 of definition 2.3.1 ensures that a "covering" step sequence exists in $S1$. The initial sequence, $a_1.a_2.b_2.b_1$, is extended by a_1, in the CSG the following step sequence $\Omega = \{a_1\}.\{a_2, b_1\}.\{a_1, b_2\}$ covers it: $1 \overset{\Omega}{\leadsto} 2$ and $1 \overset{a_1.a_2.b_2.b_1.a_1}{\Longrightarrow} 2$ with $[a_1.a_2.b_2.b_1.a_1] = [\{a_1\}.\{a_2, b_1\}.\{a_1, b_2\}]$

Section 3.2 presents examples satisfying conditions 1), 2) and 3) and violating condition 4)

2.3.2 Reachability Preservation

This section shows how general reachability properties such as liveness or deadlock may be checked directly on CSG.

Liveness

$\Sigma = < S, s_0, T, \rightarrow >$, is E-Live for $E \subset T$ iff $\forall s \in S, \forall e \in E, \exists \sigma \in T^*$ such that $s \overset{\sigma.e}{\Longrightarrow}$

$\Sigma = < S', s_0, \mathbb{T}, \leadsto >$ is E-Live iff $\forall s \in S, \forall e \in E, \exists \sigma \in \mathbb{T}^*, \exists u \in \mathbb{T}$ such that $s \overset{\sigma.u}{\leadsto}$ with $e \in u$

property If Σ is a CSG for Σ then Σ is E-Live iff Σ is E-Live

($\Leftarrow$) (resp ($\Rightarrow$)) is a direct consequence of condition 3 (resp 4) of definition 2.3.1

Deadlock preservation Let $Sink(\Sigma) =_{def} \{s \in S : s \overset{t}{\not\rightarrow} \quad \forall t \in T\})$

Property: If Σ is a CSG of Σ then $Sink(\Sigma) = Sink(\Sigma)$

$Sink(\Sigma) \subset Sink(\Sigma)$ (resp $Sink(\Sigma) \subset Sink(\Sigma)$): By absurd by considering a state $\in S' - Sink(\Sigma)$ (resp $\in S' - Sink(\Sigma)$) and by applying condition 4 (resp 3) of definition 2.3.1, this leads to a contradiction

3 On the Fly Covering Step Graph Derivation

This section introduces a basic algorithm for the covering step graph derivation directly from the description of the system. The general conditions to be fulfilled are first described. These conditions are shown to be sufficient conditions. The corresponding complete algorithm is then presented.

3.1 Basic Principles

Table 1 describes the general structure of the basic algorithm. This algorithm is similar to a standard algorithm for computing a reachable marking graph, that is $Enabled(q)$ computes the set of transitions enabled by state q, and $fire(q, t)$ computes the state obtained from state q by firing transition t .

The changes with respect to a standard algorithm are the following.

- The independence relation $\wr$ is supplied,
- The enable transitions are split in two subsets by means of functions T_U and T_M :
  - T_u, defines transitions to be explored in a standard way,
  - T_m defines transitions whose exploration will be conducted within a step. In the sequel, such transitions will be referred as "mergeable" transitions.
- The set of transition steps Π_{T_M} built by means of function Π, whose domain

```
1. Initialise: Stack is empty ; push s₀ onto Stack
                H is empty ; enter s₀ in H
                A is empty ;

2. Loop : while Stack ≠ ∅ loop
          { pop(q) from stack
          T ← Enabled(q);
          IF T = ∅ Then Print "Deadlock"
          ELSE{

          Tᵤ ← Tᵤ(T,l)
          Tₘ ← Tₘ(T,l)

          Πₜₘ ← Π(Tₘ,l)
          ∀π ∈ Πₜₘ do
             {q" ← q;
             ∀p ∈ π do {q' = fire(q",p); q" ← q'};
             enter < q, s, q" > in A;
               if q" ∈ H then {enter q" in H; put q' onto Stack}
             }

          ∀t ∈ Tᵤ do
          { q' = fire(q,t); enter < q, {t}, q' > in A
            if q' ∈ H then {enter q' in H; put q' onto Stack}
          }
          }
```

Table 1. General algorithm

is the set of mergeable transitions T_m

  • The implementation of a firing step of transitions.

3.2 Sufficient Conditions

Some sufficient conditions are now defined on the sets T_u, T_m, and Π_{T_m}. These conditions have to be satisfied in order to ensure that the algorithm of Table 1 produces indeed a CSG. These conditions are parameterised by a transitive conflict relation weaker than the initial one. This transitive relation, denoted $\#\!\#$, is such that $\# \subset \#\!\#$, consequently $\#\!\#^C \subset l$.

3.2.1 Sufficient Conditions

Table 2 describes the conditions which lead to a CSG derivation. Conditions CA_1, CA_2 deal with elements T_u and T_m, respectively. Conditions CB_1, CB_2 concern the set of transition steps Π_{T_m}.

Condition CA_1 defines a partition of the enable transition set. Condition CA_2 implies that no (initial) conflict may occur between a mergeable transition and a disable transition.

$\forall q \in S :$

> (CA_1) $T_m \cup T_u = Enabled(q)$ and $T_m \cap T_u = \emptyset$
>
> (CA_2) If $t \in T_m$ then $t'\#t \Rightarrow t' \in Enabled(q)$
>
> (CB_1) $\forall \pi \in \Pi_{T_m} : \pi \in Step(T, \#^C)$
>
> (CB_2) $\forall P \in Step(Enabled(q), \#^C), \exists \pi \in \Pi_{T_m} : P \subseteq \pi$

Table 2. Sufficient conditions on T_u, T_m and Π_{T_m}

Condition CB_1 implies that no (direct or indirect) conflict exists between transitions within a step. Condition CB_2 implies that the considered computation steps *subsume* any subset of strongly independent transitions. Conditions CA_2, CB_1, and condition CB_2 are illustrated in subsections 3.2.2, 3.2.3, respectively.

3.2.2 Confusion Case

Petri net N and its associated LTS S depicted by Figure 2 represent a *confusion* case [Rei 85]: Transition D is in conflict with transitions A_1, B_1. These latter transitions are independent. From state 1 of the LTS, transitions A_1 et B_2 may occur. When B_2 first occurs, D may then occur; however, when A_1 first occurs, D cannot happen.

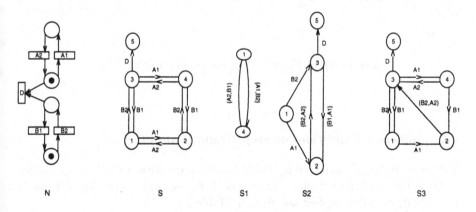

Fig. 2. Confusion Case: Illustration of Conditions CA_2 et CB_1

Motivation for CA_2:

Consider LTS $S1$ (Fig 2). From state 1, computation step $\{A1,B2\}$ is built, transition A1 is in a (potential) conflict with (still disable) transition D. The atomic computation step A_1, B_2 *hides* some internal states, namely states 2 and

3. In state 3, transitions A_1, B_1, and D are simultaneously enable. Step $\{A1,B2\}$ would prevent the firing of transition D, and a deadlock state would be omitted. In this case, "covering" condition (def 2.3.1.4) is made false.

Motivation for CB_1: Consider LTS $S2$ (Fig 2). From state 3, computation step $\{A1,B1\}$ might be derived, whereas A1, B1 are in an un-direct conflict, through D (i.e. $\{A1, B1\} \in Step(T, \iota)$ but $A1\#D$ and $B1\#D$)

The sequence $1 \overset{B2.B1.B2.D}{\Rightarrow} 5$ in S (cf. Fig 2) is not "covered" in $S2$, in the sense of definition 2.3.1. Because state 5 is a deadlock state, a sequence from state 1 to state 5 must exist. The single sequence of length 4 in $S2$ is the following[2]:
$1 \overset{\{A1\}.\{B2,A2\}.\{D\}}{\leadsto} 5$ or $[\{A1\}.\{B2, A2\}.\{D\}] \neq [B2.B1.B2.D]$.
$S3$ is a CSG for S.

3.2.3 Maximal Step Coverage

The net depicted by Figure 3 consists of two independent conflict sets. The associated LTS is represented by LTS S. CSG S_2 is a correct one with respect to the net, but step graph S_1 is an incorrect one, because it does not satisfy condition CB_2. With respect to S, sequence $0 \overset{e1.e4}{\Rightarrow} 2$ must be a possible sequence of any correct associated CSG. This is not the case for step graph S_1: $0 \overset{\{e1,e3\}}{\leadsto} 2$ and $0 \overset{\{e2,e4\}}{\leadsto} 2$ but $[\{e1, e3\}] \neq [e1.e4]$ and $[\{e2, e4\}] \neq [e1.e4]$

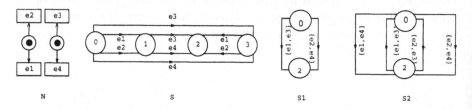

Fig. 3. Wrong Coverage: Illustration of condition CB_2

3.3 Sufficient Conditions of Step Graph Computation

The basic algorithm, depicted by Table 1 leads to the correct CSG computation, as far as the conditions given by Table 2 hold. By using the notations of Table 1, the conditions to be verified are given by Table 3.

3.3.1 Proof of conditions (1), (2), and (3)

(2) and (3) are immediate consequences of conditions CA_1, CB_1 and of the relation $\#^C \subset \iota$

---

[2] Trace equality ensures that the two sequences have the same length

$$s \overset{E}{\leadsto} s' \text{ means } < s, E, s' > \in A \text{ and } T = \{E \in \mathcal{P}(T) :< s, E, s' > \in A\}$$

(1) $H \subset S$ (2) $T \subset Step(T, \iota)$ (3) $s \overset{E}{\leadsto} s'$ implies $s \overset{E}{\Rightarrow} s'$

$$\exists s" \in, \exists \omega' \in T^*, \exists P_{\omega.\omega'} \in \boldsymbol{T^*} :$$

(4) $s \overset{\omega}{\Rightarrow} s'$ implies

$$s' \overset{\omega'}{\Rightarrow} s", s \overset{P_{\omega.\omega'}}{\leadsto} s" \text{ where } [w.w']_{(T,\iota)} = [P_{\omega.\omega'}]_{(T,\iota)}$$

Table 3. Verification Conditions

When $s \overset{E}{\leadsto} s'$, two cases have to be considered:

either $E = \{t\}$ then $s \overset{t}{\longrightarrow} s'$ and $\{t\} \in Step(T, \iota)$ or $E = \{t_1, t_2, \ldots t_k\}$ then $E \in$

$Step(Enabled(s), \iota)$ $(cf C B_1)$ and $s \overset{E}{\Rightarrow} s'$ (cf prop 2.2.3)

The enumeration starts with $H = \{s_0\}$ $(s_0 \in S)$, (2) et (3) imply that any state derived from a state in H is indeed a S state, condition (1) consequently holds.

3.3.2   **Proof of condition (4)** The following lemma is required; the proof of this lemma is shown in section 3.3.3.

Factorisation Lemma $\forall s \in S \cap S', \forall w \in T^* : s \overset{w}{\Rightarrow} s' \Rightarrow$ (a) or (b) where

    (a) $\exists t \in T_u, s \overset{t.w_1}{\Rightarrow} s'$ with $[w] = [t.w_1]$

    (b) $\exists F \in \Pi_{T_m}, \exists E \subset F : s \overset{\mathcal{E}.w_1}{\Rightarrow} s'$ with $[w] = [\mathcal{E}.w_1]$ and $(F \setminus E) \wr \|w_1\|$

Proof of (4) is performed by induction on $|\omega|$ (obvious when $|\omega| = 0$)
the property is assumed to hold for rank k, and let sequence ω be such that $|\omega| = k + 1$. Factorisation lemma is valid and two cases occur:

Case (a) is obvious: $t \in t_u$ implies $s \overset{\{t\}}{\leadsto} s_1$ and $s_1 \overset{w_1}{\Rightarrow} s'$, the induction hypothesis is then applied on s_1.

Case (b): $\exists F \in \Pi_{T_m}, \exists E \subset F : s \overset{E.w_1}{\Rightarrow} s'$ with $[w] = [\mathcal{E}.w_1]$ and $(F \setminus E) \wr \|w_1\|$

Let $R = F \setminus E$, then $s_E \overset{R}{\rightarrow}, s_E \overset{w_1}{\longrightarrow} s'$ with $R \wr w_1$ and $E \wr R$

The diamond property (2.2.5_3) implies: for $r \in Seq(R)$, then $s_E \overset{r.w_1}{\longrightarrow} s"$ and $s_E \overset{w_1.r}{\longrightarrow}$
$s"$ (cf fig 4.a)

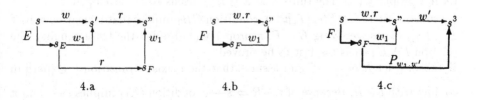

       4.a   4.b   4.c

Fig. 4. Case (b)

The case depicted by Figure 4.b is a consequence of $E \oplus R = F$, where $[w.r] = [F.w_1]$. From state s_F, $s_F \xrightarrow{w_1} s"$ where $|w_1| \leq k$ and the induction hypothesis may be used: $\exists w' \in T^*, \exists P_{w_1.w'} \in Step(T)^* : s_F \xrightarrow{w_1.w'} s^3, s_F \xrightarrow{P_{w_1.w'}} s^3$ and $[w_1.w'] = [P_{w_1.w'}]$ (cf fig 4.c).

Finally, $s \xrightarrow{w.r.w'} s^3$ and $s \xrightarrow{F.P_{w_1.w'}} s^3$. Because of $[w.r] = [F.w_1]$ and $[w_1.w'] = [P_{w_1.w'}]$ the result , $[w.r.w'] = [F.P_{w_1.w'}]$, holds.

3.3.3 Proof of Factorisation Lemma

For technical reasons, *factorisation* operator π is first introduced. This operator extracts from any transition sequence the maximal computation step, or the maximal step prefix. Operator π is defined according to the selected conflict relation.

π *Definition:* Let π be the mapping $S \times Step(T) \times T^* \times T^* \mapsto Step(T) \times T^*$ such that

$$\pi(s, E, w, \epsilon) = (E, w)$$
$$\pi(s, E, w_1, w_2) = (E, w_1.w_2) \text{ if } E \in \Pi_{T_m}$$

$$\pi(s, E, w_1, t.w') = \begin{cases} \pi(s, E \cup \{t\}, w_1, w') & \text{if } t \in T_m, \{t\} \wr \|w_1\| \\ & \text{and } E \cup \{t\} \in Step(T, \text{\rlap{/}{/}}^C) \\ (\{t\}, \mathcal{E}.w_1.w') & \text{if } t \in T_u, \{t\} \wr \|w_1\| \text{ and } \mathcal{E} \in Seq(E) \\ \pi(s, E, w_1.t, w') & \text{otherwise} \end{cases}$$

π Properties

Let $s \in S$, $w, w_1 \in T^*$ be such that $s \xrightarrow{w} s'$ and $\pi(s, \emptyset, \epsilon, w) = (E, w_1)$ Then (1) and (2)

(1) $s \xrightarrow{E} s_1, s_1 \xrightarrow{w_1} s'$ & where $[w] = [E.w_1]$

(2) $\forall t \in \|w_1\| : t \xrightarrow{s} \Rightarrow \exists t' \in E : t \text{\rlap{/}{/}} t'$

proof of factorisation lemma:

Let $s \in S$, $w, w_1 \in T^*$ be such that $s \xrightarrow{w'} s'$ and $\pi(s, \emptyset, \epsilon, w) = (E, w_1)$

As a result of the π construction, three cases occur: $E = \{t\}$ and $t \in T_u$, $E \in \Pi_{T_m}$, $E \notin \Pi_{T_m}$. The first two cases are similar, and the result directly follows from π property (1). The third case $E \notin \Pi_{T_m}$ needs to be developed.

By construction, $E \in Step(T, \text{\rlap{/}{/}}^C)$, condition CB_2 implies that transition subset $F \in \Pi_{T_m}$ exists, verifying $E \subset F$; π property (1) implies the trace equality, the relation $F \setminus E \wr w_1$ is the last to be proved.

By absurd, let R be $F \setminus E$ and assume that there exists transition $t' \in \|w_1\|$ in conflict with $t \in R$. Because of $t \in R \Rightarrow s \xrightarrow{t}$, condition CA_2 implies $s \xrightarrow{t'}$ and π property (2) then holds: $\exists t" \in E$ and $t' \equiv_\# t"$. Finally, $t, t" \in F$ such that $t \text{\rlap{/}{/}} t"$ and $F \in Step(T, \text{\rlap{/}{/}}^C)$, this is a contradiction.

3.4 Step Graph of Crossed Conflicts

This section presents a particular conflict relation and definitions of functions T_M, T_U, and Π providing an algorithm for the step graph derivation, which is a specific instantiation of the former basic algorithm (cf. Table 1).

3.4.1 Weak Conflict and Strong Independence

Let relation $\equiv_\#$ be the reflexive and transitive closure of relation $\#$. This relation is called *weak conflict relation*. Relation $\equiv_\#$ is an equivalence relation and $\# \subseteq \equiv_\#$. Relation $\equiv_\#^C$, i.e. the complement relation, is called strong independence relation and $\equiv_\#^C \subseteq \wr$

3.4.2 Crossed Conflict Step

An easy way to obey to rule CB_2 is to consider only maximal computation steps. Since conflict relation is an equivalence relation ($\equiv_\#$), the associated conflict sets supply a partition of the transition set. The set of maximal computation steps results from the *orthoproduct* [PF 90] of the equivalence classes of relation $\equiv_\#$

Orthoproduct : Let $I\!\!E$ be a set of sets, $I\!\!E = \{E_1, E_2, \ldots E_n\}$ $I\!\!E \in \mathcal{P}(\mathcal{P}(U))$
$\Pi_C(I\!\!E) =_{def} \{\{e_1, e_2, \ldots e_n\} : (e_1, e_2, \ldots e_n) \in E_1 \times E_2 \ldots \times E_n\}$

Crossed Conflict Step: The set $\Pi_C(T_m / \equiv_\#)$ is called Crossed Conflict Step. Properties:
- $E \in \Pi_C(T_m / \equiv_\#) \Rightarrow E \in Step(T_m, \equiv_\#^C)$ a fortiori $E \in Step(T_m, \wr)$
- $E \in \Pi_C(T_m / \equiv_\#), t \notin E \Rightarrow E \cup \{t\} \notin Step(T_m, \equiv_\#^C)$
  these properties follow from Π_C definition, and from $\equiv_\#^C \subseteq \wr$

3.4.3 Crossed Conflict Step Construction

The final algorithm is derived from the basic one (cf table 1) by considering the following functions:
$$T_M(T, \wr) = \{t \in T : t' \# t \Rightarrow t' \in T\}$$
$$T_U(T, \wr) = T \setminus T_M(T, \wr)$$
$$\Pi(T_M(T, \wr)) = \Pi_C((T, \wr) / \equiv_\#)$$
Property: : This algorithm produces a CSG. Conditions CA_1, CA_2 et CB_1 hold (trivial). Condition CB_2 results from properties of crossed conflict steps.

Examples: The former (right) CSG are computed by this algorithm. In the case of Milner's scheduler, (section 2.3.1), the gain is exponential: for n sites, the standard LTS consists of $n \times 2^n$ states and $(n^2 + n) \times 2^{n-1}$ transitions, the CSG consists of $n + 1$ states and $n + 1$ transitions.

4 Covering Step Graphs and Observational Equivalence

This section shows how to adapt the previous algorithm in order to generate, according to a set of observable events T_{Obs}, a covering step graph observationally equivalent ($\equiv$) to the standard LTS (i.e the LTS underlying the generated step graph and the standard LTS are ≡). The observational equivalence definition is recalled, an additional condition required for observational equivalence is given and shown to be sufficient. Finally, a specific algorithm for producing observational equivalent step graphs is supplied.

4.1 Observational Equivalence - Weak Bisimulation

Observation: Observational Equivalence is defined with respect to a set of observable events T_{Obs}. The definition of Observational Equivalence introduces a particular label τ representing internal (un-observable) computation and new transition relation $\Rightarrow$ is defined as follows:

$s \overset{\tau^*}{\Longrightarrow} s'$ iff $\exists w \in (T \setminus T_{Obs})^* : s \overset{w}{\Rightarrow} s'$,

and for all observable label a, $s \overset{\tau^* a \tau^*}{\Longrightarrow} s'$ iff $s \overset{\tau^*}{\Longrightarrow} s_1, s_1 \overset{a}{\longrightarrow} s_2$ and $s_2 \overset{\tau^*}{\Longrightarrow} s'$

In the sequel, $s \overset{\tau^* \epsilon \tau^*}{\Longrightarrow} s'$ stands for $s \overset{\tau^*}{\Longrightarrow} s'$

Observational equivalence (over finite LTS) may be defined as the limit of a decreasing sequence of equivalence relations over the states of the LTS ($\equiv_n \subset S_1 \times S_2$) [Mil 85]

$\equiv = \cap_{n \geq o} \equiv_n$ where $\equiv_0 = S_1 \times S_2$

$$s \overset{\tau^* a \tau^*}{\Longrightarrow} s' \Rightarrow \exists q' : q \overset{\tau^* a \tau^*}{\Longrightarrow} q' \text{ and } s' \equiv_{n-1} q' \text{ (i)}$$

and $s \equiv_n q$ iff $\forall a \in T_{Obs} \cup \{\epsilon\}$

$$q \overset{\tau^* a \tau^*}{\Longrightarrow} q' \Rightarrow \exists s' : s \overset{\tau^* a \tau^*}{\Longrightarrow} s' \text{ and } s' \equiv_{n-1} q' \text{ (ii)}$$

Weak-Bisimulation: A binary relation B over $S_1 \times S_2$ is a weak bisimulation iff

$$s \overset{\tau^* a \tau^*}{\Longrightarrow} s' \Rightarrow \exists q' : q \overset{\tau^* a \tau^*}{\Longrightarrow} q' \text{ and } (s', q') \in B \text{ (i)}$$

$(s, q) \in B$ iff $\forall a \in T_{Obs} \cup \{\epsilon\}$

$$q \overset{\tau^* a \tau^*}{\Longrightarrow} q' \Rightarrow \exists s' : s \overset{\tau^* a \tau^*}{\Longrightarrow} s' \text{ and } (s', q') \in B \text{ (ii)}$$

property: Observational Equivalence is the largest Weak-Bisimulation.

Extensions to LTS For $i \in \{1, 2\} : \Sigma_i = < S_i, s_{0i}, T_i, \rightarrow_i >$

$\Sigma_1 \equiv \Sigma_2$ iff $s_{01} \equiv s_{02}$

Σ_1 and Σ_2 are weak-bisimilar iff there exists a weak-bisimulation B containing (s_{01}, s_{02})

4.2 Sufficient conditions for observational equivalence

The basic algorithm (table 1) for step graphs generation and the associated sufficient conditions (table 2) are maintained. A new condition is introduced with respect to the set of observed events. Notations are those of table 2.

Free Transitions

Let $\#(t)$ be the transition subset which are in conflict with t:

$\#(t) =_{def} \{t' \in T : t' \# t\}$

and $Free(T, \#)$ the transition subset without conflict:

$Free(T, \#) =_{def} \{t \in T : \#(t) = \{t\}\}$

Additional condition for observational equivalence: Each "mergeable" transition t has to be included in a step transition π such that: all transitions in π (possibly except t) are conflict free and according whether t is observable or not, then either t is the single observable transition in π or any transition in π is unobservable. More precisely, condition CB_3 has to be fulfilled: $\forall t \in T_m, \exists \pi \in \Pi_{T_m} : t \in \pi, \pi \cap T_{Obs} = T_{Obs} \cap \{t\}$ & $\pi \setminus \{t\} \subset Free(T, \#)$
CB_3 admits the two following consequences:

c_1: Within a step, there exists at most a single observable transition or a conflict transition (a transition not belonging to $Free(T, \#)$).

c_2: For each unobservable (enabled) transition, there exists a step containing it, and this step is composed of only free and unobservable transitions.

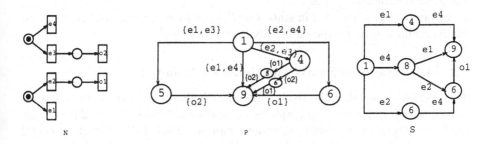

Fig. 5. Step Graph not observationally equivalent: example 1

Example 1: Figure 5 depicts a net where two "independent conflicts" are initially enabled, P, a step graph for which some steps contain two conflict transitions ($e2$ and $e4$ are independent but $e2, e4 \in Free(T_m, \iota)^C$, then c_1 is violated) and S a fragment of the complete LTS associated with the net. Observed events are o_1 and o_2.

State 8 of LTS S is considered. 8 admits the following experiments: $8 \overset{\tau^*}{\Rightarrow} 9$ and $8 \overset{\tau^* o_1 \tau^*}{\Rightarrow} 9$. State 8 does not belong to the states of the step graph P and no state of P is observationally equivalent to 8. Now, we consider state 1, a common state of S and P. In S, $1 \overset{\tau^*}{\Rightarrow} 8$ while in P $1 \rightsquigarrow q$ where $q \in \{1, 4, 5, 6, 9\}$. Since $q \not\equiv 8$ for each $q \in \{1, 4, 5, 6, 9\}$, we conclude that $S \not\equiv P$.

Example 2: Figure 6 depicts net N, step graph P associated with and fragment S of the whole LTS associated with the net. Observed events are o_1, o_2 and o_3. State 3 of LTS S is considered. 3 admits the following experiments: $3 \overset{\tau^*}{\Rightarrow} 9$ et $3 \overset{\tau^* o_3 \tau^*}{\Rightarrow} 9$. State 3 does not belong to the states of P and no state of P is observationally equivalent to 3. Now, we consider state 1, a common state of S

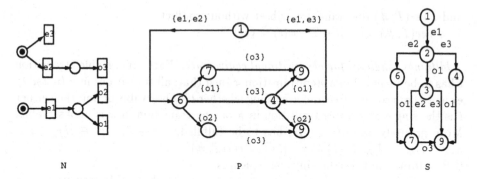

Fig. 6. Step Graph not observationally equivalent: Example 2

and P. In S, $1 \stackrel{\tau^* o_1 \tau^*}{\Longrightarrow} 3$ while in P $1 \stackrel{\tau^* o_1 \tau^*}{\rightsquigarrow} q$ where $q \in \{7, 9\}$. Since $3 \not\equiv 7$ and $3 \not\equiv 9$, we conclude that $S \not\equiv P$.

c_2 (and CB_3) is violated in state 1 of P: transition e_1 is enabled in 1 but no step containing e_1 is composed of only free transitions[3].

4.3 Observational Equivalence between Σ and $I\!\Sigma$

This section shows that the basic algorithm, depicted by table 1, furnishes an observational equivalent step graph, as far as the conditions of table 2 (sufficient conditions for a step graph generation) and condition CB_3 described in 4.2 (specific to observational equivalence) hold.

The proof is organised as follows: The two next sections show basic properties of free transitions with respect to observational equivalence. Section 4.3.3 shows that for each state, common to the standard LTS and its step graph, each "classical" experiment ($s \stackrel{w}{\Longrightarrow} q$) may be reproduced by means of $\stackrel{w}{\rightsquigarrow}$, that is $s \stackrel{w}{\rightsquigarrow} q'$ where $q \equiv q'$ and $\equiv$ is the observational equivalence defined on the standard LTS. This result is used in section 4.3.4 to build a bisimulation relation between the standard LTS and its step graph.

4.3.1 Diamond properties over free transitions

Property: For $s, s', q \in S, w \in T^* : s \stackrel{w}{\Longrightarrow} q$ and $t \in Free(T, \#)$

If $s \stackrel{t}{\longrightarrow} s'$ then $(s' \stackrel{t}{\Longrightarrow} q'$ and $q \stackrel{t}{\longrightarrow} q')$ or $(w = w_1.t.w_2$ and $s' \stackrel{w_1.w_2}{\Longrightarrow} q)$

proof: If $t \wr w$ then the result is an immediate consequence of diamond property 2.2.5. Now, consider the case where $t \# w$. Since $t \in Free(T, \#) \Rightarrow t \in \|w\|$, w may be rewritten as $w_1.t.w_2$ where $t \wr w_1$. The result follows by applying diamond property 2.2.5.

Corollary: For $s, q \in S, a \in T_{Obs} : s \stackrel{\tau^* a \tau^*}{\Longrightarrow} q$ and $t \in Free(T, \#), t \notin T_{Obs}$

If $s \stackrel{t}{\longrightarrow} s'$ Then $(s' \stackrel{\tau^* a \tau^*}{\Longrightarrow} q'$ and $q \stackrel{t}{\longrightarrow} q')$ or $(s' \stackrel{\tau^* a \tau^*}{\Longrightarrow} q)$

---

[3] ($e_1 \in \{e_1, e_2\}$ and $e_1 \in \{e_1, e_3\}$ but $\{e_2, e_3\} \subset Free(T, \#)^C$)

4.3.2 Free Transitions and Observational Equivalence

Property: If $s \xrightarrow{t} s', t \in Free(T, \#)$ and $t \notin T_{Obs}$ then $s \equiv s'$
proof: direct consequence of the former corollary.
Corollary: If $s \xRightarrow{\sigma} s'$ with $\sigma \in Free(T, \#)^*$ and $\|\sigma\| \cap T_{Obs} = \emptyset$ Then $s \equiv s'$
proof by induction on $|\sigma|$

4.3.3 Experiment Preservation

Hiding Morphisms
For $T_{Obs} \subset T$ and $\mathbb{T} \subset \mathcal{P}(T)$ verifying : $\forall E \in \mathbb{T}$ $Card(E \cap T_{Obs}) \leq 1$, we
define:

$$F_{Obs} \text{ a morphism } T^* \mapsto T^*_{Obs} \text{ by: } F_{Obs}(o.w) = \begin{cases} o.F_{Obs}(w) \text{ if } o \in T_{Obs} \\ \epsilon.F_{Obs}(w) \text{ otherwise} \end{cases}$$

$$\text{and } \mathbb{F}_{Obs} : \mathbb{T}^* \mapsto T^*_{Obs} \text{ by } \mathbb{F}_{Obs}(P_1.P_2) = \begin{cases} o.\mathbb{F}_{Obs}(P_2) \text{ if } P_1 \cap T_{Obs} = \{o\} \\ \epsilon.\mathbb{F}_{Obs}(P_2) \text{ otherwise} \end{cases}$$

Experiment preservation in Σ
For $s \in S \cap S', s \xRightarrow{w} q \Rightarrow s \xrightsquigarrow{W} q'$ with $F_{Obs}(w) = \mathbb{F}_{Obs}(W)$ and $q \equiv q'$ $in \Sigma$
Proof by induction on $|w|$. Obvious when $|w| = 0$, the property is assumed to
hold when $|w| = k$ and let sequence $t.w$, state s_1 such that $s \xrightarrow{t} s_1$
According whether $t \in T_u$ or $t \in T_m$, two cases occur.

• If $t \in T_u$ then $s \xrightsquigarrow{t} s_1$ and induction hypothesis may be applied from s_1 im-
plying that $s_1 \xrightsquigarrow{W} q'$ where $q \equiv q'$ and $F_{Obs}(w) = \mathbb{F}_{Obs}(W)$, Since $\mathbb{F}_{Obs}(t.W) = \mathbb{F}_{Obs}(t).\mathbb{F}_{Obs}(W)$ the result follows.

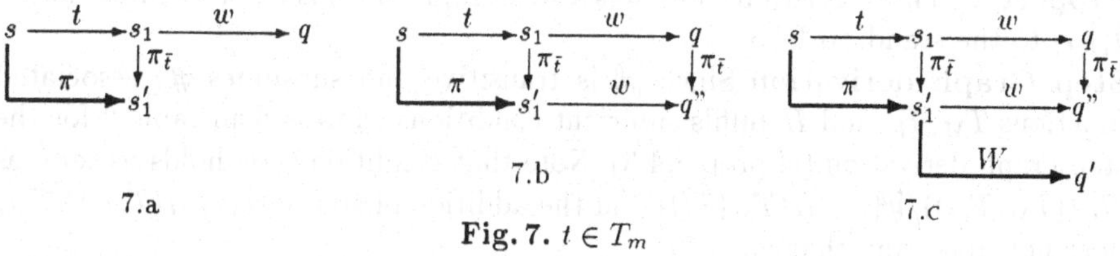

7.a   7.b   7.c

Fig. 7. $t \in T_m$

• If $t \in T_m$ then because of condition CB_3:
$$\exists \pi \in \Pi_{T_m} : t \in \pi, (\pi \setminus \{t\}) \subset Free(T, \#) \text{ \& } F_{Obs}(t) = \mathbb{F}_{Obs}(\pi)$$
(Fig 7.a) Let s_1' such that $s \xrightsquigarrow{\pi} s_1'$, the following holds: $s_1 \xRightarrow{\pi_{\bar{t}}} s_1'$ where $\pi_{\bar{t}} \in$
$Seq(\pi \setminus \{t\})$ (cf def 2.3.1). Since $\pi \setminus \{t\} \subset Free(T, \#)$, corollary 4.3.2 is valid
and it follows that $s_1 \equiv s_1'$.

(Fig 7.b) We have $s_1 \xRightarrow{w} q, s_1 \xRightarrow{\pi_{\bar{t}}} s_1'$ and $\pi_{\bar{t}} \wr w$ then $s_1' \xRightarrow{w} q'$ and $q \xRightarrow{\pi_{\bar{t}}} q$" (cf
prop 4.3.1). Moreover, $q \xRightarrow{\pi_{\bar{t}}} q$" $\Rightarrow q \equiv q$" (corollary 4.3.2)

Finally, as depicted in Fig 7.c, $s_1' \in S \cap S'$ and $s_1' \overset{w}{\Rightarrow} q"$ with $|w| = k$, and the induction hypothesis may be applied from s_1' implying that $s_1' \overset{W}{\leadsto} q'$, $q' \equiv q"$ and $F_{Obs}(w) = I\!\!F_{Obs}(W)$. Since $q" \equiv q$ and $F_{Obs}(t) = I\!\!F_{Obs}(\pi)$, the result follows.

Corollary: For $s \in S \cap S', a \in T_{Obs} \cup \{\epsilon\} : s \overset{\tau^* a \tau^*}{\Longrightarrow} q \Rightarrow s \overset{\tau^* a \tau^*}{\leadsto\leadsto} q'$ and $q \equiv q'$ $in \Sigma$

4.3.4 Σ and $I\!\!\Sigma$ are Weak-Bisimilar

Let B, be the subset of $S \times S'$ defined by $B = \{(q, q') \in S \times S' : q \equiv q' \text{ in } \Sigma\}$ property: B is a bisimulation between Σ and $I\!\!\Sigma$. (proof is a direct consequence of corollary 4.3.3 and CSG definition)

Corollary: Σ and $I\!\!\Sigma$ are observationally equivalent

4.3.5 Effective Algorithm

In this section a specific conflict relation and specific functions T_M, T_U and Π are proposed to obtain an algorithm for the derivation of an observational equivalent step graph which is a specific instantiation of the former basic algorithm (cf Table 1).

1. Functions T_M and T_U, introduced in section 3.4.3, are maintained:
   $T_M(T, \iota) = \{t \in T : t' \# t \Rightarrow t' \in T\}$ and $T_U(T, \iota) = T \setminus T_M(T, \iota)$
2. Conflict equivalence relation $\#\!\#$ is defined by the following partition of T_m:
   $T_m = \bigcup_{t \in Free(T_m, \#) \setminus T_{obs}} \{t\} \bigoplus ((T_{Obs} \cap T_m) \cup Free(T_m, \#)^C)$
   Within this partition, each un-observable free transition leads to a block reduced to itself, and a last block gathers all the transitions which are observable or involved in a conflict. As a consequence, equivalence relation $\#\!\#$ is weaker than the initial one ($\# \subset \#\!\#$).
3. $\Pi(T_M(T, \iota)) = \Pi_C((T_M(T, \iota)) / \#\!\#) \cup (Free(T_M(T, \iota)) \setminus T_{obs})$

Property: : This algorithm computes a step graph which is weak-bisimilar (w.r.t T_{Obs}) to the standard LTS

Step Graph derivation Since $\#\!\#$ is transitive and subsumes $\#$, associated functions T_M, T_U and Π fulfils sufficient conditions expressed in table 2 for the step graph derivation (cf prop 3.4.3). Note that condition CB_2 holds as soon as $\Pi_C((T_M(T, \iota)) / \#\!\#) \subset \Pi(T_M(T, \iota)$ and the addition of step $Free(T_M(T, \iota)) \setminus T_{obs}$ does not make any change.

Weak-bisimulation w.r.t T_{Obs}: $\Pi(T_M(T, \iota))$ fulfils CB_3 because of (1) the definition of partition $\#\!\#$, (2) the definition of $\Pi(T_M(T, \iota))$ by means of the orthoproduct and (3) the introduction of step $Free(T_M(T, \iota)) \setminus T_{obs}$.

5 Evaluation and Conclusion

Evaluation: The considered example is a data base system presented in [Jen 87] as illustrative example for a state space reduction using equivalent markings.

The same example has been used in [Val 89] to illustrate state space reduction method using stubborn set.

The data base system consists in $n \geq 2$ managers and a mechanism ensuring mutual exclusion for critical operations. Initially, all managers are *idle* and the base is *open*. The modification of the value by a manager is modelled by transition "update and send messages" ($usm(k)$) for which a message is sent to each partners of manager k. The sending manager (k) waits until all other managers have received his message ($rm(k,p)$), performed an update and sent an acknowledgement ($sa(k,p)$). When all acknowledgements are available, the sending manager returns to idle and the base to open ($ra(k)$).

Fig 8.a depicts the general structure of the whole state space in the case of n managers. Cube$_k$ represents the interleaved execution of transitions $rm(k,p)$ and $sa(k,p)$ for $p \in [1, N], p \neq k$. Fig 8.b depicts "cube" 1 in the case of 3 managers. For this system, pair of transitions in conflict have the following form: (t, t) or $(usm(i), usm(j))$ or $(usm(i), rm(i,j))$

Consequently each "cube" will be represented in a step graph by 3 states and 2 transition Steps ($\{rm(1,2), rm(1,3)\}$ and $\{sa(1,2), sa(1,3)\}$ for the cube depicted Fig 8.a). Finally the whole Step graph for a date base consisting of n managers is composed of $3n + 1$ states and $4n$ edges.

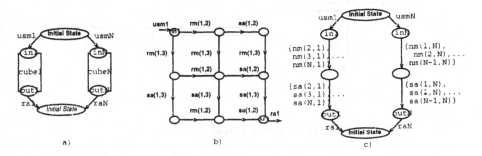

Fig. 8. Covering Step Graph for a n managers data base

Comparative Results: The following table gives the state space size for n managers, first when no reduction is performed, then with a partial exploration by means of stubborn sets [Val 89], with a symbolic exploration by means of equivalent markings [Jen 87], finally with a partial and symbolic exploration.

	states	edges
no reduction	$n3^{n-1} + 1$	$\approx n^2 \times 3^{n-2}$
Stubborn sets	$2n^2$	$2n^2$
Equivalent markings	$n^2/2$	n^2
Stubborn sets + Equivalent markings	$2n$	$2n$
Covering Step Graph	$3n + 1$	$4n$
Covering Step Graph + Equivalent markings	4	4

All the different techniques allow the size of the state space to be reduced from exponential to polynomial. Covering Step graphs take advantage over stubborn sets and equivalent markings: the size is linear for Step graphs, while it is quadratic when stubborn sets or Equivalent markings are used. Finally, as mentioned in [Val 89], "Symbolic" methods (using symmetries) and "Partial Order" methods (using independence relation) are complementary: for this particular example, an exploration by means of Step graphs and equivalent markings furnishes the optimal result since the size of the reduced state space is constant.

Behavioural analysis using Covering Step Graphs:

The same example is used to illustrate the reduction power of the proposed approach in the framework of Weak-Bisimulation. Now the step graph derivation is conducted with respect to a set of observable events. Roughly speaking, observable events are chosen according to the set of properties to be analysed [HM 85]. In the context of the present example, two kinds of observation are considered: first, observed events are relative to critical operations ($T_{Obs_1} = \{usm(k) : \forall k \in [1, N]\} \cup \{ra(k) : \forall k \in [1, N]\}$) allowing mutual exclusion property to be verified, second, the "local service" of each manager ($T_{Obs_2}(k) = \{usm(k), RM(k), SA(k)\}$ [4]) is considered.

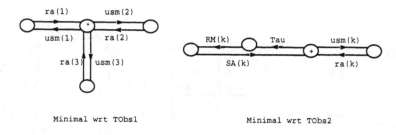

Minimal wrt TObs1   Minimal wrt TObs2

Fig. 9. Minimal Weak Equivalent

In the case of T_{Obs_1}, the obtained step graph is those obtained by general step graph derivation (Fig. 8.c). It consists in $3n + 1$ states and $4n$ edges while the minimal weak bisimilar LTS consists in $n + 1$ states and $2n$ transitions (Fig 9)

The following table summarises the obtained results for T_{Obs_2}. In this case, the minimal weak bisimilar LTS is constant (4 states and 5 edges) while the size of the obtained equivalent step graph seems quadratic in k.

K	3	4	6	10
no reduction	28/42	109/224	1459/1872	590.491/11.810.000
Equivalent step graph	10/12	13/16	19/24	58/94
minimal weak	4/5	4/5	4/5	4/5

---

[4] In order to have a better reduction with observational equivalence: events $rm(k, j)$ and $sa(k, j)$ for $k \neq j$ are respectively renamed by $RM(k)$ and $SA(k)$

Conclusion and Future works: This paper has presented Covering Step Graphs as an alternative to LTS. CSG permits the verification of general reachability properties (deadlock detection, liveness) while Weak Bisimilar CSG (derived with respect to a set of observable events) permits the verification of specific properties expressed in Hennesy-Milner Logic [HM 85]. Two algorithms are proposed for an "on the fly generation" of such graphs. Two standard examples permit us to obtain a first evaluation of our approach: the obtained results are promising, but significant experiments remain necessary. With respect to behavioural analysis, other equivalences (maximal traces, co-simulation, ...) may be also considered and associated step graph derivation algorithms proposed. Finally, the interest of step graphs in the framework of "True concurrency" semantics may be envisaged.

References

[FM 90] J. FERNANDEZ, L. MOUNIER *Verifying Bisimulation on the Fly*
3 rd. Int. Conf on Formal Description Techniques, Madrid, 1990
[GW 91] P. GODEFROID, P. WOLPER
Using Partial Orders for efficient verification of deadlock freedom and safety properties In Computer Aided Verification, 1991, LNCS 575
[GP 93] P. GODEFROID, D. PIROTIN
Refining Dependencies Improves Partial-Order verification methods
In Computer Aided Verification, 1993, LNCS 697
[Esp 93] J. ESPARZA
Model Checking using net unfoldings In TAPSOFT'93, 1993. LNCS 668
[HM 85] M. HENNESSY, R. MILNER *Algebraic Laws for Nondeterminism and Concurrency* Journal of the A.C.M Volume 32 1985
[Jen 87] K. JENSEN *Coloured Petri Nets.*
In Brauer, W., Reisig, W. & Rozenberg, G. (Ed.): Petri Nets: Central Models and their Properties. Advances in Petri Nets LNCS 254
[PF 90] D. H. PITT, D. FREESTONE *The derivation of conformance tests from lotos specifications* IEEE Transactions on Software Engineering, 16(12), 1990
[McMil 95] K. L. MCMILLAN *Trace theoretic verification of asynchronous circuits using unfoldings* In Computer Aided Verification, 1995, LNCS 939
[Maz 87] A. MAZURKIEWICZ *Trace Theory* In "Petri Nets: Applications and Relationship to other models of concurrency" LNCS 255
[Mil 85] R. MILNER *Communication and Concurrency* Prentice Hall.
[Rei 85] W. REISIG *Petri Nets: an Introduction*
EATCS, Monographs on Theoretical Computer Science, Springer Verlag, 1985
[Val 89] A. VALMARI *Stubborn Sets for reduced state space generation*
10 th Int. Conf on Application and Theory of Petri Nets, Bonn, 1989. LNCS 483
[WG 93] P. WOLPER, P. GODEFROID *Partial Order Methods for Temporal Verification*
Proceedings of CONCUR'93, LNCS 715

Performance Analysis of a Connection Management Scheme in IWU Interconnecting LANs Across ATM Networks

Jun Yuan[1], Lilin Liu and Zhanqiu Dong[2]

[1] Institute of Computer Science and Technology, Peking University,
Beijing 100871, P.R.China
Email: mmlab@748pku.pku.edu.cn
[2] Graduate School, Academia Sinica, Beijing 100039, P.R.China

Abstract. One of the challenges of interconnecting LANs with ATM based B-ISDN is that most LANs work in connectionless way and have the nature of bursty traffic; However, ATM network provides connection oriented services. This paper is concerned with performance analysis of a connection management scheme in the Interworking Units (IWUs) interconnecting LANs across ATM networks. Delayed release of established VC is modeled using stochastic Petri nets to explicitly capture the correlative characteristic of bursty traffic. The result shows that there is a tradeoff between effective use of the bandwidth and the signaling overhead caused by connection setup and release, which help us to design and configure the IWUs.

1 Introduction

Local Area Networks (LANs) have been widely developed in the 1980's because of lower cost and higher performance. However, for an isolated LAN, the resource to be shared by the users in different LANs is limited. It is an urgent requirement to interconnect the diverse LANs and form an internet that spans different areas and departments.

On the other hand, Asynchronous Transfer Mode (ATM) technology is chosen as the final transfer mode of B-ISDN and its standardization is currently progressed rapidly by ITU-T [12]. The LAN interconnection through ATM networks has attracted attention as an application in the first stage of B-ISDN era and as a driving force for advancement of the ATM networks.

One of the most challenges of the interconnection is that most LANs work in connectionless (CL) way and have the characteristics of bursty traffic, while ATM based B-ISDN offer basically connection oriented (CO) services. This means that a virtual channel has to be established before the user data can be transmitted.

To support LAN interconnection across ATM networks, two methods are proposed by ITU-T, i.e., i) end-to-end B-ISDN connection (indirect, implicit method), and ii) CLSFs (direct, explicit method) [13] . The IWUs convert the protocols between LANs and ATM network, and translate LAN local addresses or IP addresses to E.164 network addresses.

In case i), IWUs are connected by VC (virtual connection), PVC (permanent virtual connection) or semi-PVC, which may cause bandwidth waste if PVCs with peak rate are adopted, due to burstiness of CL data traffic. However, the implementation of this case is comparatively simple, which lead to the possibility of using in ATM LAN. In case ii), only semi-PVCs can be used between CLSFs, i.e., virtual overlay network between CLSFs is established on top of the ATM network. The direct method may provide efficient use of the network resource by means of statistical multiplexing of CL data traffic [14], which can capture the correlative nature of traffic source. For both methods, bandwidth allocation mechanism is required to effectively use the network resources.

Recently, Shimokoshi (see [11]) compares six bandwidth allocation schemes through simulation. The results address that the delayed release of established VC scheme is most applicable to offer the required data transfer quality to users. Heijenk (see [6]) analyzes this scheme using infinite continuous time Markov chain with matrix geometric solution. Haverkort (see [5]) define a class of infinite states SPNs which has an underlying Markov Chain of the quasi birth-death (QBD) type to model the connectionless server of ATM networks.

In this paper, we will further study the performance analysis of the delayed release of established VC scheme using stochastic Petri nets [8][7] to capture the correlative nature of bursty LAN traffic, which might be useful for bandwidth allocation and burstiness control. Stochastic Petri nets is an attractive tool, which in somewhat ease the contradiction between modeling and analysis. First, we model the LAN traffic with poisson process to find the basic action of this scheme, the underlying CTMC is somewhat alike to the model proposed in [6]; However, we consider the finite buffer size, which lead to more accurate description of real system. Then, we extend the initial model in three ways: i) MMPP traffic source is used for description of bursty nature of LAN. ii) Connection setup and release process are modeled by deterministic transitions which are nearer to the real system. iii) We introduce the connection request blocking probability into this model. The extended model is a deterministic and stochastic Petri nets (DSPN), which can be analyzed using embedded Markov chain (EMC) [9] or supplementary variables [2][3], whose complexity is almost the same. This extension can ensure that no more than one deterministic transitions are concurrently enabled in one marking, which is the tractable condition of the DSPN. The result shows that there is a tradeoff between effective use of the bandwidth and the signaling overhead caused by connection setup and release, which help us to design and configure IWUs.

2 Proposed Approach

The approach described and analyzed in this paper is based on the indirect support of connectionless data services in IWUs (see Fig.1). These IWUs are connected to each other through on demand connections.

The aim of the interconnection is to provide a transparent interconnection across ATM network. The IWU performs traffic filtering by observing the MAC

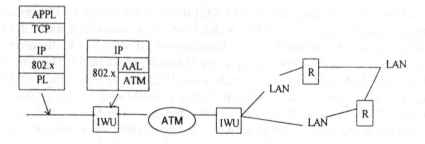

Fig. 1. System architecture

destination address. Before transmitting, an IWU performs the translation of the MAC destination address into the E.164 address of the remote IWU and the functions of the Segmentation and Reassembly (SAR) sublayer within the ATM Adaptation Layer (AAL). Each IP cell is segmented into 44 Byte blocks representing the SAR payload of an ATM cell.

The functional model of the IWU is depicted in Fig.2. The traffic monitoring is accomplished by determining the amount of incoming data within successive time intervals T. The IWU itself has to wait a certain time for the positive acknowledgment to set up a connection due to propagation delay and processing time required in the network node. During network congestion, a request may be blocked with a certain probability. Hence, the value describes the decision of a CAC algorithm used for capacity allocation. On the other hand, connection is released when there is no cell for transmission.

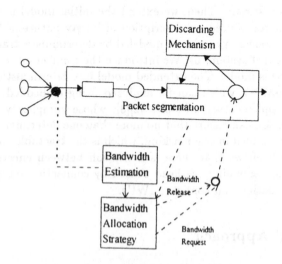

Fig. 2. Functional structure of IWU

In the proposed approach, an AAL connection will be established if the first cell of a packet has arrived, and no connection to the proper destination protocol entity is available. It can be used for consecutive cells to the same destination as well. The connection will not be released immediately after transferring the cell. The release process is delayed after the connection has not been used for a certain period of time.

This mechanism tries to exploit the expected correlation between the interarrival times of consecutive cells for the same destination entity. It assumes that the expected time until the next arrival is longer after the holding time has expired than immediately after a departure. Thus it can reduce the time that a connection has to be maintained, compared to the connection per cell mechanism, or reduce the mean delay experienced by cells, compared to the permanent connection mechanism [6]. In fact, it is a tradeoff between effective use of the bandwidth and the signaling overhead caused by connection setup and release, which help us to design and configure the IWUs. When connection request blocking probability is considered, this scheme is more attractive. While this mechanism cooperated with the selectively discard cell scheme operated by connection manager, the total performance will be promoted.

3 Performance Evaluation

For the functional model, we can derive a performance model to investigate the performance behavior of the proposed connection management mechanism.

The data stream that segmented cells enter into the IWU is bi-directional. The IWU can be modeled as sending buffer T and receiving buffer R. After the connection has been set up, the receiving buffer R receives the cells coming from the ATM network, and sends them to the destination LAN. Because the transmission rate in this direction is controlled by the source IWU, it is rarely that blocking takes place at R. Buffer T receives cells from the LAN, and sends them to the ATM network. Blocking takes place after T is full and cell will come into the IWU from LAN continuously. During blocking, the IWU simply or selectively drops down the cells until there are spaces in the buffer. We just consider the simply drop down scheme.

In the following, we establish two stochastic Petri nets model according to different traffic arrival process, i.e., Poisson arrival process and MMPP model. Petri nets is an effective mathematical tools to describe the synchronization, concurrence, collision, and other distributed action. Introducing stochastic process into Petri nets, then translating it into an equivalent continuous time Markov chain for solution, in somewhat ease the contradiction between modeling and analysis [7]. It is appropriate to model the connection establishment and release with basic store-and-forward action through this tool.

3.1 Poisson Arrival Process Based SPN Model

A data source is typically modelled as a Poisson process, in which the times between two successive cell arrivals are assumed to be independent. At the initial

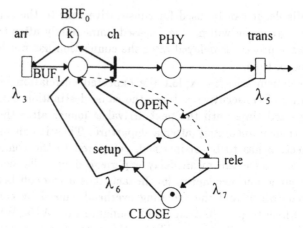

Fig. 3. SPN based model of the IWU

step, we also consider this pattern modeled by an exponential transition *arr* (see Fig.3) for easy tractability.

In general, setting up and releasing a connection will cost much larger time than transmitting a data unit. On the other hand, maintaining a connection in ATM networks during long silence period between two bursts of data transmission will also waste network resources, especially when the traffic intensity across the ATM network is low. In this simple model, we try to find the relationships among performance indices and some important system parameters, such as buffer size, traffic intensity, and connection setup and release rate.

This model comprises five places and five transitions, out of which four are exponentially timed and one is immediate. The place labeled BUF_0, which is initially marked with k token, models the idle buffer. Each exponentially arrived LAN cell moves one token from BUF_0 to place BUF_1, which models the issue of a new cell transmission request. Place $OPEN$ and $CLOSE$ model the status of a connection. The marking of place $CLOSE$ enables the firing of the exponential transition *setup* with the rate λ_6. The marking of place $OPEN$ enables the firing of exponential transition *rele* with rate λ_7, which is an important parameter to control the probability of connection release and resetup.

We define λ_7 as a function: $\lambda_7 = f(S, C_1, C_2)$, where S is the traffic source parameters and C_1 is the cost of link for data transmission, C_2 is the cost of connection setup. Given a certain workload, maintaining a longer connection yields higher costs, but releasing and establishing again a connection also lead to high costs. We choose λ_7 as the trade-off between C_1 and C_2. From the above discussion, we can see that λ_7 is an interesting parameter to control system performance and cost.

Place labeled PHY models the physical channel, whose token enables the exponential transition *trans*, representing data transmission. The volume of PHY is 1. The firing of immediate transition *ch* requires at least one token in buf_1, one token in $OPEN$, but no token in PHY. Inhibit arc from BUF_1 to *rele* represents that connection should not be released if BUF_1 is not empty.

From Fig.3, we can draw corresponding reachability graph, then acquire equivalent CTMC according to its marking process (see Fig.4).

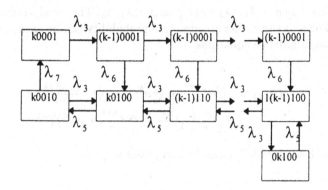

Fig. 4. Corresponding CTMC of the SPN model

We denote M_i by the state of Markov chain, i.e., $M_i = (m_4 m_3 m_2 m_1 m_0)$. Where m_4, m_3, m_2, m_1, m_0, separately represents the token numbers in place BUF_0, BUF_1, PHY, $OPEN$ and $CLOSE$. We solve Markov state equation according to (1). The total analysis process can be automatically finished by SPN analysis tools.

$$\Pi Q = 0, \quad \sum_{i=1} = 1 \tag{1}$$

Where $\Pi = (\pi_1, \pi_2, \ldots, \pi_n)$ is the steady state probability vector of CTMC. Q is state transition matrix.

However, the Markov Chain shown in Fig.4 is a QBD models as defined by Neuts in [10]. A class of infinite-state SPNs which has an underlying Markov chain of the QBD type has been defined by Haverkort in [5]. We can directly solve finite-state QBD models introducing some boundary conditions.

The following balance equations can be obtained ((2) through (7)), where (2), (3) represent the balance equations of repeating structure, (4), (5), (6) represent the boundary steady-state probability. Equation (7) is normalization condition. For simplicity, we denote π_{2i+1} by $\pi_{(K-i)i001}$, $\pi_{2(i+1)}$ by $\pi_{(K-i)i100}$ for $0 \leq i \leq$ K, and π_0 by π_{K0010}.

$$(\lambda_3 + \lambda_6)\pi_{2i+1} = \lambda_3 \pi_{2i-1}, 0 \leq i < \text{K} \tag{2}$$

$$\lambda_5 \pi_{2(i+1)} = \lambda_3(\pi_{2i} + \pi_{2i+1}), 0 \leq i \leq \text{K} \tag{3}$$

$$\lambda_7 \pi_0 = \lambda_3 \pi_1 \tag{4}$$

$$\lambda_3 \pi_{2K-1} = \lambda_6 \pi_{2K+1} \tag{5}$$

$$\lambda_5 \pi_{2(K+1)} = \lambda_3 \pi_{2K} \tag{6}$$

$$\sum_{i=0}^{K} \pi_{2i+1} + \pi_0 + \sum_{i=0}^{K} \pi_{2(i+1)} = 1 \tag{7}$$

We define σ as $\sigma = \lambda_3/(\lambda_3+\lambda_6)$, ρ as $\rho = \lambda_3/\lambda_5$. The steady-state probability π_{2K+1} is shown in (8). Other steady-state probability can be easily computed by π_{2K+1} through balance equations.

$$\pi_{2K+1} = \{\frac{1-\sigma^{K+1}}{1-\sigma} + \frac{\lambda_3}{\lambda_5}\sigma^{K-1} + \frac{\lambda_3}{\lambda_7} \cdot \frac{1-\rho^{K+2}}{1-\rho}$$
$$+ \frac{\rho}{\sigma-\rho}(\frac{\sigma-\sigma^{K+1}}{1-\sigma} - \frac{\rho-\rho^{k+2}}{1-\rho} + \rho\sigma^K)\}^{-1} \tag{8}$$

The cell loss probability can be computed as (9):

$$q = q_1 + q_2 = \pi_{2K+1} + \pi_{2(K+1)} \tag{9}$$

Where q_1 results from the fact that connection is not setup, but the buffer is full when the cells enter into the IWU. q_2 is the buffer overflow probability after the connection has been setup.

When a cell enters into the IWU, if the connection is not setup, the delay includes delay of connection setup, delay in queue and service time; If the connection has been setup, the delay consists of queue delay and service time, i.e.,

$$T = \sum_{i=0}^{K} \pi_{2i+1} \cdot \frac{1}{\lambda_6} + \frac{\sum_{i=0}^{2K+2} \pi_i \cdot m_3}{\lambda_3(1-q)} + \sum_{i=0}^{K} \pi_{2(i+1)} \cdot \frac{1}{\lambda_5} \tag{10}$$

The Average reserved bandwidth, $E[B]$, can be computed as:

$$E[B] = \lambda_5 \cdot \sum_{i=0}^{K} \pi_{2(i+1)} \tag{11}$$

As a numerical example, we consider the case where $\lambda_5 = 300$, $\lambda_6 = 10$, $\lambda_7 = 0.5$. The relationship between the cell loss probability q and traffic intensity ρ is shown in Fig.5(a).

In Fig.5(b), we depict the average delay T as a function of the traffic intensity ρ, considering different buffer size K. When ρ is small, as for increasing ρ, T_1 decreases. This results from the decreasement of delay caused by connection setup, the dominant delay. When ρ is large enough, queueing delay becomes the dominant delay, which leads to the increasing T.

Moreover, we show the effect of λ_7 to the performance indices q and T in Fig.6.

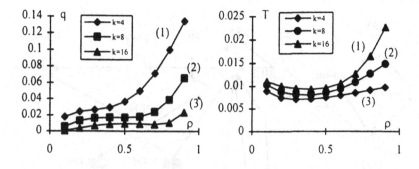

Fig. 5. cell loss probability q vs.ρ (a)   and average delay T vs. ρ(b)

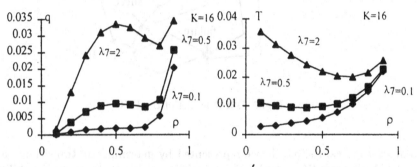

Fig. 6. cell loss probability q vs.ρ (a)   and average delay T vs. ρ(b)

3.2   MMPP LAN Traffic based DSPN Model

Due to the small and fixed size of ATM cell, sources typically produce streams of cells when active. The conventional modelling by a Poisson input process is therefore considered to be no longer valid. Instead, an ON/OFF model is used to capture the fact that sources alternate between an active (ON) period during which cells are periodically emitted and a silence(OFF) period in which no cells are produced. In case there is only one source, this yields an interrupted Poisson process(IPP). When N similar IPPs are multiplexed, one obtains a Markov Modulated Poisson process(MMPP) with $N + 1$ states[1].

In this paper, we only consider a virtual connection pair to the same destination. For simplicity, two state MMPP (IPP) traffic model is used to capture the bursty characteristics of LAN (see Fig.7). A token in place ON or OFF respectively models whether the traffic source is in a burst or not. When in a burst, cells are generated according to a Poisson process with rate λ_3. When not in a burst, no cells are generated. The time durations of ON and OFF states are subordinated to exponential distribution. That is, the exponential transition $go\text{-}on$ and $go\text{-}off$, with average rate λ_2 and λ_1 respectively model the time durations that the source stays at the OFF and ON state. Inhibit arc from OFF to arr models no cell transmission in OFF state. In real system, the connection delayed release time and connection setup time will probably be determined

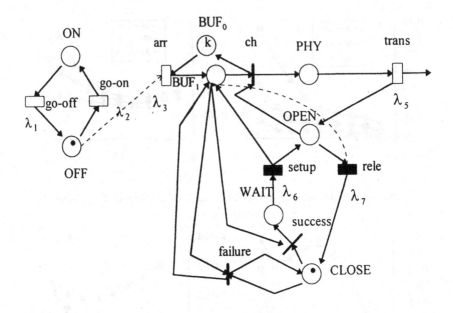

Fig. 7. DSPN model of the IWU

by fixed time values, and hence represented by deterministic transitions in our model. Moreover, We consider the connection request blocking probability in our model. This can be represented as conflict immediate transition *success* and *failure*. The firing of *failure* (whose weight is P_b representing the blocking probability) indicates the action of connection request blocking. On the other hand, The firing of *success* (whose weight is $1 - P_b$) represents successful connection establishment. In order to ensure the excution order, we define the priority of immediate transition *ch* (priority=1) lower than two immediate transitions *success* and *failure* (priority=2).

We choose EMC approach to analyze the DSPN model, whose solution is under the condition that at most one concurrent deterministic transition is enabled in each marking or there exists only several properly synchronized concurrent deterministic transitions [9].

Through P-invariance analysis, our model satisfies the above mentioned condition. The solution can be computed by first solving the EMC, by subsequently multiplying the solution of the EMC by the expected sojourn times in the markings of the DSPN and by finally normalizing the solution vertor in order to yield a vector probabilities.

Definition 1. Assume M is the set of all markings. M is partitioned into two classes MD and ME depending on the fact that deterministic transition t_d is enabled or not in them and reordered so that the state $S_i \in MD$ of the EMC come first. Define the matrix Q as (see [8])

$$Q = \begin{bmatrix} D & P \\ Q_{2,1} & Q_{2,2} \end{bmatrix}$$

The submatrix P contains the rates due to the firing of exponential transitions in conflict with the deterministic transition. The D submatrix contains the rates due to firing of exponential transitions not conflicting with the deterministic transition. $Q_{2,1}$, $Q_{2,2}$ represent the rates due to exponential transitions enabled in the markings belonging to ME.

Define the matrix Q' as

$$Q' = \begin{bmatrix} D & P \\ 0 & 0 \end{bmatrix}$$

Definition 2. Define matrix Δ as the state change probability matrix due to the firing of t.

$$\Delta^t = \begin{bmatrix} \Delta_{D,D} & \Delta_{D,E} \\ 0 & I \end{bmatrix}$$

In the following, we summarize the steps which are required for the steady-state solution of DSPN [3].

1. Generate the reduced reachability graph of DSPN
2. Obtain Ω^t, Ψ^t defined by (12),(13).

$$\Omega^t = [\omega_{ij}^t] = [Pr\{\text{state } j \text{ when } t \text{ fires}|\text{state } i \text{ at time } 0\}] = e^{Q'\tau} \quad (12)$$

$$\Psi^t = [\psi_{ij}^t] = [E\{\text{sojourn time in state } j \text{ up to the firing of } t|$$

$$\text{state } i \text{ when } t \text{ becomes enabled}\}]$$

$$= \int_0^\tau e^{Q' \cdot x} dx \quad (13)$$

3. Compute the one step transition probability matrix P of the EMC and the conversion matrix C according to (14) and (15)

$$\forall i \in ME : p_{ij} = \begin{cases} 0, & \text{if } i = j \\ \lambda_{ij}/\lambda_i, & \text{otherwise} \end{cases} \quad c_{ij} = \begin{cases} 1/\lambda_i, & \text{if } i = j \\ 0, & \text{otherwise} \end{cases} \quad (14)$$

$$\forall t \in T^D, \forall i \in MD : P(i) = \Omega_i \cdot \Delta^t, \quad c_{ij} = \begin{cases} \psi_{ij}^t, & \text{if } j \in MD \\ 0, & \text{otherwise} \end{cases} \quad (15)$$

4. Compute the steady-state solution of the EMC by solving the linear equation defined in (16)

$$\pi' \cdot (P - I) = 0, \quad \pi' \cdot e = 1 \quad (16)$$

5. Multiply the EMC solution by the matrix C of conversion factors and normalize the result in order to derive a proper vector of probabilities according to (17).

$$\gamma = \pi' \cdot C, \quad \pi = \frac{\gamma}{\sum \gamma_i} \quad (17)$$

We define the burstiness B as $B = (\lambda_1 + \lambda_2)/\lambda_2$. Then the traffic intensity ρ is computed as $\rho = \lambda_3/(\lambda_5 \cdot B)$

We solve this model numerically using the tool TimeNET [4]. At the first step, the choose the very small connection request blocking probability. In Fig.8(a), We depict the cell loss probability q as a function of the traffic intensity ρ just defined above. In Fig.8(b), we show the relationship between average delay T and ρ. Both figures consider the different buffer size K and burstiness B.

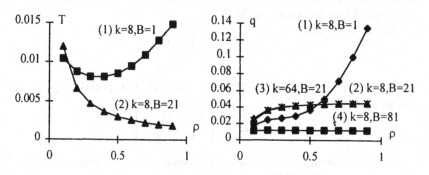

Fig. 8. cell loss probability q vs. ρ (a)   and average delay T vs. ρ (b)

When B is small, the result is similar to the SPN model. We can see this from the comparison of curve(1) in Fig.8(a) with curve(2) in Fig.5(a), and curve(1) in Fig.8(b) with curve(2) in Fig.5(b).

When B is large, delay time caused by the connection setup is the dominant delay in total delay. So the average delay decreases with the increasing ρ. But the proportion of queueing delay increases with larger ρ. Therefore, the decreasing speed of average delay is reduced. We can see this from curve(2) in Fig.8(b).

In Fig.8(a), curve(2) is close to curve(3), which reflects that it is difficult to control cell loss probability resulting from the high LAN bursty traffic only by enlarging buffer size. One solution is that reducing burstiness and control the burst length through cooperation with high level protocol (such as TCP). In our experience, with the same burstiness, effectively controlling burst length that can be captured by the connection release delay time will result in less connection resetup probability.

In Fig.9, we show the effect of the delayed release time. we choose $\rho = 0.5$, delay=50, 300 for the numerical result. The connection request blocking probability is represented by b. We can see that the cell loss probability is lower when delayed time is comparatively large. The effect is obvious especially when the blocking probability is large.

4   Conclusion

In this paper, the delayed release of established VC approach in IWU interconnecting various LANs through ATM network is analyzed using stochastic Petri

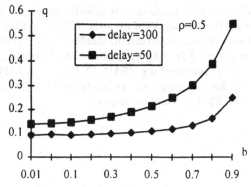

Fig. 9. Cell loss probability q vs. connection request blocking probility b

nets. The result shows that this scheme can explicitly capture the correlative characteristic of bursty traffic through carefully choosing delayed parameters. In fact, this is a tradeoff between effective use of the bandwidth and the signaling overhead caused by connection setup and release, which help us to design and configure IWU. The analysis result also shows: In highly bursty traffic, it is useless to enlarge the buffer to absorb LAN traffic. The solution is to reduce the burstiness and control the burst length through cooperation with high level protocol(such as TCP). We believe that under the traffic which have the self-similiar characteristics, the delayed release of established VC approach will be more effective. In our future work, we will further study this scheme.

References

1. Frost, V., Melamed, B.: Traffic modelling for telecommunications networks. IEEE Communications Magazine. **32** (1994) 70-81
2. German, R.: Analysis of stochastic Petri nets with non-exponentially distributed firing time. Ph.D dissertation. Berlin (1994)
3. German, R., Lindemann, C.: Analysis of stochastic Petri nets by method of supplementary variables. Performance Evaluation. **20** (1994) 70-81
4. German, R., Kelling, Ch., Zimmermann, A. and Hommel, G.: TimeNET: A toolkit for evaluating non-Markovian Petri nets Performance Evaluation. **24** (1995) 69-87
5. Haverkort, B.R.: Matrix solution of infinite stochastic Petri nets. IPDS'95. Berlin. Germany (1995) 72-81
6. Heijenk, G.: Connectionless communications using the Asynchronous Transfer Mode. Ph.D Dissertation, Twente Univ., Netherland (1994)
7. Liu, L.L., Yuan J., Dong Z.Q.: Modeling and analysis of computer networks Chinese Journal of Computers (to appear)
8. Marsan, M.A.: Stochastic Petri nets: an elementary introduction Advances in Petri nets. Lecture Notes in Computer Sicence. Springer Verlag. **424** (1989) 1-29
9. Marsan, M.A., Chiola, G.: On Petri nets with deterministic and exponentially distributed firing times. Advances in Petri nets. Lecture Notes in Computer Science. Springer Verlag. **266** (1987) 132-145
10. Neuts, M.F.: Matrix geometrix solutions in Markov models: An algorithmic approach. John Hopkins University Press. Baltimore, (1981)

11. Shimokoshi, K.: Performance comparison of bandwidth allocation mechanisms for LAN/MAN interworking through an ATM network. ICC'94. (1994) 1405-1411
12. Stallings, W.: ATM network: an introduction. MacMilan Pub. (1989)
13. Venieris, I.S., Angelopoulos, J.D., Stassinopoulos G.I.: Efficient use of protocol stacks for LAN/MAN-ATM interworking. JSAC. 11 (1993) 1160-1171
14. Yuan, J., Dong, Z.Q.: An new approach to modeling of ATM networks. IEEE Southeast Conference'96, Florida, USA (to appear)

Author Index

Lecture Notes in Computer Science

For information about Vols. 1–1018

please contact your bookseller or Springer-Verlag